Fundamentals of engineering drawing

eighth edition
FUNDAMENTALS OF

Warren J. Luzadder, p.e.
PURDUE UNIVERSITY

PRENTICE-HALL INC., ENGLEWOOD CLIFFS, NEW JERSEY 07632

ENGINEERING DRAWING

for design, product development, and numerical control

1. Mechanical drawing. I. Title.
T353.L88 1981   604'.24   80-15974
ISBN 0-13-338350-4

First–5th ed. published under title: Fundamentals of
engineering drawing for technical students and
professional draftsmen; 6th ed.: Fundamentals of
engineering drawing for design, communication, and
numerical control.

Fundamentals of Engineering Drawing
for design, product development, and numerical control
eighth edition

Warren J. Luzadder, P.E.

Editorial/production supervision by Virginia Huebner
Art direction: Jayne Conte
Cover design by Jayne Conte
Manufacturing buyer: Joyce Levatino

Printed in the United States of America
10 9 8 7 6 5 4

other books by the author:

Innovative Design with an Introduction to Design Graphics. Prentice-Hall, Inc. 1975.
Basic Graphics for Design, Analysis, Communications and the Computer.
  2nd ed., Prentice-Hall, Inc., 1968.
Fundamentos De Dibujo Para Ingenieros, 2nd ed., Compania Editoria
  Continental, 1967.
Graphics for Engineers. Prentice Hall of India Private Limited, 1964.
Technical Drafting Essentials, 2nd ed., Prentice-Hall, Inc., 1956.
Problems in Engineering Drawing, 7th ed., Prentice-Hall, Inc., 1977.
Engineering Graphics Problems for Design, Analysis and Communications.
  Prentice-Hall, Inc., 1968.
Problems in Drafting Fundamentals, Prentice-Hall, Inc., 1956.
Purdue University Engineering Drawing Films, with J. Rising et al.

PRENTICE-HALL INTERNATIONAL, INC., *London*
PRENTICE-HALL OF AUSTRALIA PTY. LIMITED, *Sydney*
PRENTICE-HALL OF CANADA, LTD., *Toronto*
PRENTICE-HALL OF INDIA PRIVATE LIMITED, *New Delhi*
PRENTICE-HALL OF JAPAN, INC., *Tokyo*
PRENTICE-HALL OF SOUTHEAST ASIA PTE. LTD., *Singapore*
WHITEHALL BOOKS LIMITED, WELLINGTON, *New Zealand*

Dedicated to my son

JAMES WILLIAM LUZADDER

Contents

PART
3 DESIGN 241

PART 4 GRAPHICS FOR DESIGN AND COMMUNICATION 277

PART 5 THE COMPUTER 397

Preface

In addition to meeting present day classroom requirements, the new edition of a text should contain added and revised material that will enable an instructor to redirect his courses to changing practices and new developments that will, in several years, justify still another edition. The changeover to the metric system continues at a slow and even pace and the widespread industrial use of computers for both design and design graphics is steadily increasing. The author has attempted to meet these changes by keeping all the computer graphics illustrations and related information up-to-date and by giving the measurement values for a majority of the illustrations and problems in millimeters, kilometers, kilograms, Newtons, and so forth. The chapter covering numerically controlled machine tools has been revised and is now all metric. Because many of the largest corporations, such as General Motors, Ford, IBM, TRW, McDonnell Douglas, Honeywell, Rockwell International, John Deere and Company, and International Harvester are converting to metric, it is time for the nation's educators to turn to the SI metric system. (Système International d'Unites). During the conversion period, now well under way, there will be widely different rates of progress toward exclusive use of this system, in various fields of industry. Some companies such as the multinational corporations that operate worldwide, and have plants and/or sales organizations in other countries, are changing over quickly and totally to metric, while others will continue to wait and watch from the sidelines until forced to change. In time, a total changeover for all is inevitable for, by necessity, this country must join the metric world to remain competitive.

During this conversion period, students in engineering colleges and schools of engineering technology must be trained to think SI (metric). Toward this end, educators can serve their students best by exerting every effort to make them feel comfortable and competent with the metric system.

Because students enter technical schools with a background in the conventional English system, the instructor can and should make a studious effort to avoid using the conventional system in lectures and for classroom problems. With increased emphasis on the use of metric, the student will gain proficiency in not one but two systems, and will also gain a decided advantage over peers not trained in the metric system.

A recent report shows that statics, dynamics, fluid mechanics, thermodynamics, and design graphics courses are now being taught using SI predominantly. In addition, target dates have been set for SI usage for steel structures (1980), machine design (1982), concrete design (1983), and construction design (1984). The fall of 1978 was the target date set for design graphics (engineering drawing) by the American Society for Engineering Education.

At the time this preface is being written Committee B1 has authorized the preparation of standard ANSI B1.13. This standard covering metric fastener threads recognizes only one series of diameter-pitch combinations ranging from M1.6 $\times$ 0.35 to M100 $\times$ 6. Committee ANSI B18 has under consideration a number of tentative standards for metric fasteners, among these are standards for hex bolts, round-head square-neck bolts, hex cap screws, socket head cap screws, hex lag screws, machine screws, hex nuts, hex slotted nuts, hex flange nuts, and plain washers. These standards have not yet been finally approved and are still subject to revision. Therefore, it was necessary to prepare Tables 12–18 in the Appendix using data gathered from a number of sources such as ISO (International Organization for Standardization) standards and Industrial Fastener Institute recommendations. Additional data was obtained from manufacturers and suppliers catalogs. The ANSI committee Y14.5 is still at work on a revised standard that will incorporate metric dimensioning. It is hoped that this standard will be approved and published early in the 1980's. In the revision of this text the author has been guided largely by the metric design and drafting standards furnished by General Motors, John Deere and Company, International Harvester, and IBM. International Harvester and IBM supplied almost complete standards, while GM furnished information needed for metric screw threads and surface quality. The thoughtfulness and kindness of the several administrators and standards engineers is deeply appreciated.

In this highly automated electronic age draftsmen and draftswomen will continue to prepare drawings by hand in the "old fashioned" way using pen or pencil on film, paper, or cloth. This statement is based upon a recent survey showing that only 23 percent of drafting was being done with automated drafting equipment. Automated drafting systems are worthwhile only if a company prepares complex drawings and has the volume to support such a system. Many companies have found that the drawings they require can be prepared by hand in much less time than it takes to prepare the programs for an automated system. For this reason, drafting departments without automated drafting far outnumber those with automation. Where automated systems do exist, draftsmen usually prepare and work on countless drawings that complement those that have been produced by the automated system. Although, admittedly, automated drafting can reduce manual drafting to a considerable extent, it can also increase the number of required drawings. Therefore, the demand for products used for manual drafting increases proportionately with the increase in sales of automated drafting equipment. The design room remains dependent on the hand-drafting skills of its design and drafting personnel.

This text has not been made totally metric because ANSI standards were not available to make this possible. Also,

since the English inch, foot, mile, acre, pound, and so forth will be with us for many years to come, it was thought to be wise to retain some drawings that show dimensions in inches and to prepare some of the illustrations showing how dimensions are to be given using both systems. In general, this was done in the chapter covering dimensioning. Where problems show inch measurements, the instructor may either require the part to be dimensioned in inches or he may ask the students to use millimeters. If a conversion to metric is made by the students, they will acquire a better understanding of the metric system. The author has had the good fortune to have been forced to face the metric system in 1960 with a Spanish translation of an earlier edition.

For convenience this text has been arranged in seven main parts: Part I, Basic Graphical Techniques; Part II, Spatial Graphics: Shape Description and Spatial Relationships; Part III, Design; Part IV, Graphics for Design and Communication; Part V, The Computer; Part VI, Graphic Methods for Engineering Communication, Design, and Computation; and Part VII, Design and Communication Drawing in Specialized Fields.

Part I covers the use of instruments, lettering, and engineering geometry. Part II provides essential information on basic graphics, spatial geometry, and pictorial representation. Part III presents the procedures of the design process and endeavors to show the interrelationship between design and manual drafting that includes both sketching and instrumental drawing. Part IV presents information needed to prepare engineering drawings. The chapters in this part are: Chapter 13, Size Description: Dimensions and Specifications; Chapter 14, Threads and Standard Machine Elements; Chapter 15, Shop Processes, Shop Terms, and Tool Drawings; and Chapter 16, Production Drawings: Preparation and Duplication. Part V (Chapters 17 and 18) covers the field of computer-aided design, automated drafting, and numerically controlled machine tools. At all institutions, particularly where computer-aided design systems are available, Chapter 17 should be assigned for reading and discussion. In addition, since practical knowledge of these systems can only be gained by either working experience or by on-the-scene observations, instructors of graphics and design should schedule plotter time and CRT (cathode-ray tube) console time for demonstrations. If this is not possible, several excellent motion pictures are available (see the

Bibliography) that may be shown in lieu of actual demonstrations.

Part VI presents the graphic methods commonly employed for communication, design, and computation. Finally, Part VII provides the information needed to prepare drawings in specialized fields so that problems may be assigned to broaden overall knowledge of the field of graphics, particularly in an area of individual and class interest.

To bring this text abreast of new technological developments, a number of leading industrial organizations have generously assisted the author by supplying appropriate illustrations that were needed in developing specific subjects. Every commercial illustration supplied by American industry has been identified using a courtesy line. The author deeply appreciates the kindness and generosity of these many companies and the busy people in their employment who found the time to select these drawings and photographs that appear in almost every chapter.

The author reaffirms his indebtedness to Professors George Shiers, W.L. Baldwin, R.H. Hammond, Byard Houck, and J.F. Zimmerly of the Ford Motor Company. The chapter on electronic drawing that was prepared by Professor Shiers for a previous edition has been used again in this edition. Professors Hammond and Houck and J.F. Zimmerly are remembered here for their contributions to the chapter on computer-aided design.

Special appreciation must be expressed for the assistance of R.B. Belford, Morris Buchanan, and John Mitchell. Mr. Belford, the Technical Director of the Industrial Fasteners Institute, supplied needed information on metric fasteners that made possible some of the new metric fastener tables in the Appendix. Mr. Buchanan presented the author with a copy of IBM Metric Standards as well as several drawings selected for use in this text. Mr. Mitchell of Cincinnati Milacron directed the preparation of the two programs appearing in the chapter on numerically controlled machine tools. Their kindness is deeply appreciated.

The author is grateful as well to Professors K.E. Botkin, R.L. Airgood and other members of the engineering graphics staff at Purdue University for their many valuable suggestions in regard to the content of this eighth edition.

Last, I would like to acknowledge two people for their outstanding work on this text. The first, Virginia Huebner, production editor, who did exceptional work in handling all aspects of production. The second, Jayne Conte, book designer, did an excellent job on the cover and interior design. My sincere thanks go to both.

W.J.L.
Purdue University

The Gateway Arch memorializes the spirit of the pioneers who settled the West after the Louisiana Purchase. Also, it might be thought of as being a testament to modern man's outstanding achievements that have resulted from innovative minds. In the years ahead still more people with daring ideas and considerable inner courage will be needed to solve the growing problems of the cities. Therefore, for this reason alone, this arch may well be called the Arch of Hope for St. Louis, for cities everywhere.

The Gateway Arch soars 630 ft upward to frame the St. Louis skyline in its new and growing splendor. Sixteen observation windows provide panoramic views to the east and west. Each leg of the Arch has a 40-passenger train, consisting of eight 5-passenger capsules that are mounted like the baskets of a Ferris wheel. In addition, for hearty souls 1076 steps lead to the observation deck at the top. Architecturally, this is creative design at its best. (*Courtesy Convention and Tourist Board of Greater St. Louis*)

introduction

1

1.1

Brief History of Drawing. For upward of twenty thousand years, a drawing has been the principal means for the portrayal of ideas through the use of lines. Its beginnings, however, are still further back in time, for our early ancestors undoubtedly explained their ideas by marking in the dust on the floors of their caves. It is deeply rooted in our instincts and, in a sense it is our one universal language even today, when some of our drawings are prepared by computers and plotters (Fig. 17.17). The earliest records of man are graphic, depicting people, deer, buffalo, and other animals of the time on the rock walls of caves. These drawings were to satisfy an elemental need for expression, long before the development of writing. However, drawing gradually freed itself from this early usage when writing was developed and it then came to be used primarily by artists and engineering designers as a means of setting forth ideas for the construction of finished works such as pyramids, war chariots, buildings, and simple mechanisms useful to man. Most of the very early drawings that still exist were made on parchment, which was very durable. Later, during the twelfth century, paper was developed in Europe and came into general use for drawings.

Only a few of the earliest drawings for fortresses, buildings, and simple mechanisms are in existence today. Those that have come down to us have been largely pictorial in nature, and they exist as carvings and paintings on walls of structures or have been woven into tapestry. One of the earliest representations shows the use of the wheel about 3200 B.C. in Mesopo-

Fig. 1.1 Idea sketch prepared by Leonardo da Vinci (1452–1519). The da Vinci "automobile" was to have been powered by two giant springs and steered by the tiller, at the left in the picture, attached to the small wheel. (*From Collection of Fine Arts Department, International Business Machines Corporation*)

tamia. The drawing presents a wheelbarrow-like structure being used by a man to transport his wife or child. The pictorial is primitive without any depth of perspective. Also, a ground-plan type of drawing for a fortress exists on a tablet prepared about 4000 B.C.

At the beginning of the Christian era, Roman architects had become skillful in preparing drawings of buildings that were to be constructed. They used straightedges and compasses to lay out the elevation and plan views and were able to prepare well-executed perspectives. However, the theory of projection of views upon imaginary planes of projection was not developed as a means of representation until sometime during the Renaissance period. Even though it is probable that Leonardo da Vinci was aware of the theory of multiview drawing (Fig. 1.6), his training as an artist prevailed and he recorded his ideas and designs for war machines and mechanical constructions by preparing perspective sketches and drawings such as the one shown in Fig. 1.1. No multiview drawings prepared by da Vinci have been found. He knew the value of a pictorial drawing, and it is interesting to note that even in this day of space travel, we prepare pictorial drawings to supplement our other design drawings (see the pictorial drawing of Skylab in Fig. 1.2).

1.2

Interrelationship of Engineering Graphics (Drawing) and Design. Since design graphics and design are interrelated in the total design process, those persons preparing the design layouts, the detailers, and the production engineer who has been assigned to the project must work closely at times with the project group leader as a part of the total design team. In general, all persons assigned to a project, both designers and those who support the design effort at any stage, should have full

Skylab

A. COMMAND & SERVICE MODULE
1. SPS Engine
2. Running Lights (8 places)
3. Scimitar Antenna
4. Docking Light
5. Pitch Control Engines
6. Crew Hatch
7. Pitch Control Engines
8. Rendezvous Window
9. EVA Handholds
10. EVA Light
11. Side Window
12. Roll Engines (2 places)
13. EPS Radiator Panels
14. SM RCS Module (4 places)
15. ECS Radiator

B. MULTIPLE DOCKING ADAPTER
1. Axial Docking Port Access Hatch
2. Docking Target
3. Exothermic Experiment
4. Infrared Spectrometer Viewfinder
5. Atmosphere Interchange Duct
6. Area Fan
7. Window Cover
8. Cable Trays
9. Inverter Lighting Control Assembly
10. L-Band Antenna
11. Proton Spectrometer
12. Infrared Spectrometer
13. Film Vault 4
14. Film Vault 1
15. Window Cover
16. SO82 (A&B) Canisters
17. M512/M479 Experiment
18. Area Fan
19. Composite Casting
20. TV Camera Input Station
21. Utility Outlet
22. TV Camera Input Station
23. M168 STS Miscellaneous Stowage Container
24. Redundant Tape Recorder
25. Radial Docking Port
26. 10-Band Multispectral Scanner
27. Proton Spectrometer
28. Temperature Thermostat
29. Radio Noise Burst Monitor
30. ATM C&D Console

C. AIRLOCK MODULE
1. Deployment Assembly Reels and Cables
2. Solar Radio Noise Burst Monitor Antenna
3. Handrails
4. DO21/DO24 Sample Panels
5. (Removed) (EVA use)
6. (Removed) (EVA use)
7. Permanent Stowage Container
8. STA IVA Station
9. Nitrogen Tanks (6 places)
10. Oxygen Tanks (6 places)
11. Molecular Sieve
12. Condensate Module
13. Electrical Feedthru Cover
14. Electronic Module 1
15. EVA Hatch
16. Airlock Instrumentation Panel
17. Molecular Sieve
18. STS C&D Console
19. ATM Deployment Assembly
20. Battery Module (2 places)
21. EVA Panel
22. Airlock Internal Hatches (2 places)
23. S193 Microwave Scatterometer Antenna
24. Running Lights (4 places)
25. Stub Antenna (2 places)
26. Thermal Blanket
27. Discone Antenna (2 places)

D. INSTRUMENT UNIT
None

E. ORBITAL WORKSHOP
1. OWS Hatch
2. Nonpropulsive Vent Line
3. VCS Mining Chamber and Filter
4. Stowage Ring Containers (24 places)
5. Light Assembly
6. Water Storage Tanks (10 places)
7. TO13 Force Measuring Unit
8. VCS Fan Cluster (3 places)
9. VCS Duct (3 places)
10. Scientific Airlock (2 places)
11. WMC Ventilation Unit
12. Emergency Egress Opening (2 places)
13. M509 Nitrogen Bottle Stowage
14. SO19 Optics Stowage Container
15. S149 Particle Collection Container
16. SO19 Optics Stowage Container
17. Sleep Compartment Privacy Curtains (3 places)
18. M13 Stowage Container
19. VCS Duct Heater (2 places)
20. M131 Rotating Chair Control Console
21. Power and Display Console
22. M131 Rotating Chair
23. WMC Drying Area
24. Trash Disposal Airlock
25. OWS C&D Console
26. Food Freezers (2 places)
27. Food Preparation Table
28. M171 Ergometer
29. MO92 Lower-Body Negative Pressure
30. Stowage Lockers
31. Experiment Support System Panel
32. M171 Gas Analyzer
33. Biomedical Stowage Cabinet
34. M171 Metabolic Analyzer
35. Meteoroid Shield
36. Nonpropulsive Vent (2 places)
37. TACS Module (2 places)
38. Waste Tank Separation Screens
39. TACS Spheres (22), Pneumatic Sphere
40. Refrigeration System Radiator
41. Acquisition Light (2 places)
42. Solar Array Wing (2 places)

F. APOLLO TELESCOPE MOUNT
1. Command Antenna
2. Telemetry Antenna
3. Solar Array Wing 1
4. Solar Array Wing 3
5. Solar Array Wing 4
6. Command Antenna
7. Telemetry Antenna
8. Sun-End Work Station Foot Restraint
9. Temporary Camera Storage
10. Quartz Crystal Microbalance
11. Acquisition Sun Sensor Assembly (2 places)
12. Acquisition Sun Sensor Assembly
13. ATM Solar Shield
14. Clothesline Attach Boom
15. EVA Lights (8 places)
16. Sun-End Film Tree Stowage
17. Handrail
18. SO82-B Experiment Aperture Door
19. He-2 Experiment Aperture Door
20. SO82-A Film Retrieval Door
21. SO82-A Experiment Aperture Door
22. SO54 Experiment Aperture Door
23. Fine Sun Sensor Aperture Door
24. SO56 Experiment Aperture Door
25. SO52 Experiment Aperture Door
26. He-1 Experiment Aperture Door
27. SO55A Experiment Aperture Door
28. SO82-B2 Experiment Aperture Door
29. SO82-B Film Retrieval Door
30. Canister Solar Shield
31. Canister Radiator
32. Canister Radiator
33. Rack
34. Charger-Battery-Regulator Modules (18 places)
35. Handrail
36. CMG Inverter Assembly (3 places)
37. Control Moment Gyro (3 places)
38. ATM Support Structure (3 places)
39. ATM Outriggers (3 places)

Fig. 1.2 Skylab—manned orbital scientific space station. The Skylab was designed to expand our knowledge of manned earth-orbital operations and to accomplish carefully selected scientific, technological, and medical investigations. (*Courtesy National Aeronautics and Space Administration*) knowledge of engineering graphics. A design engineer, to be successful as a designer, should have thorough training in this area. At the very least, he or she should be capable of preparing well-executed freehand design sketches and have a working knowledge of all of the forms of graphical expression that have been presented in this text.

Persons in the design support area, who may be expected to solve some of the design problems that arise graphically, prepare design layouts and models, and, finally, prepare the detail and assembly drawings needed for use in the production shops, must all have some basic education in the classroom and then acquire added experience where they are employed in order to become acquainted with company standards and practices. Some of these present-day "design-room professionals," in addition to their usual assignments, as mentioned, may also be expected to have some knowledge of production methods, particularly concerning numerically controlled (NC) machines, and have an understanding of computers (Chapters 17 and 18). In addition, specific

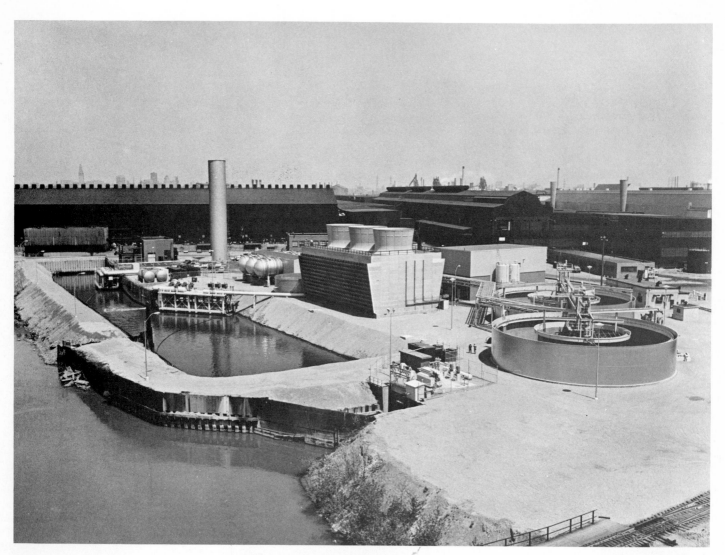

Fig. 1.3 Water treatment plant. Overall view of Republic Steel Corporation's new $18 million waste water treatment plant on the banks of the Cuyahoga River in Cleveland. This plant can treat up to 100 million gallons of water daily or the equivalent of the normal consumption of some 500 thousand people. This is the largest treatment plant of its type ever constructed by private industry in Ohio. Major components of the plant include a 350-foot long, 150-foot wide settling basin (left center in photo), two 95-foot diameter clarifiers for removal of mill scale and floating oil (right center), vacuum filters for dry solids, seven deep bed filters for removal of fine solids and traces of oil and cooling towers for heat removal. (*Courtesy Republic Steel Corp.*)

design tasks may require the use of digitizers and plotters. (See the left-hand page facing the beginning of Chapter 2.)

1.3
Engineering Graphics Today. Even though the attention of a designer may be said to be centered mainly on the problems that arise in design and development (Fig. 1.3), he must have a complete working knowledge of communication drawing, for it is often his responsibility to direct the preparation of the final drawings. These follow the preliminary design drawings and instructive sketches that he and his aids have prepared in accordance with the basic principles underlying the preparation of working drawings.

For a full and complete exchange of ideas with others, the engineering technologist must be proficient in the three means of communication that are at his disposal: (1) English, both written and oral; (2) symbols, as used in the basic sciences; and (3) engineering drawing.

As a member of the engineering team, the design draftsman must be fully capable of preparing the final drawings that will convey the information about size and shape needed to fabricate the parts and assemble the structure. This must be done in accordance with company practices and with the practices recommended in the publications of the American National Standards Institute. It is expected that the design draftsman keep up to date with the changes in standards that are constantly being made by standards engineers employed by his company and by the committees working under the sponsorship of the American National Standards Institute.

Engineering technicians, assigned to production areas or to the engineering department to aid the engineers, must have considerable knowledge of engineering drawing. Those closely assisting a design engineer or technologist may be called on to solve problems graphically, and to make working sketches relating to mechanisms, electrical circuits, and structural systems.

1.4
Organization of the Text. The purpose of this text is to present the grammar and composition of drawing so that those students in engineering colleges and technical institutes who conscientiously study the basic principles will eventually be able to prepare satisfactory industrial drawings and, after some practical experience, be capable of directing the work of others. To facilitate study, the subject matter has been separated into its various component parts: engineering geometry, multiview drawing, dimensioning, pictorial drawing, sketching, design, and so forth. Later chapters discuss the preparation of working drawings, both detail and assembly, computer-aided design, the preparation of topographic drawings, and the construction of charts and graphs. The major portion of the material presented leads up to the preparation of machine drawings, which the prospective members of some of the other branches of engineering technology think is not of interest to them. Since the methods used in the preparation of machine drawings, however, are the same methods used in the preparation of drawings in other fields, a thorough understanding of machine drawing assures a good foundation for later study in some specialized field, such as structural drawing. For those interested in specific types of drawing, some material has been presented with the assumption that the student already possesses a working knowledge of projection and dimensioning.

Proficiency in applying the principles of orthographic projection leads to easy graphical methods of solving space problems such as the determination of the clearance distance between a wheel and a fender or the true angle between a turbine blade route and the engine axis (Chapter 9).

1.5
Graphics and the Computer. The graphic language has become a means for an intimate and continuous conversational style of interchange between man and computer in the process of creative design. At present, computer systems have been developed that are in daily use and that interact with a human partner using the designer's own graphic language (Fig. 1.4). In bringing the computational power of the digital computer to bear on graphical design, special image processing systems have been built that allow the computer to both read and generate drawings. Chapter 17 covers computer graphics in considerable detail.

1.6
The Present and the Future. This is a

Fig. 1.4 An engineer working at a graphics display unit to find solutions to problems that have arisen in the design. With the electronic "light pen" the designer can revise the image and change the parameters. An engineer may work with diagrams, drawings, or printed letters and numbers directly on the face of the tube. This direct man–computer interaction leads to quick answers and permits concepts to be easily evaluated and then accepted or rejected. (*Courtesy International Business Machine Corporation*)

period of revolutionary change in the field of communication technology. At the present time, we are integrating new methods with the stylized methods of the past, using each method where it would seem to be best suited for the production of technical communications. We will continue to draw graphical representations manually and print out dimensions, notes, lists of materials, and so forth for the foreseeable future. However, plotters used on-line with computers or off-line directed by tapes will continue to gain wider utilization. Plotters are used in particular by aircraft and aerospace companies that produce products having complex contours. They are used also to a limited extent for the preparation of ordinary drawings of machine parts, for printed circuitry, for structural steel drawings, for highway route plans, and for determining the tool paths of milling (NC) machine tapes (Chapters 17 and 18).

Since nothing has happened to alter the fact that we communicate with others best in the graphic language, one advance that has come about in the use of computers has been the development of computer graphics—that is, the development of the hardware and the software that enable the computer to accept, understand, analyze, and produce engineering design data in graphic form. Persons who are responsible for the preparation of technical representations must become familiar with programming and with the use of both the computer and the plotter. These are their newest design and drafting tools. Although plotters are somewhat sophisticated as well as expensive, they save the designer and draftsman hours of tedious manual labor.

No new developments appear on the immediate horizon that will relieve the designer of the task of thinking out his design or the draftsman of the responsibility of applying the knowledge of his trade. Furthermore, recent technological breakthroughs have served to extend, rather than supplant, the direct application of the principles of engineering drawing by designers and draftsmen. The technical field needs the graphic language in all of its many forms.

As technology advances to meet the rising expectations of people all around the world, more and more drawings and other forms of graphical representation will be required. Most of these, as might be expected, will continue to be prepared manually.

Automation has already entered the field of communication technology and therefore must be accepted by those persons now in the field and by those about to enter. The only question that either a student or a technically trained man now at work can ask himself is whether or not he can adjust to new knowledge, new methods, and different requirements during his career.

1.7

Role of the Computer and Plotter in the Drawing Room (Fig. 1.5). Although words alone may not relieve entirely a person's concern about the continued importance of drafting, an understanding of the true role of the computer and plotter in the design room will answer some, if not all, of the questions that may arise in one's mind. Those who have been employed, where computers have taken over a number of graphic functions, welcome their use because they have learned that computers can accomplish those repetitive tasks, such as wire routing, schematics, and repetitious mechanical drafting, that can be very boring and time consuming to the draftsman. This leaves a large number of drawings in the field of mechanical design still to be done by draftsmen and detailers at the drawing board (see the page facing Chapter 2). In this category are drawings of piece parts that must be

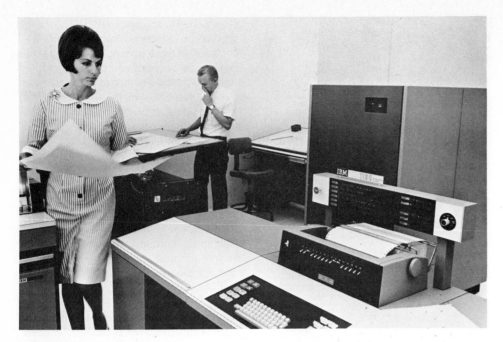

Fig. 1.5 The computer and plotter in the design room. The plotter is the newest drafting machine.

prepared manually. In the aircraft industry, where computers, digitizers, and plotters are readily available, manually prepared drawings now constitute at least 50% of the total drawing output. Each of these drawings done on the board is so unique and different as to require human intelligence for its preparation, together with indepth knowledge of graphics and shop practices. It should also be noted that the design layout, which defines the function and shape of these parts being drawn, was of necessity prepared manually in the preliminary design stage of the design project. Design layouts will continue to be drawn manually until the time comes when we are willing to forgo the idea of producing better products and commit ourselves to the continual production of products and systems already designed. Knowing the restlessness and competitive nature of humankind, this author predicts that this day is far in the future.

It should be noted, however, that at this point in time there are a few types of mechanical drawings that can be prepared other than manually through the use of a digitizer or a cathode ray tube coupled to a design-assist system (Chapter 17), the output in this case being a magnetic tape for a numerically controlled machine (Chapter 18).

Those who may be interested in learning more about the role of the computer in design should read Chapters 17 and 18. These two chapters present detailed information on computer-aided design and numerically controlled machine tools. In reading these chapters, one should come to the conclusion that there is a need to know graphics and that the computer will never make the teaching of graphics obsolete at any level of our educational system.

1.8
Major U.S. Corporations Go Metric. Although the metric system was legally approved for the United States by an Act of Congress more than 100 years ago (1866), it has not been widely used up to the present time because its usage was never made mandatory. Now, we find ourselves living alone in isolation on a nonmetric island in a world where all the major nations use the metric system. At least this statement was valid until a few years ago, when a number of our large U.S. multinational corporations started to convert to SI (Système International d'Unités) units, the primary reason for making a change at the time being more profitable international trade. These companies, with their worldwide operations, were finding it increasingly difficult to do business in countries committed to the metric system, with its different units of weights and measures. In addition, all these companies import metric machinery from their foreign subsidi-

Fig. 1.6 Metric drawing. (*Courtesy International Harvester Company*)

aries. This makes it necessary to supply spare parts, tools, and manuals for the needed interface with U.S.-designed equipment. The change to metric in the industrial field is now well under way and large manufacturing companies, such as General Motors, Ford, IBM, 3M, International Harvester, John Deere and Company, Caterpillar Tractor, Honeywell, McDonnell Douglas, Rockwell International, and TRW, are all converting to metric. See Fig. 1.6.

Total conversion will not come about immediately for much of American industry. For this reason, and because it will be a number of years before the ANSI standards will be converted to metric, this text may be said to be only about 75% metric. It must be noted that some companies who claim to be totally metric give thread specifications in the inch system. This practice will be followed until a new ANSI standard for threads has been approved. A similar situation exists for fasteners, bearings, keys, pipes, drills, and cylindrical fits. However, the student will find that the Appendix of this text contains numerous conversion tables, along with tables for metric drills, metric threads, metric fasteners, and SI metric-limits and fits. At this time a special committee set up to study and produce an Optimum Metric Fastener System has released OMFS Recommendations 2, 3, 4, and 6, which relate to screw threads. Eventually there should be a complete standard for ANSI-OMFS metric screw threads. Recently, ANSI committee Y14.5 reported that a new dimensioning standard covering metric could be approved and published as early as 1980. In the revision of this text, the author has been guided largely by the metric design and drafting standards furnished by General Motors, John Deere and Company, International Harvester, and IBM.

1.9
Educational Value of Graphics. Engineering drawing offers students an insight into the methods of attacking engineering problems. Its lessons teach the principles of accuracy, exactness, and positiveness with regard to the information necessary for the production of a nonexisting structure. Finally, it develops the engineering imagination that is so essential to the creation of successful design (Fig. 1.2).

PART 1

BASIC
GRAPHICAL
TECHNIQUES

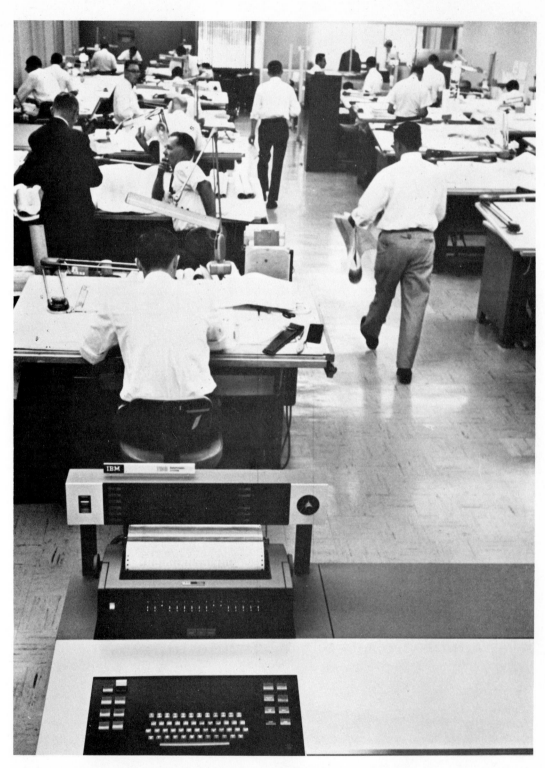

A typical design room with designers and draftsmen working at their drawing boards.
Note the computer in the foreground. At this time, computers and plotters can be
seen in most drawing rooms. Many designs are worked out using a computer, and
some of the drawings are prepared by a plotter. All engineers, technologists, and technicians should be familiar with their capabilities. (*Courtesy International Business
Machines Corporation*)

instrumental drawing— practices and techniques

2

Fig. 2.1 Essential drafting equipment.

A □ Drawing equipment and use of instruments

2.1
Introduction. The instruments and materials needed for making ordinary engineering drawings are shown in Fig. 2.1. The instruments in the plush-lined case should be particularly well made, for with inferior ones it is often difficult to produce accurate drawings of professional quality.

2.2
List of Equipment and Materials. The fol-

lowing list is a practical selection of equipment and materials necessary for making pencil drawings and ink tracings.

1. Case of drawing instruments
2. Drawing board
3. T-square
4. 45° triangle*
5. 30° —60° triangle (25 cm or 10 in.)
6. French curve
7. Scales (Secs. 2.19–2.20)
8. Drawing pencils
9. Pencil pointer (file or sandpaper pad)
10. Thumbtacks, brad machine, or Scotch tape
11. Pencil eraser
12. Cleaning eraser
13. Erasing shield
14. Dusting brush

* A 6-in. 45° Braddock lettering triangle, which may be used as either a triangle or a lettering instrument, may be substituted for this item.

Fig. 2.2 A standard set of drawing instruments.

Fig. 2.3 Purdue Riefler set.

15. Protractor
16. Lettering pens
17. Black drawing ink
18. Pad of sketching paper (plain or ruled)
19. Drawing paper
20. Tracing paper
21. Tracing cloth or film (Mylar)

To these may be added the following useful items:

22. Piece of soapstone
23. Tack lifter
24. Slide rule

2.3

Set of Instruments. A standard set of drawing instruments in a velvet-lined case and a large-bow set, which is capable of fulfilling the needs of most engineers and draftsmen, are shown in Figs. 2.2 and 2.3, respectively.

Large-bow sets are preferred by many persons, especially in the aircraft and automotive fields. A complete set might include the following: large bow (compass), beam compass with an extension bar for drawing very large circles, dividers, small bow, and a slip-handle ruling pen. Since the large bow (Fig. 2.3) is particularly suited for drawing circles of very small sizes up to 250 mm in diameter, its range covers that of both the bow instruments and the compass (without the extension bar) of the standard set (Fig. 2.2).

The large bow is preferred by its advocates because its sturdy construction permits the draftsman to exert the pressure necessary to secure black, opaque lines on pencil drawings that are to be used for making prints.

2.4

Protractor. The protractor (Fig. 2.4) is used for measuring and laying off angles.

2.5

Special Instruments and Templates. A few of the many special instruments and templates that are convenient for drawing are shown in Figs. 2.5–2.11.

Because of the time consumed in cutting back the wood to repoint an ordinary drawing pencil, some draftsmen favor the use of artist's automatic pencils (Fig. 2.5). Separate leads for these pencils may be purchased in any of the 17 degrees of hardness available for regular drawing pencils.

The flexible curves shown in Fig. 2.6, because of their limitless variations, are extremely convenient. The type shown in (*a*) is a lead bar enclosed in rubber. The more desirable one shown in (*b*) has a steel ruling edge attached to a spring with a lead core.

The drop pen (pencil) (Fig. 2.7) is designed for repeated drawing of circles of

Fig. 2.4 Protractor.

Fig. 2.5 Artist's pencils.

(a)

(b)

Fig. 2.6 Flexible curves.

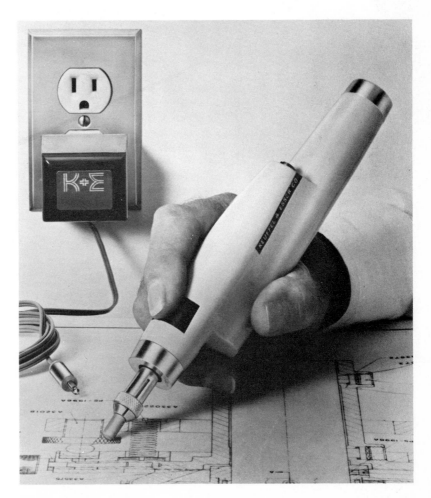

Fig. 2.7 Drop pen (pencil).

small diameter, such as the circles representing rivet heads.

The electric erasing machine (Fig. 2.8) saves valuable drafting time of those persons who prepare reproductions in ink and who find it necessary to make frequent erasures.

Several of a number of special ruling pens that are available are shown in Fig. 2.9. The contour pen (a) is used freehand for tracing contours on a map. The inking leg is free to swivel in the handle. The railroad pen (b) draws parallel lines that may represent roads or railroad tracks. The fountain ruling pen (c) is useful when much inking is to be done. It does not require frequent filling. The border pen (d) draws a line of fixed width. It is desirable to have a set of these pens available, since

Fig. 2.8 Erasing machine.

CONTOUR
PEN
(a)

RAILROAD
PEN
(b)

FOUNTAIN RULING
PEN
(c)

BORDER
PEN
(d)

Fig. 2.9 Special ruling pens.

Fig. 2.10 Proportional dividers.

a separate pen is required for each different line width.

Proportional dividers (Fig. 2.10) are used to reproduce distances to a reduced or an enlarged scale.

The use of templates (Fig. 2.11) can save valuable time in the drawing of standard figures and symbols on plans and drawings.

2.6

Drafting Machine. The drafting machine (Fig. 2.12) is designed to combine the functions of the T-square, triangles, scale, and protractor. Drafting machines are used extensively in commercial drafting rooms because it has been estimated that their use leads to a 25–50% saving in time.

2.7

Tracing Paper, Tracing Cloth, and Mylar. White lightweight tracing paper, on which pencil drawings can be made and from which blueprints can be produced, is used in most commercial drafting rooms in order to keep labor costs at a minimum.

The two general types of cloth available are ink cloth and pencil cloth. The cloth

used for ink is clear and transparent, dull on one side and glossy on the other. Pencil cloth is a white cloth with a surface specially prepared to take pencil marks readily.

The matte drawing surface of a plastic film, commonly called *Mylar,* is excellent for both pencil and ink work. The film has high dimensional stability and the printing qualities are very good. Erasures leave no ghost marks, as happens with cloth and vellum. Some users think of it as being virtually indestructible, since it is resistant to cracking and almost impossible to tear. It is rapidly replacing cloth and in some cases designers prefer it in place of vellum. It may be purchased in rolls or standard size sheets.

2.8

Pencils. The student and professional man should be equipped with a selection of good, well-sharpened pencils with leads of various degrees of hardness, such as 9H, 8H, 7H, and 6H (hard); 5H and 4H (medium hard); 3H and 2H (medium); and H and F (medium soft).

The grade of pencil to be used for various purposes depends on the type of line

(a)

(b)

Fig. 2.16 Manipulating the T-square.

deflection while drawing the line. Experienced draftsmen hold the T-square, as shown in Fig. 2.16(*b*), with the fingers pressing on the blade and the thumb on the paper. Small adjustments may be made with the hand in this position by sliding the blade with the fingers.

Horizontal lines are drawn from left to right along the upper edge of the T-square

(Fig. 2.17). (*Exception:* Left-handed persons should use the T-square head at the right side of the board and draw from right to left.) While drawing the line, the ruling hand should slide along the blade on the little finger.

2.13
Triangles. The 45° and the 30° × 60° triangles (Fig. 2.18) are the ones commonly used for ordinary work. A triangle may be checked for nicks by sliding the thumbnail along the ruling edges, as shown in Fig. 2.19.

2.14
Vertical Lines. Vertical lines are drawn upward along the vertical leg of a triangle whose other (horizontal) leg is supported and guided by the T-square blade. The blade is held in position with the palm and thumb of the left hand, and the triangle is adjusted and held by the fingers, as shown in Fig. 2.20. In the case of a right-handed person, the triangle should be to the right of the line to be drawn.

Either the 30° × 60° or the 45° triangle may be used since both triangles have a right angle. However, the 30° × 60° is generally preferred because it usually has a longer perpendicular leg.

2.15
Inclined Lines. Triangles also are used for drawing inclined lines. Lines that make angles of 30°, 45°, or 60° with the horizontal may be drawn with the 30° × 60° or the 45° triangle in combination with

Fig. 2.18 Triangles.

Fig. 2.19 Testing a triangle for nicks.

Fig. 2.17 Drawing horizontal lines.

Fig. 2.20 Drawing vertical lines.

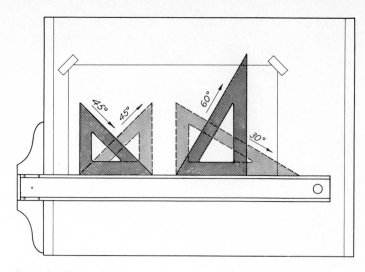

Fig. 2.21 Inclined lines.

the T-square, as shown in Fig. 2.21. If the two triangles are combined, lines that make 15° or a multiple of 15° may be drawn with the horizontal. Several possible arrangements and the angles that result are shown in Fig. 2.22.

The triangles used singly or in combination offer a useful method for dividing a circle into 4, 6, 8, 12, or 24 equal parts (Fig. 2.23). For angles other than those divisible by 15, a protractor must be used.

2.16
Parallel Lines. The triangles are used in combination to draw a line parallel to a given line. To draw such a line, place a ruling edge of a triangle, supported by a T-square or another triangle, along the given line; then slip the triangle, as shown in Fig. 2.24, to the required position and draw the parallel line along the same ruling edge that previously coincided with the given line.

2.17
Perpendicular Lines. Either the sliding triangle method [Fig. 2.25(*a*)] or the revolved triangle method [Fig. 2.25(*b*)] may be used to draw a line perpendicular to a given line. When using the sliding triangle method, adjust to the given line a side of a triangle that is adjacent to the right angle. Guide the side opposite the right angle with a second triangle, as shown in Fig. 2.25(*a*); then slide the first triangle along the guiding triangle until it is in the required position for drawing the perpendicular along the other edge adjacent to the right angle.

Although the revolved triangle method [Fig. 2.25(*b*)] is not so quickly done, it is widely used. To draw a perpendicular using this method, align along the given line the hypotenuse of a triangle, one leg of which is guided by the T-square or another triangle; then hold the guiding member in position and revolve the triangle about the right angle until the other leg is against the guiding edge. The new

Fig. 2.22 Drawing inclined lines with triangles.

Fig. 2.23 To divide a circle into 4, 6, 8, 12, or 24 equal parts.

Fig. 2.24 To draw a line parallel to a given line.

position of the hypotenuse will be perpendicular to its previous location along the given line and, when moved to the required position, may be used as a ruling edge for the desired perpendicular.

2.18
Inclined Lines Making 15°, 30°, 45°, 60°, or 75° with an Oblique Line. A line making an angle with an oblique line equal to any angle of a triangle may be drawn with the triangles. The two methods previously discussed for drawing perpendicular lines are applicable with slight modifications. To draw an oblique line using the revolved triangle method [Fig. 2.26(a)], adjust along the given line the edge that is opposite the required angle; then revolve the triangle about the required angle, slide it into position, and draw the required line along the side opposite the required angle.

To use the sliding triangle method [Fig. 2.26(b)], adjust to the given line one of the edges adjacent to the required angle, and guide the side opposite the required angle with a straight edge; then slide the triangle into position and draw the required line along the other adjacent side.

To draw a line making 75° with a given line, place the triangles together so that the sum of a pair of adjacent angles equals 75°, and adjust one side of the angle thus formed to the given line; then slide the triangle, whose leg forms the other side of the angle, across the given line into position, and draw the required line, as shown in Fig. 2.27(a).

To draw a line making 15° with a given line, select any two angles whose difference is 15°. Adjust to the given line a side adjacent to one of these angles, and guide the side adjacent with a straightedge. Remove the first triangle and substitute the

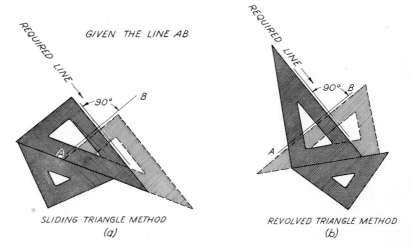

Fig. 2.25 To draw a line perpendicular to another line.

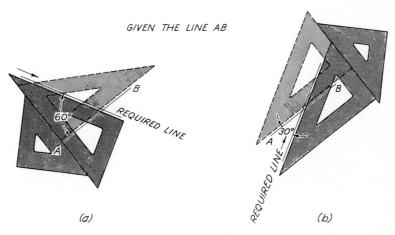

Fig. 2.26 To draw lines making 30°, 45°, or 60° with a given line.

GIVEN THE LINE AB

(a)   (b)

Fig. 2.27 To draw lines making 15° or 75° with a given line.

Fig. 2.28 Mechanical engineers' scale, full-divided.

Fig. 2.29 Engineers' decimal scale.

Fig. 2.30 Civil engineers' scale.

Fig. 2.31 Architects' scale, open-divided.

other so that one adjacent side of the angle to be subtracted is along the guiding edge, as shown in Fig. 2.27(*b*); then slide it into position and draw along the other adjacent side.

2.19

Scales—Inches and Feet. A number of kinds of scales are available for varied types of engineering design. For convenience, however, all scales may be classified according to their use as mechanical engineers' scales (both fractional and decimal), civil engineers' scales, architects' scales, or metric scales.

The mechanical engineers' scales are generally of the full-divided type, graduated proportionally to give reductions based on inches. On one form (Fig. 2.28) the principal units are divided into the common fractions of an inch (4, 8, 16, and 32 parts). The scales are indicated on the stick as eighth size (1½ in. = 1 ft), quarter size (3 in. = 1 ft), half size (6 in. = 1 ft), and full size.

Decimal scales are more widely used in industrial drafting rooms. The full size decimal scale shown in Fig. 2.29, which has the principal units (inches) divided into fiftieths, is particularly suited for use with the two-place decimal system. The half-size, three-eighths size, and quarter-size scales (Fig. 2.32) have the principal units divided into tenths.

The civil engineers' (chain) scales (Fig. 2.30) are full-divided and are graduated in decimal parts, usually 10, 20, 30, 40, 50, 60, 80, and 100 divisions to the inch.

Architects' scales (Fig. 2.31) differ from mechanical engineers' scales in that the divisions represent a foot, and the end units are divided into inches, half-inches, quarter-inches, and so forth (6, 12, 24, 48, or 96 parts). The usual scales are ⅛ in. = 1 ft, ¼ in. = 1 ft, ⅜ in. = 1 ft, ½ in. = 1 ft, 1 in. = 1 ft, 1½ in. = 1 ft, and 3 in = 1 ft.

The sole purpose of the scale is to reproduce the dimensions of an object full-size on a drawing or to reduce or enlarge them to some regular proportion, such as eighth-size, quarter-size, half-size, or double-size. The scales of reduction most frequently used are as follows:

Fractional

Mechanical engineers' scales

Full-size	(1″ = 1″)
Half-size	(½″ = 1″)
Quarter-size	(¼″ = 1″)
Eighth-size	(⅛″ = 1″)

Architects' or mechanical engineers' scales

Full-size	(12″ = 1′-0)
Half-size	(6″ = 1′-0)
Quarter-size	(3″ = 1′-0)
Eighth-size	(1½″ = 1′-0)
1″ = 1′-0	¼″ = 1′-0
¾″ = 1′-0	3/16″ = 1′-0
½″ = 1′-0	⅛″ = 1′-0
⅜″ = 1′-0	3/32″ = 1′-0

Decimal

Mechanical engineers' scales

Full-size	(1.00″ = 1.00″)
Half-size	(0.50″ = 1.00″)
Three-eighths-size	(0.375″ = 1.00″)
Quarter-size	(0.25″ = 1.00″)

Civil engineers' scales

10 scale: 1″ = 1′; 1″ = 10′;
    1″ = 100′; 1″ = 1000′
20 scale: 1″ = 2′; 1″ = 20′;
    1″ = 200′; 1″ = 2000′
30 scale: 1″ = 3′; 1″ = 30′;
    1″ = 300′; 1″ = 3000′
40 scale: 1″ = 4′; 1″ = 40′;
    1″ = 400′; 1″ = 4000′
50 scale: 1″ = 5′; 1″ = 50′;
    1″ = 500′; 1″ = 5000′
60 scale: 1″ = 60′; etc.
80 scale: 1″ = 80′; etc.

The first four scales, full-size, half-size, quarter-size, and eighth-size, are the ones most frequently selected for drawing machine parts, although other scales can be used. Since objects drawn by structural draftsmen and architects vary from small to very large, scales from full-size to 3/32 in. = 1 ft (1/128-size) are commonly encountered. For maps, the civil engineers' decimal scales having 10, 20, 30, 40, 50, 60, and 80 divisions to the inch are used to represent 10, 20, 30 ft, and so forth, to the inch.

On a machine drawing, it is considered good practice to omit the inch (″) marks in a scale specification. For example, a scale may be specified as: FULL SIZE, 1.00 = 1.00, or 1 = 1; HALF SIZE, .50 = 1.00, or ½ = 1; and so forth.

The decimal scales shown in Fig. 2.32 have been approved by the American National Standards Institute for making machine drawings when the decimal system is used.

It is essential that a draftsman always think and speak of each dimension as full-size when scaling measurements, because the dimension figures given on the finished drawing indicate full-size measurements of the finished piece, regardless of the scale used.

The reading of an open-divided scale is illustrated in Fig. 2.33 with the eighth-size (1½ in. = 1 ft) scale shown. The dimen-

sion can be read directly as 21 in., the 9 in. being read in the divided segment to the left of the cipher. Each long open division represents 12 in. (1 ft).

The reading of the full-size decimal scale is illustrated in Fig. 2.34. The largest division indicated in the illustration represents one inch, which is subdivided into tenths and fiftieths (.02 in.). In Fig. 2.32, the largest divisions on the half size, three eighths size, and quarter size decimal scales represent one inch.

Fig. 2.32 Decimal scales.

Fig. 2.33 Reading a scale.

2.20

Metric Scales. The metric scale (Figs. 2.35 and 2.36) is used in those countries where the meter is the standard of linear measurement. As of this time, the International System of Units (Systéme International d'Unités) (SI) for the measurement of length, measurement of surface, measurement of volume, measurement of weight, and so forth, is used as a world standard in every industrialized country except the United States. We stand alone with our inches, gallons, and pounds in spite of the fact that the metric system was legalized for use in this country by an Act of Congress more than one hundred years ago (1866). The metric system is legal in all states. Even though the cost of conversion to the metric system will be enormous, there can be little doubt that a changeover will be made in the near future. As a practical matter, it is probable that American industry will shift gradually to the metric system without prodding by an Act of Congress since even now the automobile industry has adopted and is using a dual-dimensioning system by which dimensions are given on drawings in both inches and millimeters. In the agricultural-machinery field both International Harvester and John Deere have gone *all metric*. Companies in other areas will probably make this step in the near future. It is for this reason that dimensional values for more than 70 percent of the problems and illustrations have been given in metric units in this text. If desired, problems given using the English system may be prepared showing metric values by utilizing the tables and other information given in the Appendix.

Since the millimeter is the standard SI unit for linear dimensions, scales for drawing are marked in millimeters so that they can be read directly. The following scales for reduction and enlargement are recommended for the drawing of machine parts.

Reduced Scales	Enlarged Scales
1:2	2:1
1:3	3:1
1:5	5:1
1:10	10:1

1:2.5 and 2.5:1 scales may be used where available.

The scales commonly used in the several fields are:

Mechanical Engineers' Scales

1:2.5	1:25
1:5	1:33⅓
1:10	1:50
1:15	1:80
1:20	1:100

Civil Engineers' Scales

1:100	1:750
1:200	1:1000
1:300	1:1250
1:400	1:1500
1:500	1:2000
1:600	1:2500
1:625	1:3000

Architects' Scales

1:1	1:125
1:2	1:200
1:5	1:250
1:10	1:300
1:20	1:400
1:25	
1:50	1:500
1:100	

The reading of distance values along the four scales of a flat scale (front and reverse sides) are shown in Fig. 2.36.

To lay off a measurement, using a scale starting at the left of the stick, align the scale in the direction of the measurement with the zero of the scale being used toward the left. After it has been adjusted to the correct location, make short marks opposite the divisions on the scale that establish the desired distance (Fig. 2.37). For ordinary work most draftsmen use the same pencil used for the layout. When extreme accuracy is necessary, however, it is better practice to use a pricker and make slight indentations (not holes) at the required points. If a regular pricker is not available, the dividers may be opened to approximately 60° and the point of one leg used as a substitute.

To ensure accuracy, place the eye directly over the division to be marked, hold the marking instrument perpendicular to the paper directly in front of the scale division, and mark the point. Always

Fig. 2.34 Reading the decimal scale.

Fig. 2.35 Flat metric scale (front and reverse sides). (*Courtesy J. S. Staedtler, Inc.*)

check the location of the point before removing the scale. If a slight indentation is made, it will be covered by the finished line; if a short mark is made and it is *very* light, it will be unnoticeable on the finished drawing.

To set off a measurement (say 68 mm) to half-size, the scale indicated either as 1:2 or 1:20 should be used. If the measurement is to be made from left to right, place the zero (0) division mark on the given line, and make an indentation (or mark) opposite the 68 division point [Fig. 2.37(*a*)]. The distance from the line to the point represents 68 mm, although it is actually 34 mm. To set off the same measurement from right to left, place the 68 mark on the given line, and make an indentation opposite the zero division mark [Fig. 2.37(*b*)].

2.21

Compass or Large Bow. The compass or large bow is used for drawing circles and circle arcs. For drawing pencil circles, the style of point illustrated in Fig. 2.38(*c*) should be used because it gives more accurate results and is easier to maintain than most other styles. This style of point is formed by first sharpening the outside of the lead on a file (Fig. 2.39) or sandpaper pad to a long flat bevel approximately 6 mm long [Fig. 2.38(*a*)] and then finishing it [Fig. 2.38(*b*)] with a slight rocking motion to reduce the width of the point. Although a hard lead (4H–6H) will maintain a point longer without resharpening, it gives a finished object line that is too light in color. Soft lead (F or H) gives a darker line but quickly loses its edge and, on larger circles, gives a thicker line at the end than at the beginning. Some draftsmen have found that a medium-grade (2H–3H) lead is a satisfactory compromise for ordinary working drawings. For design drawings, layout work, and graphical solutions, however, a harder lead will give better results.

The needle point should have the shouldered end out and should be adjusted approximately 10 mm (.38 in.) beyond the end of the split sleeve [Fig. 2.38(*a*)].

2.22

Using the Compass or Large Bow. To draw a circle, it is first necessary to draw two intersecting center lines at right angles and mark off the radius. The pivot point should be guided accurately into position at the center. After the pencil

Fig. 2.36 Reading metric scales.

Fig. 2.37 To lay off a measurement-metric scale.

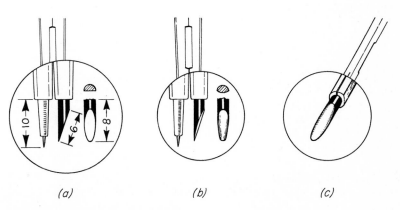

Fig. 2.38 Shaping the compass lead.

Fig. 2.39 Sharpening the compass lead.

Fig. 2.40 Using the large bow (Vemco).

point has been adjusted to the radius mark, the circle is drawn in a clockwise direction, as shown in Fig. 2.40. While drawing the circle, the instrument should be inclined slightly forward. If the pencil line is not dark enough, it may be drawn around again.

When using a compass for a radius larger than 50 mm (2 in.), the legs should be bent at the knee joints to stand approximately perpendicular to the paper (Fig. 2.41). It is particularly important that this adjustment be made when drawing ink circles, otherwise both nibs will not touch the paper. For circles whose radii exceed 125 mm (5 in.), the lengthening bar should be used to increase the capacity.

The beam compass is manipulated by steadying the instrument at the pivot leg with one hand while rotating the marking leg with the other (Fig. 2.42)

2.23
Dividers. The dividers are used principally for dividing curved and straight lines into any number of equal parts and for transferring measurements. If the instrument is held with one leg between the forefinger and second finger, and the other leg between the thumb and third finger, as illustrated in Fig. 2.43, an adjustment may be made quickly and easily with one hand. The second and third fingers are used to "open out" the legs, and the thumb and forefinger to close them. This method of adjusting may seem awkward to the beginner at first, but with practice absolute control can be developed.

2.24
Use of the Dividers. The trial method is used to divide a line into a given number of equal parts (Fig. 2.44). To divide a line into a desired number of equal parts, open the dividers until the distance between the points is estimated to be equal to the length of a division, and step off the line *lightly*. If the last prick mark misses the end point, increase or decrease the setting by an amount estimated to be equal to the error divided by the number of divisions, before lifting the dividers from the paper. Step off the line again. Repeat this procedure until the dividers are correctly set, then space the line again and indent the division points. When stepping off a line, the dividers are rotated alternately in an opposite direction on either side of the line, each half-revolution, as shown in Fig. 2.44.

Although the dividers are used to transfer a distance on a drawing, they should never be used to transfer a measurement from the scale, as the method is slow and inaccurate and results in serious damage to the graduation marks. Care should be taken to avoid pricking large unsightly holes with the divider points. It is the common practice of many expert draftsmen to draw a small freehand circle around a very light indentation to establish its location.

Fig. 2.41 Using the compass (legs bent).

Fig. 2.42 Drawing large circles (Vemco beam compass).

Fig. 2.43 To adjust the large dividers.

Fig. 2.44 Using the dividers.

2.25

Use of the Bow Instruments. The bow pen and bow pencil are convenient for drawing circles having a radius of 25 mm (1 in.) or less (Fig. 2.45). The needle points should be adjusted slightly longer than the marking points, as in the case of the compass.

Small adjustments are made by the fingers of the hand holding the instrument, with the pivot point in position at the center of the required circle or arc.

2.26

Use of the French Curve. A French curve is used for drawing irregular curves that are not circle arcs. After sufficient points have been located, the French curve is applied so that a portion of its ruling edge passes through at least three points, as shown in Fig. 2.46. It should be so placed that the increasing curvature of the section of the ruling edge being used follows the direction of that part of the curve which is changing most rapidly. To ensure that the finished curve will be free of humps and sharp breaks, the first line drawn should start and stop short of the first and last points to which the French curve has been fitted. Then the curve is adjusted in a new position with the ruling edge coinciding with a section of the line previously drawn. Each successive segment should stop short of the last point matched by the curve. In Fig. 2.46, the curve fits the three points, A, 1, and 2. A line is drawn from between point A and point 1 to between point 1 and point 2. Then the curve is shifted, as shown, to fit again points 1 and 2 with an additional point 3, and the line is extended to between point 2 and point 3.

Some people sketch a smooth continuous curve through the points in pencil before drawing the mechanical line. This procedure makes the task of drawing the curve less difficult, since it is easier to adjust the ruling edge to segments of the freehand curve than to the points.

2.27

Use of the Erasing Shield and Eraser. An erasure is made on a drawing by placing an opening in the erasing shield over the work to be erased and rubbing with a pencil eraser (never an ink eraser) until it is removed (Fig. 2.47). Excessive pressure

Fig. 2.45 Using the bow pencil.

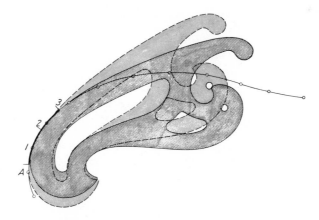

Fig. 2.46 Using the irregular curve.

Fig. 2.47 Using the erasing shield.

should not be applied to the eraser because, although the lines will disappear more quickly, the surface of the paper is likely to be permanently damaged. The fingers holding the erasing shield should rest partly on the drawing paper to prevent the shield from slipping.

2.28

Use of the Ruling Pen. The ruling pen is used to ink mechanical lines. It is always guided by the working edge of a T-square, triangle, or French curve and is never used freehand.

When ruling a line, the pen should be in a vertical plane and inclined slightly (approximately 60°) in the direction of the movement. It is held by the thumb and forefinger, as illustrated in Fig. 2.48, with the blade against the second finger and the adjusting screw on the outside away from the ruling edge. The third and fourth fingers slide along the T-square blade and help control the pen. Short lines are drawn with a hand movement, long lines with a free arm movement that finishes with a finger movement. While drawing, the angle of inclination and speed must remain constant to obtain a line of uniform width and straightness. Particular attention should be given to the position of the pen, as practically all faulty lines are due to incorrect inclination or to leaning the pen so that the point is too close to the straightedge or too far away from it. The

Fig. 2.48 Holding the pen.

correct position of the pen for drawing a satisfactory line is illustrated in Fig. 2.48.

If the pen is held so that it leans outward, as shown in Fig. 2.49(a), the point will be against the straightedge, and ink will run under and cause a blot; or, if it leans inward, as in Fig. 2.49(b), the outer nib will not touch the paper and the line will be ragged.

Unnecessary pressure against the straightedge changes the distance between the nibs, which in turn may either reduce the width of the line along its entire length or cause its width to vary, as in Fig. 2.49(d).

It will not take any beginner long to discover that care must be taken when removing a T-square or triangle away from a wet ink line [Fig. 2.49(c)].

The ruling pen is filled by inserting the quill or dropper device of the stopper between the nibs. Care must be taken, while filling, to see that there is enough ink to finish the line and that none of the ink from the filler gets on the outside of the blades. No more than 6 mm (.25 in.) should ever be put in; there is a danger of blotting if the pen is used with a greater amount.

The special ink-bottle holder and filler shown in Fig. 2.50 is a convenience that saves time when filling a pen, because only one hand is needed to raise the stopper and insert the ink.

The width of a line is determined by the distance between the nibs, which is regulated by the adjusting screw. When setting the pen, a series of test lines should be drawn with a straightedge on a small piece of the same kind of paper or cloth to establish the setting to the desired width of the line. The draftsman's line gauge, shown in Fig. 2.51, is convenient for testing the widths of trial lines as illustrated. If the first trial line is not of the desired width, another and another should be drawn until the final one agrees with the selected width as given on the gauge.

If the ink refuses to flow when the pen is

(a)   (b)

(c) ────────────────────────
T-SQUARE SLIPPED INTO LINE

(d) ────────────────────────
PEN PRESSED AGAINST T-SQUARE BLADE

(e) ────────────────────────
NOT ENOUGH INK IN PEN

(f) ────────────────────────
INK ON OUTSIDE OF PEN BLADE

(g) ────────────────────────
PEN AWAY FROM GUIDING EDGE NEAR END OF LINE

Fig. 2.49 Common faults in handling a ruling pen.

Fig. 2.50 Special ink-bottle holder and filler.

touched to the paper, either the ink has thickened or the opening between the nibs has become clogged. To start it flowing, draftsmen often touch the point to the back of a finger or pinch the blades together. Whenever this fails to produce an immediate flow, the pen should be cleaned and refilled.

A dirty pen in which the ink has been allowed to thicken will not draw any better than a dull one. To avoid changing the setting of the pen when cleaning it, fold the pen wiper twice at 90° and draw the corner of the fold between the ends of the blades (Fig. 2.52).

2.29

Tracing. Often, when it is necessary to make duplicate copies (blueprints) of important drawings for a machine or structure, the original pencil drawings are traced in ink on a tracing medium, usually tracing cloth. Contrary to the practice of old-time draftsmen and the intention of the early manufacturers of this medium, the dull side is now almost universally used for the inking surface instead of the slick side because it produces less light glare, will take both pencil and ink lines better, and will withstand more erasing. The fact that the dull side will take pencil lines is important because, on some occasions, in order to save time, drawings are made directly on the cloth and then traced. On completion of the tracing, all pencil lines including the guide lines and slope lines for the lettering may be removed by wiping the surface of the cloth with a rag moistened with a small amount of gasoline, benzene, or cleaning fluid.

When the tracing cloth has been fastened down over the drawing, a small quantity of tracing cloth powder may be sprinkled over the surface to make it take the ink evenly and smoothly. After it has been well rubbed in, the excess must be thoroughly removed by wiping with a clean cloth, for even a small amount of loose powder left on the surface can cause clogging of the pen. Powder is also used by some persons over a spot where an erasure has been made; but a better practice is to use a piece of soapstone, which will put a smooth, slick finish over the damaged area. In applying the soapstone, rub the spot and then wipe a finger over it a few times. Following this treatment, the erased area will take ink almost as well as the original surface.

Since ink lines are made much wider than pencil lines, in order to get a good

Fig. 2.51 Testing a trial line on a "try sheet."

contrast on a blueprint, they should be carefully centered over the pencil lines when tracing. The center of an ink line should fall directly on the pencil line, as shown correctly in Fig. 2.53. For ink work it might be said that ink lines are tangent when their center lines touch. In this same illustration, note the poor junctures obtained when ink lines are not centered so that their center lines are tangent.

Figure 2.54(a) shows the filled-in corner effect that frequently appears, to the disgust of the draftsman, when an ink line is drawn from or to another previously drawn line that is still wet.

When a working drawing is traced in ink on either paper or cloth, the lines should be "inked" in a definite order. Otherwise, the necessity of waiting for the ink to dry after every few lines not only wastes time

Fig. 2.52 Cleaning the ruling pen.

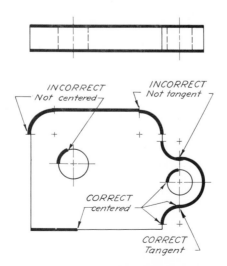

Fig. 2.53 Inking over pencil lines.

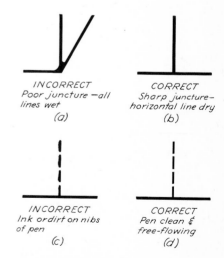

Fig. 2.54 Ink lines.

but often results in a line here and there being left out. Furthermore, hit-and-miss inking may produce lines of unequal width. It is therefore recommended that the student make a conscientious attempt to follow the order of inking suggested in this chapter.

After the paper or cloth has been fastened down over the drawing, and before the inking is begun, each tangent point should be marked and all centers should be indented.

Order of inking

1. Curved lines
   a. Circles and circle arcs (small circles first) in the order of 1, 2, and 3 below.
      (1) Visible
      (2) Invisible
      (3) Circular center lines and dimension lines
   b. Irregular curves
      (1) Visible
      (2) Invisible
2. Straight lines
   a. Visible
      (1) Horizontal, from the top of the sheet down
      (2) Vertical, from the left side of the sheet to the right
      (3) Inclined, from the left to the right
   b. Invisible
      (1) Horizontal
      (2) Vertical
      (3) Inclined
   c. Auxiliary (center, extension, dimension lines, etc.)
      (1) Horizontal
      (2) Vertical
      (3) Inclined
      (4) Section lines

(a) (b) (c) (d)

Fig. 2.55 Worn, incorrectly shaped, and correctly shaped blades.

3. Arrowheads and dimension figures
4. Notes and titles
5. Border

2.30

Sharpening a Ruling Pen. Ruling pens, whether they have been used continuously or intermittently, eventually show signs of wear. When a pen's nibs have lost their elliptical shape and have become so dull that the ink spreads under their tips, fine lines cannot be drawn until the pen has been sharpened. The way to detect this condition is to examine the tips. If bright spots may be seen on the tips, the pen is too dull for satisfactory work. An example of a worn point is illustrated in Fig. 2.55(a).

Since even new pens are seldom sharpened properly and often require retouching, every draftsman should be able to reshape and sharpen his own pen.

Incorrectly sharpened points are illustrated in Fig. 2.55(b) and (c). Points shaped as shown in either (b) or (c) are aggravating, because if a point is rounded, as in (b), the ink flows too freely; if it is too pointed, as in (c), the ink cups up and the flow is difficult to start. Only a pen that is correctly shaped and sharpened, as shown in Fig. 2.55(d), will give the results that should be expected.

Although blades should be sharpened to a thin edge, care should be taken not to make them sharp enough to cut the surface of the paper. A pen should never be sharpened, even the slightest amount, on the inside of the blades.

A fine-grained Arkansas oilstone is the best all-purpose stone for sharpening a ruling pen. The first step in sharpening is to equalize the length of the nibs and correct their shape. This may be done by bringing the blades together so that they barely touch and then drawing them lightly back and forth across the stone while swinging the pen through an arc of approximately 120° each stroke (Fig. 2.56). During this operation, it is essential that the pen be held in a vertical plane and that an even pressure be maintained against the stone. When an inspection, under a magnifying glass, reveals that the nibs have been restored to their correct shape, the blades should be opened and each nib sharpened all around the outside to a thin edge. A blade is sharpened by holding the pen as shown in Fig. 2.57 and sliding it back and forth across the stone with a slight rolling motion to preserve its original convex shape. The pen should be

Fig. 2.56 Shaping the nibs.

Fig. 2.57 Sharpening the blades.

examined from time to time so that the sharpening can be stopped as soon as the bright point disappears.

Finally, a test should be made by filling the pen and drawing a series of lines of various weights on a piece of tracing paper or tracing cloth. If the instrument is capable of drawing satisfactory lines of any weight, particularly very fine lines, it has been correctly sharpened.

2.31
Conventional Line Symbols. Symbolic lines of various weights are used in making technical drawings. The recommendations of the American National Standards Institute, as given in ANSI Y14.2M–1979, are the following:

Two widths of lines—thick and thin—are recommended for use on drawings (Fig. 2.58). Pencil lines in general should be in proportion to the ink lines except that the thicker pencil lines will be necessarily thinner than the corresponding ink lines but as thick as practicable for pencil work. Exact thicknesses may vary according to the size and type of drawing. For example, where lines are close together, the lines may be slightly thinner. The ratio of line thickness should be about two-to-one. The thin line width should be approximately 0.35 mm (.015 in.) and the thick line width approximately 0.7 mm (.03 in.).

Ink lines on drawings prepared for catalogs and books may be drawn using three widths—thick, medium, and thin. This provides greater contrast between types of lines and gives a better appearance.

The lines illustrated in Fig. 2.58 are shown full-size. When symbolic lines are used on a pencil drawing they should not vary in color. For example, center lines, extension lines, dimension lines, and sec-

tion lines should differ from object lines only in width. The resulting contrast makes a drawing easier to read. All lines, except construction lines, should be very dark and bright to give the drawing the "snap" that is needed for good appearance. If the drawing is on tracing paper the lead must be "packed on" so that a satisfactory print can be obtained. Construction lines should be drawn *very* fine so as to be unnoticeable on the finished drawing. The lengths of the dashes and spaces shown in Figs. 5.29 and 7.21 are recommended for the hidden lines, center lines, and cutting-plane lines on average-size drawings.

B □ Freehand technical lettering

2.32
Introduction. To impart to the workers in the shops all the necessary information for the complete construction of a machine or structure, the shape description, which is conveyed graphically by the views, must be accompanied by size descriptions and instructive specifications in the form of figured dimensions and notes (Fig. 2.59).

All dimensions and notes should be lettered freehand in a plain, legible style that can be rapidly executed. Poor lettering detracts from the appearance of a drawing and often impairs its usefulness, regardless of the quality of the line work.

2.33
Single-stroke Letters (Reinhardt). Single-stroke letters are now used universally for technical drawings. This style is suitable for most purposes because it possesses the qualifications necessary for

NAME	PENCIL DRAWING	INK DRAWING
VISIBLE OUTLINE (Approximately 0.5)	———————— THICK	———————— THICK
HIDDEN OUTLINE (Approximately 0.3)	– – – – – – THIN	– – – – – – MEDIUM
CENTER LINE	——–——–—— THIN	——–——–—— THIN
CROSS SECTION LINES	///////// THIN	///////// THIN
DIMENSION LINES, EXTENSION LINES, AND LEADERS	|← —— 98 —— →| THIN	|← —— 98 —— →| THIN
CUTTING PLANE LINES (THICK)	SAE STANDARD / ASA STANDARD	— — — / ▬ ▬ ▬
BREAK LINES	THICK — SHORT BREAK / THIN ∿ LONG BREAK ∿	∿ SHORT BREAK / ∿ LONG BREAK ∿
PHANTOM LINE	THIN — ADJACENT PARTS / ALTERNATE POSITION	REPEATED DETAIL / LINES OF MOTION

– Short Break Line

– Long Break Line

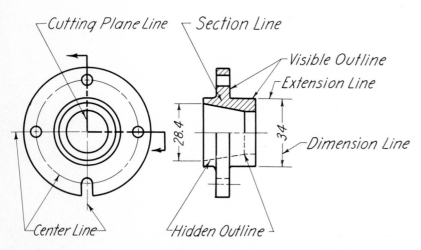

Cutting Plane Line   Section Line

Visible Outline

Extension Line

Dimension Line

Hidden Outline

Center Line

Phantom Line (Alternate Position)

Fig. 2.58 Alphabet of lines (finished weight).

53.30%   NORTH AMERICA

15.22%   SOUTH AMERICA

9.85%   EUROPE

0.38%   AFRICA

21.25%   ASIA

WORLD PRODUCTION
OF
CRUDE PETROLEUM
(a) Chart

3 RIBS - EQUISPACED

FILLETS & ROUNDS 1.5 R

CLAMP WASHER
C.I   I REQ'D.

(b) Machine Drawing

(c) Structural Drawing

66 KV - 3∅ - 60 ∿
INCOMING LINE NO. 6603

1200A-69KV

3-69KV 115V

(d) Electrical Diagram
From ASA Y32.2-1954

Fig. 2.59 Technical drawings.

legibility and speed. On commercial drawings it appears in slightly modified forms, however, since each person finally develops a style that reflects his own individuality.

The expression *single-stroke* means that the width of the straight and curved lines that form the letters are the same width as the stroke of the pen or pencil.

2.34
General Proportions of Letters. Although there is no fixed standard for the proportions of the letters, certain definite rules must be observed in their design if one wishes to have his lettering appear neat and pleasing. The recognized characteristics of each letter should be carefully studied and then thoroughly learned through practice.

It is advisable for the beginner, instead of relying on his untrained eye for proportions, to follow the fixed proportions given in this chapter. Otherwise, his lettering will most likely be displeasing to the trained eye of the professional. Later, after he has thoroughly mastered the art of lettering, his individuality will be revealed naturally

by slight variations in the shapes and proportions of some of the letters.

It is often desirable to increase or decrease the width of letters in order to make a word or group of words fill a certain space. Letters narrower than normal letters of the same height are called *compressed letters;* those that are wider are called *extended letters* (Fig. 2.60).

2.35
Lettering Pencils and Pens. Pencil lettering is usually done with a medium-soft pencil. Since the degree of hardness of the lead required to produce a dark opaque line will vary with the type of paper used, a pencil should be selected only after drawing a few trial lines. In order to obtain satisfactory lines, the pencil should be sharpened to a long conical point and then rounded slightly on a piece of scratch paper. To keep the point symmetrical while lettering, the pencil should be rotated a partial revolution before each new letter is started.

The choice of a type of pen point for lettering depends largely upon the personal preference and characteristics of

NORMAL   LETTERS

COMPRESSED LETTERS

EXTENDED   LETTERS

Fig. 2.60 Compressed and extended letters.

Fig. 2.61 Henry tank pen.

BARCH-PAYZANT

(a)

EDCO

(b)

LEROY

(c)

SPEEDBALL

(d)

Fig. 2.62 Special lettering pens.

the individual. The beginner can learn only from experience which of the many types available are best suited to him.

A pen that makes a heavy stroke should be used for bold letters in titles, and so forth, while a light-stroke pen is required for the lighter letters in figures and notes.

A very flexible point should never be used for lettering. Such a point is apt to shade the downward stem strokes as well as the downward portions of curved strokes. A good point has enough resistance to normal pressure to permit the drawing of curved and stem strokes of uniform width.

There are ordinary steel pen points with special ink-holding devices that make them especially suitable for lettering. The *Henry tank* pen, for example, shown in Fig. 2.61, has an ink reservoir that holds the ink above the point so that it feeds down the slit in an even flow. This device further assists in maintaining a uniform line by preventing the point from spreading.

Four of the many special pens designed for single-stroke letters are illustrated in Fig. 2.62. The *Barch–Payzant pen* (*a*) is available in graded sizes from No. 000 (very coarse) to No. 8 (very fine). The very fine size is suitable for lettering 3–5 mm (.12–.20 in.) high on technical drawings. The *Edco* (*b*) has a patented holder into which any one of a graded set of lettering nibs (ranging in sizes from No. 0 to No. 6) may be screwed. The tubular construction of the point makes it possible to draw uniform lines regardless of the direction of the stroke. Also of tubular construction is the *Leroy* (*c*). The *Speedball* (*d*) may be obtained in many graded sizes.

2.36

Devices for Drawing Guide Lines and Slope Lines. Devices for drawing guide lines (Fig. 2.63) are available in a variety of forms. The two most popular are the *Braddock lettering triangle* (Fig. 2.64), and the *Ames lettering guide* (Fig. 2.65).

The Braddock lettering triangle is provided with sets of grouped countersunk holes that may be used to draw guide lines by inserting a sharp-pointed pencil (4H or 6H) into the holes and sliding the triangle back and forth along the guiding edge of a T-square or a triangle supported by a T-square (Fig. 2.64). The holes are grouped to give guide lines for capitals and lower-case letters. The numbers below each set indicate the height of the capitals in thirty-seconds of an inch. For example, the No. 3 set is for capitals $\frac{3}{32}$ in. high, the No. 4 set is for capitals $\frac{1}{8}$ in. high, the No. 5 is for capitals $\frac{5}{32}$ in. high, and so on. For a metric drawing, using a metric guide, the lettering for the drawing number and drawing title should be about 7 mm high. Dimension values, tolerances, and subtitles may vary in height from 3.5 mm to 5 mm depending on the size of the drawing.

2.37

Uniformity in Lettering. Uniformity in height, inclination, spacing, and strength of line is essential for good lettering (Fig. 2.66). Professional appearance depends as much on uniformity as on the correctness of the proportion and shape of the individual letters. Uniformity in height and inclination is assured by the use of guide lines and slope lines and uniformity in weight and color, by the skillful use of the

Fig. 2.63 Guide lines and slope lines.

pencil and proper control of the pressure of its point on the paper. The ability to space letters correctly becomes easy after continued thoughtful practice.

2.38
Composition. In combining letters into words, the spaces for the various combinations of letters are arranged so that the areas appear to be equal (Fig. 2.67). For standard lettering, this area should be about equal to one-half the area of the letter M. If the adjacent sides are stems, this area is obtained by making the distance between the letters slightly greater than one-half the height of a letter, and a smaller amount, depending on the contours, for other combinations. Examples of good and poor composition are shown in Fig. 2.67.

The space between words should be equal to or greater than the height of a letter but not more than twice the height. The space between sentences should be somewhat greater. The distance between lines of lettering may vary from one-half the height of the capitals to $1\frac{1}{2}$ times their height.

2.39
Stability. If the areas of the upper and lower portions of certain letters are made equal, an optical illusion will cause them to appear to be unstable and top-heavy. To overcome this effect, the upper portions of the letters B, E, F, H, K, S, X, and Z and the figures 2, 3, and 8 must be reduced slightly in size.

An associated form of illusion is the phenomenon that a horizontal line drawn across a rectangle at the vertical center will appear to be below the center. Since the letters B, E, F, and H are particularly subject to this illusion, their central horizontal strokes must be drawn slightly above the vertical center in order to give them a more balanced and pleasing appearance.

The letters K, S, X, Z and the figures 2, 3, and 8 are stabilized by making the width of the upper portion less than the width of the lower portion.

2.40
Technique of Freehand Lettering. Any prospective engineer or technologist can learn to letter if he practices intelligently and is persistent in his desire to improve. The necessary muscular control, which must accompany the knowledge of letter-

To draw horizontal guide lines, insert the pencil point in the proper holes and slide the triangle back and forth. Use side of slot for drawing inclined slope lines.

Fig. 2.64 Braddock lettering triangle.

ing, can be developed only through constant repetition.

Pencil letters should be formed with strokes that are dark and sharp, never with strokes that are gray and indistinct. Beginners should avoid the tendency to form letters by sketching, as strokes made in this manner vary in color and width.

When lettering with ink, the results obtained depend largely on the manner in which the pen is used. Many beginners complain that the execution of good freehand lettering is impossible with an ordi-

Fig. 2.65 Ames lettering instrument.

UNIFORMITY IN HEIGHT, INCLINATION, AND STRENGTH OF LINE IS ESSENTIAL FOR GOOD LETTERING

Fig. 2.66 Uniformity in lettering.

POOR

GOOD

Fig. 2.67 Letter areas.

Fig. 2.68 "Inking" the pen.

INK ON POINT

Fig. 2.69

Fig. 2.70 Holding the pen.

nary pen point, although their own incorrect habits may have resulted in the inability to make strokes of uniform width. This lack of uniformity may be due to one of four causes: (1) excessive pressure on the pen point; (2) an accumulation of lint, dirt, or dried ink on the point; (3) tilting the point while forming a stroke; or (4) fresh ink on the point. The latter cause requires some explanation, since very few persons know the proper method of "inking" the pen. The pen should be wiped thoroughly clean, and the ink should be deposited on the underside over the slot, well above the point (Fig. 2.68). When the pen is filled in this manner, the ink feeds down the slit in an even flow, making possible the drawing of uniform curved and straight lines. If ink is placed on the point or allowed to run to the point, an excessive amount of ink will be deposited on the first letters made, and the width of the strokes will be somewhat wider than the strokes of, say, the sixth or seventh letter (Fig. 2.69).

When lettering, the pen is held as shown in Fig. 2.70. It should rest so loosely between the fingers that it can be slid up and down with the other hand.

The thin film of oil on a new point must be removed by wiping before the point is used.

2.41
Inclined and Vertical Capital Letters. The letters shown in Figs. 2.71 and 2.72 have been arranged in related groups. In laying out the characters, the number of widths has been reduced to the smallest number consistent with good appearance; similarities of shape have been emphasized and minute differences have been eliminated. Each letter is drawn to a large size on a cross-section grid that is 2 units wider, to facilitate the study of its characteristic shape and proportions. Arrows with numbers indicate the order and direction of the strokes. The curves of the inclined capital letters are portions of ellipses, while the curves of the vertical letters are parts of circles.

The I, T, L, E, and F

The letter I is the basic or stem stroke. The horizontal stroke of the T is drawn first, and the stem starts at the exact center of the bar. The L is 5 units wide, but it is often desirable to reduce this width when an L is used in combination with such letters as A and T. It should be observed that the letter L consists of the first two strokes of the E. The middle bar of the E is $3\frac{1}{2}$ units long and is placed slightly above the center for stability. The top bar is $\frac{1}{2}$ unit shorter than the bottom bar. The letter F is the E with the bottom bar omitted.

The H and N

Stroke 3 of the H should be slightly above the center, for stability. The outside parallel strokes of the N are drawn first to permit an accurate estimate of its width. The inclined stroke should intersect these accurately at their extremities.

The Z and X

The top of the Z should be 1 unit narrower than the bottom, for stability. In the smaller sizes, this letter may be formed without lifting the pen. The X is similar to the Z in that the top is made 1 unit narrower than the bottom. The inclined strokes cross slightly above center.

The A, V, M, and W

The horizontal bar of the A is located up from the bottom a distance equal to one-third the height of the letter. The V is the letter A inverted without the crossbar, and is the same width. The letters M and W are the widest letters of the alphabet. The

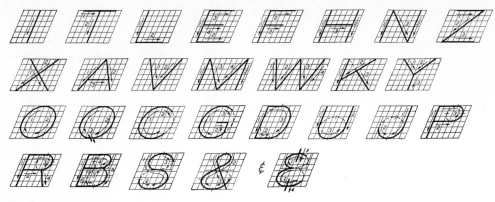

Fig. 2.71 Inclined capital letters.

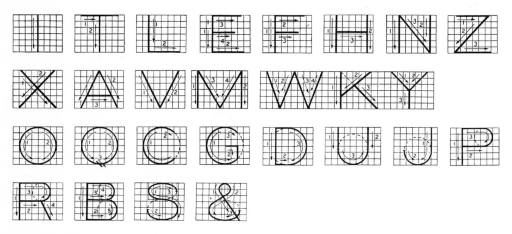

Fig. 2.72 Vertical capital letters.

outside strokes of the M are drawn first, so that its width may be judged accurately. The inside strokes of this letter meet at the center of the base. The W is formed by two modified V's. Alternate strokes are parallel.

The K and Y

The top of the letter K should be made 1 unit narrower than the bottom, for stability. Stroke 2 intersects the stem one-third up from the bottom. Stroke 3 is approximately perpendicular to stroke 2, and, if extended, would touch the stem at the top. The strokes of the Y meet at the center of the enclosing parallelogram or square.

The O, Q, C, and G

Stroke 1 of the letter O starts just to the right of the top and continues to the left around the side to a point beyond the bottom. Thus stroke 1 forms more than half of the ellipse or circle. The Q is the letter O with the added kern, which is a straight line located near the bottom tangent point. The C is based on the O, but since it is not a complete ellipse or circle, it is narrower than either the O or Q. The top extends 1 unit down and the bottom 1 unit up on the right side. G is similar to C. The horizontal portion of stroke 2 starts at the center.

The D, U, and J

The first two strokes of the D form an incomplete letter L. Stroke 3 starts as a horizontal line. The bottom third of the U is one-half of an ellipse or circle. J is similar to the letter U.

The P, R, and B

The middle horizontal bar of the P is located at the center of stroke 1. The curved portion of stroke 3 is one-half of a perfect ellipse or circle. The R is constructed similarly to the P. The tail joins at the point of tangency of the curve and middle bar. To stabilize the letter B, the top is made ½ unit narrower than the bottom and the middle bar is placed slightly above the center. The curves are halves of ellipses or circles.

The S and &

The upper and lower portions of the S are perfect ellipses with one-quarter removed. The top ellipse should be made ½ unit narrower than the lower one, for stability. In the smaller sizes this letter may be made with one or two strokes, depending upon its size. The true ampersand is made with three strokes. Professionals, however, usually represent an ampersand with a character formed by using portions of the upper and lower ellipses of the numeral 8 with the addition of two short bars (Fig. 2.71).

Although many favor the inclined letters, recent surveys indicate that vertical letters are used more generally.

2.42
Inclined and Vertical Numerals. The numerals shown in Figs. 2.73 and 2.74 have been arranged in related groups in accordance with the common characteristics that can be recognized in their construction.

The 1, 7, and 4

The stem stroke of the 4 is located 1 unit in from the right side. The bar is 1½ units above the base. The stem of the 7 terminates at the center of the base.

The 0, 6, and 9

The cipher, which is 1 unit narrower than the letter O, is the basic form for this group. In the figure 6, the right side of the large ellipse ends 1 unit down from the top, and the left side ends at the center of the base. The small loop is slightly more than three-fourths of an ellipse. The 9 is the 6 inverted.

The 8, 3, and 2

Each of these figures is related to the

letter S, and the same rule of stability should be observed in their construction. The top portion of the figure 8 is shorter and ½ unit narrower than the lower portion. Each loop is a perfect ellipse. The figure 3 is the 8 with the lower-left quarter of the upper loop and the upper-left quarter of the lower loop omitted. The 2 is simply three-quarters of the upper loop of the 8 and the upper-left quarter of the lower loop of the 8 with straight lines added.

The 5

This figure is a modification of the related groups previously described. The top is ½ unit narrower than the bottom, for stability. The curve is a segment of a perfect ellipse, ending one unit up from the bottom.

2.43
Single-stroke Lowercase Letters. Single-stroke lowercase letters, either vertical or inclined, are commonly used on map drawings, topographic drawings, structural drawings, and in survey field books. They are particularly suitable for long notes and statements because, first, they can be executed much faster than capitals and, second, words and statements formed with them can be read more easily.

The construction of inclined lowercase letters (Fig. 2.75) is based on the straight line and the ellipse. This basic principle of forming letters is followed more closely for lowercase letters than for capitals. The body portions are two-thirds the height of the related capitals. As shown in Fig. 2.76, ascenders extend to the cap line, and descenders descend to the drop line. For lowercase letters based on a capital letter 6 units high, the waistline is 2 units down from the top and the drop line 2 units below the base line.

The order of stroke, direction of stroke, and formation of the letters follow the same principles as for the capitals. The letters are presented in family groups having related characteristics, to enable the beginner to understand their construction. The vertical lowercase letters, illustrated in Fig. 2.77, are constructed in the same manner as inclined letters.

The i, l, k, and t

All letters of this group are formed by

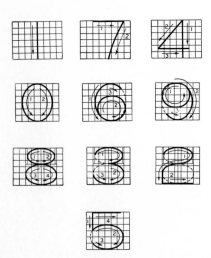

Fig. 2.73 Inclined numerals.

Fig. 2.74 Vertical numerals.

straight lines of standard slope. The i is drawn 4 units high, and the dot is placed halfway between the waistline and cap line. Stroke 2 of the k starts at the waistline and intersects stroke 1 at a point 2 units above the base. Stroke 3, extended should intersect stroke 1 at the top. The t is 5 units high, and the crossbar is on the waistline.

The v, w, x, and z

All of these letters are similar to the capitals. Alternate strokes of the w are parallel. The width of the top of both the x and the z is made ½ unit less than the width across the bottom, for stability.

The o, a, b, d, p, and q

The bodies of the letters in this group are formed by the letter o, and they differ only in the position and length of the stem stroke. The o is made with two strokes, and the first stroke should form more than half of the character.

The g

The g is related to the letters o and y. Stroke 3 starts at the waistline and ends slightly beyond the point of tangency of the curve with the drop line.

The c and e

The c is a modified letter o. It is not a complete form, and, therefore, its width is less than its height. Stroke 1 ends 1 unit up on the right side, stroke 2 ends 1 unit down. The e is similarly constructed, except for the fact that stroke 2 continues as a curve and finishes as a horizontal line that terminates at the middle of the back.

The h, n, r, and m

The curve of the h is the upper portion of the letter o. Stroke 2 starts 2 units above the bottom of the stem and finishes parallel to stroke 1. The n differs from the h in that the stem stroke extends only from the waistline to the base line. The r is a portion of the letter n, stroke 2 ending 1 unit down from the top. The m consists of two modified letter n's. The straight por-

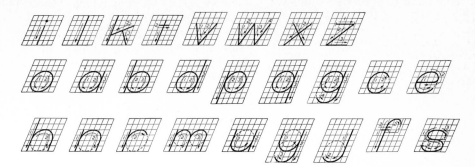

Fig. 2.75 Inclined lowercase letters.

Fig. 2.76

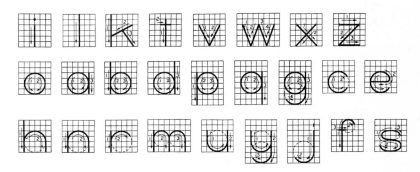

Fig. 2.77 Vertical lowercase letters.

tions of strokes 2 and 3 are parallel to stroke 1.

The u and y

The letter u is an inverted n, and the curve is a portion of the letter o. It should be noted that stroke 2 extends to the base line. The y is a partial combination of the letters u and g.

The j and f

The portion of the j above the base line is the letter i. The curve is the same as that which forms the tail of the g. The curved portion of the letter f is 2½ units

Fig. 2.78

*HANDLE PIN
C.R.S. 1 REQ'D.*

Fig. 2.79 Use of large and small caps.

APPROVED BY CUSTOMER	*ROSS-SHOER CO.*
ADAMS MFG. CO. AURORA, ILL.	
ASSEMBLY STANDARD DUTY FLANGE UNIT	

SCALE *FULL SIZE*	DATE *JAN. 3, 1959*
DRAWN BY *CARL KING*	CHECKED BY *J.B. JONES*
TRACED BY *K. BODINE*	APPROVED BY *W.L. KOE*
ORDER NO. *M-L-24631*	DRG. *5-19-9612*

Fig. 2.80 Machine-drawing title block.

wide, stroke 1 starting slightly to the right of the point of tangency with the cap line.

The s

The lowercase s is almost identical to the capital S.

2.44
Fractions. The height of the figures in the numerator and denominator is equal to three-fourths the height of the whole number, and the total height of the fraction is twice the height of the whole number. The division bar should be horizontal and centered between the fraction numerals, as shown in Fig. 2.78. It should be noted that the sloping center line of the fraction bisects both the numerator and denominator and is parallel to the sloping center line of the whole number.

2.45
Large Caps and Small Caps in Combination. Many commercial draftsmen use a combination of large caps and small caps in forming words, as illustrated in Fig. 2.79. When this style is used, the height of the small caps should be approximately three-fifths the height of the first capital letter of the word.

2.46
Titles. Every drawing, sketch, graph, chart, or diagram has some form of descriptive title to impart certain necessary information and to identify it. On machine drawings, where speed and legibility are prime requirements, titles are usually single-stroke. On display drawings, maps, and so on, which call for an artistic effect, the titles are usually composed of "built-up" ornate letters.

Figure 2.80 shows a title block that

Fig. 2.81 Leroy lettering device.

might be used on a machine drawing. It should be noted that the important items are made more prominent by the use of larger letters formed with heavier lines. Less important data, such as the scale, date, drafting information, and so on, are given less prominence.

To be pleasing in appearance, a title should be symmetrical about a vertical center line and should have some simple geometric form. An easy way to ensure the symmetry of a title is first to count the letter and word spaces, then, working to the left and right from the middle space or letter on the vertical center line, to sketch the title lightly in pencil, before lettering it in finished form. An alternative method is to letter a trial line along the edge of a piece of scrap paper and place it in a balanced position just above the location of the line to be lettered.

2.47
Mechanical Lettering Devices and Templates. Although mechanical lettering devices produce letters that may appear stiff to an expert, they are used in many drafting rooms for the simple reason that they enable even the unskilled to do satisfactory lettering with ink. The average draftsman rightly prefers stiff uniformity to wavy lines and irregular shapes. One of the oldest instruments of this sort on the market is the *Wrico* outfit. With a satisfactory set of Wrico pens and templates, letters ranging in size from $3/32$ to $1/2$ in. in height may be executed. The letters are formed by a stylographic pen that is guided around the sides of openings in a template made of transparent pyralin.

The *Leroy* device, shown in Fig. 2.81, is possibly even more efficient than the Wrico, for it does not require the sliding of a template to complete a letter.

2.48
Commercial Gothic (Fig. 2.82). Gothic letters are comparatively easy to construct. They are drawn first in outline, in skeleton form, and then filled in with a ballpoint pen. This style, from which the single-stroke engineering letters are derived, is used to some extent for titles on machine drawings, but more frequently it appears on maps, charts, and graphs that are prepared for display purposes.

2.49
Modern Roman. The modern Roman letters shown in Fig. 2.83 are used extensively by engineers for names and titles

on maps. Students should be familiar with this alphabet, for it appears in all modern publications. As in the case of the gothic style, these letters must be drawn first in outline, in skeleton form, and then filled in with a ballpoint or wide-line pen. The straight lines are usually drawn with a ruling pen. The curved lines, except for the serifs, may be formed either mechanically or freehand. When attempting to construct this type of letter it is wise to bear in mind the following facts: (1) All vertical strokes are heavy except those forming the letters M, N, and U. (2) All horizontal strokes are light. (3) The width of the heavy strokes may vary from one-eighth to one-sixth the height of the letter.

Roman letters may be drawn either extended or compressed, depending on the area to be covered or the space restrictions determined by other lettering, such as on maps. If a draftsman using the roman-style letter would have his work look professional, he must pay particular attention to detail. Care must be taken to keep the proportions of a letter appearing more than once in a name or title identical.

7 ABCDEFG
HIJKLMN
OPQRSTU
VWXYZ &
1234567
890

SCALE
0   10

Fig. 2.82 Gothic letters and numerals.

ABCDEFG
HIJKLMN
OPQRSTU
VWXYZ &
12345678
90

abcdefghijk
lmnopqrstu
vwxyz

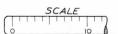

SCALE

Fig. 2.83 Modern Roman letters and numerals.

Exercises in instrumental drawing

The following elementary exercises have been designed to offer experience in the use of the drafting instruments. The designs should be drawn *lightly* with a hard pencil. After making certain that all constructions shown on a drawing are correct, the lines forming the designs should be heavied with a medium-hard pencil. The light construction lines need not be erased if the drawing is relatively clean.

1. (Fig. 2.84). On a sheet of drawing paper reproduce the line formations shown. If the principal border lines have not been printed on the sheet, they may be drawn first so that the large 130 × 195 mm rectangle can be balanced horizontally and vertically within the border. To draw the inclined lines, first draw the indicated measuring lines through the lettered points at the correct angle, and mark off the specified spacing. These division points establish the locations of the required lines of the formation. The six squares of the formation are equal in size.

2. (Fig. 2.85). Reproduce the line formations shown.

3. (Fig. 2.86). This exercise is designed to give the student practice with the bow pencil and compass by drawing some simple geometric figures. The line work within each large circle

may be reproduced with only the knowledge that the diameter is $3\frac{1}{4}$ in. (82 mm). All circles and circle arcs are to be made finished-weight when they are first drawn, since retracing often produces a double line. Do not "overrun" the straight lines or stop them too short.

Fig. 2.85

Fig. 2.84

Fig. 2.86

3.00 DIA.

Fig. 2.87

76.2 [3.00] DIA

Dimension in [ ] is in inches

Fig. 2.88

[76.2]
3 DIA.

[38.1]
1½ DIA.

*Dimensions in [ ] are in
millimeters*

Fig. 2.89

4–6. (Figs. 2.87–2.89). Reproduce the following designs according to the instructions given for problem 3, making the dashes of the arcs approximately ⅛ in. (3 mm) long.

7–9. (Figs. 2.90–2.92). Reproduce the line work within each square, using the dimensions given. (The dimensions shown, however, are for the student's use only and should not appear on the finished drawing.) Arcs should be made

9

72

72

Fig. 2.90

3

3

Fig. 2.91

3

3

Fig. 2.92

finished-weight when first drawn. The straight lines of each design may be drawn with a hard pencil and later heavied with a softer pencil. Do not erase the construction lines.

10–13. (Figs. 2.93–2.96). Reproduce the geometric shapes.

Dimension in [ ] is in inches

Fig. 2.93   Oval.

Dimension in [ ] is in inches

Fig. 2.94   Ellipse (approximate).

Fig. 2.95   Ellipse (approximate).

Fig. 2.96   Ellipse (pictorial).

Exercises

It should be noted that while these exercises are offered to give the student practice in letter forms and word composition, they also contain statements of important principles of drawing, shop notes, and titles with which every engineer should be familiar. Each lettering exercise should be submitted to the instructor for severe criticism before the student proceeds to the next. Section 2.34 should be reread before starting the first exercise.

1. Letter the statement given in Fig. 2.97 in ⁵⁄₃₂-in. (4 mm) capital letters using an appropriate pencil that is suited to the type of paper being used and that will produce uniform opaque lines. The necessary guide lines should be drawn with a hard pencil.

2, 3. Letter the statements given in Figs. 2.98 and 2.99 in ⅛-in. (3 mm) capital letters using an appropriate pencil that is suited to the type of paper being used and that will produce uniform opaque lines. The necessary guide lines should be drawn with a hard pencil.

4. Letter the statement given in Fig. 2.100 in ⁵⁄₃₂-in. (4 mm) capital letters using an appropri-

ate pencil that is suited to the type of paper being used and that will produce uniform opaque lines. The necessary guide lines should be drawn with a hard pencil.

5–7. Letter the statements given in Figs. 2.101 and 2.103 in ⅛-in. (3 mm) capital letters using an appropriate pencil that is suited to the type of paper being used and that will produce uniform opaque lines. The necessary guide lines should be drawn with a hard pencil.

8–14. Letter the following statements in ⅛-in. (3 mm) capital letters using an appropriate pencil that is suited to the type of paper being used and that will produce uniform opaque lines. The necessary guide lines should be drawn with a hard pencil.

8. A GOOD STUDENT REALIZES THE IMPORTANCE OF NEAT AND ATTRACTIVE LETTERING.

9. THE POSITION OF THE VIEWS OF AN ORTHOGRAPHIC DRAWING MUST BE IN STRICT ACCORDANCE WITH THE UNIVERSALLY RECOGNIZED ARRANGEMENT ILLUSTRATED IN FIG. 5.6.

POOR LETTERING DETRACTS FROM THE APPEARANCE OF A DRAWING

Fig. 2.97

IN LEARNING TO LETTER, CERTAIN DEFINITE RULES OF FORM & DESIGN MUST BE OBSERVED

Fig. 2.98

WHEN LETTERING WITH INK, THE INK SHOULD BE WELL ABOVE THE TIP OF THE POINT

Fig. 2.99

#26 (.1470) DRILL AND REAM FOR #1×1 TAPER PIN WITH PC #41 IN POSITION

Fig. 2.100 Lettered statement. Draw guide lines as shown.

10. THE VIEWS OF AN ORTHOGRAPHIC DRAWING SHOULD SHOW THE THREE DIMENSIONS: WIDTH, DEPTH, AND HEIGHT.

11. AN INVISIBLE LINE SHOULD START WITH A SPACE WHEN IT FORMS AN EXTENSION OF A SOLID LINE.

12. THE FRONT VIEW OF AN ORTHOGRAPHIC DRAWING SHOULD BE THE VIEW THAT SHOWS THE CHARACTERISTIC SHAPE OF THE OBJECT.

13. AN AUXILIARY VIEW SHOWS THE TRUE SIZE AND SHAPE OF AN INCLINED SURFACE.

14. A SECTIONAL VIEW SHOWS THE INTERIOR CONSTRUCTION OF AN OBJECT.

15. Draw horizontal and inclined guide lines and letter the following detail titles. Use $\frac{5}{32}$-in. (4 mm) capitals for the part names and $\frac{1}{8}$-in. (3 mm) capitals for the remainder of the titles.

BASE
C.I.   1 REQ'D

BUSHING
BRO.   1 REQ'D

SPINDLE
C.R.S.   1 REQ'D

16–20. Using a hard pencil, draw two or more sets of horizontal and inclined guide lines for $\frac{5}{32}$-in. letters; then letter the following exercises in lowercase letters.

16. The front and top views are always in line vertically.

17. The front and side views are in line horizontally.

18. The depth of the top view is the same as the depth of the side view.

19. If a line is perpendicular to a plane of projection, its projection will be a point.

20. If a line is parallel to a plane of projection, its projection on the plane is exactly the same length as the true length of the line.

S.A.E. 1020 –COLD DRAWN STEEL BAR
1–12 UNF–2B   1–8 UNC–2A   1–5 SQUARE
BREAK ALL SHARP CORNERS UNLESS OTHERWISE SPECIFIED

Fig. 2.101 Lettered statement. Draw guide lines as shown.

METRIC NOTES: TRUE R 18±0.1   Ø10±0.1–18 DEEP
3×45° CHAMFER
8±0.1–6 SLOTS EQL SP

METRIC

Fig. 2.102 Lettered statement. Draw guide lines as shown.

NECK 3 WIDE ×1.5 DEEP   M22×2.5–6H (20 DRILL)
Ø6±0.1–9.5 CBORE ×3 DEEP
6.5 THRU–10 DIA ×82° CSK

METRIC

Fig. 2.103 Lettered statement. Draw guide lines as shown.

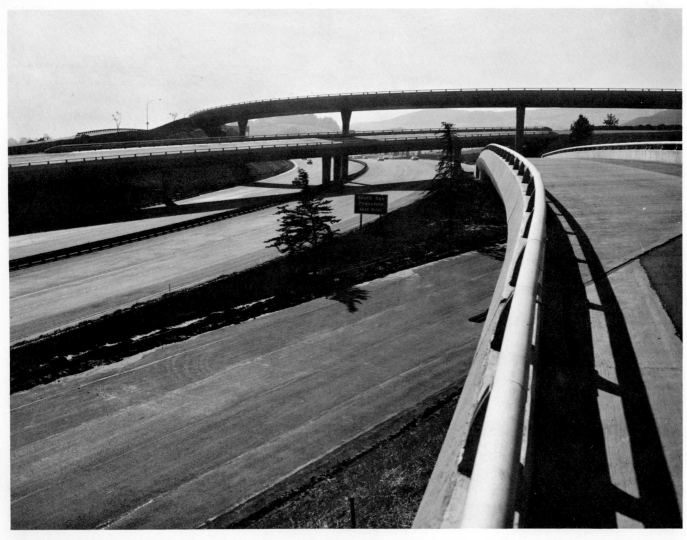

The design of highway interchanges involves the application of the geometry of circle arcs. The up-and-down curves of the concrete roadway are parabolic. In the mechanical engineering field, geometry is used in the design of an endless number of machine parts as well as for gears, cams, and linkages. (*Courtesy Department of Public Works, State of California*)

3.9

To Construct an Angle Equal to a Given Angle (Fig. 3.8). Given the angle BAC and the line $A'C'$ that forms one side of the transferred angle. Use any convenient radius with the vertex A as a center, and strike the arc that intersects the sides of the angle at D and E. With A' as a center, strike the arc intersecting $A'C'$ at E'. With E' as a center and the chord distance DE as a radius, strike a short intersecting arc to locate D'. $A'B'$ drawn through D' makes angle $B'A'C'$ equal angle BAC.

3.10

To Draw a Line Through a Given Point and the Inaccessible Intersection of Two Given Lines (Fig. 3.9). Given the lines KL and MN, and the point P. Construct any triangle, such as PQR, having its vertices falling on the given lines and the given point. At some convenient location construct triangle STU similar to PQR, by drawing SU parallel to PR, TU parallel to QR, and ST parallel to PQ. PS is the required line.

3.11

To Construct an Angle, Tangent Method (Fig. 3.10). Draftsmen and designers often find it necessary to draw long lines having an angle between them that is not equal to an angle of a triangle. Such an angle may be laid off with a protractor, but it should be remembered that as the lines are extended any error is multiplied. To avoid this situation, the tangent method may be used. The tangent method involves trigonometry but, since it is frequently used, a discussion of it here is pertinent. (See Table 1 of the Appendix.)

In this method, a distance D_1 is laid off along a line that is to form one side of the angle, and a distance D_2, equal to D_1 times the natural tangent of the angle, is marked off along a perpendicular through point P. A line through point X is the required line, and angle A is the required angle. In laying off the distance D_1, unnecessary multiplication will be eliminated if the distance is arbitrarily made 10 full-size units. However, in order to make the construction large enough for accuracy and at the same time keep all the lines on the drawing, it may be necessary to lay off the 10 units to some scale other than full-size. When a decimal-inch scale is being used, the construction may be drawn using either a half-size or quarter-size scale. If a metric scale is used one might decide upon the 1 : 50 or some other selected scale.

This method is also used for angles formed by short lines whenever a protractor is not available.

3.12

To Construct an Angle, Chord Method (Fig. 3.11). Engineers and draftsmen frequently select the chordal method for constructing an angle accurately. This method, as applied in laying out a given angle, involves the use of an easily determined chord length for a selected length of radius laid off in units. Given the angle (say, $29°$), the procedure is as follows: First, lay off any convenient distance, usually 10 units, along a line that is to form one side of the angle and strike an arc of indefinite length using this distance as a radius. Second, obtain the unit chord value for the given angle from Table 2 of the Appendix ($29° = .5008$) and multiply this value by 10. With the compass or dividers lay off the chord length ($.5008 \times 10 = 5.008$) along the arc from the starting line and complete the angle.

3.13

To Construct a Triangle, Given Its Three Sides (Fig. 3.12). Given the three sides AB, AC, and BC. Draw the side AB in its correct location. Using its end points A and B as centers and radii equal to AC and BC, respectively, strike the two intersecting arcs locating point C. ABC is the required triangle. This construction is particularly useful for developing the surface of a transition piece by triangulation.

3.14

To Construct an Equilateral Triangle (Fig. 3.13). Given the side AB.

(a) Using the end points A and B as centers and a radius equal to the length of AB, strike two intersecting arcs to locate C. Draw lines from A to C and C to B to complete the required equilateral triangle.

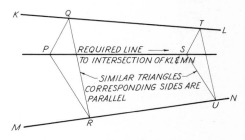

Fig. 3.9 To draw a line through a given point and the inaccessible intersection of two given lines.

Fig. 3.10 To construct an angle, tangent method.

Fig. 3.11 To construct an angle, chord method.

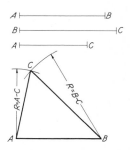

Fig. 3.12 To construct a triangle, given its three sides.

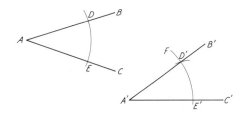

Fig. 3.8 To construct an angle equal to a given angle.

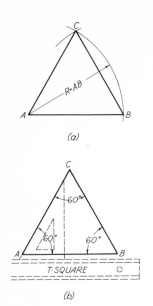

Fig. 3.13 To construct an equilateral triangle.

(b) Using a 30°–60° triangle, draw through A and B lines that make 60° with the given line. If the line AB is inclined, the 60° lines should be drawn as shown in Fig. 2.26

3.15
To Transfer a Polygon (Fig. 3.14). Given the polygon ABCDE.

(a) Enclose the polygon in a rectangle. Draw the "enclosing rectangle" in the new position and locate points A, B, C, D, and E along the sides by measuring from the corners of the rectangle. A compass may be used for transferring the necessary measurements.

(b) To transfer a polygon by the triangle method, divide the polygon into triangles and, using the construction explained in Sec. 3.13, reconstruct each triangle in its transferred position.

3.16
To Construct a Square (Fig. 3.15).

(a) Given the side AB. Using a T-square and a 45° triangle, draw perpendiculars to line AB through points A and B. Locate point D at the intersection of a 45° construction line through A and the perpen-

dicular from B. Draw CD parallel to AB through D to complete the square. To eliminate unnecessary movements the lines should be drawn in the order indicated.

(b) Given the diagonal length EF. Using a T-square and a 45° triangle, construct the square by drawing lines through E and F at an angle of 45° with EF in the order indicated.

(c) The construction of an inscribed circle is the first step in one method for drawing a square when the location of the center and the length of one side are given.

Using a T-square and a 45° triangle, draw the sides of the square tangent to the circle. This construction is used in drawing square bolt heads and nuts.

3.17
To Construct a Regular Pentagon (Fig. 3.16). Given the circumscribing circle. Draw the perpendicular diameters AB and CD. Bisect OB and, with its midpoint E as a center and EC as a radius, draw the arc CF. Using C as a center and CF as a radius, draw the arc FG. The line CG is one of the equal sides of the required pentagon. Locate the remaining vertices by striking off this distance around the circumference.

If the length of one side of a pentagon is given, the construction described in Sec. 3.20 should be used.

3.18
To Construct a Regular Hexagon (Fig. 3.17).

(a) Given the distance AB across corners. Draw a circle having AB as a diameter. Using the same radius and with points A and B as centers, strike arcs intersecting the circumference. Join these points to complete the construction.

(b) Given the distance AB across corners. Using a 30°–60° triangle and a

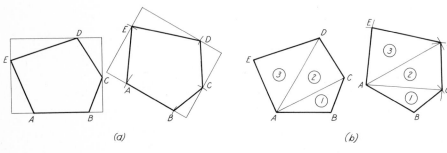

(a)   (b)

Fig. 3.14 To transfer a polygon.

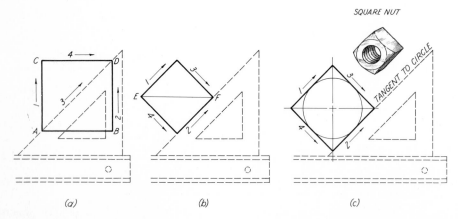

(a)   (b)   (c)

Fig. 3.15 To construct a square.

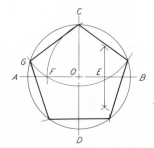

Fig. 3.16 To construct a regular pentagon.

Fig. 3.17 To construct a regular hexagon.

T-square, draw the lines in the order indicated by the numbers on the figure.

(*c*) Given the distance across flats. Draw a circle whose diameter equals the distance across flats. Using a 30°–60° triangle and a T-square, as shown, draw the tangents that establish the sides and vertices of the required hexagon.

This construction is used in drawing hexagonal bolt heads and nuts.

3.19
To Construct a Regular Octagon (Fig. 3.18).

(*a*) Given the distance across flats. Draw the circumscribed square and its diagonals. Using the corners as centers and one-half the diagonal as a radius, strike arcs across the sides of the square. Join these points to complete the required octagon.

(*b*) Given the distance across flats. Draw the inscribed circle; then, using a 45° triangle and T-square, draw the tangents that establish the sides and vertices of the required octagon.

3.20
To Construct Any Regular Polygon, Given One Side (Fig. 3.19).

Given the side LM. With LM as a radius, draw a semicircle and divide it into the same number of equal parts as the number of sides needed for the polygon. Suppose the polygon is to be seven-sided. Draw radial lines through points 2, 3, and so forth. Point 2 (the second division point) is always one of the vertices of the polygon, and line $L2$ is a side. Using point M as a center and LM as a radius, strike an arc across the radial line $L6$ to locate point N. Using the same radius with N as a center, strike another arc across $L5$ to establish O on $L5$. Although this procedure may be continued with point O as the next center, more

accurate results will be obtained if point R is used as a center for the arc to locate Q, and Q as a center for P.

3.21
To Divide the Area of a Triangle or Trapezoid into a Given Number of Equal Parts (Fig. 3.20).

(*a*) Given the triangle ABC. Divide the side AC into (say, five) equal parts, and draw a semicircle having AC the diameter. Through the division points (1, 2, 3, and 4) draw perpendicular lines to points of intersection with the semicircle (5, 6, 7, and 8). Using C as a center, strike arcs through these points (5, 6, 7, and 8) that will cut AC. To complete the construction, draw lines parallel to AB through the points (9, 10, 11, and 12) at which the arcs intersect the side AC.

(*b*) Given the trapezoid $DEBA$. Extend the sides of the trapezoid to form the triangle ABC and draw a semicircle on AC with AC as a diameter. Using C as a center and CD as a radius, strike an arc cutting the semicircle at point P. Through P draw a perpendicular to AC to locate point Q. Divide QA into the same number of equal parts as the number of equal

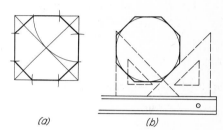

Fig. 3.18 To construct a regular octagon.

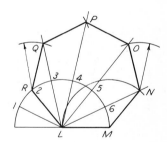

Fig. 3.19 To construct any regular polygon, given one side.

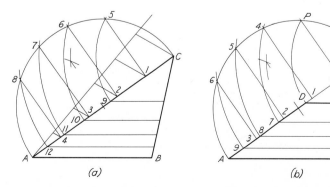

Fig. 3.20 To divide the area of a triangle or trapezoid into a given number of equal parts.

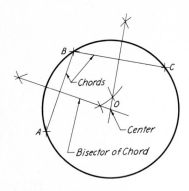

Fig. 3.21   To find the center of a circle through three points.

areas required (in this case, four), and proceed using the construction explained in (*a*) for dividing the area of a triangle into a given number of equal parts.

3.22

To Find the Center of a Circle Through Three Given Points Not in a Straight Line (Fig. 3.21).   Given the three points A, B, and C. Join the points with straight lines (which will be chords of the required circle), and draw the perpendicular bisectors. The point of intersection O of the bisectors is the center of the required circle, and OA, OB, or OC is its radius.

3.23

Tangent Circles and Arcs.   Figure 3.22 illustrates the geometry of tangent circles. In (*a*) it can be noted that the locus of centers for circles of radius R tangent to AB is a line that is parallel to AB at a distance R from AB. The locus of centers for circles of the same radius tangent to

CD is a line that is parallel to CD at distance R (radius) from CD. Since point O at which these lines intersect is distance R from both AB and CD, a circle of radius R with center at O must be tangent to both AB and CD.

In (*b*) the locus of centers for circles of radius R_3 that will be tangent to the circle with a center at O and having a radius R_1 is a circle that is concentric with the given circle at distance R_3. The radius of the locus of centers will be $R_1 + R_3$. In the case of the circle with center at point P, the radius of the locus of centers will be $R_2 + R_3$. Points Q and Q_1, where these arcs intersect, are points that are distance R_3 from both circles. Therefore, circles of radius R_3 that are centered at Q and Q_1 will be tangent to both circles with centers at O and P.

3.24

To Draw a Circular Arc of Radius R Tangent to Two Lines (Fig. 3.23).

(*a*) Given the two lines AB and CD at right angles to each other, and the radius of the required arc R. Using their point of intersection X as a center and R as a radius, strike an arc cutting the given lines at T_1 and T_2 (tangent points). With T_1 and T_2 as centers and the same radius, strike the intersecting arcs locating the center O of the required arc.

(*b*),(*c*) Given the two lines AB and CD, not at right angles, and the radius R. Draw lines EF and GH parallel to the given lines at a distance R. Since the point of intersection of these lines is distance R from both given lines, it will be the center O of the required arc. Mark the tangent points T_1 and T_2 that lie along perpendiculars to the given lines through O.

These constructions are useful for drawing fillets and rounds on views of machine parts.

3.25

To Draw a Circular Arc of Radius R_1 Tangent of a Given Circular Arc and a Given Straight Line (Fig. 3.24).   Given the line AB and the circular arc with center O.

(*a*),(*b*) Draw line CD parallel to AB at a distance R_1. Using the center O of the given arc and a radius equal to its radius plus or minus the radius of the required arc (R_2 plus or minus R_1), swing a parallel arc intersecting CD. Since the line CD and the intersecting arc will be the loci of centers for all circles of radius R_1, tangent respectively to the given line AB and the given arc, their point of intersection P will

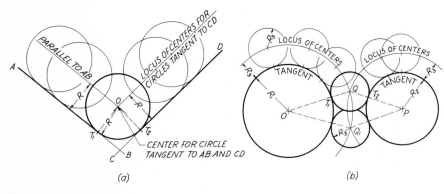

Fig. 3.22   Tangent circles.

(*a*)   (*b*)

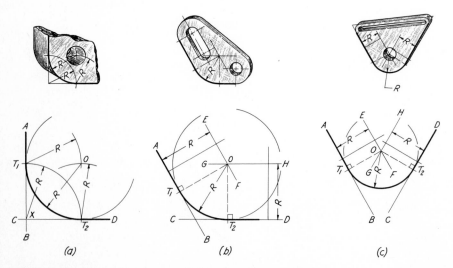

(*a*)   (*b*)   (*c*)

Fig. 3.23   To draw a circular arc tangent to two lines.

be the center of the required arc. Mark the points of tangency T_1 and T_2. T_1 lies along a perpendicular to AB through the center P, and T_2 along a line joining the centers of the two arcs.

This construction is also useful for drawing fillets and rounds on views of machine parts.

3.26
To Draw a Circular Arc of a Given Radius R_1 Tangent to Two Given Circular Arcs (Fig. 3.25). Given the circular arcs AB and CD with centers O and P, and radii R_2 and R_3, respectively.

(*a*) Using O as a center and R_2 plus R_1 as a radius, strike an arc parallel to AB. Using P as a center and R_3 plus R_1 as a radius, strike an intersecting arc parallel to CD. Since each of these intersecting arcs is the locus of centers for all circular arcs of radius R_1 tangent to the given arc to which it is parallel, their point of intersection S will be the center for the required arc that is tangent to both. Mark the points of tangency T_1 and T_2 that lie on the lines of centers PS and OS.

(*b*) Using O as a center and R_2 plus R_1 as a radius, strike an arc parallel to AB. Using P as a center and R_3 minus R_1 as a radius, strike an intersecting arc parallel to CD. The point of intersection of these arcs is the center for the required arc.

3.27
To Draw a Reverse (Ogee) Curve (Fig. 3.26).

(*a*) *Reverse (ogee) curve connecting two parallel lines.* Given the two parallel lines AB and CD. At points B and C, the termini and tangent points of the reverse curve, erect perpendiculars. Join B and C with a straight line and assume a point E as the point at which the curves will be tangent to each other. Draw the perpendicular bisectors of BE and EC. Since an arc tangent to AB at B must have its center on the perpendicular BP, the point of intersection P of the bisector and the perpendicular is the center for the required arc that is to be tangent to the line at B and the other required arc at point E. For the same reason, point Q is the center for the other required arc.

This construction is useful to engineers in laying out center lines for railroad tracks, pipelines, and so forth.

(*b*) *Reverse (ogee) curve connecting two nonparallel lines.* Given the two nonparallel lines AB and CD. At points B and C, the termini and tangent points,

erect perpendiculars. Along the perpendicular at B lay off the given (or selected) radius R and draw the arc having P as its center. Then draw a construction line through point P perpendicular to CD to establish the location of point X. With the position of X known, join points X and C with a straight line along which will lie the chords of the arcs forming the ogee curve between points X and C. The broken line XY (not a part of the construction) has

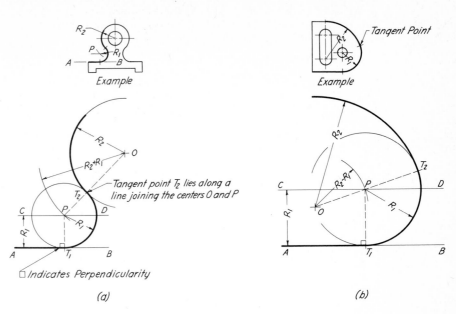

Fig. 3.24   To draw a circular arc tangent to a given circular arc and a line.

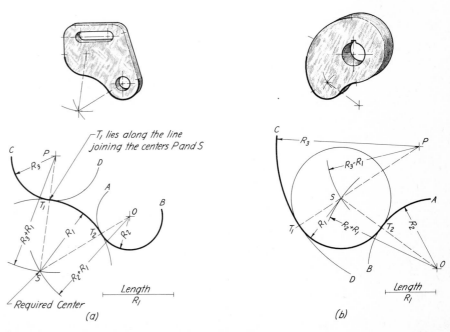

Fig. 3.25   To draw a circular arc tangent to two given arcs.

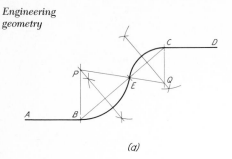

(a)

Fig. 3.26   To draw a reverse curve.

(b)

Fig. 3.27   To draw a reverse curve tangent to three lines.

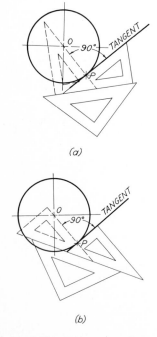

(a)

(b)

Fig. 3.28   To draw a line tangent to a circle at a point on the circumference.

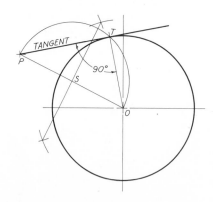

Fig. 3.29   To draw a line tangent to a circle through a given point outside.

been added to show that the procedure to be followed in completing the required curve will be as previously explained for drawing a reverse curve joining two parallel lines. In this case the parallel lines are XY and CD, instead of the lines AB and CD as in (a).

An alternative method for establishing the needed center for the required arc has been added to the illustration in (b). In this method the radius distance R is laid off upward along a perpendicular to CD through C. With point S established by this measurement, the line PS, as drawn, becomes the chord of an arc (not shown) that will have the same center as the required arc EC. The intersection of the perpendicular bisector of PS with the perpendicular erected downward from C will establish the position of point Q, the center of concentric arcs having chords PS and EC.

3.28
To Draw a Reverse Curve Tangent to Three Given Lines (Fig. 3.27). Given the lines AB and CD that are intersected by a third line BC at points B and C. Assume the position of point E (point of tangency) along BC and locate the termini points T_1

and T_2 by making CT_1 equal to CE and BT_2 equal to BE. The intersections of the perpendiculars erected at points T_1, E, and T_2 establish the centers P and Q of the arcs that form the reversed curve.

3.29
To Draw a Line Tangent to a Circle at a Given Point on the Circumference (Fig. 3.28). Given a circle with center O and point P on its circumference. Place a triangle supported by a T-square or another triangle in such a position that one side passes through the center O and point P. When using the method illustrated in (a), align the hypotenuse of one triangle on the center of the circle and the point of tangency; then, with the guiding triangle held in position, revolve the triagnle about the 90° angle and slide into position for drawing the required tangent line.

Another procedure is shown in (b). To draw the tangent by this method, align one leg of a triangle, which is adjacent to the 90° angle, through the center of the circle and the point of tangency; then slide it along the edge of a guiding triangle into position.

This construction satisfies the geometric requirement that a tangent must be perpendicular to a radial line drawn to the point of tangency.

3.30
To Draw a Line Tangent to a Circle Through a Given Point Outside the Circle (Fig. 3.29). Given a circle with center O and an external point P. Join the point P and the center O with a straight line, and bisect it to locate point S. Using S as a center and SO (one-half PO) as a radius, strike an arc intersecting the circle at point T (point of tangency). Line PT is the required tangent.

3.31
To Draw a Tangent Through a Point *P* on a Circular Arc Having an Inaccessible Center (Fig. 3.30). Draw the chord PB; then erect a perpendicular bisector. With point P as a center swing an arc through point C where the perpendicular bisector cuts the given arc. With C as a center and a radius equal to the chord distance CE, draw an arc to establish the location of point F. A line drawn through points P and F is the required tangent.

3.32
To Draw a Line Tangent to a Circle Through a Given Point Outside the Circle (Fig. 3.31). Place a triangle supported by

a T-square or another triangle in such a position that one leg passes through point P tangent to the circle, and draw the tangent. Slide the triangle along the guiding edge until the other leg coincides with the center O, and mark the point of tangency. Although this method is not as accurate as the geometric one explained in Sec. 3.30, it is frequently employed by commercial draftsmen.

3.33
To Draw a Line Tangent to Two Given Circles (Fig. 3.32). Given two circles with centers O and P and radii R_1 and R.

   (a) *Open belt.* Using P as a center and a radius equal to R minus R_1, draw an arc. Through O draw a tangent to this arc using the method explained in Sec. 3.30. With the location of tangent point T established, draw line PT and extend it to locate T_1. Draw OT_2 parallel to PT_1. The line from T_2 to T_1 is the required tangent to the given circles.

   (b) *Crossed belt.* Using P as a center and a radius equal to R plus R_1, draw an arc. With the location of tangent point T determined through use of the method shown in Fig. 3.29, locate tangent point T_1 on line TP and draw OT_2 parallel to PT. The line $T_1 T_2$, drawn parallel to OT, is the required tangent.

3.34
To Approximate a Curve with Tangent Circular Arcs (Fig. 3.33). Draftsmen often find it desirable to approximate a noncircular curve with a series of tangent arcs. If the curve consists of a number of points, a pleasing curve should be sketched lightly through these points before starting to draw the arcs. The centers and radii are selected by trial, but it must be remembered after the first arc has been drawn (as far as it coincides with the sketched curve) that when arcs are tangent, the centers are on a common normal through their point of tangency. Sometimes draftsmen use this method to draw curves in ink instead of using a French curve.

3.35
To Lay Off the Approximate Length of the Circumference of a Circle (Fig. 3.34). Draw a line through point A tangent to the circle and lay off along it a distance AB equal to three times the diameter ($3D$). Using point E on the circumference as a center and a radius equal to the radius of the circle, strike an arc to establish the

Fig. 3.30 To draw a tangent to a circular arc having an inaccessible center.

Fig. 3.31 To draw a line tangent to a circle through a given point outside.

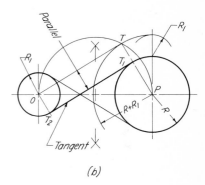

     (a)   (b)

Fig. 3.32 To draw a line tangent to two given circles.

location of point C. Draw CD perpendicular to the vertical center line through point A. DB is the rectified length of the circumference; however, it is slightly longer than the true circumference by a negligible amount (approximate error $1/21,800$).

3.36
To Lay Off the Approximate Length of a Circular Arc on Its Tangent (Fig. 3.35). Given the arc AB.

   (a) Draw the tangent through A, and extend the chord BA. Locate point C by laying off AC equal to one-half the length of the chord AB. With C as a center and a

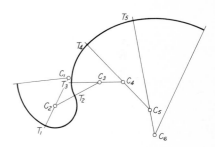

Fig. 3.33 To approximate a curve with tangent circular arcs.

Fig. 3.34 To lay off the approximate length of the circumference of a circle.

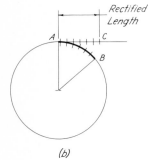

Fig. 3.35 To lay off the approximate length of a circular arc on its tangent.

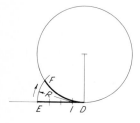

Fig. 3.36 To lay off a specified length along an arc.

radius equal to CB, strike an arc intersecting the tangent at D. The length AD along the tangent is slightly shorter than the true length of the arc AB by an amount that is so minute it may be disregarded.

(b) Draw the tangent through A. Using the dividers, start at B and step off equal chord distances around the arc until the point nearest A is reached. From this point (without raising the dividers) step off along the tangent an equal number of distances to locate point C. If the point nearest A is indented into the tangent instead of the arc, the almost negligible error in the length of AC will be still less.

Since the small distances stepped off are in reality the chords of small arcs, the length AC will be slightly less than the true length of the arc. For most practical purposes the difference may be disregarded.

When the central angle (θ) and the radius of an arc are known, the length of the arc may be computed by the formula $L = 2\pi R(\theta/360°) = 0.01745R\theta$.

3.37
To Lay Off a Specified Length Along a Given Circle Arc (Fig. 3.36). On the tangent to the arc, lay off the distance DE representing the specified length of arc. Divide DE into four equal parts. Then, using point 1 as a center and with the length $1–E$ as the radius R, strike an arc intersecting the given arc at F. The arc DF is approximately equal in length to the line DE. For large angles, it is advisable to make the construction for one-half of DE.

3.38
Conic Sections (Fig. 3.37). When a right circular cone of revolution is cut by planes at different angles, four curves of intersection are obtained that are called *conic sections*.

When the intersecting plane is perpen-

dicular to the axis, the resulting curve of intersection is a *circle*.

If the plane makes a greater angle with the axis than do the elements, the intersection is an *ellipse*.

If the plane makes the same angle with the axis as the elements, the resulting curve is a *parabola*.

Finally, if the plane makes a smaller angle with the axis than do the elements or is parallel to the axis, the curve of intersection is a *hyperbola*.

The geometric methods for constructing the ellipse, parabola, and hyperbola are discussed in succeeding sections.

3.39
Ellipse (Fig. 3.37). Mathematically the ellipse is a curve generated by a point moving so that at any position the sum of its distances from two fixed points (foci) is a constant (equal to the major diameter). It is encountered very frequently in orthographic drawing when holes and circular forms are viewed obliquely. Ordinarily, the major and minor diameters are known.

3.40
To Construct an Ellipse, Foci Method (Fig. 3.38). Draw the major and minor axes (AB, CD) and locate the foci F_1 and F_2 by striking arcs, centered at C and having a radius equal to OA (one-half of the major diameter). The construction is as follows: Determine the number of points needed along the circumference of each quadrant of the ellipse for a relatively accurate layout (say, four) and mark off this number of division points (P, Q, R, and S) between O and F_1 on the major axis. In many cases it may be desirable to use additional points spaced closer together nearer F_1 in order to form accurately the sharp curvature at the end of the ellipse. Next, with F_1

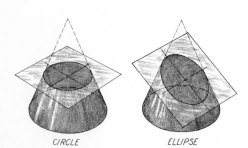

CIRCLE   ELLIPSE   PARABOLA   HYPERBOLA

Fig. 3.37 Conic sections.

Fig. 3.38 To construct an ellipse, foci (definition) method.

and F_2 as centers and the distances AP and BP as radii, respectively, strike intersecting arcs to locate P' on the circumference of the ellipse. Distances AQ and BQ are radii for locating points Q'. Locate points R' and S' in a similar manner and complete the ellipse using a French curve.

This method is sometimes known as the definition method, since it is based on the mathematical definition of the ellipse as given in Sec. 3.39.

3.41
To Construct an Ellipse, Trammel Method (Fig. 3.39). Given the major axis AB and the minor axis CD. Along the straight edge of a strip of paper or cardboard, locate the points O, C, and A so that the distance OA is equal to one-half the length of the major axis, and the distance OC is equal to one-half the length of the minor axis. Place the marked edge across the axes so that point A is on the minor axis and point C is on the major axis. *Point O will fall on the circumference of the ellipse.* Move the strip, keeping A on the minor axis and C on the major axis, and mark at least five other positions of O on the ellipse in each quadrant. Using a French curve, complete the ellipse by drawing a smooth curve through the points. The ellipsograph, which draws ellipses mechanically, is based on this same principle. The trammel method is an accurate method.

An alternative method for marking off the location of points A, O, and C is given in Fig. 3.39.

3.42
To Construct an Ellipse, Concentric Circle Method (Fig. 3.40). Given the major axis AB and the minor axis CD. Using the center of the ellipse (point O) as a center, describe circles having the major and minor axes as diameters. Divide the circles into equal central angles and draw diametrical lines such as P_1P_2. From point P_1 on the circumference of the larger circle, draw a line parallel to CD, the minor axis, and from point P_1' at which the diameter P_1P_2 intersects the inner circle, draw a line parallel to AB, the major axis. The point of intersection of these lines, point E, is on the required ellipse. At points P_2 and P_2' repeat the same procedure and locate point F. Thus, two points are established by the line P_1P_2. Locate at least five points in each of the four quadrants. The ellipse is completed by drawing a smooth curve through the points.

This is one of the most accurate methods used to form ellipses.

3.43
To Construct an Ellipse, Four-Center Method (Fig. 3.41). Given the major axis AB and the minor axis CD. Draw the line AC. Using the center of the ellipse O as a center and OC as a radius, strike an arc intersecting OA at point E. Using C as a center and EA as a radius, strike an arc intersecting the line AC at F. Draw the perpendicular bisector of the line AF. The points G and H, at which the perpendicular bisector intersects the axes AB and CD (extended), are the centers of two of the arcs forming the ellipse. Locate the other two centers, J and K, by laying off OJ equal to OH and OK equal to OG. To determine the junction points (tangent points) T, T_1, T_2, and T_3, for the arcs, draw lines through the centers of the tangent arcs. The figure thus formed by the four circle arcs approximates a true ellipse. When an accurate ellipse is required, this method should not be used.

3.44
To Construct an Ellipse, Parallelogram Method (Fig. 3.42). Given the major axis AB and the minor axis CD. Construct the circumscribing parallelogram. Divide AO and AE into the same number of equal parts (say, four) and number the division points from A. From C draw a line through point 3 on line AE, and from D draw a line through point 3 on line AO. The point of intersection of these lines is on the required ellipse. Similarly, the intersections of lines from C and D through points numbered 1 and 2 are on the ellipse. A similar construction will locate points in the other three quadrants of the ellipse. Use of a French curve will permit a

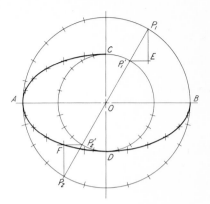

Fig. 3.40 To construct an ellipse, concentric circle method.

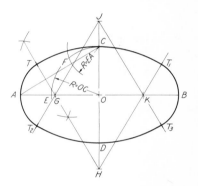

Fig. 3.41 To construct an ellipse, center method.

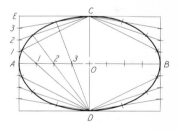

Fig. 3.42 To construct an ellipse, parallelogram method.

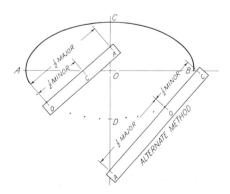

Fig. 3.39 To construct an ellipse, trammel method.

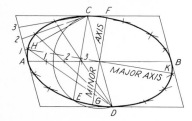

Fig. 3.43 To draw the major and minor axes of an ellipse, given the conjugate diameters.

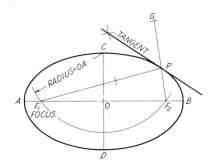

Fig. 3.44 To draw a tangent to an ellipse.

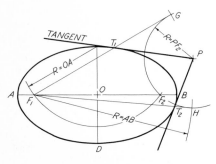

Fig. 3.45 To draw a tangent to an ellipse through a point outside the ellipse.

Fig. 3.46 To construct a parabola.

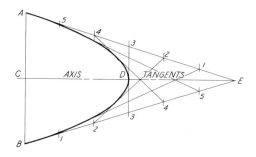

Fig. 3.47 To construct a parabola, tangent method.

smooth curve to be drawn through the points.

Had the circumscribing parallelogram not been a rectangle as in Fig. 3.42, the completed construction would appear as in Fig. 3.43, and AB and CD would be conjugate axes. To establish the major and minor axes, draw a semicircle on CD as a diameter, intersecting the ellipse at E. FG, running parallel to CE through the center of the ellipse, will be the required minor axis. HK, running through the center of the ellipse parallel to DE and perpendicular to FG, will be the major axis.

3.45

To Draw a Tangent to an Ellipse at Any Given Point (Fig. 3.44). Given any point, such as P, on the perimeter of the ellipse $ABCD$. Using C as a center and a radius equal to OA (one-half the major diameter), strike arcs across the major axis at F_1 and F_2. From these points, which are foci of the ellipse, draw F_1P and F_2G. The bisector of the angle GPF_1 is the required tangent to the ellipse.

3.46

To Draw a Tangent to an Ellipse From a Given Point P Outside the Ellipse (Fig. 3.45). With the end of the minor axis as a center and a radius R equal to one-half the length of the major axis, strike an arc to find the foci F_1 and F_2. With point P as a center and the distance PF_2 as a radius, draw an arc. Using F_1 as a center and the length AB as a radius, strike arcs cutting the arc with center P at points G and H. Draw lines GF_1 and HF_1 to establish the location of the tangent points T_1 and T_2. Draw the required tangent.

3.47

Parabola (Fig. 3.37). Mathematically the parabola is a curve generated by a point moving so that at any position its distance from a fixed point (the focus) is always

exactly equal to its distance to a fixed line (the directrix). The construction shown in Fig. 3.46 is based on this definition.

In engineering design, the parabola is used for parabolic sound and light reflectors, for vertical curves on highways, and for bridge arches.

3.48

To Construct a Parabola (Fig. 3.46). Given the focus F and the directrix AB. Draw the axis of the parabola perpendicular to the directrix. Through any point on the axis, for example, point C, draw a line parallel to the directrix AB. Using F as a center and the distance OC as a radius, strike arcs intersecting the line at points P_4 and P_4'. Repeat this procedure until a sufficient number of additional points have been located to determine a smooth curve. The vertex V is located at a point midway between O and F.

To construct a tangent to a parabola, say, at point P_6, draw the line P_6D parallel to the axis; then bisect the angle DP_6F. The bisector of the angle is the required tangent.

3.49

To Construct a Parabola, Tangent Method (Fig. 3.47). Given the points A and B and the distance CD from AB to the vertex. Extend the axis CD, and set off DE equal to CD. EA and EB are tangents to the parabola at A and B, respectively.

Divide EA and EB into the same number of equal parts (say, six) and number the division points as shown. Connect the corresponding points 1 and 1, 2 and 2, 3 and 3, and so forth. These lines, as tangents of the required parabola, form its envelope. Draw the tangent curve.

3.50

To Construct a Parabola, Offset Method (Fig. 3.48). Given the enclosing rectangle $A'ABB'$. Divide VA' into any number of equal parts (say, ten) and draw from the division points the perpendiculars parallel to VC, along which offset distances are to be laid off. The offsets vary as the square of their distances from V. For example, since V to 2 is two-tenths of the distance from V to A', 2-2' will be $(.2)^2$ or .04 of $A'A$. Similarly, 6-6' will be $(.6)^2$ or .36 of $A'A$; and 8-8' will be .64 of $A'A$. To complete the parabola, lay off the computed offset values along the perpendiculars and form the figure with a French curve.

The entire construction can be done graphically (as illustrated) by first calculating the values for the squared distances and then dividing the depth distance (along the axis) proportionally using these values. The graphical method shown in Fig. 3.7 was used.

The offset method is preferred by civil engineers for laying out parabolic arches and computing vertical curves for highways. The parabola shown in Fig. 3.48 could represent a parabolic reflector.

3.51
To Construct a Curve of Parabolic Form Through Two Given Points (Fig. 3.49). Given the points A and B. Assume a point C. Draw the tangents CA and CB, and construct the parabolic curve using the tangent method shown in Fig. 3.47. This method is frequently used in machine design to draw curves that are more pleasing than circular arcs.

3.52
To Construct a Parabola, Parallelogram Method. Since the dimensions for a parallelogram that will enclose a given parabola are generally known, a parabola may be constructed by the parallelogram method illustrated in Fig. 3.50. With the enclosing rectangle drawn to the given width and depth dimensions, divide VA and AB into the same number of equal divisions (say, five) and number the division points as shown in (a). Draw light construction lines from point V to each of the division points along AB. Then draw lines parallel to the axis from points 1, 2, 3, and 4 on VA. The intersection of the construction lines from points numbered 1 is on the parabola.

Likewise, the intersection of the lines from points numbered 2 is on the parabola. The complete parabolic outline passes through the additional points at the intersection of the lines from points numbered 3 and the lines from points numbered 4. The method as explained for a parabola enclosed in a rectangle may be applied to a nonrectangular shape, as shown in (b).

3.53
To Locate the Directrix and Focus of a Given Parabolic Curve (Fig. 3.51). With the location of the axis known [see (a)] draw the tangent PY by locating point Y on the axis extended at a distance D_1 from the vertex V. Draw PX parallel to the axis and construct angle FPY equal to YPX. The point at which the line forming the newly constructed angle cuts the given axis is the focus F. The directrix through O

is perpendicular to the axis at the same distance from the vertex V as the focus F.

When the position of the axis is not known, the procedure illustrated in (b) will establish the location of the axis, focus, and directrix. As the initial step, draw two parallel chords at random and locate the midpoint of each, points R and S. Draw line RS to establish the direction of the axis that will be parallel to RS. Next, draw a chord perpendicular to RS at any location and through the midpoint M draw the axis of the parabola as required. The position of the vertex is now known and one might follow the procedure given in (a) to locate the focus and directrix. However, the position of the directrix may be found easily and quickly merely by drawing a tangent to the parabola at 45° with the axis. The directrix passes through O at the intersection of the tangent and the axis extended.

3.54
Hyperbola (Fig. 3.37). Mathematically the hyperbola can be described as a curve generated by a point moving so that at any position the difference of its distances

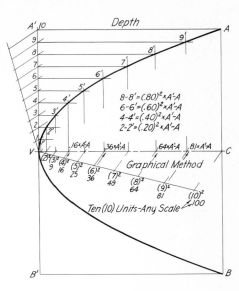

Fig. 3.48 To construct a parabola, offset method.

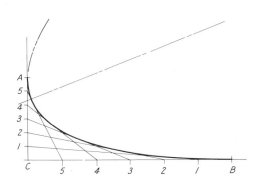

Fig. 3.49 To construct a curve of parabolic form.

Fig. 3.50 To construct a parabola, parallelogram method.

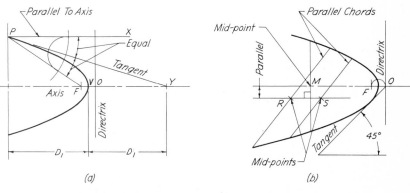

Fig. 3.51 To locate the directrix and focus of a given parabolic curve.

Fig. 3.52 To construct a hyperbola.

Fig. 3.53 To construct an equilateral hyperbola.

from two fixed points (foci) is a constant (equal to the transverse axis of the hyperbola). This definition is the basis for the construction shown in Fig. 3.52.

3.55
To Construct a Hyperbola (Fig. 3.52). Given the foci F_1 and F_2, and the transverse axis AB. Using F_1 and F_2 as centers, and any radius R_1 greater than F_1B, strike arcs. With these same centers and a radius equal to $R_1 - AB$, strike arcs intersecting the first arcs. These intersecting arcs establish the positions of four symmetrically located points (P_1, P_2, P_3, and P_4) using a single pair of radii. Additional sets of four points are obtained by assuming a different initial radius each time. Repeat this procedure, as outlined, until a sufficient number of points have been located to determine a smooth curve.

The tangent to the hyperbola at any point, such as P_1, is the bisector of the angle between the focal radii F_1P_1 and F_2P_1.

As hyperbolic curves extend toward infinity they gradually approach two straight lines known as *asymptotes*. These may be located by drawing a circle having the distance $F_1 - F_2$ as a diameter and erecting perpendiculars to the transverse axis through points A and B. The points at which these perpendicular lines intersect the circle are points (D_1, D_2, D_3, and D_4) on the asymptotes.

3.56
To Construct an Equilateral, or Rectangular, Hyperbola Through a Given Point P —Asymptotic to Given Axes (Fig. 3.53). Given the asymptotes CA and CB and any point P on the curve. Bisect the angle ACB to establish the position of the axis of symmetry of the hyperbola. Then, through point P, draw the lines PD and PF perpendicular to CA and CB, respectively. Along PF extended, mark off points 1, 2,

3, 4, and so forth, as needed. Through these points draw rays converging at C and intersecting DE (DP extended) at points 1', 2', 3', 4', 5', and so forth. From each of the numbered points along DE draw perpendiculars to DP and through the numbered points along FG draw perpendiculars to PF. The intersections of corresponding lines (perpendiculars) establish the location of the points 1'', 2'', 3'', 4'', and so forth, along the required curve.

3.57
Involute. The spiral curve traced by a point on a cord as it unwinds from around a circle or a polygon is an *involute curve*. Figure 3.54 (*a*) shows an involute of a circle, while (*b*) shows that of a square. The involute of a polygon is obtained by extending the sides and drawing arcs using the corners, in order, as centers. The circle in (*a*) may be considered to be a polygon having an infinite number of sides.

3.58
To Draw an Involute of a Circle [Fig. 3.54(*a*)]. Divide the circumference into a number of equal parts. Draw tangents through the division points. Then, along each tangent, lay off the rectified length of the corresponding circular arc, from the starting point to the point of tangency. The involute curve is a smooth curve through these points. The involute of a circle is used in the development of tooth profiles in gearing.

3.59
To Draw the Involute of a Polygon [Fig. 3.54(*b*)]. Extend the sides of the polygon as shown in (*b*). With the corners as centers, in order around the polygon, draw arcs terminating on the extended sides. The first radius is equal to the length of one side of the polygon. The radius of each successive arc is the distance from the center to the terminating point of the previous arc.

3.60
Cycloid. A cycloid is the curve generated by a point on the circumference of a moving circle when the circle rolls in a plane along a straight line, as shown in Fig. 3.55.

3.61
To Draw a Cycloid (Fig. 3.55). Draw the generating circle and the line AB tangent

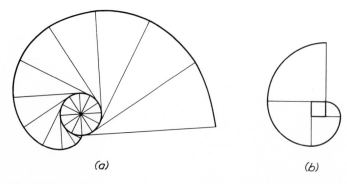

(*a*)   (*b*)

Fig. 3.54 Involute.

to it. The length AB should be made equal to the circumference of the circle. Divide the circle and the line AB into the same number of equal parts. With this much of the construction completed, the next step is to draw the line of centers CD through point O and project the division points along AB to CD by drawing perpendiculars. Using these points as centers for the various positions of the moving circle, draw circle arcs. For the purpose of illustration, assume the circle is moving to the left. When the circle has moved along CD to x, point P will have moved to point P_x. Similarly, when the center is at y, P will be at P_y. To locate positions of P along the cycloidal curve, project the division points of the divided circle in their proper order, across to the position circles. A smooth curve through these points will be the required cycloid.

Fig. 3.55 Cycloid.

3.62
Epicycloid (Fig. 3.56). An epicycloid is the curve generated by a point on the circumference of a circle that rolls in a plane on the outside of another circle. The method used in drawing an epicycloid is similar to the one used in drawing the cycloid.

3.63
Hypocycloid (Fig. 3.57). A hypocycloid is the curve generated by a point on the circumference of a circle that rolls in a plane on the inside of another circle. The method used to draw a hypocycloid is similar to the method used to draw the cycloid.

Additional information on the use of cycloidal curves to form the outlines of cycloidal gear teeth may be found in Chapter 21.

3.64
Spiral of Archimedes. Archimedes' spiral is a plane curve generated by a point moving uniformly around and away from a fixed point. In order to define this curve more specifically, it can be said that it is generated by a point moving uniformly along a straight line while the line revolves with uniform angular velocity about a fixed point.

The definition of the spiral of Archimedes is applied in drawing this curve as illustrated in Fig. 3.58. To find a sufficient number of points to allow the use of an irregular curve for drawing the spiral it is the practice to divide the given circle into a number of equal parts (say, 12) and draw radial lines to the division points. Next,

divide a radial line into the same number of equal parts as the circle and number the division points on the circumference of the circle beginning with the radial line adjacent to the divided one. With the center of the circle as a center, draw concentric arcs that in each case will start at a numbered division point on the divided radial line and will end at an intersection with the radial line that is numbered correspondingly. The arc starting at point 1 gives a point on the curve at its intersection with radial line 1, the arc starting at 2 gives an intersection point on radial line 2, etc. The spiral is a smooth curve drawn through these intersection points.

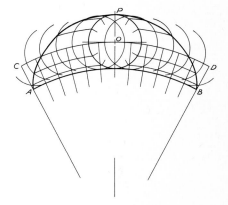

Fig. 3.56 Epicycloid.

3.65
Helix (Fig. 3.59). The cylindrical helix is a space curve that is generated by a point moving uniformly on the surface of a cylinder. The point must travel parallel to the axis with uniform linear velocity while at the same time it is moving with uniform angular velocity around the axis. The curve can be thought of as being generated by a point moving uniformly along a straight line while the line is revolving with uniform angular velocity around the axis of the given cylinder. Study the pictorial drawing, Fig. 3.59.

The first step in drawing a cylindrical helix is to lay out the two views of the cylinder. Next, the lead should be measured along a contour element and divided into a number of equal parts (say, 12). Divide the circular view of the cylinder into the same number of parts and number the division points.

The division lines of the lead represent the various positions of the moving point as it travels in a direction parallel to the axis of the cylinder along the moving line. The division points on the circular view are the related position of the moving line. For example, when the line has moved from the 0 to the 1 position, the point has traveled along the line a distance equal to one-twelfth of the lead; when the line is in

Fig. 3.57 Hypocycloid.

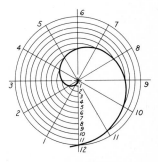

Fig. 3.58 Spiral of Archimedes.

Fig. 3.59 Helix.

the 2 position, the point has traveled one-sixth of the lead. (See pictorial drawing, Fig. 3.59.) In constructing the curve the necessary points are found by projecting from a numbered point on the circular view to the division line of the lead that is numbered similarly.

A helix may be either right-hand or left-hand. The one shown in Fig. 3.59 is a left-hand helix.

When the cylinder is developed, the helix becomes a straight line on the development, as shown. It is inclined to the base line at an angle known as the "helix angle." A screw thread is an example of a practical application of the cylindrical helix.

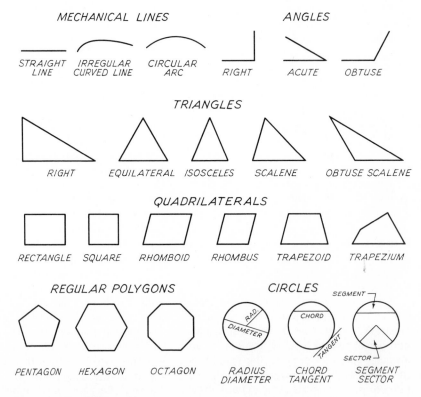

Fig. 3.60 Geometric shapes.

PRISMS

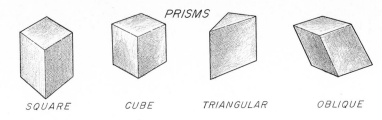

SQUARE   CUBE   TRIANGULAR   OBLIQUE

CYLINDERS   CONES

RIGHT   OBLIQUE   RIGHT   OBLIQUE

PYRAMIDS

SQUARE   TRIANGULAR   OBLIQUE   TRUNCATED

DOUBLE-CURVED SURFACES

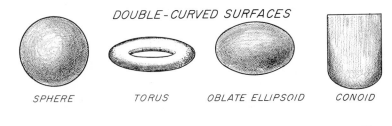

SPHERE   TORUS   OBLATE ELLIPSOID   CONOID

SERPENTINE   PARABOLOID   HYPERBOLOID

Fig. 3.61   Geometric solids.

Problems

The following exercises not only require the student to study and use certain common geometric constructions but also furnish additional practice in applying good line technique to the drawing of instrumental figures and practical designs. All work should be very accurately done. Tangent points should be indicated by a light, short dash across the line.

1. Draw a line 80 mm (3.15 in.) long. Divide it proportionally in the ratio 1:2:3. Use the method shown in Fig. 3.7.

2. Construct a regular hexagon having a 66 mm (2.6 in.) distance across flats. Select the most practical procedure.

3. Construct a regular hexagon having a 80 mm (3.15 in.) distance across corners. Select the most practical method.

4. Construct a regular pentagon having 30 mm (1.18 in.) sides. Use the method illustrated in Fig. 3.19.

5. Draw two horizontal lines 50 mm (1.97 in.) apart. Locate two points 75 mm (2.95 in.) apart horizontally, one on each line. Draw an ogee curve tangent to these lines. Study the procedure illustrated in Fig. 3.26(*a*).

6. Construct an ellipse having a major diameter of 108 mm (4.25 in.) and a minor diameter of 70 mm (2.76 in.) Use the trammel method illustrated in Fig. 3.39.

7. Construct an ellipse having a major diameter of 100 mm (3.94 in.) and a minor diameter of 70 mm (2.75 in.). Use the concentric circle method illustrated in Fig. 3.40. Find a sufficient number of points to obtain a smooth curve.

8. Construct a parabola with vertical axis. Make the focus 19 mm (.75 in.) from the directrix. Select a point on the curve and draw a line tangent to the parabola. Study Sec. 3.48 and Fig. 3.46.

9. Construct a hyperbola having a transverse axis of 25 mm (.98 in.) and foci 41 mm (1.62 in.) apart. Study Sec. 3.55 and Fig. 3.52.

10. Construct the involute of an equilateral triangle with 25 mm (.98 in.) sides. Study Secs. 3.57 and 3.59.

11. Construct the cycloid generated by a 38 mm (1.50 in.) circle. Study Sec. 3.61 and Fig. 3.55.

12. Construct the epicycloid generated by a 38 mm (1.50 in.) circle rolling on a 127 mm (5.00 in.) circle. Study Sec. 3.62 and Fig. 3.56.

13. Construct the hypocycloid generated by a 38 mm (1.50 in.) circle rolling on a 114 mm (4.50 in.) circle. Study Sec. 3.63 and Fig. 3.57.

14. Reconstruct the view of the wrench and hexagonal nut shown in Fig. 3.62. Mark all tangent points with short lines.

15. Construct the shape of the slotted guide shown in Fig. 3.63. Show all construction for locating centers, and mark points of tangency.

16. Construct the adjustable Y-clamp shown in Fig. 3.64. Show all construction for locating centers and mark points of tangency.

17. Reconstruct the end view of the dolly block shown in Fig. 3.65.

18. Reconstruct the view of the electrode shown in Fig. 3.66.

19. Reconstruct the plat of a land survey shown in Fig. 3.67. Use the tangent method, as explained in Sec. 3.11, to determine the direction of the center line of State Road 26. The triangles used in combination will produce the other angles.

20. Reconstruct the view of the spline plate shown in Fig. 3.68.

21. Reconstruct the view of the adjustment plate shown in Fig. 3.69.

22. (Fig. 3.70). Part *A* is free to pivot about a shaft. If this part should be revolved in a coun-

Fig. 3.62 Wrench.

Fig. 3.63 Slotted guide.

Fig. 3.64 Adjustable Y-clamp.

Fig. 3.65 End view, dolly block.

Fig. 3.66 Electrode. (Numerical values shown in [ ] are in millimeters.)

Fig. 3.67 Plat of a land survey.

Fig. 3.68 Spline plate. (Numerical values shown in [ ] are in millimeters.)

Fig. 3.69 Adjustment plate.

terclockwise direction as indicated by the arrows, it would contact surface *C*. Reproduce the drawing as given and show part *A* revolved until it is in contact with surface *C*. Use the symbolic line for showing an alternative position for part *A* in this new position. Show all geometric constructions clearly and do not erase construction lines.

23. (Fig. 3.71). Part *A* revolves about shaft *B* in a clockwise direction from the position shown until surface *C* comes into contact with the cylindrical surface of the roller. Reproduce the drawing as given and show part *A* in its revolved position using the symbolic alternative position line for this new position. Show all geometric constructions clearly and do not erase construction lines.

24. (Fig. 3.72). Draw the top of the arch through point *C*. Use an exact method. Scale and record the rise of the arch (one-half the length of the minor axis). Locate the foci. Show all construction.

25. (Fig. 3.73). The design of a counterweight system is such that point *A* (pivot point) of the counterweight and point *C* (end view of the axis of the shaft and roller) are in line horizontally. A point *B* on the counterweight is 105 from *A* on

Fig. 3.70 Geometric construction.

Fig. 3.71 Geometric construction.

Fig. 3.72 Elliptical arch.

Fig. 3.73 Design of a counterweight system.

Fig. 3.74 Highway interchange.

Fig. 3.75 Cam and follower.

a line making an angle of 41° with *AC* through *A*. Points *A* and *B* are the centers of R40 and R25 arcs, respectively.

Draw an arc of R80 tangent to the 25 and 40 arcs to form the lower portion of the outline of the counterweight. Complete the upper part of the outline by drawing a reverse curve tangent to the top of the 25 and 40 arcs. The radius of the curve tangent to the 40 arc is to be 43% of the total length of the two chords of the reverse curve. Using geometry, determine the location of the point of tangency of the counterweight and roller when the counterweight swings clockwise into contact with the outside surface of the stop-roller.

26. Prepare a plan view drawing showing the center lines of the highway interchange (Fig. 3.74). Select a suitable scale.

27. It is desired to know the angular displacement of the center line of the cam (Fig. 3.75) when the follower moves to a position 18.8 below the position shown. Show the cam and follower in their new positions in phantom outline. Use only an approved geometrical method to find the location of the centers of the arcs. A trial-and-error method is not acceptable. Show all construction and mark all points of tangency.

Determine the angle through which the center line of the cam has moved. Dimension the angle between the center lines.

28. (Fig. 3.76). It is desired to construct the elliptical lunar trajectories of an unmanned lunar spacecraft after it has been fired into space by an Atlas–Agena D rocket. As it approaches along the path indicated, its control rockets are fired at firing-point FP (on the tangent) in such a manner as to place the spacecraft in an elliptical orbit, which has the center of the moon as one focus and the major axis in the direction shown by the center line. After exactly one and one-half orbits, the control rockets are fired again to place the spacecraft in a smaller orbit, which also has the center of the moon as one focus. The major axis of the smaller orbit makes an angle of 30° with the major diameter of the first orbit. Plot the first orbit using the foci method and plot the second orbit by the trammel method. Determine the location of the second firing.

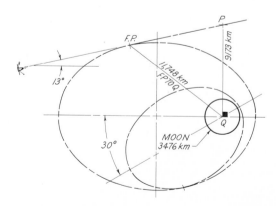

Fig. 3.76 Elliptical lunar orbits.

PART

2

SPATIAL
GRAPHICS:
SHAPE
DESCRIPTION
AND SPATIAL
RELATIONSHIPS

Recovery forces move in as a command module impacts into the waters of the Pacific.
(*Courtesy National Aeronautics and Space Administration*)

the theory of projection

4

Fig. 4.1 Perspective projection.

4.1
Introduction. Since engineers are confronted with the task of recording the shapes and sizes of three-dimensional objects on the plane of a sheet of drawing paper, it is obvious that recognized procedures must be followed if their drawings and sketches are to be easily understood. Size description and shape description are equally important, but, in order to simplify the presentation of the fundamentals of making drawings and sketches, this chapter is concerned entirely with the methods commonly employed in describing shape.

A later chapter will discuss size description.

Each of the different methods—axonometric, oblique, and orthographic—is based on some form of projection. The theory governing a method should be understood thoroughly before it is used.

4.2
Perspective (Scenographic) Projection. In perspective projection, the projecting lines (visual rays) converge to a point, as shown in Fig. 4.1. The representation on the

transparent picture plane may be considered the view that would be seen by one eye located at a definite point in space. The picture is established on the imaginary plane by the piercing points of the projecting lines from the eye to the object. The size of the view depends on the distance from the observer to the plane and the distance from the plane to the object.

Perspective projection is not suitable for working drawings because a perspective view does not reveal exact size and shape. It is used to some extent by engineers in preparing preliminary sketches.

4.3

Orthographic Projection (Parallel Projection). If the observer in Fig. 4.1 moves straight back from the picture plane until he is an infinite distance from it, the projecting lines (projectors) from the eye to the object become parallel to each other and perpendicular to the picture plane. The resulting projection (Fig. 4.2) will then be the same shape and size as the front surface of the object. From a practical viewpoint, the projection may be thought of as being formed by perpendicular projectors extended from the object to the plane. The view is called an *orthographic projection*.

Since the view shown in Fig. 4.2 does not reveal the thickness of the object, one or more additional projections (Fig. 4.3) are necessary to complete the description. Two projections are usually sufficient to describe simple objects, but three or more are necessary for complicated ones.

The picture planes are customarily called the *principal* or *coordinate planes of projection*, and the perpendiculars, *projectors*. In engineering drawing, the planes are usually arranged as shown in Fig. 4.4. Since all three are mutually perpendicular, they are called the horizontal, frontal, and profile coordinate planes. To maintain this mutual relationship when laying out views, it is the usual practice to consider the frontal plane as lying in the plane of the paper and the horizontal and profile planes as being revolved into position (Fig. 4.6). Note in Fig. 4.5 the manner in which the planes are revolved. This theoretical treatment of the coordinate planes establishes an absolute relationship between the views. Visualizing an object would be considerably more difficult than it is, if it were not for this fixed relationship, for it would be impossible to determine quickly the direction of sight for a particular view.

4.4

One-plane Projection. If the object is turned and then tilted so that three faces are inclined to the plane of projection, the resulting projection is a special type of

Fig. 4.3 Planes of projection.

OBSERVER AT INFINITY

Fig. 4.2 Orthographic projection.

Fig. 4.4 Planes of projection.

orthographic projection known as *axonometric projection*. Figure 4.7 illustrates an axonometric projection of a cube. Note that the projectors from the object to the plane are perpendicular to the plane. The three recognized subdivisions of axonometric projection, namely, isometric, dimetric, and trimetric, are explained in Chapter 11.

Another form of one-plane projection is known as *oblique projection*. This form differs from orthographic projection in that, although one face is imagined to be parallel to the plane of projection, the projectors make an angle other than 90° with it (Fig. 4.8). Obviously an infinite number of different views are possible, depending on the angle the parallel projectors make with the plane of projection (picture plane). The various subdivisions are cavalier projection, cabinet projection, and clinographic projection (Chapter 11).

Axonometric projection, oblique projection, and perspective projection may all be classed together as *one-plane pictorial projection*.

Fig. 4.7 Theory of axonometric projection.

Fig. 4.5 Revolution of the planes of projection.

Fig. 4.6 Planes revolved into the plane of the paper.

Fig. 4.8 Oblique projection.

4.5

First- and Third-angle Projection. If the horizontal and frontal planes are assumed to extend indefinitely on one side of the profile plane, four dihedral angles are formed and are designated as the *first, second, third,* and *fourth angles* (Fig. 4.9). The lines of intersection of these planes are called coordinate axes. Their point of intersection is called the origin. In this discussion of first- and third-angle projection, it should be remembered that no matter in which angle the object is placed, the observer views it from in front of the frontal plane and from above the horizontal plane. To avoid misunderstandings, the directions for revolving the horizontal and profile planes into the frontal plane are illustrated in Fig. 4.10. Note that the first and third quadrants are "opened" and the second and fourth are "closed" in revolving the horizontal plane into the frontal plane.

If an object, such as the one shown in Fig. 4.10, is placed so that its main faces are parallel to the principal planes, the respective projections on each plane will show the true size and shape of all surfaces that are parallel to that principal plane. Theoretically, the object could have

been shown in any one of the four quadrants. It has been placed in the third quadrant simply because engineering custom in the United States dictates the use of the third. This quadrant is used because the views, when revolved into the frontal plane, are in their natural positions. That is, the top view appears *above* the front view, as is expected, and the profile view, showing the *right side,* falls on the *right* of the front view.

In some foreign countries, first-angle projection is used for working drawings (study Fig. 4.11). Observe that the top view is projected on the horizontal plane and the front view on the frontal plane. For this reason, the top view falls below the front view when the coordinate planes are revolved.

In this country, the use of first-angle projection for working drawings was abandoned by engineering draftsmen some ninety years ago, although it is still used by architects and structural designers.

4.6

Systems of Projection. The different systems of projection may be conveniently classified as follows:

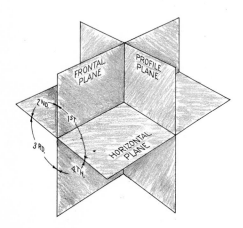

Fig. 4.9 Planes of projection.

Fig. 4.10 Third-angle projection.

Fig. 4.11 First-angle projection.

In the Gulf of Mexico, EXXON is testing a prototype of a system that makes it possible to reach oil and natural gas deeper beneath the sea than has ever before been possible. The prototype is called the Submerged Production System (SPS). After the SPS unit has been lowered and anchored in position on the sea floor, a drilling ship (not shown) will complete a number of wells through openings on the unit. When the drilling has been finished, the ship will depart. After production starts, the SPS will carry out the complete operation. All functions can be monitored remotely controlled either from aboard a ship or from a shore station. Punching the right button will either start up or shut down parts of the unit. The SPS is designed to pipe oil and gas to shore or to a tanker, as shown. The SPS is environmentally safe. (*Courtesy EXXON Corporation*)

multiview representation for design and product development

5

A □ Principal views

5.1

Introduction. Engineers use the orthographic system of projection for describing the shape of machine parts and structures (Fig. 5.1). Practical application of this method of describing an object results in a drawing consisting of a number of systematically arranged views that reproduce the object's exact shape. It was explained in Chapter 4, Sec. 4.3 that a set of views showing the object from different positions is always taken. These views, positioned in strict accordance with the universally recognized arrangement, must show the three dimensions, width, height, and depth. Although three views (Fig. 5.2) are usually required to describe an ordinary object, only two may be needed for a particularly simple one. A very complicated object may require four or more views. A view projected on an auxiliary plane also may be desirable (Fig. 8.3). Such a view often makes possible the elimination of one of the principal views. Therefore, it is up to the individual to determine the number and type of views needed to produce a satisfactory drawing. He will soon develop a knack for this, if he bears in mind that the number of views required depends entirely on the complexity of the shape to be described.

5.2

Definition. Multiview (multiplanar) projection is a method by means of which the exact shape of an object can be represented by two or more separate views produced on projection planes that are usually at right angles to each other.

75

Fig. 5.1 A multiview drawing. (*Courtesy International Business Machines Corporation*)

5.3
Methods of Obtaining the Views. The views of an object may be obtained by either of two methods: (1) the natural method; (2) the "glass box" method.

Since the resulting views will be the same in either case, the beginner should adopt the method he finds the easiest to understand. Both methods are explained here in detail.

5.4
Natural Method. In using this method, each of the necessary views is obtained by looking directly at the particular side of the object the view is to represent.

Figure 5.2 shows three of the principal views of an object: the front, top, and side views. They were obtained by looking directly at the front, top, and right side, respectively. In the application of this method, some consider the position of the object as fixed and the position of the observer as shifted for each view; others find it easier to consider the observer's position as fixed and the position of the object as changed for each view (Fig. 5.2). Regardless of which procedure is followed,

Fig. 5.2   Obtaining three views of an object.

Fig. 5.3   Position of views.

the top and side views must be arranged in their natural positions relative to the front view.

Figure 5.3 illustrates the natural relationship of views. Note that the top view is *vertically above* the front view, and the side view is *horizontally in line with* the front view. In both of these views *the front of the block is toward the front view.*

5.5
"Glass Box" Method.   An imaginary glass box is used widely by instructors to explain

the arrangement of orthographic views. An explanation of this scheme can best be made by reviewing the use of planes of projection (Chapter 4). It may be considered that planes of projection placed parallel to the six faces of an object form an enclosing glass box (Fig. 5.4). The observer views the enclosed object from the outside. The views are obtained by running projectors from points on the object to the planes. This procedure is in accordance with the theory of orthographic projection explained in Sec. 4.3, as well as the

Fig. 5.4   "Glass box."

definition in Sec. 5.2. The top, front, and right side of the box represent the *H* (horizontal), *F* (frontal), and *P* (profile) projection planes.

Since the projections on the sides of the three-dimensional transparent box are to appear on a sheet of drawing paper, it must be assumed that the box is hinged (Fig. 5.5) so that, when it is opened outward into the plane of the paper, the planes assume the positions illustrated in Figs. 5.5 and 5.6. Note that all of the planes, except the back one, are hinged to the frontal plane. In accordance with this universally recognized assumption, the top projection must take a position directly above the front projection, and the right-side projection must lie horizontally to the right of the front projection. To identify the separate projections, engineers call the one on the frontal plane the *front view* or *front elevation,* the one on the horizontal plane the *top view* or *plan,* and the one on the side or profile plane the *side view, side elevation,* or *end view.* Figure 5.6 shows the six views of the same object as they would appear on a sheet of drawing paper. Ordinarily, only three of these views are necessary (front, top, and right side). A bottom or rear view will be required in comparatively few cases.

Fig. 5.5 Opening the glass box.

Fig. 5.6 Six views of an object on a sheet of drawing paper.

Fig. 5.7 "Second position" for the side view.

5.6

"Second Position." Sometimes, especially in the case of a broad, flat object, it is desirable to hinge the sides of the box to the horizontal plane so that the side view will fall to the right of the top view, as illustrated in Fig. 5.7. This arrangement conserves space on the paper and gives the views better balance.

5.7

Principles of Multiview Drawing. The following principles should be studied carefully and understood thoroughly before any attempt is made to prepare an orthographic drawing:

1. The front and top views are *always* in line vertically (Fig. 5.3).

2. The front and side views are in line horizontally, except when the second position is used (Fig. 5.3).

3. The front of the object in the top view faces the front view (Fig. 5.5).

4. The front of the object in the side view faces the front view (Fig. 5.5).

5. The depth of the top view is the same as the depth of the side view (or views) (see Fig. 5.8).

6. The width of the top view is the same as the width of the front view (Fig. 5.8).

7. The height of the side view is the same as the height of the front view (Fig. 5.8).

Fig. 5.8 View terminology.

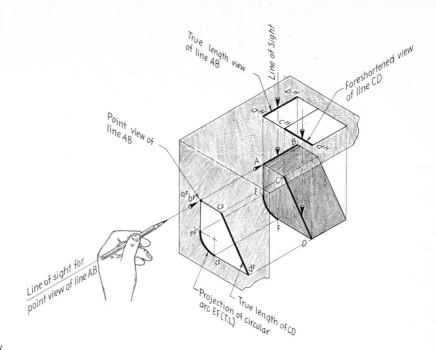

Fig. 5.9 Projected views of lines.

8. A view taken from above is a top view and *must* be drawn above the front view (Fig. 5.6).

9. A view taken from the right, in relation to the selected front, is a right-side view and *must* be drawn to the right of the front view (Fig. 5.6).

10. A view taken from the left is a left-side view and *must* be drawn to the left of the front view (Fig. 5.6).

11. A view taken from below is a bottom view and *must* appear below the front view (Fig. 5.6).

5.8
Projection of Lines. A line may project either in true length, foreshortened, or as a point in a view depending on its relationship to the projection plane on which the view is projected (see Fig. 5.9). In the top view, the line projection $a^H b^H$ shows the true length of the edge AB (see pictorial) because AB is parallel to the horizontal plane of projection. Looking directly at the frontal plane, along the line, AB projects as a point ($a^F b^F$). Lines, such as CD, that are inclined to one of the planes of projection, will show a foreshortened projection in the view on the projection plane to which the line is inclined and true length in the view on the plane of projection to which the line is parallel. The curved line projection $e^F f^F$ shows the true length of the curved edge.

The student should study Fig. 5.10 and attempt to visualize the space position of each of the given lines. It is very necessary both in preparing and reading graphical representations to recognize the position of a point, line, or plane and to know whether the projection of a line is true length or foreshortened, and whether the projection of a plane shows the true size and shape. The indicated reference lines may be thought of as representing the edges of the glass boxes shown. The projections of a line are identified as being on either a frontal, horizontal, or profile plane by the use of the letters F, H, or P with the lowercase letters that identify the end points of the line. For example, in Fig.

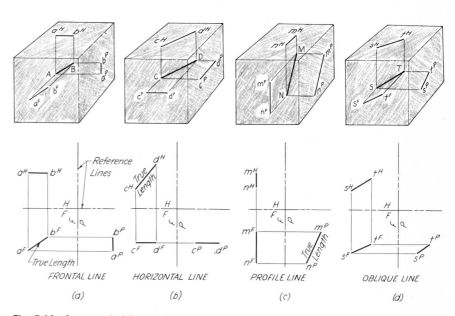

FRONTAL LINE   HORIZONTAL LINE   PROFILE LINE   OBLIQUE LINE

(a)   (b)   (c)   (d)

Fig. 5.10 Some typical line positions.

5.10(a), $a^H b^H$ is the horizontal projection of line AB, $a^F b^F$ is the frontal projection, and $a^P b^P$ is the profile projection.

It is suggested that the student hold a pencil before him and move it into the following typical line positions to observe the conditions under which the pencil representing a line, appears in true length.

1. *Vertical line.* The vertical line is perpendicular to the horizontal and will therefore appear as a point in the H (top) view. It will appear in true length in the F (frontal) view and in the P (profile) view.

2. *Horizontal line* [Fig. 5.10(b)]. The horizontal line will appear in true length when viewed from above because it is parallel to the H-plane of projection and its end points are theoretically equidistant from an observer looking downward.

3. *Inclined line* [Fig. 5.10(c)]. The inclined line is any line not vertical or horizontal that is parallel to either the frontal plane or the profile plane of projection. An inclined line will show true length in the F (frontal) view or P (profile) view.

4. *Oblique line* [Fig. 5.10(d)]. The oblique line will not appear in true length in any of the principal views because it is inclined to all of the principal planes of projection. It should be apparent in viewing the pencil alternately from the directions used to obtain the principal views, namely, from the front, above, and side, that one end of the pencil is always farther away from the observer than the other. Only when looking directly at the pencil from such

a position that the end points are equidistant from the observer can the true length be seen. On a drawing, the true length projection of an oblique line will appear in a supplementary A (auxiliary) view projected on a plane that is parallel to the line (Sec. 9.7).

5.9
Meaning of Lines. On a multiview drawing a visible or invisible line may represent either the intersection of two surfaces, the edge view of a surface, or it may be the limiting element of a surface. These three different meanings of a line are illustrated in Fig. 5.11. In the top view, the curved line is an edge view of surface C, while a straight line is the edge view of surface A. The full circle in the front view may be considered as the edge view of the cylindrical surface of the hole. In the side view, the top line, representing the contour element of the cylindrical surface, indicates the limits for the surface and therefore can be thought of as being a surface limit line. The short vertical line in this same view represents the intersection of two surfaces. In reading a drawing, one can be sure of the meaning of a line on a view only after an analysis of the related view or views. All views must be studied carefully.

5.10
Projection of Surfaces. The components of most machine parts are bounded by either plane or single-curved surfaces. Plane surfaces bound cubes, prisms, and pyramids, while single-curved surfaces, ruled by a moving straight line, bound cylinders and cones. The projected representations (lines or areas) of both plane and single-curved surfaces are shown in Fig. 5.12. From this illustration the student should note that (1) when a surface is parallel to a plane of projection, it will appear in true size in the view on the plane of projection to which it is parallel; (2) when it is perpendicular to the plane of projection, it will project as a line in the view; and (3) when it is positioned at an angle, it will appear foreshortened. A surface will always project either as a line or an area on a view. The area representing the surface may be either a full-size or foreshortened representation.

In Fig. 5.12 the cylindrical surface A appears as a line in the side (profile) view and as an area in the top and front views. Surface B shows true size in the top view and as a line in both the front and side

Fig. 5.11 **Meaning of lines.**

views. Surface C, a vertical surface, will appear as a line when observed from above.

5.11
Analysis of Surfaces, Lines, and Points in Three Principal Views. An analysis of the representation of the surfaces of a mutilated block is given pictorially in Fig. 5.13. It can be noted that each of the surfaces A, B, and C appears in true size and shape in one view and as a line in each of the other two related views. Surface D, which is inclined, appears with foreshortened length in the top and side views and as an inclined line in the front view.

Three views of each of the visible points are shown on the multiview drawing. At the very beginning of an elementary course in drawing, a student will often find it helpful to number the corners of an object in all views.

5.12
Selection of Views. Careful study should be given to the outline of an object before the views are selected; otherwise there is no assurance that the object will be described completely from the reader's viewpoint (Fig. 5.14). Only those views that are necessary for a clear and complete description should be selected. Since the repetition of information only tends to confuse the reader, superfluous views should be avoided.

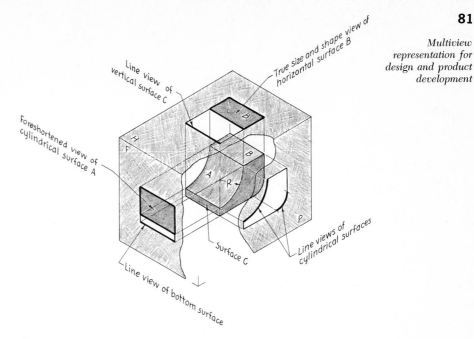

Fig. 5.12 Projected views of surfaces.

Although some objects, such as cylinders, bushings, bolts, and so forth, require only two views (front and side), more complicated pieces may require an auxiliary or sectional view in addition to the ordinary three views.

The space available for arranging the views often governs the choice between the use of a top or side view. The difference between the descriptive values of the

Fig. 5.13 Analysis of surfaces, lines, and points.

View does not reveal the contour shape of either the hole or the slot nor does it show that the corners are cut at 45°

Good Choice
(a)

Poor Choice
(b)

View does not show the characteristic shape

Good Choice
(c)

Poor Choice
(d)

Fig. 5.14   Choice of views.

TOP VIEW

Border

FRONT VIEW

Top view offers best arrangement with front view of a long object

(a)

FRONT VIEW   SIDE VIEW

Border

Side view offers best arrangement with front view of a short object

(b)

Fig. 5.15   Choice of views.

Should show the characteristic shape

(a)

Should show the characteristic shape

(b)

Fig. 5.16   Principal view of an object.

two frequently is not great. For example, a draftsman often finds that the views of a long object will have better balance if a top view is used [see Fig. 5.15(a)]; while in the case of a short object [see (b)], the use of a side view may make possible a more pleasing arrangement. It should be remembered that the choice of views for many objects is definitely fixed by the contour alone, and no choice is offered as far as spacing is concerned. It is more important to have a set of views that describes an object clearly than one that is artistically balanced.

Often there is a choice between two equally important views, such as between a right-side and left-side view or between a top and bottom view (Fig. 5.6). In such cases, one should adhere to the following rule: *A right-side view should be used in preference to a left-side view and a top view in preference to a bottom view.* When this rule is applied to irregular objects, the front (contour) view should be drawn so that the most irregular outline is toward the top and right side.

Another rule, one that must be considered in selecting the front view, is as follows: *Place the object to obtain the smallest number of hidden lines.*

5.13

Principal (Front) View. The principal view is the one that shows the characteristic contour of the object [see Fig. 5.16(a) and (b)]. Good practice dictates that this be used as the front view on a drawing. It should be clearly understood that the view of the natural front of an object is not always the principal view, because frequently it fails to show the object's characteristic shape. Therefore, another rule to be followed is: *Ordinarily, select the view showing the characteristic contour shape as the front view, regardless of the normal or natural front of the object.*

When an object does have a definite normal position, however, the front view should be in agreement with it. In the case of most machine parts, the front view can assume any convenient position that is consistent with good balance.

5.14

Invisible Lines. Dotted lines are used on an external view of an object to represent surfaces and intersections invisible at the point from which the view is taken. In Fig. 5.17(a), one invisible line represents a line of intersection or edge line, while the other invisible line may be considered to represent either the surface or lines of inter-section. On the side view in (b) there are invisible lines, which represent the contour elements of the cylindrical holes.

5.15

Treatment of Invisible Lines. The short dashes that form an invisible line should be drawn carefully in accordance with the recommendations in Sec. 5.24. An invisible line always starts with a dash in contact with the object line from which it starts, unless it forms a continuation of a visible line. In the latter case, it should start with a space, in order to establish at a glance the exact location of the endpoint of the visible line (see Fig. 5.18C). Note that the effect of definite corners is secured at points A, B, E, and F, where, in each case, the end dash touches the intersecting line. When the point of intersection of an invisible line and another object line does not represent an actual intersection on the object, the intersection should be open as at points C and D. An open intersection tends to make the lines appear to be at different distances from the observer.

Parallel invisible lines should have the breaks staggered.

The correct and incorrect treatment for starting invisible arcs is illustrated at G and G'. Note that an arc should start with a dash at the point of tangency. This treatment enables the reader to determine the exact end points of the curvature.

5.16

Omission of Invisible Lines. Although it is common practice for commercial draftsmen to omit hidden lines when their use tends to confuse further an already overburdened view or when the shape description of a feature is sufficiently clear in another view, it is not advisable for a beginning student to do so. The beginner, until he has developed the discrimination that comes with experience, will be wise to show all hidden lines.

5.17

Precedence of Lines. When one discovers in making a multiview drawing that two lines coincide, the question arises as to which line should be shown or, in other words, which line must have precedence if the drawing is to be read intelligently. For example, as revealed in Fig. 5.19, a solid line may have the same position as an invisible line representing the contour element of a hole, or an invisible line may

Fig. 5.17 Invisible lines.

Fig. 5.18 Correct and incorrect junctures of invisible outlines.

occur at the same place as a center line for a hole. In these cases the decision rests on the relative importance of each of the two lines that can be shown. The precedence of lines is as follows:

Solid lines (visible object lines) take precedence over all other lines.

Dashed lines (invisible object lines) take precedence over center lines, although evidence of center lines may be indicated as shown in both the top and side views of Fig. 5.19.

A *cutting-plane line* takes precedence over a center line where it is necessary to indicate the position of a cutting plane.

5.18

Projection of Angles. When an angle lies in a plane that is parallel to one of the

Fig. 5.19 Precedence of lines.

Fig. 5.20 Projection of angles.

Fig. 5.21 Treatment of tangent surfaces.

planes of projection, the angle will show in true size in the view on that particular plane of projection to which the angle is parallel. In Fig. 5.20 those angles indicated as actual show in their true size. The 60° angle, which lies in a surface that is not parallel to the *H*-plane, appears at less than 60° in the top view. The 30° angle for the sloping line on the component portion that is inclined backward projects at greater than 30° in the front view. It may be said that, except for a 90° angle having one leg as a normal line, angles lying on inclined planes will project either larger or smaller than true size, depending on the position of the plane in which the angle lies. A 90° angle always projects in true size, even on an inclined plane, if the line forming one side of the angle is parallel to the plane of projection and a normal view of the line results. The normal view of a line is any view of a line that is obtained

with the direction of sight perpendicular to the line.

What has been stated concerning angles on inclined surfaces can be easily verified by the student if he will observe what happens to the angles of a 30° × 60° triangle resting on the long leg, as it is revolved from a vertical position downward onto the surface of his desk top.

5.19

Treatment of Tangent Surfaces. When a curved surface is tangent to a plane surface, as illustrated in several ways on the pictorial drawing in Fig. 5.21, no line should be shown as indicated at *A* and *B* in the top view and as noted for the front and side views. At *C* in the top view the line represents a small vertical surface that must be shown even though the upper and lower lines for this surface may be omitted in the front view, depending on the decision of the draftsman. In the top view a line has been drawn to represent the intersection of the inclined and horizontal surfaces at the rear, even though they meet in a small round instead of a sharp edge. The presence of this line emphasizes the fact that there are two surfaces meeting here that are at a definite angle, one to the other. Several typical examples of tangencies and intersections have been illustrated in Fig. 5.22.

5.20

Parallel Lines. When parallel surfaces are cut by a plane, the resulting lines of intersection will be parallel, as shown by the pictorial drawing in Fig. 5.23(*b*), where the near corner of the object has been removed by the oblique plane *ABC*. It can be observed from the multiview drawing in (*c*) that "when two lines are parallel in space, their projections will be parallel in all of the views," even though at times both lines may appear as points on one view.

In Fig. 5.23, three views are to be drawn that show the block after the near front corner has been removed [see (*b*)]. Several of the required lines of intersection can be readily established through the given points *A*, *B*, and *C* that define the oblique plane. For example, $c^F b^F$ can be drawn in the front view and the line through a^F can be drawn parallel to it. In the top view, $a^H b^H$ should be drawn first and the intersection line through c^H should then be drawn parallel to this *H*-view of *AB*. The drawing can now be com-

Fig. 5.22 Treatment of tangent surfaces.

pleted by working back and forth from view to view while applying the rule that a plane intersects parallel planes along lines of intersection that are parallel. The remaining lines are thus drawn parallel to either AB or CB [see the pictorial in (b)].

5.21

Plotting an Elliptical Boundary. The actual intersection of a circular cylinder or cylindrical hole with a slanting surface (inclined plane) is an ellipse (Fig. 5.24). The elliptical boundary in (a) appears as an ellipse in the top view, as a line in the front view, and as a semicircle in the side view. The ellipse was plotted in the top view by projecting selected points (such as points A and B) from the circle arc in the side view, as shown. For example, point A was projected first to the inclined line in the front view and then to the top view. The mitre line shown was used to project the depth distance for A in the top view for illustrative purposes only. Ordinarily, dividers should be used to transfer measurements to secure great accuracy.

In (b) the intersection of the hole with the sloping surface is represented by an ellipse in the side view. Points selected around the circle in the top view (such as points C and D) projected to the side view as shown permit the draftsman to form the elliptical outline. It is recommended that a smooth curve be sketched freehand through the projected points before the French curve is applied to draw the finished ellipse, because it is easier to fit a curved ruling edge to a line than to scattered points.

5.22

Projecting a Curved Outline (Space Curve). When a boundary curve lies in an inclined plane, the projection of the curve may be found in another view by projecting points along the curve, as illustrated in Fig. 5.25. In the example, points selected along the arcs forming the curve in the top view were first located in the side view, using distances taken from the top view, as shown by the X and Y measurements. Then the front view positions of these points, through which the front view of the curve must pass, were established by projecting horizontally from the side view and downward from the top view.

5.23

Treatment of Intersecting Finished and Unfinished Surfaces. Figure 5.26 illustrates the removal of material when machining surfaces, cutting a slot, and drilling a hole in a small part. The italic f on a surface of a pictorial drawing in this text indicates that the surface has been machined. The location of sharp and rounded corners, as illustrated in (b) and (c), are noted on the multiview drawing. A discussion covering rounded internal and external corners is given in Sec. 5.44. At this time, the f is not recommended for use on a detail drawing in any ANSI standard. See Sec. 13.7.

Fig. 5.23 Parallel lines.

Fig. 5.24 Representation of an elliptical boundary.

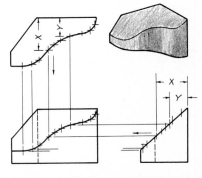

Fig. 5.25 Projecting a space curve.

Fig. 5.26 Rough and finished surfaces on a casting.

Unfinished Casting (a)

Base and Slot-Machined (b)

Hole-Drilled (c)

5.24
To Make an Orthographic Drawing. The location of all views should be determined before a drawing is begun. This will ensure balance in the appearance of the finished drawing. The contour view is usually started first. After the initial start, the draftsman should construct his views simultaneously by projecting back and forth from one to the other. It is poor practice to complete one view before starting the others, as much more time will be required to complete the drawing. Figure 5.27 shows the procedure for laying out a three-view drawing. The general outline of the views

Fig. 5.27 Steps in making a three-view drawing of an object.

first should be drawn lightly with a hard pencil and then heavied with a medium-grade pencil. Although experienced persons sometimes deviate from this procedure by drawing in the lines of known length and location in finished weight while constructing the views, it is not recommended that beginners do so (see Fig. 5.27, step III).

Although a 45° mitre line is sometimes used for transferring depth dimensions from the top view to the side view, or vice versa, as shown in Fig. 5.28(*b*), it is better practice to use dividers, as in (*a*). Continuous lines need not be drawn between the views and the mitre line, as in the illustration, for one may project from short dashes across the mitre line. The location of the mitre line may be obtained by extending the construction lines representing the front edge of the top view and the front edge of the side view to an intersection.

When making an orthographic drawing in pencil, the beginner should endeavor to use the line weights recommended in Sec. 2.31. The object lines should be made very dark and bright, to give snap to the drawing as well as to create the contrast necessary to cause the shape of the object to stand out. Special care should be taken to gauge the dashes and spaces in invisible object lines. On ordinary drawings, 3 mm (.12 in.) dashes and 0.8 mm (.03 in.) spaces are recommended (Fig. 5.29).

Center lines consist of alternate long and short dashes. The long dashes are from 20 to 40 mm (.80 to 1.60 in.) long, the short dashes 3 mm (.12 in.), and the spaces 0.8 mm (.03 in.) (Fig. 5.29). The following technique is recommended in drawing center lines:

1. Where center lines cross, the short dashes should intersect symmetrically (Fig. 5.29). (In the case of very small circles the breaks may be omitted.)

2. The breaks should be so located that they will stand out and allow the center line to be recognized as such.

3. Center lines should extend approximately 3 mm (.12 in.) beyond the outline of the part whose symmetry they indicate (Fig. 5.29).

4. Center lines should not end at object lines.

5. Center lines that are aligned with object lines should have not less than a 1.5 mm (.06 in.) space between the end of the center line and the object line.

For a finished drawing to be pleasing in appearance, all lines of the same type must be uniform, and each type must have proper contrast with other symbolic types. The contrast between the types of pencil lines is similar to that of ink lines (Fig. 2.58), except that pencil lines are never as wide as ink lines (read Sec. 2.31). On commercial drawings, the usual practice is to "burn in" the object lines by applying heavy pressure.

If reasonable care is taken not to soil a drawing, it will not be necessary to clean any part of it with an eraser. Since the practice in most commercial drawing rooms is not to erase construction lines if they have been drawn lightly, the student, at the very beginning of his first course, should try to acquire habits that ensure cleanliness.

When constructing a two-view drawing of a circular object, the pencil work must start with the drawing of the center lines, as shown in Fig. 5.30. This is necessarily the first step, because the construction of the circular (contour) view is based on a horizontal and a vertical center line. The horizontal object lines of the rectangular view are projected from the circles.

5.25

Visualizing an Object from Given Views. Most students in elementary graphics courses find it difficult to visualize an object from two or more views. This trouble is largely due to the lack of systematic procedure for analyzing complex shapes.

The simplest method of determining shape is illustrated pictorially in Fig. 5.31. This method of "breaking down" may be applied to any object, since all objects may be thought of as consisting of elemental geometric forms, such as prisms, cylinders, cones, and so on. These imaginary component parts may be additions in the form of projections or subtractions in the form of cavities. Following such a detailed geometric analysis, a clear picture of an entire object can be obtained by mentally assembling a few easily visualized forms.

It should be realized, when analyzing component parts, that it is impossible ordinarily to determine whether a form is an addition or a subtraction by looking at one view. For example, the small circles in the top view in Fig. 5.31 indicate a cylindrical form, but they do not reveal whether the form is a hole or a projection. By consulting the front view, however, the

Fig. 5.28 Methods for transferring depth dimensions.

Transfer measurements between top and side views using dividers

PRACTICAL METHOD
(Most Accurate)

(a)

45° Mitre Line

(b)

SHORTEN DASHES FOR LIMITED DISTANCES

SHORT DASHES AT CENTER

Fig. 5.29 Invisible lines and center lines.

BORDER

LOCATE CENTER LINES &
MARK RADIAL DISTANCES

STEP I

MARK DISTANCES

DRAW CIRCLES IN FINISHED WEIGHT

STEP II

"BLOCK-IN" VIEW

STEP III

DARKEN OBJECT LINES

STEP IV

Fig. 5.30 Steps in making a two-view drawing of a circular object.

Fig. 5.31 "Breaking down" method.

form is shown to be a hole (subtracted cylinder).

The graphic language is similar to the written language in that neither can be read at a glance. A drawing must be read patiently by referring systematically back and forth from one view to another. At the same time the reader must imagine a three-dimensional object and not a two-dimensional flat projection.

A student usually will find that a pictorial sketch will clarify the shape of a part that is difficult to visualize. The method for preparing quick sketches in isometric is explained in Secs. 6.14 and 6.15.

5.26
Interpretation of Adjacent Areas of a View. To obtain a full understanding of the true geometric shape of a part, all the areas on a given view must be carefully analyzed, because each area represents a surface on the part. For example, in reading a drawing it must be determined whether a particular area in a top view

represents a surface that is inclined or horizontal and whether the surface is higher or lower than adjacent ones. Five distinctly different objects are shown in Fig. 5.32, all having the same top view. In determining the actual shape of these objects, memory and previous experience can be a help, but one can easily be misled if he does not approach the analysis with an open mind, for it is only by trial-and-error effort and by referring back and forth from view to view that a drawing can be read. In considering area A in (a) it might be thought that the triangular surface could be high and horizontal, which would be correct because of the arrangement of lines in the front view. However, in considering the top view alone, A could be either sloping, as in (c) and (e), or low and horizontal, as in (b). An analysis of the five parts reveals that the surface represented by area B can also be either sloping, high and horizontal, or low and horizontal. Area C offers even a wider variety of possibilities in that it may be either low and horizontal (a), sloping (b), cylindrical [(c) and (d)], or high and horizontal (e).

The student must realize at this point that since there are infinite possibilities for the shape, position, and arrangement of surfaces that form objects, he must learn to study tediously the views of any object with which he is not familiar until he is sure of the exact shape. Multiview drawings cannot be read with ease of our written language, which lists all the components in a dictionary.

5.27
True-length Lines. Students who, lacking a thorough understanding of the principles of projection (Sec. 5.8), find it difficult to determine whether or not a projection of a line in one of the principal views

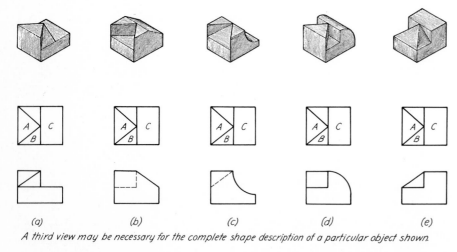

(a)   (b)   (c)   (d)   (e)

A third view may be necessary for the complete shape description of a particular object shown.

Fig. 5.32 Meaning of areas.

shows the true length of the line, should study carefully the following facts:

1. If the projection of a line shows the true length of the line, one of the other projections must appear as a horizontal line, a vertical line, or a point, on one of the other views of the drawing.

2. If the top and front views of a line are horizontal, then both views show the true length.

3. If the top view of a line is a point, the front and side views show the true length.

4. If the front view of a line is a point, the top and side views show the true length.

5. If the top and front views of a line are vertical, the side view shows the true length.

6. If the side projection of a line is a point, the top and front views show the true length.

7. If the front view of a line is horizontal and the top view is inclined, the top inclined view shows the true length.

8. If the top view of a line is horizontal and the fron view is inclined, the front inclined view shows the true length.

5.28
Representation of Holes. In preparing drawings of parts of mechanisms, a draftsman finds it necessary to represent machined holes, which most often are either drilled, drilled and reamed, drilled and countersunk, drilled and counterbored, or drilled and spotfaced. Graphically, a hole is represented to conform with the finished form. The form may be completely specified by a note attached to the view showing the circular contour (Fig. 5.33). The shop note, as prepared by the draftsman, usually specifies the several shop operations in the order that they are to be performed in the shop. For example, in (d) the hole, as specified, is drilled before it is counterbored. When depth has not been given in the note for a hole, it is understood to be a through hole; that is, the holes goes entirely through the piece [(a), (c), (d), and (e)]. A hole that does not go through is known as a *blind hole* (b). For such holes, depth is the length of the cylindrical portion. Drilled, bored, reamed, cored, or punched holes are always specified by giving their diameters, never their radii. Drill diameters for number- and letter-size drills may be found in Table 49 in

the Appendix. Metric drills have been listed in Table 23.

In drawing the hole shown in (a), which must be drilled before it is reamed, the limits are ignored and the diameter is scaled to the nearest regular inch or millimeter size. In (b) the 30° × 60° triangle is used to draw the approximate representation of the conical hole formed by the drill point. In (c) a 45° triangle has been used to draw an approximate representation of the outline of the conical enlargement. The actual angle of 82° is ignored in order to save time in drawing. The spotface in (e) is most often cut to a depth of 1.5 mm (.06 in.); however, the depth is usually not specified.

The beginner should now scan the several sections in the chapter on shop processes to obtain some general information on the production of holes. Complete information on the preparation of shop notes for holes may be found in Chapters 13, 14, and 15.

B □ Revolution

5.29
To Find the True Length of a Line by Revolution. In engineering drafting, it is frequently necessary to determine the true length of a line when constructing the development of a surface (Chapter 10). The true lengths must be found of those

(For representations of threaded holes see chapter covering screw threads and fasteners)

Fig. 5.33 Representation of holes.

lines that are not parallel to any coordinate plane and therefore appear foreshortened in all the principal views. (see Sec. 5.8). The practical as well as theoretical procedure is to revolve any such oblique line into a position parallel to a coordinate plane such that its projection on that particular plane will be the same length as the line. In Fig. 5.34(*a*), this is illustrated by the edge AB on the pyramid. AB is oblique to the coordinate planes, and its projections are foreshortened. If this edge line is imagined to be revolved until it becomes parallel to the frontal plane, then the projection ab_r in the front view will be the same length as the true length of AB.

A practical application of this method is shown in Fig. 5.34(*c*). The true length of the edge AB in Fig. 5.34(*a*) would be found by revolving its top projection into the position ab_r, representing AB revolved parallel to the frontal plane, and then projecting the end point b_r down into its new position along a horizontal line-through b. The horizontal line represents the horizontal plane of the base, in which the point B travels as the line AB is revolved.

Commercial draftsmen who are unfamiliar with the theory of coordinate planes find the true-length projection of a line by visualizing the line's revolution. They think of an edge as being revolved until it is in a plane perpendicular to the line of sight of an observer stationed an infinite distance away (Figs. 10.10, 10.11, and 10.12). The process corresponds to that used in drawing regular orthographic views (Sec. 5.4). Usually this method is more easily understood by a student.

Note in Fig. 5.34(*a*) and (*b*) that the true length of a line is equal to the hypotenuse of a right triangle whose altitude is equal to the difference in the elevation of the end points and whose base is equal to the top projection of the line. With this fact in mind, many draftsmen determine the true length of a line by constructing a true-length triangle similar to the one illustrated in Fig. 5.34(*d*).

**5.30
Revolution of an Object.** Although in general the views on a working drawing represent a machine part satisfactorily when shown in a natural position, it is sometimes desirable to revolve an element until it is parallel to a coordinate plane in order to improve the respresentation or to reveal the true size and shape of a principal surface, or true length of a line.

The distinguishing difference between this method and the method of auxiliary projection (Chapter 8) is that, in the procedure of revolution, the observer turns (revolves) the object with respect to the customary planes of projection, instead of shifting his viewing position with respect to either an inclined or an oblique surface of the object.

Despite the fact that the revolution of an entire object, as illustrated in Fig. 5.35, rarely has a practical application in industry, the making of such a drawing provides excellent drill in projection. Therefore, since the several articles that follow are intended primarily for training students, the practical applications have been omitted, while the procedures for revolving simple objects are explained in detail.

**5.31
Simple (Single) Revolution.** When the regular views are given, an object may be shown in another position, as may be required, by imagining it to be revolved about an axis perpendicular to one of the principal (horizontal, frontal, or profile) planes. A single revolution about such an axis is known as a "simple revolution." The three general cases are shown in Fig. 5.35, spaces II, III, and IV.

**5.32
To Determine the True-shape Projection (Normal View) of a Plane Surface—Revolution Method.** If a surface is parallel to one of the principal planes of projection, its projection on that plane will show its true shape, as has been explained in Sec. 5.10. In Fig. 5.35, space I, the inclined

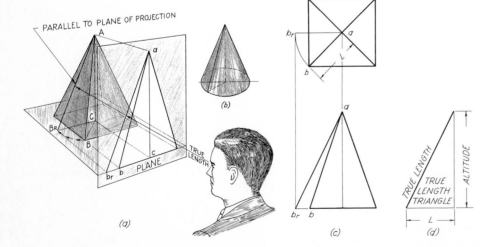

Fig. 5.34 True length of a line, revolution method.

surface $ABCD$ appears as a line ($a^F b^F c^F d^F$) in the front view and is shown foreshortened in the top and side views. Line A–B was taken as the axis of rotation and the surface was revolved into a position parallel to the H-plane, as shown by the pictorial drawing ($ABC_R D_R$). First, the edge view of the surface, as represented by the line $a^F b^F c^F d^F$, was revolved about $a^F b^F$ into a horizontal position, as shown by the line $a^F b^F c_r^F d_r^F$. The complete H-view ($a^H b^H c_r^H d_r^H$) of the surface as revolved shows the true shape of $ABCD$.

5.33
Revolution About a Vertical Axis Perpendicular to the Horizontal Plane. A simple revolution about an axis perpendicular to the horizontal plane is illustrated in Fig. 5.35, space II. The object is first revolved about the assumed imaginary axis until it is in the desired position (see pictorial). The views of the part in its revolved position then are obtained by orthographic projection, as in the case of any ordinary multiview drawing. Since the top view will not be changed in shape by the revolution, it must be drawn first in its revolved position (at 30° in this case) and the front and side views should be projected from it. Since the heights to all of the points on the object also remain unchanged by the revolution, height distances could be conveniently projected from the initial views in I.

The top view may be drawn directly in revolved position, without first drawing the usual orthographic views. If this procedure is followed, the height distances for the front and side views may be set off to known dimensions.

5.34
Revolution about a Horizontal Axis Perpendicular to the Frontal Plane. If an object is revolved about an imaginary axis perpendicular to the frontal plane, as shown in Fig. 5.35, space III, the front view changes in position but not in shape. The front view, therefore, should be drawn first in its revolved position, and the top and side views should be projected from it. The depth of the top view and the side view remain unchanged since the depth distance is parallel to the axis. If the usual unrevolved views are not drawn first, the front view may be drawn directly in its revolved position and depth distances can then be laid off to known dimensions.

Fig. 5.35 Single revolution about an axis perpendicular to a principal (horizontal, frontal, or profile) plane.

5.35
Revolution about a Horizontal Axis Perpendicular to the Profile Plane. A single revolution of an object about an axis perpendicular to the profile plane is illustrated in Fig. 5.35, space IV. Since in this case it is the side view that is perpendicular to the axis and revolves parallel to the coordinate plane of projection, it is the side view that remains unchanged in its shape. The width of the top and front views is not affected by the revolution. Therefore, horizontal dimensions for these views may be set off by using known measurements.

From these general cases of simple revolution, two principles have emerged that can be stated as follows:

1. The view that is perpendicular to the axis of revolution changes only in position.

2. The lengths of the lines parallel to the axis do not change during the revolution and, therefore, may be either laid off to known measurements or projected from the usual orthographic views of the object.

5.36
Clockwise and Counterclockwise Revolution. An object may be revolved either

COUNTERCLOCKWISE
(a)

CLOCKWISE
(b)

Fig. 5.36   Direction of revolution.

clockwise or counterclockwise about an axis of revolution. The direction is indicated by the view to which the axis is perpendicular. For example, front views, when revolved as in Fig. 5.36(b), show a clockwise revolution. When revolved as in (a), their revolution is counterclockwise. Top views show a clockwise revolution when revolved to the right. Right-side views indicate a clockwise direction of revolution when they have been revolved to the right and a counterclockwise direction when revolved to the left.

5.37

Successive (Multiple) Revolution. Since it is possible to show an object in any position relative to the coordinate planes of projection, it can be drawn as may be required by making a series of successive simple revolutions. Usually such a series is limited to three or four stages. Figure 5.37 shows an object revolved successively about two separate axes. The usual orthographic views of the given object are shown at the left in space I. In space II, the object has been revolved (counterclockwise) about a vertical axis to obtain an edge view (EV) of the oblique surface ABC (see pictorial). Note that the line AC projects as a point ($a^F c^F$) in the front view. The identification of each corner may not be necessary in the case of a simple object. However, if the object is in the least complex, possible confusion is avoided if each corner point is either

numbered or identified by a letter, as was done for the oblique surface in the illustration. In space III at the right, the object is represented after it has been revolved from its previous position in space II. In this case, since the edge view (EV) of surface ABC is now horizontal, a true-shape projection (TSP) of the surface will appear in the top view.

A convenient method of copying a view in a new revolved position is first to trace it on a small piece of tracing paper and then to place the traced view in correct position and draw the new view. An alternative method is to place the traced view in correct position and prick the corner points, using the dividers opened out. With the traced view removed, the revolved view on the final drawing may be completed by joining the prick-points.

C □ Conventional practices

5.38

To reduce the high cost of preparing engineering drawings and at the same time to convey specific and concise information without a great expenditure of effort, some generally recognized systems of symbolic representation and conventional practices have been adopted by American industry.

A standard symbol or conventional representation can express information that might not be understood from a true-line representation unless accompanied by a lettered statement. In many cases, even though a true-line representation would convey exact information, very little more would be gained from the standpoint of better interpretation. Some conventional practices have been adopted for added clearness. For instance, they can eliminate awkward conditions that arise from strict adherence to the rules of projection.

These idioms of drawing have slowly developed with the graphic language until at the present time they are universally recognized and observed and appear in the various standards of the American National Standards Institute.

Professional men and skilled workmen have learned to accept and respect the use of the symbols and conventional practices, for they can interpret these representations accurately and realize that their use saves valuable time in both the drawing room and the shop.

Normal Position   Revolution For Edge View   Revolution To Obtain TSP

Fig. 5.37   Successive revolution to show the true shape of an oblique surface.

5.39

Half Views and Partial Views. When the available space is insufficient to allow a satisfactory scale to be used for the representation of a symmetrical piece, it is considered good practice to make one view either a half view or a partial view, as shown in Fig. 5.38. The half view, however, must be the top or side view and not the front view, which shows the characteristic contour. The half view should be the front half of the top or side view. In the case of the partial view shown in (b), a break line is used to limit the view.

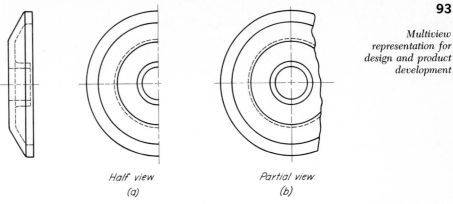

Half view
(a)

Partial view
(b)

Fig. 5.38. Half views and partial views.

5.40

Accepted Violations of True Projection in the Representation of Boltheads, Slots, and Holes for Pins. A departure from true projection is encountered in representing a bolthead. For example, on a working drawing, it is considered the best practice to show the head across corners in both views, regardless of the fact that in true projection one view would show "across flats." This method of treatment eliminates the possibility of a reader's interpreting a hexagonal head to be a square head (Fig. 5.39). Furthermore, the showing of a head the "long way" in both views clearly reveals the space needed for proper clearance.

In the case of the slotted head fasteners, the slots are shown at 45° in the end views in order to avoid placing a slot on a center line, where it is usually difficult to draw so that the center line passes accurately through the center (Fig. 5.40). This practice does not affect the descriptive value of the drawing, because the true size and shape of the slot is shown in the front view. The hole for a pin is shown at 45° for the same reason. In such a position it may be more quickly observed.

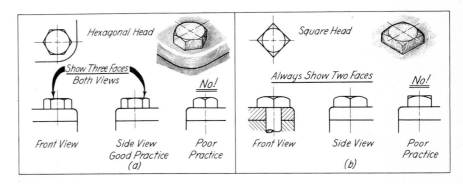

Fig. 5.39 Treatment of bolt heads.

Fig. 5.40 Treatment of slots and holes in fasteners and pins.

5.41

Treatment of Unimportant Intersections. The conventional methods of treating various unimportant intersections are shown in Fig. 5.41. To show the true line of intersection in each case would add little to the value of the drawing. Therefore, in the views designated as preferred, true projection has been ignored in the interest of simplicity. On the side views, in (a) and (b), for example, there is so little difference between the descriptive values of the true and approximate representations of the holes that the extra labor necessary to draw the true representation is unwarranted.

Fig. 5.41 Treatment of unimportant intersections.

5.42

Aligned Views. Pieces that have arms, ribs, lugs, or other features at an angle are shown aligned or "straightened out" in one view, as illustrated in Fig. 5.42. By this method, it is possible to show the true shape as well as the true position of such features. In Fig. 5.43, the front view has been drawn as though the slotted arm had been revolved into alignment with the element projecting outward to the left. This practice is followed to avoid drawing an element—that is at an angle—in a foreshortened position.

5.43

Conventional Treatment of Radially Arranged Features. Many objects that have radially arranged features may be shown more clearly if true projection is violated, as in Fig. 5.42(*b*). Violation of true projection in such cases consists of intentionally showing such features swung out of position in one view to present the idea of symmetry and show the true relationship of the features at the same time. For example, while the radially arranged holes in a flange (Fig. 5.44) should always be shown in their true position in the circular view, they should be shown in a revolved position in the other view in order to show their true relationship with the rim.

Radial ribs and radial spokes are similarly treated [Fig. 5.42(*a*)]. The true projection of such features may create representations that are unsymmetrical and misleading. The preferred conventional method of treatment, by preserving symmetry, produces representations that are more easily understood and that at the same time are much simpler to draw. Figure 5.45 illustrates the preferred treatment for radial ribs and holes.

5.44

Representations of Fillets and Rounds. Interior corners, which are formed on a casting by unfinished surfaces, are always filled in (filleted) at the intersection in order to avoid possible fracture at that point. Sharp corners are also difficult to obtain and are avoided for this reason as well. Exterior corners are rounded for appearance and for the comfort of persons who must handle the part when assembling or repairing the machine on which the part is used. A rounded internal corner is known as a *fillet;* a rounded external corner is known as a *round* (Fig. 5.46).

When two intersecting surfaces are ma-

Fig. 5.42 Conventional practice of representing ribs and lugs.

Fig. 5.43 Aligned views.

Fig. 5.44 Radially arranged holes.

Fig. 5.45 Conventional treatment of radially arranged ribs.

FILLET   ROUND

Fig. 5.46 Fillets and rounds.

chined, however, their intersection will
become a sharp corner. For this reason,
all corners formed by unfinished surfaces
should be shown "broken" by small
rounds, and all corners formed by two
finished surfaces, or one finished surface
and one unfinished surface, should be
shown "sharp." Although in the past it
has been the practice to allow pattern-
makers to use their judgment about the
size of fillets and rounds, many present-
day companies require their designers and
draftsmen to specify their size even
though their exact size may not be important.

Since fillets and rounds eliminate the
intersection lines of intersecting surfaces,
they create a special problem in orthographic representation. To treat them in
the same manner as they would be treated
if they had large radii results in views that
are misleading. For example, the true-pro-
jection view in Fig. 5.47(c) confuses the
reader, because at first glance it does not
convey the idea that there are abrupt
changes in direction. To prevent such a
probable first impression and to improve
the descriptive value of the view, it is necessary to represent these theoretically
nonexisting lines. These characteristic
lines are projected from the approximate
intersections of the surfaces, with the fillets disregarded.

Figure 5.48 illustrates the accepted
conventional method of representing the
"run-out" intersection of a fillet in cases
where a plane surface is tangent to a
cylindrical surface. Although run-out arcs
such as these are usually drawn freehand,
a French curve or a bow instrument may
be used. If they are drawn with the bow
instrument, a radius should be used that
is equal to the radius of the fillet, and the
completed arc should form approximately
one-eighth of a circle.

Fig. 5.47 Conventional practice of representing nonexisting lines of intersection.

Fig. 5.48 Conventional treatment for fillets.

Fig. 5.49 Approximate methods of representing run-outs for intersecting fillets and rounds.

POINT OF TANGENCY

(a) *(b)* *(c)*

Fig. 5.50 Broken-out view.

38 8
380
76

Over-all dimension not to scale.
Break permits ends to be drawn closer together

ROUND SECTION-SOLID

ROUND SECTION-TUBULAR

RECTANGULAR SECTION-SOLID

RECTANGULAR SECTION-HOLLOW

RECTANGULAR SECTION-WOOD

CONVENTIONAL BREAK

Fig. 5.51 Conventional breaks.

detail or assembly drawings. The break representations for indicating the broken ends of rods, shafts, tubes, and so forth, are designed to reveal the characteristic shape of the cross section in each case. Although break lines for round sections may be drawn freehand, particularly on small views, it is better to draw them with either an irregular curve or a bow instrument. The breaks for wood sections, however, always should be drawn freehand.

5.46
Ditto Lines. When it is desirable to minimize labor in order to save time, ditto lines may be used to indicate a series of identical features. For example, the threads on the shaft shown in Fig. 5.52 are just as effectively indicated by ditto lines as by a completed profile representation. When ditto lines are used, a long shaft of this type may be shortened without actually showing a conventional break.

5.47
A Conventional Method for Showing a Part in Alternative Positions. A method frequently used for indicating an alternative position of a part or a limiting position of a moving part is shown in Fig. 5.53. The dashes forming the object lines of the view showing the alternative position should be of medium weight. The phantom line shown in Fig. 2.58 is recommended for representing an alternative position.

5.48
Conventional Representation. Symbols are used on topographic drawings, architectural drawings, electrical drawings, and machine drawings. No engineer serving in a professional capacity can very well escape their use.

Most of the illustrations that are shown in Fig. 5.54 should be easily understood. However, the crossed-lines (diagonals) symbol has two distinct and different meanings. First, this symbol may be used on a drawing of a shaft to indicate the position of a surface for a bearing or, second, it may indicate that a surface perpendicular to the line of sight is flat. These usages are illustrated with separate examples.

The generally accepted methods of representing intersecting fillets and rounds are illustrated in Fig. 5.49. The treatment, in each of the cases shown, is determined by the relationship existing between the sizes of the intersecting fillets and rounds.

5.45
Conventional Breaks. A relatively long piece of uniform section may be shown to a larger scale, if a portion is broken out so that the ends can be drawn closer together (Fig. 5.50). When such a scheme is employed, a conventional break is used to indicate that the length of the representation is not to scale. The American National Standard conventional breaks, shown in Fig. 5.51, are used on either

Fig. 5.52 Ditto lines.

CHAIN

CHAIN

ROPE-CABLE   INSULATED CABLE

KNURL

BEARING

FLAT SURFACE

Fig. 5.53 Alternative positions.   **Fig. 5.54 Conventional symbols.**

Problems

The problems that follow are intended primarily to furnish study in multiview projection through the preparation of either sketches or instrumental drawings. Many of the problems in this chapter, however, may be prepared in more complete form. Their views may be dimensioned as are the views of working drawings, if the student will study carefully the beginning of the chapter covering dimensioning before attempting to record size description (Chapter 13). All

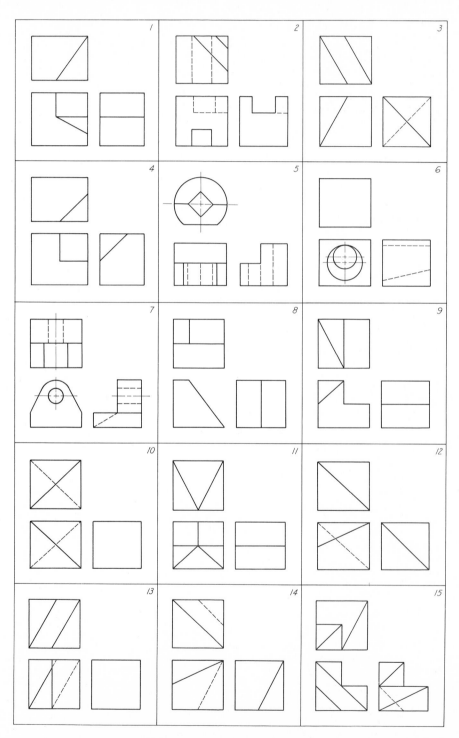

Fig. 5.55 Missing-line (or lines) exercises.

dimensions should be placed in accordance with the general rules of dimensioning. The problems given at the end of Chapter 6 offer further study in multiview representation.

The views shown in a sketch or drawing should be spaced on the paper with aim for balance within the borderlines. Ample room should be allowed between the views for the necessary dimensions. If the views are not to be dimensioned, the distance between them may be made somewhat less than would be necessary otherwise.

Before starting to draw, the student should reread Sec. 5.24 and study Fig. 5.27, which shows the steps in making a multiview drawing. The preparation of a preliminary sketch always proves helpful to the beginner.

All construction work should be done in light lines with a sharp hard pendil. A drawing should be checked by an instructor before the lines are "heavied in," unless the preliminary sketch was checked beforehand.

A part showing linear dimensions given in millimeters may be dimensioned in inches by converting the values using Table 5 in the Appendix.

1. (Fig. 5.55). Add the missing line or lines in one view of each of the three-view drawings. When the missing line or lines have been determined, the three views of each object will be consistent with one another.

2. (Fig. 5.56). Draw or sketch the third view for each of the given objects.

3–6. (Figs. 5.57–5.60). Reproduce the given views and draw the required view. Show all hidden lines.

7. (Fig. 5.61). Make an orthographic drawing or sketch of the bench stop. The views may be dimensioned. The shaft portion that fits into the hole in the bench top is 19 mm. in diameter and 50 mm. long.

8–39. (Figs. 5.62 and 5.93). Make multiview drawings of the given objects. The views of a drawing may or may not be dimensioned.

40. (Fig. 5.94). Make a complete orthographic drawing of the anchor bracket.

41. (Fig. 5.95). Make a complete orthographic drawing of the tube holder.

Fig. 5.56 Third-view problems.

Fig. 5.57

Fig. 5.58

Fig. 5.59

Fig. 5.60

Fig. 5.61 Bench stop.

Fig. 5.62 Corner block.

Fig. 5.63 Stop block.

Fig. 5.64 Rest block.

Fig. 5.65 Adjustment block.

Fig. 5.66 Locating block.

Fig. 5.67 Safety block.

Fig. 5.68 Angle block.

Fig. 5.69   Cross stop.

Fig. 5.70   End block.

Fig. 5.71   Bevel block.

Fig. 5.72   Stabilizer block.

FILLETS & ROUNDS R3

Fig. 5.73   Mounting bracket.

.594DIA FINISH ALL OVER

Fig. 5.74   Index guide.

FILLETS & ROUNDS R3

Fig. 5.75   Guide bracket.

Fig. 5.76   Control guide.

Fig. 5.77 Shoe block.

Fig. 5.78 Holder block.

Fig. 5.79 Slotted guide.

Fig. 5.80 Pivot guide.

Fig. 5.81 Control guide.

Fig. 5.82 Corner bracket.

Fig. 5.83 Auxiliary fork.

Fig. 5.84 Jaw block.

Fig. 5.85 Shifter.

Fig. 5.86 Stud guide.

Fig. 5.87 Ejector bracket. (Dimension values shown in [ ] are in millimeters.)

Fig. 5.88 Guide clip.

Fig. 5.89 Control rod guide.

Fig. 5.90 Shaft bracket. (Dimension values shown in [ ] are in millimeters.)

Fig. 5.91 Lathe leg.

FILLETS & ROUNDS R3

Fig. 5.92 Feed guide.

FILLETS & ROUNDS R3

Fig. 5.93 Control bracket.

FILLETS & ROUNDS R3

Fig. 5.94 Anchor bracket.

TUBING 2.750 O.D.

.18 SAWCUT
.94
.38
.62 R
7 - 14UNC-2B
16 HEX NUT
1.38
3.50
.75
3.61
.62
.88
.50
.24 R
.62
.30
.50
2.50
3.74
3.50
½-UNC CAP SCREW

FILLETS & ROUNDS .12^R

Fig. 5.95   Tube holder.

The ideas of a designer may be sketched in pictorial form as they are visualized. Later during a period of preliminary study, a combination of orthographic and pictorial sketches are prepared as problems are recognized and possible solutions are considered. Sales engineers frequently include pictorial sketches along with orthographic sketches when preparing field reports. Sketches, when used in combination with the written and spoken language, lead to a full understanding by all persons with whom one finds it necessary to communicate.

The pictorial sketch above shows a designer's idea for a quick acting bench clamp.

freehand sketching for visualization and communication

6

A □ Sketching and design

6.1

Introduction. From the earliest times pictorial representations have been the means of conveying the ideas of one person to another person and of one group to another group (Fig. 6.2). There is little doubt that our ancestors traced out in the dust on the cave floor many crude pictures to supplement their guttural utterances. On their cave walls these same primitive men and women drew pictures, which

today convey to others the stories of their lives. They used the only permanent means they were aware of at that time.

At present, we have at our command the spoken languages, which no doubt developed from limited semiintelligent throat sounds, and the written languages, graphical and symbolic in form. The descriptive powers of the various forms of presentation may be compared in Fig. 6.1. The use of sign language representation, (*a*), is rather easy to learn and may be quickly executed but interpretation is restricted to persons who understand the

particular language in which it is presented. The multiview representation, shown in (*b*), may be understood universally by persons who have been trained in its use. However, the given views will prove to be almost meaningless to the many who have not had the advantage of needed training. What then is the one form of representation that can be understood and used by all? It is the pictorial form shown in (*c*).

It is necessary that a designer or engineering technologist be capable of executing well-proportioned and understand-

控駛火箭

(a)   (b)   (c)

Fig. 6.1 Graphic methods for presenting ideas (symbolic, multiview, and pictorial).

Fig. 6.2 Designer's idea sketch. In the future heli-lifted classrooms may be transported to every part of the earth. The lessons would be projected in 3-D. Transmittal would be carried by laser beams, relayed by satellite. (*Courtesy Raymond Loewy/William Smith, Inc., and Charles Bruning Company*)

able freehand pictorial sketches, for it is one of the most important modes of expression that he has available for use. Like other means for conveying ideas, when depended on alone, it will usually prove to be inadequate. However, when used in combination with the written or spoken language and related graphical representations, it makes a full understanding by others become sure and not just possible (Fig. 6.3). Each method of expression is at

hand to supplement another to convey the intended idea.

The designer should be capable of speaking his own language and possibly one foreign language fluently; he should be able to write to present his ideas clearly and accurately; he should be familiar with the graphical method of presenting shape through the use of multiviews; and, finally, he should be competent to execute well-proportioned and understandable pictorial sketches, which are needed to clarify and ensure complete transfer of his ideas to others.

The designer is a creative person living in a world where all that he creates must exist in space. He must visualize space conditions, space distances, and movement in space. In addition he must be able to retain as well as alter the image of his idea, which will at the very start exist only in his mind.

As his idea forms the designer nearly always resorts to sketching to organize his thoughts quickly and to more clearly visualize the problems that appear. The first ideas may be sketched in pictorial form as they are visualized. Later, in making a preliminary study, a combination of orthographic design sketches and pictorial sketches may quickly pile up on his desk as problems are recognized and possible solutions are recorded for reference and for conferences with others.

The designer's use of sketches, both pictorial and orthographic (Fig. 6.4), continues throughout the preliminary design stages and into the development and detailing stages. This comes about because the designer is usually called on to serve as both planner and director. Throughout all stages in the development of a product he must solve problems and clarify instructions. Very often a pictorial sketch of some detail of construction will prove to be more intelligible and will convey the idea much better than an orthographic sketch, even when dealing with an experienced draftsman or detailer (Fig. 6.22).

Design sketches may be done in the quiet of the designer's office or amid the confusion of the conference table. To meet the requirement of speed of preparation, one must resist all temptation to use instruments of any type and rely on the pencil alone, for the true measure of the quality of a finished sketch is neatness and good proportion rather than the straightness of the lines. A pictorial sketch need not be an artistic masterpiece to be useful.

6.8

Straight Lines. The pencil should rest on the second finger and be held loosely by the thumb and index finger about 25–40 mm (1–1½ in.) above the point.

Horizontal lines are sketched from left to right with an easy arm motion that is pivoted about the muscle of the forearm. The straight line thus becomes an arc of infinite radius. When sketching a straight line, it is advisable first to mark the end points with light dots or small crosses (Fig. 6.8).

The complete procedure for sketching a straight line is as follows:

1. Mark the end points.

2. Make a few trial motions between the marked points to adjust the eye and hand to the contemplated line.

3. Sketch a *very* light line between the points by moving the pencil in two or three sweeps. When sketching the trial line, the eye should be on the point toward which the movement is directed. With each stroke, an attempt should be made to correct the most obvious defects of the stroke preceding, so that the finished trial line will be relatively straight.

4. Darken the finished line, keeping the eye on the pencil point on the trial line. The final line, replacing the trial line, should be distinct, black, uniform, and straight.

It is helpful to turn the paper through a *convenient angle* so that the horizontal and vertical lines assume a slight inclination (Fig. 6.9). A horizontal line, when the paper is in this position, is sketched to the right and upward, thus allowing the arm to be held slightly away from the body and making possible a free arm motion.

Short vertical lines may be sketched either downward or upward, without changing the position of the paper. When sketching downward, the arm is held slightly away from the body and the movement is toward the sketcher (Fig. 6.10). To sketch vertical lines upward, the arm is held well away from the body. By turning the paper, a long vertical line may be made to assume the position of a horizontal line and can be sketched with the same general movements used for the latter.

Inclined lines running upward from lower left to upper right may be sketched upward with the same movements used for horizontal lines, but those running

Fig. 6.7 Pencil points and sketch lines.

MECHANICAL LINE

VISIBLE OBJECT LINE — THICK AND DENSE BLACK

INVISIBLE OBJECT LINE — MEDIUM AND DENSE BLACK

CONSTRUCTION LINE — THIN AND LIGHT

DIMENSION LINE — THIN AND BLACK

CENTER LINE — THIN AND BLACK

Fig. 6.8 Steps in sketching a straight line.

Fig. 6.9 Sketching horizontal lines.

downward from upper left to lower right are sketched with the general movements used for either horizontal or vertical lines, depending on their inclination (Fig. 6.11). Inclined lines may be more easily sketched by turning the paper to make them conform to the direction of horizontal lines.

Fig. 6.10   Sketching vertical lines.

Fig. 6.11   Sketching inclined lines.

6.9
Circles.   Small circles may be sketched by marking radial distances on perpendicular center lines. When additional points are needed, the distances can be marked off either by eye or by measuring with a marked strip of paper (Fig. 6.12). Larger circles may be constructed more accurately by sketching two or more diagonals, in addition to the center lines, and by sketching short construction lines perpendicular to each, equidistant from the center. Tangent to these lines, short arcs are drawn perpendicular to the radii. The circle is completed with a light construction line, and all defects are corrected before darkening (Fig. 6.13).

C □ Multiview sketches

6.10
Making a Multiview Sketch (Fig. 6.14).
When making orthographic working sketches a systematic order should be followed, and all the rules and conventional practices used in making working drawings should be applied. The following procedure is recommended:

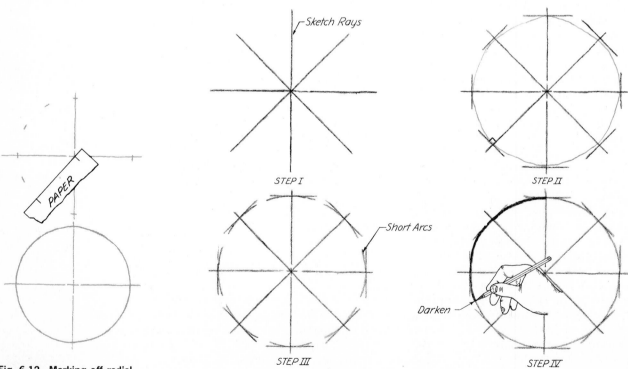

Fig. 6.12   Marking off radial distances.

Fig. 6.13   Sketching large circles.

1. Examine the object, giving particular attention to detail.

2. Determine which views are necessary.

3. "Block-in" the views, using light construction lines.

4. Complete the detail and darken the object lines.

5. Sketch extension lines and dimension lines, including arrowheads.

6. Complete the sketch by adding dimensions, notes, title, date, sketcher's name or initials, and so on.

7. Check the entire sketch carefully to see that no dimensions have been omitted.

The beginning student should read Part A of Chapter 5 before he attempts to make a multiview sketch.

6.11

Proportions. The beginner must recognize the importance of being able to estimate comparative relationships between the width, height, and depth of an object being sketched. The complete problem of proportioning a sketch also involves relating the estimated dimensions for any component parts, such as slots, holes, and projections, to the over-all dimensions of the object. It is not the practice to attempt to estimate actual dimensions, for sketches are not usually made to scale. Rather one must decide, for example, that the width of the object is twice its height, that the width of a given slot is equal to one-half the width of the object, and that its depth is approximately one-fourth the overall height.

To become proficient at sketching one must learn to recognize proportions and be able to compare dimensions "by eye." Until one is able to do so, he can not really "think with his pencil." Some people can develop a keen eye for proportion with only a limited amount of practice and can maintain these estimated proportions when making the views of a sketch. Others have alternately discouraging and encouraging experiences. Discouragement comes when one's knowledge of sketching is ahead of his ability and he has not had as much practice as he needs. The many who find it difficult to make the proportions of the completed sketch agree with the estimated proportions of the object may begin by using the graphical method shown in Figs. 6.15(a), (b), and (c). This

Fig. 6.14 Steps in sketching.

method is based on the fact that a rectangle (enclosing a view) may be divided to obtain intermediate distances along any side that are in such proportions to the total length as one-half, one-fourth, one-third, and so on. Those who start with this rectangle method as an aid in proportioning should abandon its use when they have developed their eye and sketching skills so that it is no longer needed.

Sketching must be done rapidly, and the addition of unnecessary lines consumes much valuable time. Furthermore, the addition of construction lines distracts the reader, and it is certain that they do not contribute to the neatness of the sketch.

The midpoint of a rectangle is the point of intersection of the diagonals, as shown in Fig. 6.15(a). A line sketched through

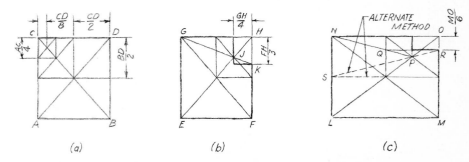

Fig. 6.15 Methods of proportioning a rectangle representing the outline of a view.

this point that is perpendicular to any side will establish the midpoint of that side. Should it be necessary to determine a distance that is equal to one-fourth the length of a side, (say, *AC*), the quarter-point may be located by repeating this procedure for the small rectangle representing the upper left-hand quarter of *ABCD*.

With the midpoint *J* located by the intersecting diagonals of the small rectangle [representing one-fourth of the larger rectangle *EFGH*, as in (*b*)], the one-third point along *FH* may be located by sketching a line from point *G* through *J* and extending it to line *FH*. The point *K* at the intersection of these lines establishes the needed one-third distance.

To determine one-sixth the length of a side of a rectangle, as in (*c*), sketch a line from *N* through point *P*, as was done in (*b*), to determine a one-third distance. Point *Q* at which the line *NP* crosses the center line of the rectangle establishes the one-sixth distance along the center line.

Figure 6.16 shows how this method for dividing the sides of a rectangle might be used to proportion an orthographic sketch.

The square may be used to proportion a view after one dimension for the view has been assumed. In this method, additional squares are added to the initial one having the assumed length as one side (Fig. 6.17). As an example, suppose that it has been estimated that the front view of an object should be three times as long as it is high. In Fig. 6.17 the height of the view has been represented by the line *AB* sketched to an assumed length. The first step in making the construction is to

sketch the initial square *ABCD* and extend *AC* and *BD* to indefinite length, being certain that the overall length from *A* and *B* will be slightly greater than three times the length of *AB*. Then the center line must be sketched through the intersection of *AD* and *BC*. Now *BX* extended to *E* locates *EF* to form the second square, and *DY* extended to point *G* locates the line *GH*. Line *AG* will be three times the length of *AB*.

D □ Sketching in isometric oblique and perspective

6.12
Pictorial Sketching. Students may employ pictorial sketches to advantage as an aid in visualizing and organizing problems. Sales engineers may frequently include pictorial sketches with orthographic sketches when preparing field reports on the needs and suggestions of the firm's customers.

With some training anyone can prepare pictorial sketches that will be satisfactory for all practical purposes. Artistic ability is not needed. This fact is important, for many persons lack only the necessary confidence to start making pictorial sketches.

6.13
Mechanical Methods of Sketching. Many engineers have found that they can produce satisfactory pictorial sketches by using one of the so-called mechanical methods. They rely on these methods because of their familiarity with the procedures used in making pictorial drawings with instruments.

The practices presented in Chapter 11 for the mechanical methods, axonometric, oblique, and perspective, are followed generally in pictorial sketching, except that angles are assumed and lengths are estimated. For this reason, one must develop an eye for good proportion before he

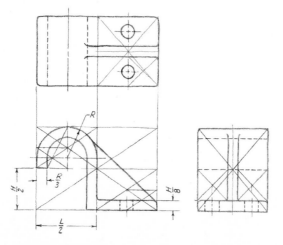

Fig. 6.16 Rectangle method applied in making an orthographic sketch.

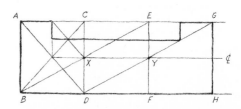

Fig. 6.17 Build-up method.

will be able to create a satisfactory pictorial sketch that will be in no way misleading.

A student having difficulty in interpreting a multiview drawing usually will find that a pictorial sketch, prepared as illustrated in Fig. 6.18, will clarify the form that he is trying to visualize, even before the last lines of the sketch have been drawn.

6.14

Isometric Sketching. Isometric sketching starts with three isometric lines, called axes, which represent three mutually perpendicular lines. One of these axes is sketched vertically, the other two at 30° with the horizontal. In Fig. 6.18 (step I), the near front corner of the enclosing box lies along the vertical axis, while the two visible receding edges of the base lie along the axes receding to the left and to the right.

If the object is of simple rectangular form, as in Fig. 6.18, it may be sketched by drawing an enclosing isometric box (step I) on the surfaces of which the orthographic views may be sketched (step II). Care must be taken in assuming lengths and distances so that the finished view (step III) will have relatively correct proportions. In constructing the enclosing box (step I), the vertical edges are parallel to the vertical axis, and edges receding to the right and to the left are parallel to the right and left axes, respectively.

Objects of more complicated construction may be "blocked in," as shown in Fig. 6.19. Note that the projecting cylindrical features are enclosed in "isometric" prisms and that the circles are sketched within isometric squares. The procedure in Fig. 6.19 is the same as in Fig. 6.18, except that three enclosing isometric boxes are needed in the formation of the final representation instead of one.

In sketching an ellipse to represent a circle pictorially, an enclosing "isometric square" (rhombus) is drawn having sides equal approximately to the diameter of the true circle (step I, Fig. 6.20). The ellipse is formed by first drawing arcs tangent to the midpoints of the sides of the isometric square in light sketchy pencil lines (step II). In finishing the ellipse (step III) with a dark heavy line, care must be taken to obtain a nearly elliptical shape.

Figure 6.21 shows the three positions for an isometric circle. Note that the major axis is horizontal for an ellipse on a horizontal plane (I).

An idea sketch prepared in isometric is shown in Fig. 6.22.

6.15

Proportioning. As stated in Sec. 6.11, one should eventually be able to judge lengths and recognize proportions. Until this ultimate goal has been reached, the graphical method presented in Fig. 6.15 may be used with pictorial sketching (Fig. 6.23).

Fig. 6.18 Steps in isometric sketching.

Fig. 6.19 Blocking in an isometric sketch.

SKETCH "ISOMETRIC SQUARE" STEP I

SKETCH SHORT ARCS STEP II

COMPLETE ELLIPSE STEP III

Fig. 6.20 Isometric circles.

Fig. 6.21 Isometric circles.

Fig. 6.22 Idea sketch in isometric.

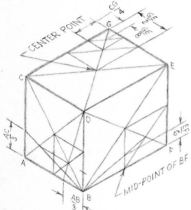

Fig. 6.23 Method for proportioning a rhomboid.

Fig. 6.24 Proportioning method applied.

The procedures as used are identical, the only recognizable difference being that the rectangle in the first case now becomes a rhomboid. Figure 6.24 illustrates how the method might be applied in making a sketch of a simple object. The enclosing box was sketched first with light lines, and then the graphical method was applied as shown to locate the points at one-quarter and one-half of the height. To establish the line of the top surface that is at a distance equal to one-third of the length from the end, a construction line was sketched from A to the midpoint B to locate C at the point of intersection of AB with the diagonal. Point C will fall on the required line.

Fig. 6.25 Blocking in an oblique sketch.

6.16

Sketches in Oblique. A sketch in oblique shows the front face without distortion, in its true shape. It has this one advantage over a representation prepared in isometric, even though the final result usually will not present so pleasing an appearance. It is not recommended for objects having circular or irregularly curved features on any but the front plane or in a plane parallel to it.

The beginner who is familiar with axonometric sketching will have very little difficulty in preparing a sketch in oblique, for, in general, the methods of preparation presented in the previous sections apply to both. The principal difference between these two forms of sketching is in the position of the axes, oblique sketching being unlike the isometric in that two of the axes are at right angles to each other. The third axis may be at any convenient angle, as indicated in Fig. 6.25.

Figure 6.26 shows the steps in making an oblique sketch using the proportioning methods previously explained for dividing a rectangle and a rhomboid. The receding lines are made parallel when a sketch is made in oblique projection.

The distortion and illusion of extreme elongation in the direction of the receding axis may be minimized by foreshortening to obtain proportions that are more realistic to the eye and by making the receding lines converge slightly. The resulting sketch will then be in a form of pseudoperspective, which resembles parallel perspective to some extent.

6.17

Perspective Sketching. A sketch that has been prepared in accordance with the concepts of perspective will present a somewhat more pleasing and realistic effect than one in oblique or isometric. A perspective sketch actually presents an object as it would appear when observed from a particular point. The recognition of this fact, along with an understanding of the concepts that an object will appear smaller at a distance than when it is close and that horizontal lines converge as they recede until they meet at a vanishing point, should enable one to produce sketches having a perspective appearance. In sketching an actual object, a position should be selected that will show it to the best advantage. When the object exists only in one's mind or on paper in orthographic form, then the object must be visualized and the viewing position assumed.

At the start, the principal lines should be sketched in lightly, each line extending for some length toward its vanishing point. After this has been accomplished, the enclosing perspective squares for circles should be blocked in and the outline for minor details added. When the object lines have been darkened, the construction lines extending beyond the figure may be erased.

Figure 6.27 shows a parallel or one-point perspective that bears some resemblance to an oblique sketch. All faces in planes parallel to the front show their true shape. All receding lines should meet at a single vanishing point. Figure 6.28 is an angular or two-point perspective.

As stated previously, a two-point perspective sketch shows an object as it would appear to the human eye at a fixed point in space and not as it actually exists. All parallel receding lines converge (Fig. 6.29). Should these receding lines be horizontal, they will converge at a vanishing point on the eyeline. Those lines extending toward the right converge to a vanishing point to the right (VP_R), and those to the left converge to the left (VP_L). These vanishing points are at the level of the observer's eye (Fig. 6.29). A system of lines that is neither perpendicular nor horizontal will converge to a VP for inclined lines.

In one-point perspective one of the principal faces is parallel to the picture plane. All of the vertical lines will appear as vertical, and the receding horizontal lines will converge to a single vanishing point (Fig. 6.30).

Those interested in a complete discussion of the geometry of perspective drawing should read Secs. 11.27–11.40. The beginner should make two or three mechanically drawn perspectives at the start to fix the fundamentals of the methods of perspective projection in his mind, even though there is some difference between

sketching what one sees or imagines and true geometrical perspective.

In making a sketch in artist's perspective, several fundamental concepts must be recognized.

First, a circle sketched in perspective will resemble an ellipse (Fig. 6.29). The long diameter of the repesentation of a circle on a horizontal plane is always in a horizontal direction (Fig. 6.31).

Second, if an object or a component part of an object is above the eyeline, it will be seen from below. Should the object be below the eyeline, it will be seen from above (Fig. 6.29). The farther the object is removed below or above the eyeline, the more one can see of the top or the bottom surface.

Third, the nearest vertical edge of an object will be the longest vertical line of the view, as shown in Fig. 6.29. When two or more objects of the same actual height appear in a perspective sketch, their represented heights will decrease in the view as they near the vanishing point.

Fig. 6.27 Sketch in parallel perspective.

Sketch Box
STEP I

Block-in Outline   *Complete Pictorial*
STEP II   *STEP III*

Fig. 6.26 Steps in oblique sketching.

**Fig. 6.28 Sketch in angular
perspective.**

Fig. 6.29 Preparing a perspective sketch.

6.18

Preparing a Perspective Sketch. The application of proportioning methods to the construction and division of an enclosing box is shown in Fig. 6.29. Read Secs. 6.11 and 6.15. The construction of a required perspective by steps is as follows:

Step I. Sketch the eyeline. This line should be well toward the top of the sheet of sketch paper.

Step II. Locate VP_L and VP_R on the eyeline. These vanishing points should be placed as far apart as possible.

Step III. Assume the position and length for the near front edge AB. The length of this line, along with the spacing of the vanishing points, establishes the size of the finished sketch. The position of AB determines how the visible surfaces are to appear. For instance, if the line AB had been moved downward from the position shown in (a), much more of the top surface would be seen in (b) and (c). Should AB have been moved to the right from its position shown in (a), the left side would have become more prominent. If AB were placed midway between the two vanishing points, then both the front and left-side surfaces would be at 45° with the picture plane for the perspective. As it has been placed in (a), the side face is at 60° to the picture plane, while the front face is at 30°. Before establishing the position for the near front edge, one must decide which surfaces are the most important surfaces and how they may best be displayed in the sketch.

Step IV. Sketch light construction lines from points A and B to each vanishing point.

Step V. Determine the proportions for the enclosing box, in this case $4\frac{1}{2}$, 2, and $1\frac{1}{2}$, and mark off two equal units along AB.

Step VI. Sketch perspective squares, representing the faces of 1-unit cubes, starting at AB and working toward each vanishing point. In cases where an overall length must be completed with a partial unit, a full-perspective square must be sketched at the end.

Step VII. Subdivide any of the end squares if necessary and sketch in the enclosing box (a).

Step VIII. Locate and block in the details, subdividing the perspective squares as required to establish the location of any detail. When circles are to be sketched in perspective by a beginner, it is advisable to sketch the enclosing box first using light lines (b).

Step IX. Darken the object lines of the sketch. Construction lines may be removed and some shading added to the surfaces as shown in (c), if desired.

A sketch in one-point perspective might be made as shown in Fig. 6.30. For this particular sketch the enclosing box was made to the assumed overall proportions for the part. Then the location of the details was established by subdividing the regular rectangle of the front face of the enclosing box and the perspective rectangle of the right side.

Fig. 6.30 Sketch made in one-point perspective.

Pencil Shading. The addition of some shading to the surfaces of a part will force its form to stand out against the white surface of the sketching paper and will increase the effect of depth in a view that might otherwise appear to be somewhat flat.

Seldom are technologists and engineers able to do creditable work in artistic shading, with cast shadows included, as they could many years ago when training in art was part of an engineer's education. It is unfortunate that they lack this training at the present time, for art and design go hand in hand. This is especially true today, for a pleasing and appealing styling sells more products than a good mechanical design.

Within the scope of this chapter, written for beginning students, it will only be possible to present a few simple rules as a guide for those making a first attempt at surface shading. However, continued practice and some thought should lead one to the point where he can do a creditable job of shading and definitely improve a pictorial sketch.

When shading an imaginary part, a designer may consider the source of the light to be located in a position to the left, above, and in front of the object. Of course, if the part actually exists and is being sketched by viewing it, then the sketcher should attempt to duplicate the degrees of shade and shadows as they are observed.

With the light source considered to be to the left, above, and in front of the object, a rectangular part would be shaded as shown in Fig. 6.31(a). The use of gradation of tone on the surfaces gives additional emphasis to the depth. To secure this added effect by shading, the darkest tone on the surface that is away from the light must be closest to the eye. As the surface recedes, the tone must be made lighter in value with a faint trace of reflected light showing along the entire length of the back edge. On the lighted side, the lightest area must be closest to the eye, as indicated by the letter L_1 in (a). To make this lighted face appear to go into the distance, it is made darker as it recedes, but it should never be made as dark as the lightest of the dark tones on the dark surface.

Shading a cylindrical part is not as easy to do as shading a rectangular part but, if it is realized that practically half of the cylinder is in the light and half in the dark

and that the lightest light and the darkest dark fall along the elements at the quarter-points of each half, then one should not find the task too difficult (b). The two extremes are separated by lighter values of shade. The first quarter on the lighted side must be made lighter than the last quarter on the dark side. In starting at the left and going counterclockwise there is a dark shade of light blending into the full light at the first quarter-point. From this point and passing the center to the dark line, the tone should become gradually darker. If vertical lines are used for shading, they should be spaced closer and closer together as they approach the dark line. The extreme-right-hand quarter should show the tones of reflected light.

There are two ways that pencil shading may be applied. If the paper has a medium-rough surface, solid tone shading may be used with one shade blending into the other. For the best results, the light tones are put on first over all areas to be shaded. The darker tones are then added by building up lighter tones to the desired intensity for a particular area. For this form of shading, a pencil with flattened point is used.

The other form of shading, and the one that is best suited for quick sketches, is produced with lines of varied spacing and weight. Light lines with wide spacing are used on the light areas and heavy lines that are closely spaced give the tone for the darkest areas. No lines are needed for the lightest of the light areas.

6.20
Conventional Treatment of Fillets, Rounds, and Screw Threads. Sketches that are not given full pencil shading may be given a more or less realistic appearance by representing the fillets and rounds of the unfinished surfaces as shown in Fig. 11.41. The conventional treatment for screw threads is shown in (b) and (c) of the same illustration.

Fig. 6.31 **Shading rectangular and cylindrical parts.**

6.21

Use of an Overlay Sheet. In preparing a design sketch, an overlay sheet may be used to advantage in making a sketch that is complicated by many details (Fig. 6.32). In this case, a quick sketch showing the general outline of the principal parts is made first in a rather rough form. Then an overlay sheet is placed over this outline sketch and the lines are retraced. In doing so, slight corrections can be made for any errors existing in the proportions of the parts or in the position of any of the lines of the original rough sketch. When this has been done, the representation of the related minor parts are added. If at any time one becomes discouraged with a sketch (multiview or pictorial) that he is making and feels that he should make a new start, he should use an overlay sheet, for there are usually many features on his existing sketch that may be retraced with a great saving of time.

6.22

Illustration Sketches Showing Mechanisms Exploded. A sketch of a mechanism showing the parts in exploded positions along the principal axes is shown in Fig. 6.33. Through the use of such sketches those who have not been trained to read multiview drawings may readily understand how a mechanism should be assembled, for both the shapes of the parts and their order of assembly, as denoted by their space relationship, is shown in pictorial form.

Illustration sketches may be made for discussions dealing with ideas for a design, but more frequently they are prepared for explanatory purposes to clarify instructions for preparing illustration drawings in a more finished form as assembly illustrations, advertising illustrations, catalogue illustrations, and illustrations for service and repair charts.

Many persons find that it is desirable, when preparing sketches of exploded mechanisms, to first block in the complete mechanism with all of the parts in position. At this initial stage of construction, the parts are sketched in perspective in rough outline and the principal axes and object lines are extended partway toward their vanishing points.

When the rough layout has been completed to the satisfaction of the person preparing the sketch, an overlay sheet is placed over the original sketch and the parts are traced directly from the sketch beneath in exploded positions along the principal axes and along the axes of holes. Frequently some beginners make a traced sketch of each individual part and place the sketches in exploded positions before preparing the finished sketch. Others with more experience accomplish the same results by first tracing the major part along with the principal axes and then moving the overlay sheet as required to trace off the remaining parts in their correct positions along the axes.

It should be recognized that in preparing sketches of exploded mechanisms in this manner the parts are not shown perspectively reduced although they are

USE OF AN OVERLAY SHEET

A VIEW FROM THE ROUGH SKETCH

A VIEW FROM THE FINISHED SKETCH ON OVERLAY SHEET

Fig. 6.32 Use of an overlay sheet for creating a final and complete sketch of a mechanism.

removed from their original place in the pictorial assembly outward and toward the vanishing points. To prepare a sketch of this type with all of the parts shown in true geometrical perspective would result in a general picture that would be misleading and one that would be apt to confuse the nontechnical person because some parts would appear much too large or too small to be mating parts.

6.23
Pictorial Sketching on Ruled Paper. Although one should become proficient in sketching on plain white bond paper, a specially ruled paper, shown in Fig. 6.34, can be used by those who need the help of guidelines.

Block-in assembly in outline

Retrace parts on overlay sheet in exploded positions from assembly sketch

Fig. 6.33 Sketch showing the parts of a mechanism in exploded positions.

Fig. 6.34 Sketches on isometric paper.

Problems

The problems presented with this chapter have been selected to furnish practice in freehand drawing. The individual pieces that appear in pictorial form have been taken from a wide variety of mechanisms used in different fields of engineering. The student may be required to prepare complete working sketches of these parts if his instructor desires. Such an assignment, however, would presuppose an understanding of the fundamentals of dimensioning as they are presented in the beginning sections of Chapter 13.

The problems presented in Fig. 6.53–6.60 were selected to give practice in preparing pictorial sketches—isometric, oblique, and perspective. In addition to developing proficiency in sketching, these problems offer the student further opportunity to gain experience in reading drawings. Additional problems that are suitable for pictorial sketching may be found in Chapter 11.

1. (Fig. 6.35). Reproduce the one-view sketch on a sheet of sketching paper.

2–4. (Figs. 6.36–6.38). Sketch, freehand, the necessary views of the given objects as assigned. The selected length for the unit will determine the size of the views. Assume any needed dimensions that are not given in units.

5–14. (Figs. 6.39–6.48). These problems are designed to give the student further study in multiview representation and, at the same time, offer him the opportunity to apply good line technique to the preparation of sketches.

Only the necessary views on which all of the hidden lines are to be shown should be drawn.

If dimensions are to be given, ample space must be allowed between the views for their placement. The beginning sections of Chapter 13 present the basic principles of size description.

15. (Fig. 6.49). Make a complete three-view sketch of the motor base. The ribs are 10 mm thick. At points A, four holes are to be drilled for 12 mm bolts that are to be 64 mm center to center in one direction and 80 mm in the other. At points B four holes are to be drilled for 12 mm bolts that fasten the motor base to a steel column. Fillets and rounds are $R3$.

16. (Fig. 6.50). Make a three-view orthographic sketch of the motor bracket.

17. (Fig. 6.51). Make a complete three-view sketch of the tool rest and/or the tool rest bracket. The rectangular top surface of the tool rest is to be 30 mm above the center line of the hole for the 12 mm bolt. The overall dimensions

GASKET,
PUNCH FROM $\frac{1}{16}$ CORK

Fig. 6.35

Fig. 6.36 Wedge block.

Fig. 6.37 End block.

Fig. 6.38 Corner block.

FILLETS & ROUNDS R3

Fig. 6.39 Shifter.

Fig. 6.40 Tool rest.

Fig. 6.41 Offset trip lever.

FILLETS & ROUNDS R3

Fig. 6.42 Guide link.

Fig. 6.43 Idler lever (weldment).

Fig. 6.44 Control bracket.

Fig. 6.45 Index guide.

Fig. 6.46 Arm bracket.

Fig. 6.47 Bearing bracket.

Fig. 6.48 Rear support bracket.

Fig. 6.49 Motor base.

Fig. 6.50 Motor bracket.

of the top are 30 × 65 mm. It is to be 6 mm thick. The overall dimensions of the rectangular pad of the bracket are 32 × 48 mm. The center line of the adjustment slot is 15 mm above the center line of the top holes in the rectangular pad and the distance from center line to center line of the slot is 35 mm. The bracket is to be fastened to a housing with 6 mm pan head machine screws.

18. Make a freehand exploded pictorial representation of the ladder shoe shown in Fig. 6.52. Show the parts as if all intersecting surfaces were sharp, thus eliminating any consideration of representing fillets and rounds.

Fig. 6.51 Tool rest and tool rest bracket.

Fig. 6.52 Ladder shoe.

19–24. (Figs. 6.53–6.58). Make freehand isometric sketches of the objects as assigned.

25–28. (Figs. 6.57–6.60). Make freehand oblique sketches of the objects as assigned.

Fig. 6.53

Fig. 6.54

Fig. 6.55

Fig. 6.56

Fig. 6.57

Fig. 6.58

Fig. 6.59

Fig. 6.60

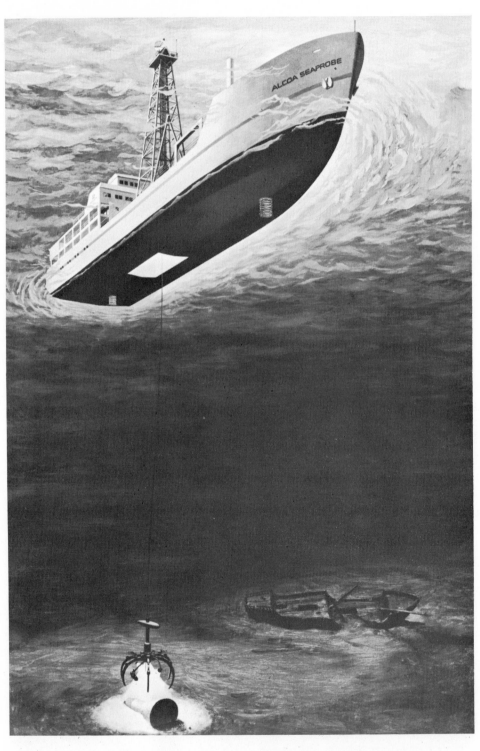

Alcoa Seaprobe at work. The design of this recovery ship required the application of the principles of multiview drawing in the development of creative ideas. (*Courtesy Aluminum Corporation of America*)

sectional views 7

7.1

Sectional Views (Fig. 7.1). Although the invisible features of a simple object usually may be described on an exterior view by the use of hidden lines, it is unwise to depend on a perplexing mass of such lines to describe adequately the interior of a complicated object or an assembled mechanism. Whenever a representation becomes so confused that it is difficult to read, it is customary to make one or more of the views "in section" (Fig. 7.2). A view "in section" is one obtained by imagining the object to have been cut by a cutting plane, the front portion being removed to reveal clearly the interior features. Figure 7.3 illustrates the use of an imaginary cutting plane. At this point it should be understood that a portion is shown removed only in a sectional view, not in any of the other views (Fig. 7.3).

When the cutting plane cuts an object lengthwise, the section obtained is commonly called a longitudinal section; when crosswise, it is called a cross section. It is designated as being either a full section, a half section, or a broken section. If the plane cuts entirely across the object, the section represented is known as a *full section*. If it cuts only halfway across a symmetrical object, the section is a *half section*. A *broken section* is a partial one, which is used when less than a half section is needed (Fig. 7.3).

On a completed sectional view, fine section lines are drawn across the surface cut by the imaginary plane, to emphasize the contour of the interior (see Sec. 7.8).

Fig. 7.1 Working drawing with sectional views. (*Courtesy Warner and Swasey Co.*)

SECTIONAL VIEW

Fig. 7.2 Sectional view.

EXTERIOR VIEW

7.2

Full Section. Since a cutting plane that cuts a full section passes entirely through an object, the resulting view will appear as illustrated in Fig. 7.3. Although the plane usually passes along the main axis, it may be offset (Fig. 7.4) to reveal important features.

A full-sectional view, showing an object's characteristic shape, usually replaces an exterior front view; however, one of the other principal views, side or top, may be converted to a sectional view if some interior feature thus can be shown to better advantage or if such a view is needed in addition to a sectioned front view.

The procedure in making a full-sectional view is simple, in that the sectional view is an orthographic one. The imaginary cut face of the object simply is shown as it would appear to an observer looking directly at it from a point an infinite distance away. In any sectional view, it is considered good practice to omit all invisible lines unless such lines are necessary to clarify the representation. Even then they should be used sparingly.

7.3

Half Section. The cutting plane for a half section removes one-quarter of an object. The plane cuts halfway through to the axis or center line so that half the finished sectional view appears in section and half appears as an external view (Fig. 7.3). This type of sectional view is used when a view is needed showing both the exterior and interior construction of a symmetrical object. Good practice dictates that hidden lines be omitted from both halves of the view unless they are absolutely necessary for dimensioning purposes or for explaining the construction. Although the use of a solid object line to separate the two halves of a half section has been approved by the Society of Automotive Engineers and has been accepted by the American National Standards Institute [Fig. 7.5(a)], many draftsmen prefer to use a center line, as

Fig. 7.3 Types of sectional views.

FULL SECTION

HALF SECTION

BROKEN SECTION

Fig. 7.4 Offset cutting plane.

FULL SECTION

Fig. 7.5 Half section.

shown in Fig. 7.5(*b*). They reason that the removal of a quarter of the object is theoretical and imaginary and that an actual edge, which would be implied by a solid line, does not exist. The center line is taken as denoting a theoretical edge.

7.4

Broken Section. A broken or partial section is used mainly to expose the interior of objects so constructed that less than a half section is required for a satisfactory description (Fig. 7.6). The object theoretically is cut by a cutting plane and the front portion is removed by breaking it away. The "breaking away" gives an irregular boundary line to the section.

7.5

Revolved Section. A revolved section is useful for showing the true shape of the cross section of some elongated object,

such as a bar, or some feature of an object, such as an arm, spoke, or rib (Figs. 7.1 and 7.7).

To obtain such a cross section, an imaginary cutting plane is passed through the member perpendicular to the longitudinal axis and then is revolved through 90° to bring the resulting view into the plane of the paper (Fig. 7.8). When revolved, the section should show in its true shape and in its true revolved position, regardless of the location of the lines of the exterior view. If any lines of the view interfere with the revolved section, they should be omitted (Fig. 7.9). It is sometimes advisable to provide an open space for the section by making a break in the object (Fig. 7.7).

7.6

Removed (Detail) Sections. A removed section is similar to a revolved section,

Fig. 7.6 Broken section.

Fig. 7.7 Revolved section.*

Fig. 7.8 Revolved section and cutting plane.

Fig. 7.9 Correct and incorrect treatment of a revolved section.

*ANSI Y14.2-1957.

except that it does not appear on an external view but instead is drawn "out of place" and appears adjacent to it (Fig. 7.10). There are two good reasons why detail sections frequently are desirable. First, their use may prevent a principal view of an object, the cross section of which is not uniform, from being cluttered with numerous revolved sections (Fig. 7.11). Second, they may be drawn to an enlarged scale in order to emphasize detail and allow for adequate dimensioning.

Whenever a detail section is used, there must be some means of identifying it. Usually this is accomplished by showing the cutting plane on the principal view and then labeling both the plane and the resulting view, as shown in Fig. 7.11.

7.7

Phantom Sections. A phantom or hidden section is a regular exterior view on which the interior construction is emphasized by crosshatching an imaginary cut surface with dotted section lines (Fig. 7.12). This type of section is used only when a regular section or a broken section would remove some important exterior detail or, in some instances, to show an accompanying part in its relative position with regard to a particular part (Fig. 7.13). Instead of using a broken line with dashes of equal length, the phantom line shown in Fig. 2.58 could have been used to represent the outline of the adjacent parts shown in Fig. 7.13.

7.8

Section Lining. Section lines are light continuous lines drawn across the imaginary cut surface of an object for the purpose of emphasizing the contour of its interior. Usually they are drawn at an angle of 45° except in cases where a number of adjacent parts are shown assembled (Fig. 7.17).

To be pleasing in appearance, these lines must be correctly executed. While on ordinary work they are spaced about 2 mm (.09 in.) apart, there is no set rule governing their spacing. They simply should be spaced to suit the drawing and the size of the areas to crosshatched. For example, on small views having small areas, the section lines may be as close as 0.8 mm (.03 in.), while on large views having large areas they may be as far apart as 3 mm (.12 in.). In the case of very thin plates, the cross section is shown solid black (Fig. 7.14).

The usual mistake of the beginning stu-

dent is to draw the lines too close together. This, plus the unavoidable slight variations, causes the section lining to appear streaked. Although several forms of mechanical section liners are available, most draftsmen do their spacing by eye. The student is advised to do likewise, being careful to see that the initial pitch, as set by the first few lines, is maintained across the area. To accomplish this, he should check back from time to time to make sure there has been no slight gen-

Fig. 7.10 Removed sections.*

Fig. 7.11 Removed (detail) sections.*

Fig. 7.12 Phantom section.

Fig. 7.13 Phantom sectioning— adjacent parts.

Fig. 7.14 Thin sections.

*ANSI Y14.2-1957.

133

eral increase or decrease in the spacing. An example of correct section lining is shown in Fig. 7.15(a), and, for comparison, examples of faulty practice may be seen in Fig. 7.15(b), (c), and (d). Experienced draftsmen realize that nothing will do more to ruin the appearance of a drawing than carelessly executed section lines.

As shown in Fig. 7.16, the section lines on two adjacent pieces should slope at 45° in opposite directions. If a third piece adjoins the two other pieces, as in Fig. 7.17(a), it ordinarily is section-lined at 30°. An alternative treatment that might be used would be to vary the spacing without changing the angle. On a sectional view showing an assembly of related parts, *all portions of the cut surface of any part must be section-lined in the same direction, for a change would lead the reader to consider the portions as belonging to different parts. Furthermore, to allow quick identification, each piece (and all identical pieces) in every view of the assembly drawing should be section-lined in the same direction.*

Shafts, bolts, rivets, balls, and so on, whose axes lie in the plane of section, are not treated the same as ordinary parts. Having no interior construction to be shown, they are drawn in full and thus tend to make the adjacent sectioned parts stand out to better advantage (Fig. 7.18).

Whenever section lines drawn at 45° with the horizontal are parallel to part of the outline of the section (see Fig. 7.19), it is advisable to draw them at some other angle (say, 30° or 60°). Those drawn as in (a) and (c) produce an unusual appearance that is contrary to what is expected. Note the more natural effect obtained in (b) and (d) by sloping the lines at 30° and 75°.

7.9

Outline Sectioning. Very large surfaces may be section-lined around the bounding outline only, as illustrated in Fig. 7.20.

7.10

Symbolic Representation for a Cutting Plane. The symbolic lines that are used to represent the edge view of a cutting plane are shown in Fig. 7.21. The line is as heavy as an object line and is composed of either alternate long and short dashes or a series of dashes of equal length. The latter form is used in the automobile industry and has been approved by the SAE (Society of Automotive Engineers) and the American National Standards Institute. On drawings of ordinary size, when alternate long and short dashes are used for the cutting-plane line, the long dashes are 20 mm (.80 in.) long, the short dashes 3 mm (.12 in.) long, and the spaces 0.8 mm (.03 in.) wide, depending on the size of the drawing. When drawn in pencil on manila paper, they are made with a medium pencil.

Arrowheads are used to show the direction in which the imaginary cut surface is

Poor Practice
No contrast
(b)

Poor Practice
Uneven spacing-
lines of varying widths
(c)

Good Practice
(a)

Poor Practice
Lines too close
(d)

Fig. 7.15 Faults in section lining.

Fig. 7.16 Two adjacent pieces.

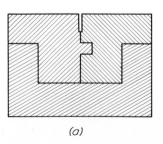

(a)

Fig. 7.17 Three adjacent pieces.

(b)

POOR PRACTICE
(a)

PREFERRED
(b)

POOR PRACTICE
(c)

PREFERRED
(d)

Fig. 7.19 Section lining at 30°, 60°, or 75°.

Fig. 7.20 Outline sectioning.

Fig. 7.18 Treatment of shafts, fasteners, ball bearings, and other parts. (*Courtesy New Departure Division, General Motors Corporation*)

viewed, and reference letters are added to identify it (Fig. 7.22).

Whenever the location of the cutting plane is obvious, it is common practice to omit the edge-view representation, particularly in the case of symmetrical objects. But if it is shown, and coincides with a center line, it takes precedence over the center line.

7.11
Summary of the Practices of Sectioning

1. A cutting plane may be offset in order to cut the object in such a manner as to reveal an important detail that would not be shown if the cutting plane were continuous (Fig. 7.4).

2. All visible lines beyond the cutting plane for the section are usually shown.

3. Invisible lines beyond the cutting plane for the section are usually not

shown, unless they are absolutely necessary to clarify the construction of the piece. In a half section, they are omitted in the unsectioned half, and either a center line or a solid line is used to separate the two halves of the view (Figs. 7.3 and 7.5).

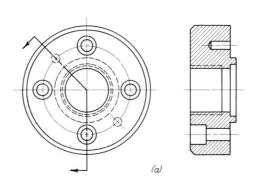

(a)

Fig. 7.21 Cutting plane lines (AN Standard).

SAE STANDARD

ASA STANDARD
(c)

APPROXIMATE
AMERICAN STANDARD
(b)

SECT B-B

SECT. A-A

Fig. 7.22 Sectional view.*

Auxiliary View

Auxiliary Section A-A

Cutting Plane A-A

Line of Sight

Fig. 7.23 Auxiliary section.

4. On a view showing assembled parts, the section lines on adjacent pieces are drawn in opposite directions at an angle of 45° (Fig. 7.16).

5. On an assembly drawing, the portions of the cut surface of a single piece in the same view or different views always should be section-lined in the same direction, with the same spacing (Fig. 7.18).

6. The symbolic line indicating the location of the cutting plane may be omitted if the location of the plane is obvious (Fig. 7.1).

7. On a sectioned view showing assembled pieces, an exterior view is preferred for shafts, rods, bolts, nuts, rivets, and so forth, whose axes are in the plane of section (Fig. 7.18).

7.12

Auxiliary Sections. A sectional view, projected on an auxiliary plane, is sometimes necessary to show the shape of a surface cut by a plane or to show the cross-sectional shape of an arm, rib, and so forth, inclined to any two or all three of the principal planes of projection (Fig. 7.23). When a cutting plane cuts an object, as in Fig. 7.23, arrows should show the direction in which the cut surface is viewed. Auxiliary sections are drawn by the usual method for drawing auxiliary views. When the bounding edge of the section is a curve, it is necessary to plot enough points to obtain a smooth one. Section 8.11 explains in detail the method for constructing the required view. A section view of this type usually shows only the inclined cut surface.

7.13

Conventional Sections. Sometimes a less confusing sectioned representation is obtained if certain of the strict rules of projection are violated, as explained in Chapter 5. For example, an unbalanced and confused view results when the sectioned view of the pulley shown in Fig. 7.24 is drawn in the true projection, as in (*a*). It is better practice to preserve symmetry by showing the spokes as if they were aligned into one plane, as in (*c*). Such treatment of unsymmetrical features is not misleading, since their actual arrangement is revealed in the circular view. The spokes are not sectioned in the preferred view. If they were, the first impression would be that the wheel had a solid web (*b*). See also Fig. 7.25.

When there are an odd number of holes in a flange, as is the case with the part in Fig. 7.26, they should be shown aligned in the sectioned view to reveal their true location with reference to the rim and the axis of the piece. To secure the so-called aligned section, one usually considers the

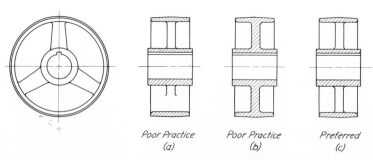

Poor Practice
(a)

Poor Practice
(b)

Preferred
(c)

Fig. 7.24 Conventional treatment of spokes in section.

cutting plane to be bent to pass through the angled hole, as shown in the pictorial drawing. Then, the bent portion of the plane (with the hole) is imagined to be revolved until it is aligned with the other portion of the cutting plane. As straightened out, the imaginary continuous plane produces the preferred section view shown in (a).

Figure 7.27 shows another example of conventional representation. The sectional view is drawn as though the upper projecting lug had been swung until the portion of the cutting plane through it formed a continuous plane with the other portion (Sec. 5.42). It should be noted that the hidden lines in the sectioned view are necessary for a complete description of the construction of the lugs.

7.14

Ribs in Section. When a machine part has a rib cut by a plane of section (Fig. 7.28), a "true" sectional view taken through the rib would prove to be false and misleading, because the crosshatching on the rib would cause the object to appear "solid." The preferred treatment is to omit arbitrarily the section lines from the rib, as illustrated by Fig. 7.28(a). The resulting sectional view may be considered the view that would be obtained if the plane were offset to pass just in front of the rib (b).

An alternative conventional method, approved but not as frequently used, is illustrated in Fig. 7.29. This practice of omitting alternate section lines some-

Fig. 7.25 Spokes in section.*

Fig. 7.26 Drilled flanges.

Fig. 7.27 Revolution of a portion of an object.

Fig. 7.28 Conventional treatment of ribs in section.

*ANSI Y14.2-1957.

Fig. 7.29 Alternative treatment of ribs in section.

times is adopted when it is necessary to emphasize a rib that might otherwise be overlooked.

7.15

Half Views. When the space available is insufficient to allow a satisfactory scale to be used for the representation of a symmetrical piece, it is considered good practice to make one view a half view, as shown in Fig. 7.30. The half view, however, must be the top or side view and not the front view, which shows the characteristic contour. The half view should be the rear half.

7.16

Material Symbols. The section-line symbols recommended by the American National Standards Institute for indicating various materials are shown in Fig. 7.31. Code section lining ordinarily is not used

on a working (detail) drawing of a separate part. It is considered unnecessary to indicate a material symbolically when its exact specification must be given as a note. For this reason, and in order to save time as well, the easily drawn symbol for cast iron is commonly used on detail drawings for all materials. Contrary to this general practice, however, some few chief draftsmen insist that symbolic section lining be used on all detail drawings prepared under their supervision.

Code section lining usually is employed on an assembly section showing the various parts of a unit in position, because a distinction between the materials causes the parts to "stand out" to better advantage. Furthermore, a knowledge of the type of material of which an individual part is composed often helps the reader to identify it more quickly and understand its function.

Fig. 7.30 Half view.

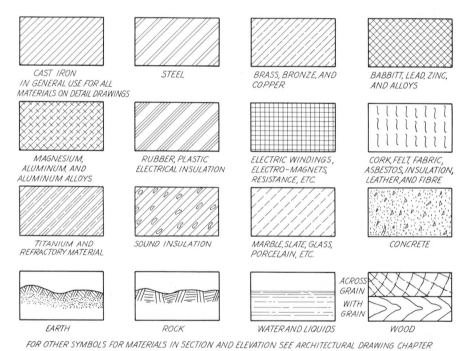

Fig. 7.31 Material symbols (AN Standard).

Problems

The following problems were designed to emphasize the principles of sectioning. Those drawings that are prepared from the pictorials of objects may be dimensioned if the elementary principles of dimensioning (Chapter 13) are carefully studied.

1. (Fig. 7.32). Reproduce the top view and change the front view to a full section view in accordance with the indicated cutting plane.

2. (Fig. 7.33). Reproduce the top view and change the front and side views to sectional views that will be in accordance with the indicated cutting planes.

3. (Fig. 7.34). Draw a front view of the pulley (circular view) and a side view in full section.

4. (Fig. 7.35). Reproduce the top view of the rod support and draw the front view in full section. Read Sec. 5.43 before starting to draw.

5. (Fig. 7.36). Draw a front view of the V pulley (circular view) and a side view in full section.

6. (Fig. 7.37). Reproduce the two view of the hand wheel and change the right-side view to a full section.

7. (Fig. 7.38). Reproduce the top view of the control housing cover and convert the front view to a full section.

8. (Fig. 7.39). Reproduce the circular front view of the pump cover and convert the right-side view to a full section.

9. (Fig. 7.40). Reproduce the front and top views of the steady brace. Complete the top view and draw the required side view and auxiliary section. Since this is a structural drawing, the figure giving the value of a distance appears above an unbroken dimension line in accordance with the custom in this field of engineering. The slope (45°) of the inclined member to which the plates are welded is indicated by a slope triangle with 12-in. legs.

This problem has been designed to make it necessary for the student to test his power of visualization if he is to determine the shape of the inclined structural member. Good judgment must be exercised in determining the location of the third hole in each plate. All hidden object lines should be shown.

10–19. (Figs. 7.41–7.50). These problems may be dimensioned, as are working drawings. For each object, the student should draw all the views necessary for a working drawing of the part. Good judgment should be exercised in deciding whether the sectional view should be a full section or a half section. After the student has made his decision, he should consult his class instructor.

Fig. 7.35 Rod support.

Fig. 7.33 Mutilated block.

Fig. 7.32 Mutilated block.

Fig. 7.34 Pulley.

Fig. 7.36 V pulley.

Fig. 7.37 Hand wheel.

Fig. 7.38 Control housing cover.

Fig. 7.39 Pump cover.

$1\frac{5}{16}$ DRILL-3 HOLES $1\frac{5}{16}$ DRILL-3HOLES

DRAW AUXILIARY SECTION A-A

$2\frac{1}{16}$ DRILL

DRAW SIDE VIEW

Fig. 7.40 Steady brace.

ϕ60 ϕ76 5 ϕ35

4 HOLES EQUISPACED

ϕ5

1.5 × 45° CHAMFER

R4

45°

18°

18°

ϕ25

75

1.5 × 45° CHAMFER

FINISH ALL OVER

Fig. 7.41 Cover.

13 DRILL-3HOLES EQUALLY SPACED

ϕ130

95 B.C

10

ϕ70

35

ϕ50

50

3 RIBS EQUALLY SPACED

25.5 BORE-38 C'BORE × 12.7 DEEP

FILLETS & ROUNDS R3

Fig. 7.43 Rod yoke.

1.31 .44 R

.25 R

.30 R

R

1.81

VIEW 'A'

FILLET & ROUNDS .12 R

ϕ120

ϕ110

88

R3

ϕ78

ϕ50

28

28

ϕ32

25

ϕ44

65

13

15

KEYWAY 5 WIDE × 2.5 DEEP

Fig. 7.42 Cone pulley.

1.00 DIA. 'A'

2.00 DIA.

2.50 DIA.

.38

.50

.96

1.50

.312 DIA.-.34 DEEP

1.88 DIA.

.06 × 45° CHAMFER

.516 DIA.4 HOLES

3.62 C TO C

.50

1.62

.40 R

.17

.88 DIA.-SPOTFACE × .03 DEEP

Fig. 7.44 Control housing cover.

Fig. 7.45 Centering bearing.

Fig. 7.46 Slotted guide link.

Fig. 7.47 Shifter link.

Fig. 7.48 Cover. (Dimension values shown in [ ] are in millimeters.)

auxiliary views

8

Fig. 8.1 Theory of projecting an auxiliary view.

A □ Primary auxiliary views

8.1
Introduction. When it is desirable to show the true size and shape of an irregular surface, which is inclined to two or more of the coordinate planes of projection, a view of the surface must be projected on a plane parallel to it. This imaginary projection plane is called an *auxiliary plane*, and the view obtained is called an *auxiliary view* (Fig. 8.1).

The theory underlying the method of projecting principal views applies also to auxiliary views. In other words, an auxiliary view shows an inclined surface of an object as it would appear to an observer stationed an infinite distance away (Fig. 8.2).

8.2
Use of Auxiliary Views. In commercial drafting, an auxiliary view ordinarily is a partial view showing only an inclined surface. The reason for this is that a projec-

tion showing the entire object adds very little to the shape description. The added lines are likely to defeat the intended purpose of an auxiliary view. For example, a complete drawing of the casting in Fig. 8.3 must include an auxiliary view of the inclined surface in order to show the true shape of the surface and the location of the holes. Compare the views in (*a*) and (*b*) and note the confused appearance of the view in (*b*). In technical schools, some instructors require that an auxiliary view show the entire object, including all invisible lines. Such a requirement, though impractical commercially, is justified in the classroom, for the construction of a complete auxiliary view furnishes excellent practice in projection.

A partial auxiliary view often is needed to complete the projection of a foreshortened feature in a principal view. This second important function of auxiliary views is illustrated in Fig. 8.16 and explained in Sec. 8.13.

8.3

Types of Auxiliary Views. Although auxiliary views may have an infinite number of positions in relation to the three principal planes of projection, primary auxiliary views may be classified into three general types in accordance with position relative to the principal planes. Figure 8.4 shows the first type, where the auxiliary plane is perpendicular to the frontal plane and inclined to the horizontal plane of projection. Here the auxiliary view and top view have one dimension that is common to both: the depth. Note that the auxiliary plane is hinged to the frontal plane and that the auxiliary view is projected from the front view.

In Fig. 8.5 the auxiliary plane is perpendicular to the horizontal plane and inclined to the frontal and profile planes of projection. The auxiliary view is projected from the top view, and its height is the same as the height of the front view.

The third type of auxiliary view, as shown in Fig. 8.6, is projected from the side view and has a common dimension with both the front and top views. To construct it, distances may be taken from either the front or top view.

Fig. 8.2 Auxiliary view.

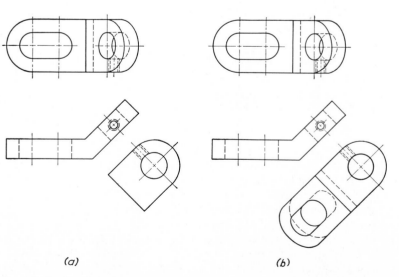

(a)   (b)

Fig. 8.3 Partial and complete auxiliary views.

Transfer the D distances in the top, taken from R-L in the direction of the arrow, to the auxiliary view. Both of these views show related distances that are equal in the direction of depth. (See pictorial drawing)

Fig. 8.4 Auxiliary view projected from front view.

Height distances (D_1 and D_2) for the auxiliary view are equal to the height distances (D_1 and D_2) in the front view.

Fig. 8.5 Auxiliary view projected from top view.

All three types of auxiliary views are constructed similarly. Each is projected from the view that shows the slanting surface as a line, and the distances for the view are taken from the other principal view that has a common dimension with the auxiliary. A careful study of the three illustrations will reveal the fact that the inclined auxiliary plane is always hinged to the principal plane to which it is perpendicular.

8.4
Symmetrical and Unsymmetrical Auxiliary Views. Since auxiliary views are either symmetrical or unsymmetrical about a center line or reference line, they may be termed (1) symmetrical, (2) unilateral, or

Width distances in the auxiliary view are equal to corresponding width distances in the front (or top) view

— *This portion of the bracket would not be drawn when a partial front view is used as indicated*

Direction of Sight

Line(edge) view of inclined surface

PARTIAL FRONT VIEW

SIDE VIEW

Fig. 8.6 Auxiliary view projected from side view.

DISTANCES D_1 D_2 & D_3 TAKEN FROM TOP VIEW

TOP VIEW

FRONT VIEW

Fig. 8.7 Symmetrical auxiliary view of an inclined surface.

(3) bilateral, according to the degree of symmetry. A symmetrical view is drawn symmetrically about a center line, the uni-lateral view entirely on one side of a reference line, and the bilateral view on both sides of a reference line.

8.5

To Draw a Symmetrical Auxiliary View. When an inclined surface is symmetrical, the auxiliary view is "worked" from a center line (Fig. 8.7). The first step in drawing such a view is to draw a center line parallel to the inclined line that represents an edge view of the surface. If the object is assumed to be enclosed in a glass box, this center line may be considered the line of intersection of the auxiliary plane and an imaginary vertical center plane. There are professional draftsmen who, not acquainted with the glass box, proceed without theoretical explanation. Their method is simply to draw a working center line for the auxiliary view and a corresponding line in one of the principal views.

Although theoretically, this working center line may be drawn at any distance from the principal view, actually it should be so located to give the whole drawing a balanced appearance. If not already shown, it also must be drawn in the principal view showing the true width of the inclined surface.

The next step is to draw projection lines from each point of the sloping face, remembering that the projectors make an angle of 90° with the inclined line representing the surface. With the projectors drawn, the location of each point in the auxiliary can be established by setting the dividers to each point's distance from the center line in the principal view and transferring the distance to the auxiliary view. For example, point X is projected to the auxiliary by drawing a projector from point X in the front view perpendicular to the center line. Since its distance from the center line in the top view is the same as it is from the center line in the auxiliary view, the point's location along the projector may be established by using the distance taken from the top view. In the case of point X, the distance is set off from the center line toward the front view. Point Y is set off from the center line away from the front view. A careful study of Fig. 8.7 reveals the fact that if a point lies between the front view and the center line of the top view, it will lie between the front view and the center line of the auxiliary view, and, conversely, if it lies away from the front view with reference to the center line of the top view, it will lie away from the front view with reference to the center line of the auxiliary view.

8.6

Unilateral Auxiliary Views. When constructing a unilateral auxiliary view, it is necessary to work from a reference line that is drawn in a manner similar to the working center line of a symmetrical view. The reference line for the auxiliary view may be considered to represent the line of intersection of a reference plane, coinciding with an outer face, and the auxiliary plane (Fig. 8.8). The intersection of this plane with the top plane establishes the reference line in the top view. All the points are projected from the edge view of the surface, as in a symmetrical view, and it should be noted in setting them off that they all fall on the same side of the reference line.

Figure 8.9 shows an auxiliary view of an entire object. In constructing such a view, it should be remembered that the projectors from all points of the object are perpendicular to the auxiliary plane, since the observer views the entire figure by looking directly at the inclined surface. The distances perpendicular to the auxiliary reference line were taken from the front view.

8.7

Bilateral Auxiliary Views. The method of drawing a bilateral view is similar to that of drawing a unilateral view, the only difference being that in a bilateral view the inclined face lies partly on both sides of the reference plane, as shown in Fig. 8.10.

8.8

Curved Lines in Auxiliary Views. To draw a curve in an auxiliary view, the draftsman must plot a sufficient number of points to ensure a smooth curve (Fig. 8.11). The points are projected first to the inclined line representing the surface in the front view and then to the auxiliary view. The distance of any point from the center line in the auxiliary view is the same as its distance from the center line in the end view.

8.9

Projection of a Curved Boundary. In Fig. 8.11, the procedure is illustrated for plotting the true size and shape of an inclined surface bounded by a curved outline. A similar procedure can be followed to plot a

Fig. 8.8 Unilateral auxiliary view.

Fig. 8.9 Auxiliary view of an object.

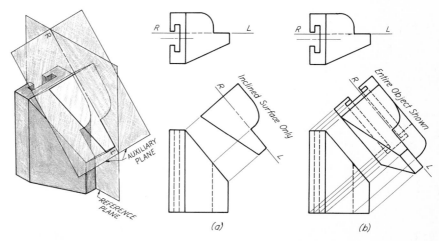

Fig. 8.10 Bilateral auxiliary view.

Fig. 8.11 Curved-line auxiliary view.

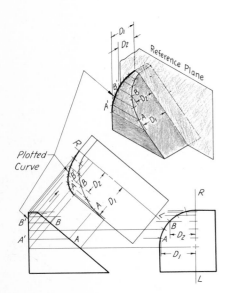

Fig. 8.12 Plotted boundary curves.

curve, such as the one on the left end of the object shown in Fig. 8.12, in an auxiliary view. Since the points on the curved outline of the vertical surface are being viewed from the same direction as those on the curved boundary of the inclined surface, the projectors from points on both surfaces will be parallel. From the pictorial drawing, it can be observed that points A and A' and B and B' lie on elements of the cylindrical surface. Also it should be noted that A and A' are the same distance from the reference plane as are B and B'. It is for this reason that

Fig. 8.13 To determine the true dihedral angle between inclined surfaces.

point A' in the auxiliary view is at the intersection of the projector from A' in the front view and a line through A in the auxiliary view, drawn parallel to the reference line RL.

8.10
Dihedral Angles. Frequently, an auxiliary view may be needed to show the true size of a dihedral angle—that is, the true size of the angle between two planes. In Fig. 8.13, it is desirable to show the true size of the angle between the planes forming the V-slot by means of a partial auxiliary view, as shown. The direction of sight (see pictorial) must be taken parallel to the edge lines 1–2 and 3–4 so that these lines will appear as points and the surfaces forming the dihedral angle will project as line views in the auxiliary view. The reference line for the partial auxiliary view would necessarily be drawn perpendicular to the lines 1–2 and 3–4 in the top view. Since the plane on which the auxiliary view is projected is a vertical one, height dimensions were used—that is, distances in the direction of the dimension D in the auxiliary view were taken from the front view.

8.11
To Construct an Auxiliary View, Practical Method. The usual steps in constructing an auxiliary view are shown in Fig. 8.14. The illustration should be studied carefully, as each step is explained in the drawing.

Fig. 8.14 Steps in constructing an auxiliary view.

Fig. 8.15 Partial views.

8.12

Auxiliary and Partial Views. Often the use of an auxiliary view allows the elimination of one of the principal views (top or side) or makes possible the use of a partial principal view. The shape description furnished by the partial views shown in Fig. 8.15 is sufficient for a complete understanding of the shape of the part. The use of partial views simplifies and drawing, saves valuable drafting time, and tends to make the drawing easier to read.

A break line is used at a convenient location to indicate an imaginary break for a partial view.

8.13

Use of an Auxiliary View to Complete a Principal View. As previously stated, it is frequently necessary to project a foreshortened feature in one of the principal views from an auxiliary view. In the case of the object shown in Fig. 8.16, the foreshortened projection of the inclined face in the top view can be projected from the auxiliary view. The elliptical curves are plotted by projecting points from the auxiliary view to the front view and from there to the top view. The location of these points in the top view with respect to the center line is the same as their location in the auxiliary view with respect to the auxiliary center line. For example, the distance

Fig. 8.16 Use of auxiliary to complete a principal view.

D_1 from the center line in the top view is the same as the distance D_1 from the auxiliary center line in the auxiliary view.

The steps in preparing an auxiliary view and using it to complete a principal view are shown in Fig. 8.17.

8.14

Line of Intersection. It is frequently necessary to represent a line of intersection between two surfaces when making a

DRAW
PRINCIPAL
VIEWS

STEP I

DRAW
REFERENCE
LINES (℄)

STEP II

DRAW PROJECTION
LINES FROM
EDGE VIEW OF
INCLINED FACE

STEP III

COMPLETE
AUXILIARY
VIEW

STEP IV

COMPLETE
TOP
VIEW

STEP V

Fig. 8.17 Steps in preparing an auxiliary view and completing a principal view.

multiview drawing involving an auxiliary view. Figure 8.18 shows a method for drawing the line of intersection on a principal view. In this case the scheme commonly used for determining the intersection involves the use of elements drawn on the surface of the cylindrical portion of the part, as shown on the pictorial drawing. These elements, such as *AB*, are common to the cylindrical surface. Point *B*, where the element pierces the flat surface, is a point that is common to both surfaces and therefore lies on the line of intersection.

On the orthographic views, element *AB* appears as a point on the auxiliary view and as a line on the front view. The loca-

tion of the projection of the piercing point on the front view is visible upon inspection. Point *B* is found in the other principal view by projecting from the front view and setting off the distance *D* taken from the auxiliary view. The distance *D* of point *B* from the center line is a true distance for both views. The center line in the auxiliary view and side view can be considered as the edge view of a reference plane or datum plane from which measurements can be made.

8.15
True Length of a Line. The true length of an oblique line may be determined either by means of an auxiliary view or by revo-

ELEMENT OF
CYLINDRICAL SURFACE

POINT COMMON TO BOTH SURFACES

PROJECTION OF PIERCING POINT

Fig. 8.18 Line of intersection.

lution of the line. Separate discussions of the procedure to be followed in the application of these methods are given in Chapters 5 and 9. To determine the true length of a line by revolution, see Sec. 5.29. To find the true length through the use of an auxiliary view, read Sec. 9.7.

B □ Secondary auxiliary views

8.16

Secondary (Oblique) Auxiliary Views. Frequently an object will have an inclined face that is not perpendicular to any one of the principal planes of projection. In such cases it is necessary to draw a primary auxiliary view and a secondary auxiliary or oblique view (Fig. 8.19). The primary auxiliary view is constructed by projecting the figure on a primary auxiliary plane that is perpendicular to the inclined surface and one of the principal planes. This plane may be at any convenient location. In the illustration, the primary auxiliary plane is perpendicular to the frontal plane. Note that the inclined face appears as a straight line in the primary auxiliary view. Using this view as a regular view, the secondary auxiliary view may be projected on a plane parallel to the inclined face. Figure 8.19(*b*) shows a practical application of the theoretical principles shown pictorially in (*a*).

It is suggested that the student read Sec. 9.14 in which the procedure for drawing the normal (true-shape) view of an oblique surface is presented step by step.

Figure 8.20 shows the progressive steps in preparing and using a secondary auxiliary view of an oblique face to complete a principal view. Reference planes have been used as datum planes from which to take the necessary measurements. Step II shows the partial construction of the primary auxiliary view in which the inclined surface appears as a line. Step III shows the secondary auxiliary view projected from the primary view and completed, using the known measurements of the lug. The primary auxiliary view is finished by projecting from the secondary auxiliary view. Step IV illustrates the procedure for projecting from the secondary auxiliary view to the top view through the primary auxiliary in order to complete the foreshortened view of the lug. It should be noted that distance D_1 taken from refer-

(*a*)

(*b*)

Fig. 8.19 Secondary auxiliary view of an oblique face.

ence R_2P_2 in the secondary auxiliary is transferred to the top view because both views show the same width distances in true length. A sufficient number of points should be obtained to allow the use of an irregular curve. Step V shows the projec- tion of these points on the curve to the front view. In this case the measurements are taken from the primary auxiliary view because the height distances from refer- ence plane R_1P_1 are the same in both views.

Fig. 8.20 Steps in drawing a secondary auxiliary view and using it to complete a principal view.

Problems

The problems shown in Fig. 8.21 are designed to give the student practice in constructing auxiliary views of the inclined surfaces of simple objects formed mainly by straight lines. They will provide needed drill in projection if, for each of the objects in Fig. 8.21, an auxiliary is drawn showing the entire object. Complete drawings may be made of the objects shown in Figs. 8.22–8.30. If the views are to be dimensioned, the student should adhere to the rules of dimensioning given in Chapter 13 and should not take too seriously the locations for the dimensions on the pictorial representations.

1. (Fig. 8.21). Using instruments, reproduce the given views of an assigned object and draw an auxiliary view of its inclined surface.

2. (Fig. 8.22). Draw the views that would be necessary on a working drawing of the dovetail bracket.

3. (Fig. 8.23). Draw the necessary views of the anchor bracket. Make partial views for the top and end views.

4. (Fig. 8.24). Draw the views that would be necessary on a working drawing of the feeder bracket.

5. (Fig. 8.25). Draw the necessary views of the anchor clip. It is suggested that the top view be a partial one and that the auxiliary view show only the inclined surface.

6. (Fig. 8.26). Draw the views that would be necessary on a working drawing of the angle bracket. Note that two auxiliary views will be required.

7. (Fig. 8.27). Draw the necessary views of the offset guide. It is suggested that partial views be used, except in the view where the inclined surface appears as a line.

8. (Fig. 8.28). Draw the views that would be needed on a working drawing of the gear cover. The opening on the inclined face is circular.

9. (Fig. 8.29). Draw the necessary views of the cutoff clip.

10. (Fig. 8.30). Draw the necessary views of the ejector clip.

11. (Fig. 8.31). Draw the views as given. Complete the top view.

12. (Fig. 8.32). Draw the views as given. Complete the auxiliary view and the front view.

13. (Fig. 8.33). Draw the views that would be necessary on a working drawing of the 45° elbow.

14. Make a multiview drawing of the airplane engine mount shown in Fig.. 8.34. The engine mount is formed of three pieces of steel plate welded to a piece of steel tubing. The completed drawing is to consist of four views. It is suggested that the front view be the view obtained by looking along and parallel to the axis of the

Fig. 8.21

Fig. 8.22 Dovetail bracket.

Fig 8.23 Anchor bracket.

Fig. 8.24 Feeder bracket.

Fig. 8.25 Anchor clip.

Fig. 8.26 Angle bracket. (Dimension values shown in [ ] are in millimeters.)

Fig. 8.27 Offset guide.

Fig. 8.28 Gear cover.

Fig. 8.29 Cut-off clip.

Fig. 8.30 Ejector clip.

6.75 DRILL – 8 HOLES
EQUALLY SPACED

THESE HOLES FOR
INSPECTION ADJUSTMENT,
& LUBRICATION

6.75 DRILL – 2 HOLES

25.5 DRILL

CASTING OUTLINE

DUST CAP

CAST 38

R20

R10

44 CORE

FILLETS & ROUNDS R3

Fig. 8.31 Housing cover.

FILLETS & ROUNDS R3

Ø10 – 2 HOLES

COMPLETE
VIEWS

SYM. ABOUT ₵

16 ± 0.02 – THRU

Fig. 8.32 Sliding tool base.

tube. The remaining views that are needed are an auxiliary view showing only the inclined lug, a side view that should be complete with all hidden lines shown, and a partial top view with the inclined lug omitted.

15. (Fig. 8.35). Draw the views given and add the required primary and secondary auxiliary views.

16. (Fig. 8.36). Draw the necessary views of the tool holder.

17. (Fig. 8.37). Draw the necessary views of the angle block. One view should be drawn to show the true size and shape of the oblique surface.

18. (Fig. 8.38). Draw the necessary views of the locating slide.

19. (Fig. 8.39). Using instruments, draw a secondary auxiliary view that will show the true size and shape of the inclined surface of an as-signed object. The drawing must also show the given principal views.

20. (Fig. 8.40). Draw the layout for the support anchor as given and then, using the double-auxiliary-view method, complete the views as required.

The plate and cylinder are to be welded. Since the faces of the plate show as oblique surfaces in the front and top views, double auxiliary views are necessary to show the thickness and the true shape.

Start the drawing with the auxiliary views that are arranged horizontally on the paper, then complete the principal views. The inclined face of the cylinder will show as an ellipse in top and front views; but do not show this in the auxiliary view that shows the true shape of the square plate.

How would you find the view that shows the true angle between the inclined face and the axis of the cylinder?

Fig. 8.33 45° elbow.

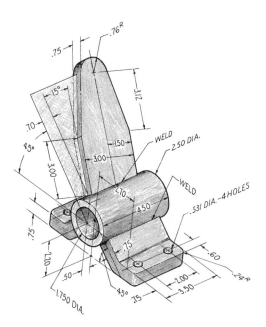

Fig. 8.34 Airplane engine mount.

Fig. 8.35 Cross anchor.

Fig. 8.36 Tool holder.

.500 DRILL THRU
AXIS OF HOLE 1.250
ABOVE SURFACE OF PAD

FILLETS & ROUNDS .12ᴿ

Fig. 8.38 Locating slide.

Fig. 8.37 Angle block.

Fig. 8.39

Fig. 8.40 Support anchor. (Dimension values shown in [ ] are in millimeters.)

The partially completed underground Judiciary Square Station on the 157.7-km Metro System in our nation's capitol. When finished 77.1 km of the line will be underground, and 81.4 km will be at the surface or on aerial ways. The system could not have been designed successfully without the application of graphic methods. (*Courtesy Washington Metropolitan Area Transit Authority*)

basic spatial geometry for design and analysis

9

A □ Basic descriptive geometry

9.1

Introduction. On many occasions, problems arise in engineering design that may be solved quickly by applying the basic principles of orthographic projection. If one thoroughly understands the solution for each of the problems presented, he should find it easy, at a later time, to analyze and solve almost any of the practical problems he may encounter.

It should be pointed out at the very beginning that to solve most types of problems one must apply the principles and methods used to solve a few basic problems, such as: (1) to find the true length of a line, (2) to find the point projection of a line, and (3) to find the true size and shape of a surface (Fig. 9.1). To find information such as the angle between surfaces, the angle between lines, or the clearance between members of a structure, one must use, in proper combination, the methods of solving these basic problems. Success in solving problems by

projection depends largely on the complete understanding of the principles of projection, the ability to visualize space conditions, and the ability to analyze a given situation. Since the ability to analyze and to visualize are of utmost importance in engineering design, the student is urged to develop these abilities by resisting the temptation to memorize step procedures.

9.2

Projection of a Point. Figure 9.2(a) shows the projection of point S on the three

161

Fig. 9.1 The frame of this satellite could not have been designed without the application of descriptive geometry methods. (*Courtesy TRW Systems Group*)

principal planes of projection and a supplementary plane A. The notation used is as explained in Sec. 5.8. Point s^F is the view of point S on the frontal plane, s^H is the view of S on the horizontal plane, and s^P is its view on the profile plane. For convenience and ease in recognizing the projected view of a point on a supplementary plane, the supplementary planes are designated as A-planes and O-planes. A (auxiliary)-planes are always perpendicular to one of the principal planes. Point s^A is the view of S on the A-plane. The view of S on an O (oblique)-plane would be designated s^O. Additional O-planes are identified as O_1, O_2, O_3, etc., in the order that they follow the first O-plane.

Since it is necessary to represent on one plane (the working surface of our drawing paper) the views of point S that lie on mutually perpendicular planes of projection, the planes are assumed to be hinged so that they can be revolved, as shown in Fig. 9.2(*b*), until they are in a single plane, as in (*c*). The lines about which the planes of projection are hinged are called *reference lines*. A reference line is identified by the use of capital letters representing the adjacent planes, as *FH*, *FP*, *FA*, *HA*, *AO*, and so forth (see Fig. 9.2).

It is important to note in (*c*) that the projections s^F and s^H fall on a vertical line, s^F and s^P lie on a horizontal line, and s^H and s^A lie on a line perpendicular to the reference line *HA*. In each case this results from the fact that point S and its projections on adjacent planes lie in a plane perpendicular to the reference line for those planes [Fig. 9.2(*a*)]. This important principal of projection determines the location of views when the relationship of lines and planes form the problem.

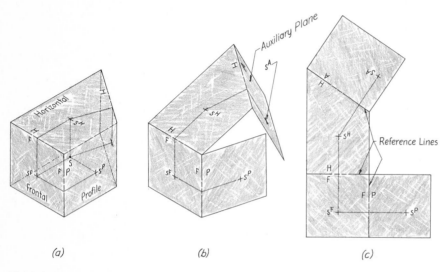

(a)   (b)   (c)

Fig. 9.2 Frontal, horizontal, profile, and auxiliary views of a point S.

9.3

Projection of a Straight Line. Capital letters are used for designating the end points of the actual line in space. In the projected views, these points are identified as shown in Fig. 9.3. The student should read Sec. 5.8, which presents the principles of multiview drawing. In particular, he should study the related illustration, which shows some typical line positions.

9.4

Projection of a Plane in Space. Theoretically, a plane is considered to be flat and unlimited in extent. A plane can be delineated graphically by: (1) two intersecting lines, (2) a line and a point not on the line, (3) two parallel lines, and, lastly, (4) three

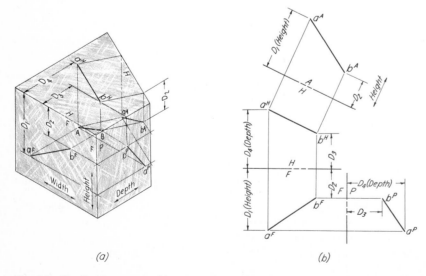

(a)   (b)

Fig. 9.3 The line in space and in successive views.

points not on a straight line. For graphical purposes and to facilitate the solution of space problems as presented in this chapter, planes will be bounded and usually triangular. The space picture and multiview representation of a plane ABC are given in Fig. 9.4. The pictorial at the left in (a) shows the projected views on the principal planes of projection. In (b) the planes are shown being opened outward to be in the plane of the paper, as in (c). It should be noted that plane ABC projects as a line (edge view) on the auxiliary plane. The three points A, B, and C of the plane are projected in the same manner as the single point S in Fig. 9.2 and are identified similarly (a^F, a^H, a^P, a^A, etc.).

9.5

Parallel Lines. Any two lines in space must be either (1) parallel, (2) intersecting, or (3) nonintersecting and nonparallel (called *skew lines*). Figure 9.18 shows intersecting lines, while Fig. 9.17 shows skew lines. Parallel lines are shown in Fig. 9.5.

It might be stated as a rule of projection, with one exception, that when two lines are parallel their projections will be parallel in every view (Fig. 9.5). In other words the lines will appear to be parallel in every view in which both appear. This is true even though in specific views they may appear as points or their projections may coincide. In either case they are still parallel because both conditions indicate that the lines have the same direction. The exception that has been mentioned occurs when the F- and H-projections of two inclined profile lines are shown. For proof of parallelism a supplementary view should be drawn, which may or may not be the profile view.

The true or shortest distance between two parallel lines can be determined on the view that will show these lines as points. The true distance can be measured between the points (Fig. 9.5).

9.6

To Determine the True Length of a Line. An observer can see the true length of a line when he looks in a direction perpendicular to it. It is suggested that the student hold a pencil before him and move it into the following typical line positions to observe the conditions under which the pencil, representing a line, appears in true length.

1. *Vertical line.* The vertical line is perpendicular to the horizontal and will

Fig. 9.4 The plane in space and in successive views.

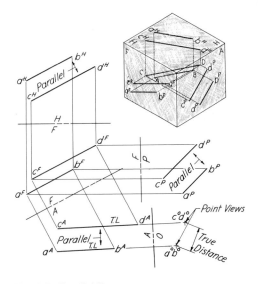

Fig. 9.5 Parallel lines.

therefore appear as a point in the H (top)-view. It will appear in true length in the F (frontal)-view, in true length in the P (profile)-view, and in true length in any auxiliary view that is projected on an auxiliary plane that is perpendicular to the horizontal plane of projection.

2. *Horizontal line.* The horizontal line will appear in true length when viewed from above because it is parallel to the *H*-plane of projection and its end points are theoretically equidistant from an observer looking downward.

3. *Inclined line.* The inclined line will show true length in the *F*-view or *P*-view, for by definition (Sec. 5.8) an inclined line is one that is parallel to either the *F*-plane or the *P*-plane of projection. However, it cannot be parallel to both planes of projection at the same time.

4. *Oblique line.* The oblique line will not appear in true length in any of the principal views because it is inclined to all of the principal planes of projection. It should be apparent, in viewing the pencil alternately from the directions used to obtain the principal views, namely, from the front, above, and side, that one end of the pencil is always

farther away from the observer than the other. Only when looking directly at the pencil from such a position that the end points are equidistant from the observer can the true length be seen. On a drawing, the true length projection of an oblique line will appear in a supplementary *A* (auxiliary)-view on a plane that is parallel to the line.

9.7
To Determine the True Length of an Oblique Line. In order to find the true length of an oblique line, it is necessary to select an auxiliary plane of projection that will be parallel to the line (Figs. 9.6 and 9.7).

Given: The *F* (frontal)-view $a^F b^F$ and the *H* (top)-view $a^H b^H$ of the oblique line *AB* (Fig. 9.6).

Solution: (1) Draw the reference line *HA* parallel to the projection $a^H b^H$. The *A*-plane for this reference line will be parallel to *AB* and perpendicular to the *H*-plane (see pictorial drawing). (2) Draw lines of projection from points a^H and b^H perpendicular to the reference line. (3) Transfer height measurements from the *F*-view to the *A*-view to locate a^A and b^A. In making this transfer of measurements, the students should attempt to visualize the space condition for the line and understand that, since the *F*-view and *A*-view both show height and because the planes of projection for these views are perpendicular to the *H*-plane, the perpendicular distance D_1 from the reference line *HF* to point a^F must be the same as the distance D_1 from the reference line *HA* to point a^A.

The projection $a^A b^A$ shows the true length of the line *AB*.

It was not necessary to use an auxiliary plane perpendicular to the *H*-plane to find the true length of line *AB* in Fig. 9.6. The auxiliary plane could just as well have been perpendicular to either the *F*- or *P*-planes. Figure 9.7 shows the use of an auxiliary plane perpendicular to the frontal plane to find the true length of the line. In this case the auxiliary view has depth distances in common with top view, as indicated.

9.8
Perpendicular Lines. Lines that are perpendicular in space will have their projections perpendicular in any view that shows either or both of the lines in true length. A second rule of perpendicularity might be that, when a line is perpendicular to a

Fig. 9.6 To find the true length of an oblique line.

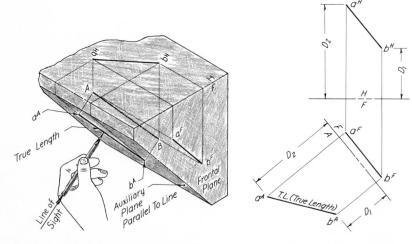

Fig. 9.7 To find the true length of a line.

plane, it will be perpendicular to every line in that plane. A careful study of Fig. 9.8 will verify these rules. For instance, it should be noted that the lines AB and CD lie in a plane that is outlined with broken lines and that $e^H f^H$ is perpendicular to $a^H b^H$ because $a^H b^H$ shows the true length of the line AB. In the A-view, we see that $e^A f^A$ is perpendicular to the line view of the plane and is therefore perpendicular to both AB and CD. The O-view shows the true shape (TSP) of the plane and the line EF as a point. This again verifies the fact that EF is perpendicular to the plane and to lines AB and CD. Otherwise line EF would not appear as a point. Note also that the O-view shows the true length (TL) of AB and CD.

9.9

Bearing of a Line. The bearing of a line is the horizontal angle between the line and a north–south line. A bearing is given in degrees with respect to the meridian and is measured from 0° to 90° from either north (N) or south (S). The bearing reading indicates the quadrant in which the line is located by use of the letters N and E, S and E, S and W, or N and W, as N 48° E or S 54° 40′ W. The bearing of a line is measured in the H-view (Fig. 9.9).

9.10

Point View of a Line (Fig. 9.10). It was pointed out in the first section of this chapter that the solutions of many types of problems depend on an understanding of a few basic constructions. One of these basic constructions involves the finding of the view showing the point view or point projection of a line. For instance, this construction is followed when it is necessary to determine the dihedral angle between two planes, for the true size of the angle will appear in the view that shows the line common to the two planes as a point.

A line will show as a point on a projection plane that is perpendicular to the line. The observer's direction of sight must be along and parallel to the line. When a line appears in true length on one of the principal planes of projection, only an auxiliary view is needed to show the line as a point. However, in the case of an oblique line both an auxiliary and an oblique view are required, for a point view must always follow a true-length view. In other words, the plane of projection for the view showing the line as a point must be adjacent to

the plane for the true view and be perpendicular to it.

Given: The F-view $a^F b^F$ and the H-view $a^H b^H$ of the oblique line AB (Fig. 9.10).

Solution: (1) Draw the view showing the TL (true length) of AB. This is an auxiliary view drawn, as explained in Sec. 9.7. (2) Draw reference line AO perpendicular to the true-length projection $a^A b^A$. This reference line is for an O-plane that is perpendicular to the A-plane. (3) Draw a projection line from $a^A b^A$ and transfer the distance D_3 from the H-view to the O-view. It should be noted from the pictorial drawing that the distance D_3 is common to both of these views, and that points a^O and b^O coincide to give a point or end view of line AB.

9.11

To Find the Shortest Distance from a Point to a Line. The shortest distance between a given point and a given straight line must be measured along a perpendicular

Fig. 9.8 Perpendicular lines.

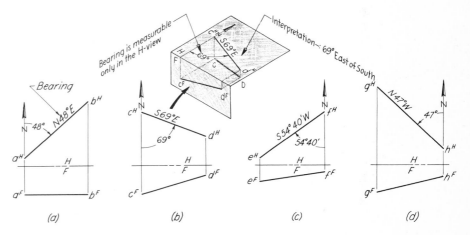

Fig. 9.9 Bearing of a line.

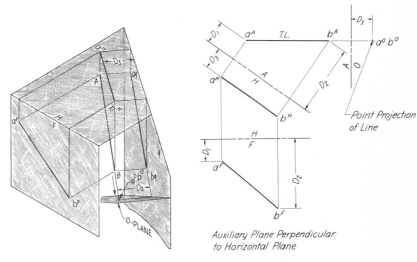

Fig. 9.10 Point view of a line.

drawn from the point to the line. Since lines that are perpendicular will have their projections show perpendicular in any view showing either or both lines in true length (Sec. 9.8), the perpendicular must be drawn in the view showing the given line in true length.

Given: The *F*- and *H*-views of the line *AB* and point *C* (Fig. 9.11).

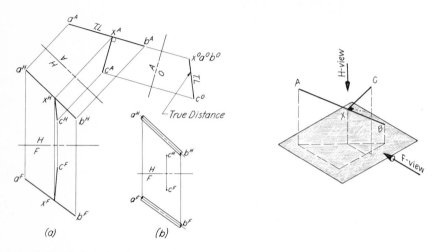

Fig. 9.11 To find the shortest distance from a point to a line.

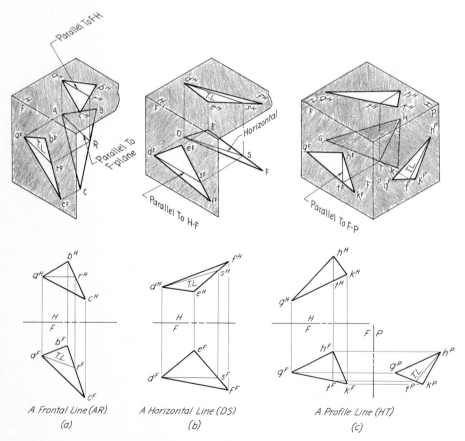

A Frontal Line (AR)
(a)

A Horizontal Line (DS)
(b)

A Profile Line (HT)
(c)

Fig. 9.12 Location of a principal line in a plane.

Solution: (1) Draw the *A* (auxiliary)-view showing the true-length view $a^A b^A$ of line *AB* and view c^A of point *C*. (2) Draw $c^A x^A$ perpendicular to $a^A b^A$. Line $c^A x^A$ is a view of the required perpendicular from point *C* to its juncture with line *AB* at point *X*. (3) Draw reference line *AO* parallel to $c^A x^A$. This reference line locates an *O*-plane, which will be parallel to the perpendicular *CX* and perpendicular to the *A*-plane. The *O*-view will show the true length of *CX*. Line *CX* does not show true length in any of the other views.

**9.12
Principal Lines of a Plane (Fig. 9.12).** Those lines that are parallel to the principal planes of projection are the *principal lines of a plane.* A principal line may be either a horizontal line, a frontal line, or a profile line. Principal lines are true-length lines and one such line may be drawn in any plane to appear true length in any one of the principal views, as desired. This is an important principle that is the basis for the solution of many problems involving lines and planes.

**9.13
To Obtain the Edge View of a Plane.** When a plane is vertical, an edge view of it will be seen from above and it will be represented by a line in the top view. Should a plane be horizontal, it will appear as an edge in the frontal view. However, planes are not always vertical or horizontal; frequently they are inclined or oblique to the principal planes of projection.

Finding the edge view of a plane is a basic construction that is used to determine the slope of a plane (dip), to determine clearance, and to establish perpendicularity. The method presented here is part of the construction used to obtain the true size and shape of a plane, to determine the angle between a line and plane, and to establish the location at which a line pierces a plane.

The edge view of an oblique plane can be obtained by viewing the plane with direction of sight parallel to it. The edge view will then appear in an auxiliary view. When the auxiliary view shows height, the slope of the plane is shown (Fig. 9.13).

Given: The *F*- and *H*-views of plane *ABC.*

Solution: (1) Draw the horizontal line *AX* in the plane. Because *AX* is parallel to the horizontal, $a^F x^F$ will be horizontal and must be drawn before the position of $a^H x^H$ can be established. (2) Draw refer-

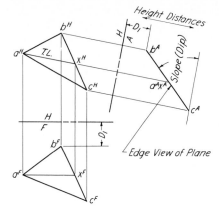

Fig. 9.13 To find the edge view of a plane.

ence line *HA* perpendicular to $a^H x^H$. (3) Construct the *A*-view, which will show the plane *ABC* as a straight line. Since the *F*-view and *A*-view have height as a common dimension, the distances used in constructing the *A*-view were taken from the *F*-view. It should be noted by the reader that an edge view found by projecting from the front view will show the angle that the plane makes with the *F*-plane. Similarly, the edge view found by projecting from the side view will show the angle with the *P*-plane.

9.14
To Find the True Shape (TSP) of an Oblique Plane. Finding the true shape of a plane by projection is another of the basic constructions that the student must understand, for it is used to determine the solution of two of the problems that are to follow. In a way, the construction shown in Fig. 9.14 is a repetition of that shown in Fig. 8.19. However, repetition in the form of another presentation should help even those students who feel they understand the method for finding the true shape of an oblique surface of an object.

To see the true size and shape of an oblique plane an observer must view it with a line of sight perpendicular to it. To do this he must, as the first step in the construction, obtain an edge view of the plane. An *O*-view taken from the *A*-view will then show the true shape of the plane.

Given: The *F*- and *H*-views of plane *ABC*.

Solution: (1) Draw a frontal line *CD* in plane *ABC*. (2) Draw reference line *FA*

perpendicular to $c^F d^F$ and construct the *A*-view showing the edge view of the plane. The auxiliary view has depth in common with the *H*-view. (3) Draw reference line *AO* parallel to the edge view in the *A*-view and construct the *O*-view, which will show the true size and shape of plane *ABC*. The needed distances from the reference line to points in the *O*-view are found in the *F*-view.

9.15
To Find the Piercing Point of a Line and a Plane. Determining the location of the point where a line pierces a plane is an-

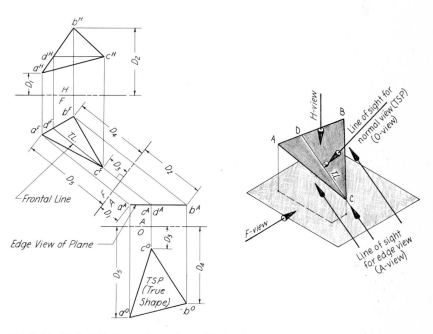

Fig. 9.14 To find the true shape of an oblique plane.

other fundamental operation with which one must be familiar. A line, if it is not parallel to a plane, will intersect the plane at a point that is common to both. In a view showing the plane as an edge, the piercing point appears where the line intersects (cuts) the edge view. This method is known as the edge-view method to distinguish it from the cutting-plane method, which may be found in Chapter 10.

The simple cases occur when the plane appears as an edge in one of the principal views. The general case, for which we use an auxiliary plane, is given in Fig. 9.15.

Given: Plane ABC and line ST.

Solution: (1) Draw the horizontal line BX in ABC. (2) Draw reference line HA perpendicular to $b^H x^H$ and construct the A-view showing an edge view of the plane as line $a^A b^A c^A$ and the view of the line $s^A t^A$. Point p^A where the line cuts the edge view of the plane is the A-view of the piercing point. (3) Project point P back from the A-view first to the H-view and then to the F-view.

**9.16
To Determine the Angle Between a Line and a Given Plane.** The true angle between a given line and a given plane will be seen in the view that shows the plane as an edge (a line) and the line in true length. The solution shown in Fig. 9.16 is based on this premise. The solution as presented might be called the *edge-view method.*

Given: The F- and H-views of plane ABC and line ST.

Solution: (1) Draw the frontal line BX in the plane ABC. (2) Draw reference line FA perpendicular to $x^F b^F$ and construct the A-view. This view will show plane ABC as line $a^A b^A c^A$; however, since this view does not show ST in true length, the true angle is not shown. (3) Draw reference line AO parallel to the edge view of the plane and construct the O-view that will show line ST viewed obliquely and plane ABC in its true size and shape. (4) Draw reference line OO_1 parallel to $s^O t^O$ and construct the second oblique view.

Line $s^{O_1} t^{O_1}$ will show the true length of ST in this second oblique view. Plane ABC will be seen again as an edge (line), for it now appears on an adjacent view taken perpendicular to the view showing

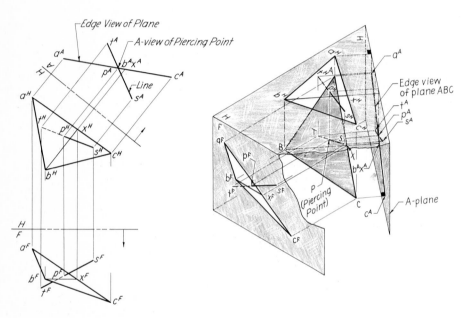

Fig. 9.15 To find the piercing point of a line and a plane.

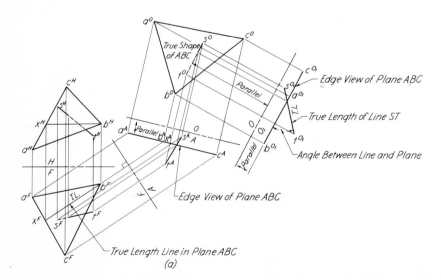

(a)

Fig. 9.16 To find the angle between a line and a plane.

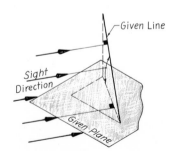

(b)

True angle can be seen in a view which shows the plane as an edge and the line in true length true shape (TSP). The required angle can now be measured between the true-length view of line ST and the edge view of plane ABC.

In the illustration of Fig. 9.16 three supplementary views were required to obtain the true angle. If the plane had appeared as an edge in one of the given views, only two supplementary views would have been needed; if it had appeared in true shape in a given view only one additional view, properly selected to show the plane as an edge and the line in true length, would be needed.

9.17
To Determine the Angle Between Two Nonintersecting (Skew) Lines.
The angle between two nonintersecting lines is measurable in a view that shows both lines in true length. The analysis and construction that might be followed to obtain the needed view are shown in Fig. 9.17.

> **Given:** The two nonintersecting lines AB and CD.
>
> **Solution:** (1) Draw the reference line HA parallel to $a^H b^H$ and construct the A-view that shows AB in true length and the view of the line CD. (2) Draw reference line AO perpendicular to $a^A b^A$ and draw the O-view in which the line AB will appear as a point ($a^o b^o$). (3) Draw reference line OO_1, parallel to $c^o d^o$ and construct the O-view showing the true length of CD. Line AB must also show true length in this view since AB, showing as a point in the O-view, is parallel to the OO_1-plane. With both lines shown in true length in the O_1-view, the required angle may be measured in this view.

9.18
To Determine the True Angle Between Two Intersecting Oblique Lines.
Since, as previously stated, two intersecting lines establish a plane, the true angle between the intersecting lines may be seen in a true-shape view of a plane containing the lines (Fig. 9.14). In Fig. 9.18 line AC completes plane ABC containing the given lines AB and BC. It is necessary to find the true angle between AB and BC.

> **Solution:** (1) Draw the frontal line XC. (2) Draw reference line FA perpendicular to $c^F x^F$ and construct the A-view showing an edge view of plane ABC. (3) Draw reference line AO parallel to the edge view $a^A c^A b^A$ and construct the O-view, which will show the TSP of plane ABC. In this view it is desirable to show only the given lines. The true angle between AB

and BC is shown by $a^o b^o c^o$. As a practical application this method might be used to determine the angle between two adjacent sections of bent rod, as shown in (b).

9.19
Distance Between Two Parallel Planes.
When two planes are parallel their edge views will appear as parallel lines in the same view. The clearance or perpendicular distance between them can be measured in this view (Fig. 9.19). The existence of planes as parallel lines in a view is another proof that they are parallel.

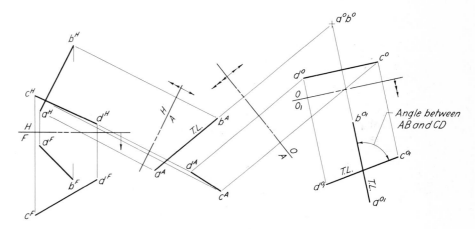

Fig. 9.17   To find the angle between two skew lines.

Fig. 9.18   To find the true angle between two intersecting lines.

Fig. 9.19   Distance between parallel planes.

Edge View of Planes

Fig. 9.20 To find the dihedral angle between two planes.

9.20

To Find the Dihedral Angle Between Two Planes. The angle between two planes is known as a *dihedral angle*. The true size of this angle between intersecting planes may be seen in a plane that is perpendicular to both. For this condition as set forth, the intersecting planes will appear as edges and the line of intersection of the two planes as a point. The true angle may be measured between the edge views of the planes (Fig. 9.20).

Given: The intersecting planes $ABCD$ and $CDEF$. The line of intersection is line CD, as shown in the pictorial drawing.

Solution: (1) Draw reference line FA parallel to $c^F d^F$ and construct the A-view. This view will show CD in true length ($c^A d^A$). (2) Draw reference line AO perpendicular to $c^A d^A$ and construct the adjacent O-view. Since this view was taken looking along line CD, points C and D are coincident and appear as a single point identified as $c^O d^O$. The intersecting planes show as edge views and the true angle between the given planes may be measured between these edge-view lines. When two planes are given that do not intersect, the dihedral angle may be found after the line of intersection has been determined.

9.21

To Find the Shortest Distance Between Two Skew Lines. As was stated in Sec. 9.5, any two lines that are not parallel and do not intersect are called *skew lines*. The shortest distance between any such lines must be measured along one line and only one line that can be drawn perpendicular

to both. This common perpendicular can be drawn in a view that is taken to show one line as a point. Its projection will be perpendicular to the view of the other line and will show in true length (Fig. 9.21).

Given: The F- and H-views of two skew lines AB and CD.

Solution: (1) Draw an A-view adjacent to the H-view to show line AB in true length ($a^A b^A$). Line CD should also be shown in this same view ($c^A d^A$). (2) Draw reference line AO perpendicular to $a^A b^A$ and draw the O-view in which line AB will appear as a point ($a^O b^O$). It is in this view that the exact location of the required perpendicular can be established. (3) Draw the line $e^O f^O$ through point $a^O b^O$ perpendicular to $c^O d^O$. The shortest distance between the skew lines now appears in true distance as the length of $e^O f^O$. Although the CD does not appear in true length in the O-view, $e^O f^O$ does, and hence $c^O d^O$ and $e^O f^O$ will appear perpendicular. (4) Complete the A-view by first locating point e^A on $c^A d^A$ and then draw $e^A f^A$ parallel to reverence line AO. (5) Locate points E and F in the H- and F-views remembering that point E is located on line CD and point F on line AB.

In engineering design an engineer frequently has to locate and find the length of the shortest line between two skewed members in order to determine clearance or the length of a connecting member. In underground construction work one might use this method to locate a connecting tunnel.

In Fig. 9.21(b) and (c) we see the two rods for which the clearance distance was determined in (a). Lines AB and CD represent the center lines of the rods.

B □ Vector geometry

9.22

Vector Methods. In order to be successful in solving some types of problems that arise in design, a well-trained technologist should have a working knowledge of *vector geometry*. The methods presented in this chapter should furnish the student with some background knowledge for solving force problems as they appear in the study of mechanics, strength of materials, and design. Through discreet use of the methods of vector geometry, as well as mathematical methods, it is possible to solve engineering problems quickly within a fully acceptable range of accuracy. Since any quantity having both magnitude and

(a)

Shortest Distance

(b)

(c)

Fig. 9.21 To find the shortest distance between two skew lines.

direction may be represented by a fixed or rotating vector, vector operations are commonly used for problems in the design of frame structures, problems dealing with velocities in mechanisms, and for problems arising in the study of electrical properties. Because a student in a beginning course in engineering graphics should have basic principles rather than specialized cases presented to him, the methods given in this chapter for solving both two-dimensional and three-dimensional force problems deal mainly with static structures or, in other words, structures with forces acting so as to be in equilibrium. In a study of physics, graphical methods are useful for the composition and resolution of forces.

It is hoped that as a student progresses through his other undergraduate courses, he will desire to learn more about the use of vector methods for solving problems, and that he will become able to recognize the cases when he may have a choice between a graphical and an algebraic method. The graphical method is the better for many cases because it is much quicker and can be checked more easily.

An example of a vector addition is shown in Fig. 9.22. An airplane is flying north with a cross wind from the west. If the speed of the plane is 250 km per hr (250 kilometers per hour) and the wind is blowing toward the east at 98 km per hr, the plane will be flying NE (northeast) at 270 km per hr. Vectors can be used for problems of this type because forces acting on a body have both magnitude and direction.

9.23
Force. In our study of vector methods, a force may be defined as a cause which tends to produce motion in an object.

A force has four characteristics which determine it. First, a force has "magnitude." The value of this magnitude is expressed in terms of some standard unit. Second, a force has a "line of action." This is the line along which the force acts. Third, a force has "direction." This is the direction in which it tends to move the object upon which it acts. Fourth, and last, a force has a "point of application." This is the place at which it acts upon the object, often assumed to be a point at the center of gravity.

When values are given for magnitude, it is essential that consistent units be used. The British gravitational system uses feet, pounds, and seconds as basic units while

the metric(SI) absolute system, now being adopted in the U.S., uses the kilogram, meter, and second as units of mass, length, and time. Since the student will find it necessary to be familiar with both systems during this period of change to SI, both systems have been used in this chapter.

The SI unit of force (Fig. 9.23), called a newton (N), is the force that must be applied to a mass of one kilogram (kg) to produce an acceleration of one meter per second per second. The weight of a body in newtons is equal to its mass in kilograms multiplied by $9.81 \, m/s^2$. See Appendix Table 9.

Conversion Factors

English to SI	SI to English
Force 1 lb = 4.448 N = 4.448 kg·m/s²	1 N = 0.2248 lb

9.24
Vector. A force can be represented graphically by a straight line segment with an arrowhead at one end. Such an arrow when used for this purpose is known as a *vector* (Fig. 9.23). The position of the body of the arrow represents the line of action of the force while the arrowhead points out the direction. The magnitude is represented to some selected scale by the overall length of the arrow itself.

When a force acts in a two-dimensional plane, only one view of the vector is needed. However, if the force is in space, two views of the vector must be given.

9.25
Addition of Vector Forces—Two Forces. For a thorough understanding of the principles of vector addition, two simple examples will be considered first.

If one of two men who find it necessary to move a supply cabinet pushes on it with 60 lb of force while the other pushes in the same direction with 40 lb of force, the total force exerted to move the cabinet is 100 lb. The representation of two or more such forces in the manner shown in Fig. 9.24 amounts to a vector addition. Should these men be in a prankish mood and decide to push in opposite directions, as illustrated in Fig. 9.25, the cabinet might move provided the 20-lb resultant force were sufficient to overcome friction. The 20-lb resultant comes from a graphical addition.

Now let it be supposed that force A represented by F_A and force B represented by F_B in Fig. 9.26 act from a point

Fig. 9.22 Vector problem.

Fig. 9.23 Vector.

Fig. 9.24 Vector addition (English Gravitational System).

Fig. 9.25 Forces in opposite directions (English Gravitational System).

Fig. 9.26 Parallelogram of forces.

Fig. 9.27 Vector triangle.

Fig. 9.28 Resultant of two forces.

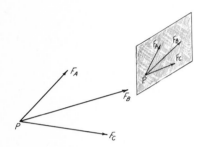

Fig. 9.29 Coplanar, concurrent forces.

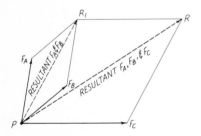

Fig. 9.30 Resultant of three or more forces with a common point of application (parallelogram method).

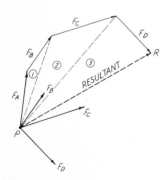

Fig. 9.31 Resultant of forces (polygon of forces).

P, the point of application. The resultant force on the body will not now be the sum of forces *A* and *B*, but instead will be the graphical addition of these forces as represented by the diagonal of a parallelogram having sides equal to the scaled length of the given forces. This single force *R* of 470 N would produce an effect upon the body that would be equivalent to the combined forces F_A and F_B. The single force which could replace any given force system, is known as the *resultant* (*R*) for the force system.

Figure 9.26 shows that the resultant *R* divides the parallelogram into two equal triangles. Therefore, *R* could have been found just as well by constructing a single triangle as shown in Fig. 9.27 provided that the vector F_B is drawn so that its tail-end touches the tip-end of F_A, and *R* is drawn with its arrow-end of to the tip-end of F_B. Since either of the triangles shown in Fig. 9.26 could have been drawn to determine *R*, it should be obvious the resultant is the same regardless of the order in which the vectors are added. However, it is important that they be added tip-end to tail-end and that the vector arrows show the true direction for the action of the concurrent forces in the given system.

To find the resultant of two forces, which are applied as shown in Fig. 9.28, it is first necessary to move the vector arrows along their lines of action to the intersection point *P* before one can apply the parallelogram method.

The forces of a system whose lines of action all lie in one plane are called *coplanar forces*. Should the lines of action pass through a common point, the point of application, the forces are said to be *concurrent*. Figure 9.29 shows a system of forces that are both concurrent and coplanar.

9.26
Addition of Vectors—Three or More Forces. The parallelogram method may be used to determine the resultant for a system of three or more forces that are concurrent and coplanar. In applying this method to three or more forces, it is necessary to draw a series of parallelograms, the number depending upon the number of vector quantities that are to be added graphically. For example, in Fig. 9.30 two parallelograms are required to determine the resultant *R* for the system. The resultant R_1 for forces F_A and F_B is determined by the first parallelogram to be drawn,

and then R_1 is combined in turn with F_C by forming the second and larger parallelogram. By combining the forces in this way *R* becomes the resultant for the complete system.

Where a considerable number of vectors form a system, a somewhat less complicated diagram results, and less work is required when the triangle method is extended and applied to the formation of a vector diagram such as the one shown in Fig. 9.31. In this case the diagram is formed by three vector triangles, one adjacent to the other, and the resultant of forces F_A and F_B is combined with F_C to form the second triangle. Finally, by combining the resultant of the three forces F_A, F_B, and F_C with F_D, the vector *R* is obtained, which represents the magnitude and direction of the resultant of the four forces. In the construction F_B, F_C, and F_D in the diagram must be drawn so as to be parallel respectively to their lines of action in the system. However, the order in which they are placed in the diagram is optional as long as one vector joins another tip to tail.

9.27
Vector Components. A component may be defined as one of two or more forces into which an original force may be resolved. The components, which together have the same action as the single force, are determined by a reversal of the process for vector addition, that is, the original force is resolved using the parallelogram method (see Fig. 9.32). The resolution of a plane vector into two components, horizontal and vertical, is illustrated in (*a*). In (*b*), the resolution of a force into components of specified direction is shown.

9.28
Forces in Equilibrium. A body is said to be in equilibrium when the opposing forces acting upon it are in balance. In such a state the resultant of the force system will be zero. The concurrent and coplanar force system shown in Fig. 9.33 is in a state of equilibrium, for the vector triangle closes and each vector follows the other tip to tail.

An *equilibrant* is the force that will balance two or more forces and produce equilibrium. It is a force that would equal the resultant of the system but would necessarily have to act in an opposite direction.

Figure 9.34 shows a weight supported by a short steel cable. The force to be determined is that needed to hold the weight in a state of equilibrium when it is swung from the position indicated by the broken lines into the position shown by solid lines.

This may be done by drawing a vector triangle with the forces in order from tip to tail. The 396 N force vector represents the equilibrant, the force that will balance the 686.7 N force and the 795 N tension force now in the cable. The reader may wonder at the increase in the tension force in the cable from a 686.7 N force when hanging straight down to a 795 N force when the cable is at an angle of 30° with the vertical. It might help to realize that as the weight is swung outward toward a position where the cable will be horizontal, both the tension force and the equilibrant will increase. Theoretically it would require forces infinitely large to hold the system in equilibrium with the cable in a horizontal position.

In solving a force system graphically it is possible to determine two unknowns in a coplanar system.

Now suppose that it is desired to determine the forces acting in the members of a simple truss as shown in Fig. 9.35(a). To determine these forces graphically, one should isolate the joint supporting the weight and draw a diagram, known as a free-body diagram to show the forces acting at the joint (b). Although the lines of this diagram may have any length, they must be parallel to the lines in the space diagram in (a). Since the boom will be in compression, a capital letter C has been placed along the line that represents the boom in the diagram. A letter T has been placed along the line for the cable because it will be in tension. Although the diagram may not have been essential in this particular case, such a diagram does play an important part in solving more complex systems.

In constructing the force polygon, it is necessary to start by drawing the vertical vector, for the load is the only force having a known magnitude and direction. After this vertical vector has been drawn to a length representing 4.9 kN, using a selected scale, the force polygon (triangle) may be completed by drawing the remaining lines representing the unknown forces parallel to their known lines of action as shown in (a). The force polygon will close since the force system is in equilibrium.

The magnitude of the unknown forces in the members of the truss can now be determined by measuring the lines of the

Fig. 9.32 Components.

Fig. 9.33 Forces in equilibrium.

Fig. 9.34 Determination of forces (graphically).

Fig. 9.35 Determination of the forces at the joint of a simple truss.

diagram using the same scale selected to lay out the length of the vertical vector. This method might be used to determine the forces acting in the members at any point in a truss.

9.29
Coplanar, Noncurrent Force Systems. Forces in one plane having lines of action that do not pass through a common point are said to be *coplanar, nonconcurrent forces* (Fig. 9.36).

9.30
Two Parallel Forces. When two forces are parallel and act in the same direction, their resultant will have a line of action that is parallel to the lines of action of the

Fig. 9.36 Coplanar, nonconcurrent forces.

Fig. 9.37 Parallel forces.

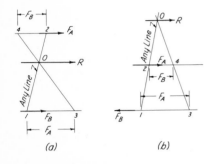

Fig. 9.38 Determination of the position of the resultant of parallel forces (graphical method).

Fig. 9.39 Moment of a force.

Fig. 9.40 Force couples.

given forces and it will be located between them. The magnitude of the resultant will be equal to the sum of the two forces [Fig. 9.37(a)], and it will act through a point that divides any perpendicular line joining the lines of action of the given forces inversely as the forces.

Should the two forces act in opposite directions, as shown in (b), the resultant will be located outside of them and will have the same direction as the greater force. Its magnitude will be equal to the difference between the two given forces. The proportion shown with the illustration in (b) may be used to determine the location of the point of application of the resultant. Those who prefer to determine graphically the location of the line of action for the resultant may use the method illustrated in Fig. 9.38. This method is based on well-known principles of geometry.

With the two forces F_A and F_B given, any line 1–2 is drawn joining their lines of action. From this line two distances must be laid off along the lines of action of the given forces. If the given forces act in the same direction, then the distances are laid off in opposite directions from the line 1–2, (a). If they act in opposite directions, the distances must be laid off on the same side of line 1–2, (b). In Fig. 9.38(a), a length equal by scale to F_A was laid off from point 1 on the line of action of F_B. Then from point 2 a length equal to F_B was marked off in an opposite direction. These measurements located points 3 and 4, the end points of the line intersecting line 1–2 at point O. Point O is on the line of action of the resultant R. In Fig. 9.38(b) this method has been applied to establish the location of the resultant for two forces acting in opposite directions.

9.31

Moment of a Force. The *moment of a force* with respect to a point is the product of the force and the perpendicular distance from the given point to the line of action of the force. In the illustration, Fig. 9.39, the moment of the force F_A about point P is Mom. $= F_A \times d$. The perpendicular distance d is known as the lever arm of the force.

9.32

Force Couples (Fig. 9.40). Two equal forces that act in opposite directions are known as a *couple*. A couple does not have a resultant, and no single force can counteract the tendency to produce rotation. The measurement of this tendency is the moment of the couple that is the product of one of the forces and the perpendicular distance between them.

To prevent the rotation of a body that is acted upon by a couple, it is necessary to use two other forces that will form a second couple. The body acted upon by these couples will be in equilibrium if each couple tends to rotate the body in opposite directions and the moment of one couple is equal to the other.

9.33

String Polygon—Bow's Notation. A system for lettering space and force diagrams, known as *Bow's notation*, is widely used by technical authors. Its use in this chapter will tend to simplify the discussions which follow.

In the space diagram, shown in Fig. 9.41(a), each space from the line of action of one force to the line of action of the next one is given a lowercase letter such as a, b, c, and d in alphabetical order. Thus the line of action for any particular force can be designated by the letters of the areas on each side of it. For example, in Fig. 9.43 the line of action for the 1080-lb force, acting downward on the beam, would be designated as line of action bc. On the force diagram, corresponding capital letters are used at the ends of the vectors. In Fig. 9.41(b), AB represents the magnitude of ab in the space diagram and BC represents the magnitude of bc.

To find the resultant of three or more parallel forces graphically, the "funicular" or "string polygon" is used. The magnitude and direction of the required resultant for the system shown in Fig. 9.41 are known. The magnitude, representing the algebraic sum of the given forces, appears

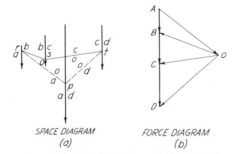

Fig. 9.41 Resultant of parallel forces—Bow's notation.

the necessity for having two unknown forces coincide in one of the views in the discussion in Sec. 9.40.

After the space diagram has been drawn to scale, the steps toward the final solution are as follows:

1. Draw a free-body diagram showing the joint at A, the joint at which the load is applied. The lines of this diagram may be of any length, but each line in it must be parallel respectively to a corresponding line in the view of the space diagram to which the free-body diagram is related. In this case, it is the horizontal (top) view. A modified form of Bow's notation was used for convenience in identifying the forces. The vertical load has been shown pulled to one side in order that this force can be made to fall within the range of the notation. This diagram is important to the solution of this problem, for it enables one to see and note the direction of all of the forces that the members exert upon the joint (note arrowheads). Capital letters were used on the free-body diagram to identify the spaces between the forces, rather than lowercase letters as is customary, so that lowercase letters could be used for the ends of the views of the vectors in the vector diagram.

2. Using a selected scale, start the two views of the vector diagram (*b*) by laying out vector RS representing the only known force, in this case the 1000-lb load. Since RS is a force acting in a vertical direction $r^F s^F$ will be in true length in the F-view and will appear as point $r^H s^H$ in the H-view.

3. Complete the H- and F-views of the vector diagram. Since each vector line in the top view must be parallel to a corresponding line in the top view of the space diagram, $s^H t^H$ must be drawn parallel to $a^H b^H$, $t^H u^H$ parallel to $a^H c^H$, and $u^H r^H$ parallel to $a^H d^H$. Since the forces acting at joint A are in equilibrium, the vector triangle will close and the vectors will appear tip to tail.

In the frontal view of the vector diagram, $s^F t^F$ will be parallel to $a^F b^F$, $t^F u^F$ will be parallel to $a^F c^F$, and $r^F u^F$ will be parallel to $d^F a^F$.

4. Determine the magnitude of the forces acting on joint A. Since vector RU shows its true length in the F-view, the true magnitude of the force represented may be determined by scaling $r^F u^F$ using the same scale used to lay

Fig. 9.47 Solution of a concurrent, noncoplanar force system—special case (English Gravitational System).

out the length of $r^F s^F$. Although it is known that vectors ST and TU are equal in magnitude, it is necessary to find the true length representation of one or the other of these vectors by some approved method before scaling to determine the true value of the force.

An arrowhead may now be added to the line of action of each force in the free-body diagram to indicate the direction of the action. Since the free-body diagram was related to the top view of the space diagram, the arrowhead for each force will point in the same direction as does the arrowhead on the corresponding vector in the H-view of the vector diagram. These arrowheads show that the forces in members AB and AC are acting away from joint A and are therefore *tension forces*. The force in AD acts toward A and thus is a *compression force*.

9.39

To Find the Three Unknown Forces of a Simple Load-bearing Frame—Composition Method (Fig. 9.48). Since the forces in AB and AC lie in an inclined plane that appears as an edge in the F-view, they may be composed into a single force that will have the same effect in the force system as the forces it replaces. This replacement force along with the force in AD and the load now become the forces that would be acting in a simple load-bearing truss (see Fig. 9.35). After the vertical vector of this concurrent coplanar force system has been drawn to a selected scale, the triangular force polygon in (b) may be completed by drawing the remaining lines representing the forces in AD and the resultant R. These lines are drawn parallel to their known lines of action as shown in the F-view of the space diagram in (a). Finally, the true length view of R must be transferred to the auxiliary view in (a) and resolved into its component forces, acting in AB and AC, that collec-

tively have the same action as the resultant R.

9.40

To Find the Three Unknown Forces of a Simple Load-bearing Truss—General Case. For the general case shown in Fig. 9.49, the known force is in a vertical position as in the previous problem, but no two of the three unknown forces appear conincident in either of the two given views. For this reason, it is necessary at the very start to add a complete auxiliary view to the space diagram that will combine with the existing top view to give a point view of one member and a line view of two of the three unknown forces. To obtain this desired situation, one should start with the following steps, which will transform the general case into the special case with which one should now be familiar.

> **Step 1.** Draw a true length line in the plane of two of the members. In Fig. 9.49(a) this line is DE, which appears in true length (TL) in the H-view.

> **Step 2.** Draw the needed auxiliary view, taken so the DE will appear as a point ($d^A e^A$) and OB and OC will be coincident (line $o^A b^A c^A$). This construction involves finding the edge view of a plane (see Sec. 9.13). In this particular case, the auxiliary view has height in common with the frontal view.

> **Step 3.** Draw the two views of the vector diagram by assuming the H-view and the A-view to be the given views of the special case. Proceed by the steps set forth for the special case in Sec. 9.38.

> **Step 4.** Determine the magnitude of the forces and add arrowheads to the free-body diagram to show the direction of action of the forces acting on point O.

9.41

In practice, engineers and technologists find wide use for methods that solve problems through the use of three-dimensional vector diagrams, for any quantity having both magnitude and direction may be represented by a vector. And, although the examples used in the chapter dealt with static structures, which are in the field of the structural engineer, vector diagram methods are used frequently by the electrical engineer for solving problems arising in his field and by the mechanical engineer for problems dealing with bodies in motion. The student will without a doubt encounter some of these methods in a textbook for a later course or will have them presented to him by his instructor.

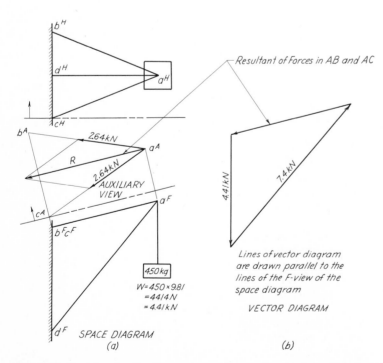

Fig. 9.48 To determine the three unknown forces of a simple load-bearing frame—composition method.

SPACE
DIAGRAM

FREE-BODY
DIAGRAM

VECTOR DIAGRAM
(b)

$W = 208 \, kg$
$W_c = 2.08 \times 9.8/ = 2040.5 \, N$
$= 2.04 \, kN$

(a)

Fig. 9.49   Solution of a concurrent, noncoplanar force system—general case.

Problems

Descriptive geometry

Problems 1 through 5 have been selected and arranged to offer the student an opportunity to apply basic principles of descriptive geometry.

The problems can be reproduced to a suitable size by transferring the needed distances from the drawing to one or the other of the scales that have been provided for each group of problems. If the inch scale is used, metric values in problem instructions should be converted to decimal inches.

1. (Fig. 9.50). Reproduce the given views of the line or lines of a problem as assigned and determine the true length or lengths using the auxiliary view method explained in Sec. 9.7. A problem may be reproduced to a suitable size by transferring the needed distances from the drawing to one of the given scales to determine values. The distances, as they are determined, should be laid off on the drawing paper using a full-size scale.

2. (Fig. 9.51). Reproduce the given views of the plane of a problem as assigned and draw the view showing the true size and shape. Use the method explained in Sec. 9.14. Determine needed distances by transferring them from the drawing to one of the given scales, by means of the dividers.

3. (Fig. 9.52). These problems are intended to give some needed practice in manipulating views to obtain certain relationships of points and lines. Determine needed distances by transferring them from the drawing to one of the accompanying scales, by means of the dividers.

1. Determine the distance between points A and B.
2. Draw the H- and F-views of a 13 mm perpendicular erected from point N of the line MN.
3. Draw the H-view of the 95 mm line ST.
4. Draw the H- and F-views of a plane represented by an equilateral triangle and containing line AB as one of the edges. The added plane ABC is to be at an angle of 30° with plane $ABDE$.
5. If the figure $MNOP$ is a plane surface, an edge view of the surface would appear as a line. Draw such a view to determine whether or not $MNOP$ is a plane.
6. A vertical pole with top O is held in place by three guy wires. Determine the slope in tangent value of the angle for the guy wire that has a bearing of N 23° W.

4. (Fig. 9.53). In this group of problems it is required to determine the shortest distance between skew lines and the angle formed by intersecting lines.

1. Show proof that the plane $ABCD$ is an oblique plane.

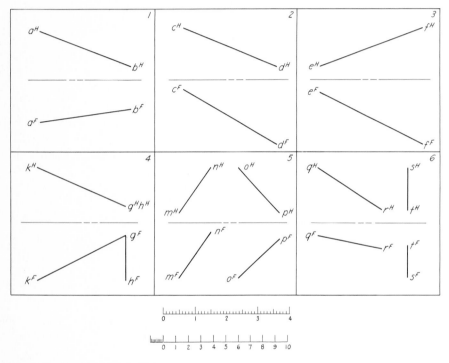

Fig. 9.50   True length of a line.

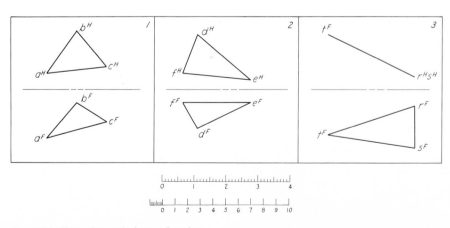

Fig. 9.51   True size and shape of a plane.

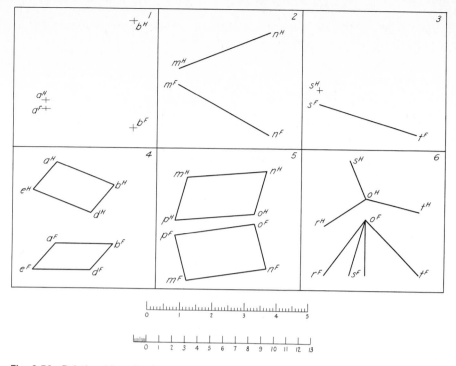

Fig. 9.52 Relationships of points and lines.

2. Determine the shortest distance between the lines *MN* and *ST*.
3. Through point *K* on line *GH* draw the *F*- and *H*-views of a line that will be perpendicular to line *EF*.
4. Determine the angle between line *AB* and a line intersecting *AB* and *CD* at the level of point *E*.

5. Through a point on line *MN* that is 32 mm from point *N*, draw the *F*- and *H*-views of a line that will be perpendicular to line *ST*.
6. Erect a 25 mm perpendicular at point *K* in the plane *EFGH*. Connect the outer end point *L* of the perpendicular with *F*. Determine the angle between *LF* and *KF*.

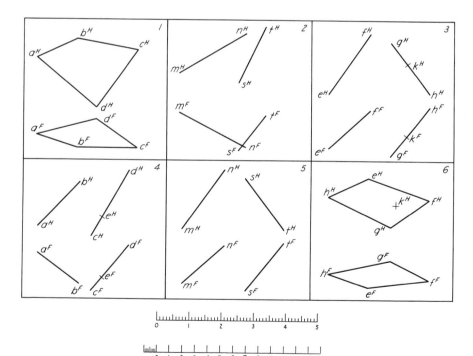

Fig. 9.53 Shortest distance between skew lines and the angle formed by intersecting lines.

5. (Fig. 9.54). These problems require that the student determine the angle between a line and a plane and the angle between two given planes.

1. Determine the angle between the planes *MNQP* and *RST*.
2. Determine the angle between the line *ST* and
   (a) The *H*-plane.
   (b) The *F*-plane.
3. The line *EF* has a bearing of N 53° E. What angle does this line make with the *P*-plane?
4. Draw the *F*- and *H*-views of a line through point *K* that forms an angle of 35° with plane *MNQR*.
5. The top and front views of planes *ABC* and *RST* are partially drawn.
   (a) Complete the views including the line of intersection.
   (b) Determine the angle between the line of intersection and the *H*-plane of projection.
6. Two views of a plane *RST* and the top view of an arrow are shown. The arrow, pointing downward and toward the left, is in a plane that forms an angle of 68° with plane *RST*.

The arrow point is 6 mm from the plane *RST*.
(a) Draw the front view of the arrow.
(b) Draw the top and front views of the line of intersection of the 68° plane and plane *RST*.

Vector geometry

The following problems have been selected to emphasize the basic principles underlying vector geometry. By solving a limited number of the problems presented, the student should find that he has a working knowledge of some vector methods that are useful for solving problems in design that involve the determination of the magnitude of forces as well as their composition and resolution. The student is to select his own scale remembering that a drawing made to a large scale usually assures more accurate results.

1. (Fig. 9.55). A force of 408 N acts downward at an angle of 60° with the horizontal. Determine the vertical and horizontal components of this force.

2. (Fig. 9.56). Determine the resultant force for the given coplanar, concurrent force system.

3. (Fig. 9.57). Determine the magnitude of the force F_C and the angle that the resultant force R makes with the horizontal for the given coplanar, concurrent force system.

4. (Fig. 9.58). Determine the magnitude and direction of the equilibrant for the given coplanar, concurrent force system.

5. (Fig. 9.59). A block weighing 45 lb is to be pulled up an inclined plane sloping at an angle of 30° with the horizontal. If the frictional resistance is 16 lb, what is the magnitude of the force F_M that is required to move the block uniformly up the plane?

Fig. 9.54 Angle between a line and a plane and the angle between two planes.

Fig. 9.57

Fig. 9.55

Fig. 9.56 Force system.

Fig. 9.58

6. (Fig. 9.60). A horizontal beam AB is hinged at B as shown. The end of the beam at A is connected by a cable to a hook in the wall at C. The load at A is 110 kg. Using the dimensions as given, determine the tension force in the cable and the reaction on the hinge at B. The weight of the beam is to be neglected.

7. (Fig. 9.61). A 270 kg load is supported by cables as shown. Determine the magnitude of the tension in the cables.

8. (Fig. 9.62). A ship that is being pulled through the entrance of a harbor is headed due east through a cross current moving at 4 knots as shown. If the ship is moving at 12 knots, what is the speed of the tugboat?

9. (Fig. 9.63). Determine the magnitude of the reactions R_1 and R_2 of the beam with loads as shown.

10. (Fig. 9.64). Determine the magnitude of the reactions R_1 and R_2 of the beam.

11. (Fig. 9.65). Determine the magnitude of the reactions R_1 and R_2 for the roof truss shown. Each of the six panels is of the same length.

12. (Fig. 9.66). Determine the magnitude of reactions R_1 and R_2 to the wind loads acting on the roof truss as shown.

13. (Fig. 9.67). A tripod with a 40 kg load is set up on a level floor as shown. Determine the stresses in the three legs due to the vertical load on the top.

14–15. (Figs. 9.68 and 9.69). Determine the stresses in the members of the space frame shown.

Fig. 9.59

Fig. 9.60

Fig. 9.61 Cable support system. (Values shown in [ ] are in the metric system.)

Fig. 9.62

Fig. 9.63

Fig. 9.64 Simple beam.

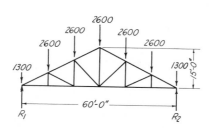

Fig. 9.65

Basic spatial geometry for design and analysis

Fig. 9.66

Fig. 9.67

Fig. 9.68 Space frame.

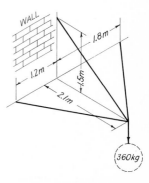

Fig. 9.69 Space frame. (Values shown in [ ] are in the metric system.)

This advanced concept train is now being built under contract. These advanced-technology cars feature an aluminum structure with exterior panels made of vandal-resistant acrylic plastic. Designed to conserve electricity, the electric propulsion system of ACT-1 includes a flywheel within the cars which store up energy that can be used to supply the electric motors when needed. In the development of the ACT-1 cars, many of the problems involving intersections and developments were solved by graphical methods. (*Courtesy U.S. Office of Transportation*)

developments and intersections 10

10.1

Introduction. Intersections and developments are logically a part of the subject of descriptive geometry. A few of the many applications that can be handled without advanced study in projection, however, are presented in this chapter.

Desired lines of intersection between geometric surfaces may be obtained by applying the principles of projection with which the student is already familiar. Although developments are laid out and are not drawn by actual projection in the manner of exterior views, their construc-

tion nevertheless requires the application of orthographic projection in finding the true lengths of elements and edges.

10.2

Geometric Surfaces. A geometric surface is generated by the motion of a geometric line, either straight or curved. Surfaces that are generated by a moving straight line are known as *ruled surfaces*, and those generated by a curved line are known as *double-curved surfaces*. Any position of the generating line, known as

a *generatrix*, is called an *element of the surface*.

Ruled surfaces include planes, single-curved surfaces, and warped surfaces.

A *plane* is generated by a straight line moving in such a manner that one point touches another straight line as it moves parallel to its original position.

A *single-curved surface* is generated by a straight line moving so that in any two of its near positions it is in the same plane.

A *warped surface* is generated by a straight line moving so that it does not lie in the same plane in any two near positions.

Double-curved surfaces include surfaces that are generated by a curved line moving in accordance with some mathematical law.

10.3

Geometric Objects. Geometric solids are bounded by geometric surfaces. They may be classified as follows:

1. Solids bounded by plane surfaces: Tetrahedron, cube, prism, pyramid, and others.

2. Solids bounded by single-curved surfaces: Cone and cylinder (generated by a moving straight line).

3. Solids bounded by warped surfaces: Conoid, cylindroid, hyperboloid of one nappe, and warped cone.

4. Solids bounded by double-curved surfaces: Sphere, spheroid, torus, paraboloid, hyperboloid, and so on (surfaces of revolution generated by curved lines).

A □ Developments

10.4

Introduction. A layout of the complete surface of an object is called a *development* or *pattern*. The development of an object bounded by plane surfaces may be thought of as being obtained by turning the object, as illustrated in Figs. 10.1 and 10.2, to unroll the imaginary enclosing surface on a plane. Practically, the drawing operation consists of drawing the successive surfaces in their true sizes with their common edges joined.

The surfaces of cones and cylinders also may be unrolled on a plane. The development of a right cylinder (Fig. 10.3) is a rectangle having a width equal to the altitude of the cylinder and a length equal to the cylinder's computed circumference (πd). The development of a right circular cone (Fig. 10.4) is a sector of a circle having a radius equal to the slant height of the cone and an arc length equal to the circumference of its base.

Warped and double-curved surfaces cannot be developed accurately, but they may be developed by some approximate method. Ordinarily, an approximate pattern will prove to be sufficiently accurate

for practical purposes if the material of which the piece is to be made is somewhat flexible.

Plane and single-curved surfaces (prisms, pyramids, cylinders, and cones), which can be accurately developed, are said to be developable. Warped and double-curved surfaces, which can be only approximately developed, are said to be nondevelopable.

10.5

Practical Developments. On many industrial drawings, a development must be shown to furnish the necessary information for making a pattern to facilitate the cutting of a desired shape from sheet metal. Because of the rapid advance of the art of manufacturing an ever-increasing number of pieces by folding, rolling, or pressing cut sheet-metal shapes, one must have a broad knowledge of the methods of constructing varied types of developments. Patterns also are used in stonecutting as guides for shaping irregular faces.

A development of a surface should be drawn with the inside face up, as it theoretically would be if the surface were unrolled or unfolded, as illustrated in Figs. 10.1–10.4. This practice is further justified because sheet-metal workers must make the necessary punch marks for folding on the inside surface.

Although in actual sheet-metal work extra metal must be allowed for lap at seams, no allowance will be shown on the developments in this chapter. Many other practical considerations have been purposely ignored, as well, in order to avoid confusing the beginner.

10.6

To Develop a Right Truncated Prism. Before the development of the lateral surface of a prism can be drawn, the true lengths of the edges and the true size of a right section must be determined. On the right truncated prism, shown in Fig. 10.5, the true lengths of the prism edges are shown in the front view and the true size of the right section is shown in the top view.

The lateral surface is "unfolded" by first drawing a "stretch-out line" and marking off the widths of the faces (distances 1–2, 2–3, 3–4, and so on, from the top view) along it in succession. Through these points light construction lines are then drawn perpendicular to the line 1_D1_D, and the length of the respective edge is set off on each by projecting from

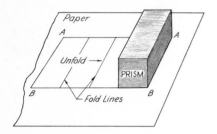

Fig. 10.1 Development of a prism.

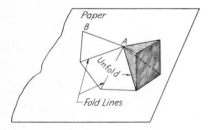

Fig. 10.2 Development of a pyramid.

Fig. 10.3 Development of a cylinder.

Fig. 10.4 Development of a cone.

Fig. 10.5 Standard method of developing the lateral surface of a right prism.

the front view. When projecting edge lengths to the development, the points should be taken in a clockwise order around the perimeter, as indicated by the order of the numbers in the top view. The outline of the development is completed by joining these points. Thus far, nothing has been said about the lower base or the inclined upper face. These may be joined to the development of the lateral surface, if so desired.

In sheet-metal work, it is usual practice to make the seam on the shortest element in order to save time and conserve solder or rivets.

10.7
To Develop an Oblique Prism. The lateral surface of an oblique prism, such as the one shown in Fig. 10.6, is developed by the same general method used for a right prism. Similarly, the true lengths of the edges are shown in the front view, but it is necessary to find the true size of the right section by auxiliary plane construction. The widths of the faces, as taken from the auxiliary right section, are set off along the stretch-out line, and perpendicular construction lines representing the edges are drawn through the division points. The lengths of the portions of each respective

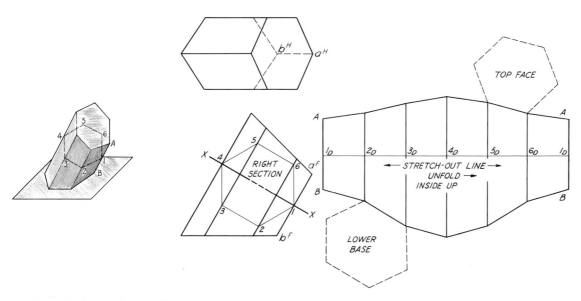

Fig. 10.6 Development of an oblique prism.

edge, above and below plane *XX*, are transferred to the corresponding line in development. Distances above plane *XX* are laid off above the stretch-out line, and distances below *XX* are laid off below it. The development of the lateral surface is then completed by joining the end points of the edges by straight lines. Since an actual fold will be made at each edge line when the prism is formed, it is the usual practice to heavy these edge (fold) lines on the development.

The stretch-out line might well have been drawn in a position perpendicular to the edges of the front view, so that the length of each edge might be projected to the development (as in the case of the right prism).

10.8

To Develop a Right Cylinder. When the lateral surface of a right cylinder is rolled out on a plane, the base develops into a straight line (Fig. 10.7). The length of this line, which is equal to the circumference of a right section ($\pi \times$ diam), may be calculated and laid off as the stretch-out line $1_D 1_D$.

Since the cylinder can be thought of as being a many-sided prism, the development may be constructed in a manner similar to the method illustrated in Fig. 10.5. The elements drawn on the surface of the cylinder serve as edges of the many-sided prism. Twelve or 24 of these elements ordinarily are used, the number depending on the size of the cylinder. Usually they are spaced by dividing the

circumference of the base, as shown by the circle in the top view, into an equal number of parts. The stretch-out line is divided into the same number of equal parts, and perpendicular elements are drawn through each division point. Then the true length of each element is projected to its respective representation on the development, and the development is completed by joining the points with a smooth curve. In joining the points, it is advisable to sketch the curve in lightly, freehand, before using the French curve. Since the surface of the finished cylindrical piece forms a continuous curve, the elements on the development are not heavied. When the development is symmetrical, as in this case, only one-half need be drawn.

A piece of this type might form a part of a two-piece, three-piece, or four-piece elbow. The pieces are usually developed as illustrated in Fig. 10.8. The stretch-out line of each section is equal in length to the computed perimeter of a right section.

10.9

To Develop an Oblique Cylinder. Since an oblique cylinder theoretically may be thought of as enclosing a regular oblique prism having an infinite number of sides, the development of the lateral surface of the cylinder shown in Fig. 10.9 may be constructed by using a method similar to the method illustrated in Fig. 10.6. The circumference of the right section becomes stretch-out line $1_D 1_D$ for the development.

Fig. 10.7   Development of a right circular cylinder.

Fig. 10.8 Two-piece elbow.

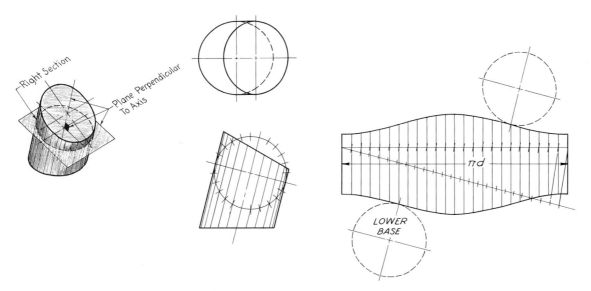

Fig. 10.9 Development of an oblique cylinder.

10.10

To Determine the True Length of a Line.
In order to construct the development of
the lateral surface of some objects, it frequently is necessary to determine the true
lengths of oblique lines that represent the
edges. The general method for determining the true lengths of lines inclined to all
of the coordinate planes of projection has
been explained in detail in Sec. 5.29.

10.11

True-length Diagrams. When it is necessary in developing a surface to find the
true lengths of a number of edges or elements, some confusion may be avoided by
constructing a true-length diagram adjacent to the orthographic view, as shown in
Fig. 10.10. The elements were revolved
into a position parallel to the F (frontal)
plane so that their true lengths show in
the diagram. This practice prevents the

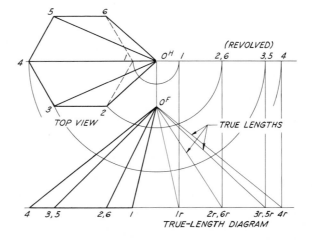

Fig. 10.10 True-length diagram (the revolution method).

front view in the illustration from being cluttered with lines, some of which would represent elements and others their true lengths.

Figure 10.12 shows a diagram that gives the true lengths of the edges of the pyramid. Each line representing the true length of an edge is the hypotenuse of a right triangle whose altitude is the altitude of the edge in the front view and whose base is equal to the length of the projection of the edge in the top view. The lengths of the top projections of the edges of the pyramid are laid off horizontally from the vertical line, which could have been drawn at any distance from the front view. Since all the edges have the same altitude, this line is a common vertical leg for all the right triangles in the diagram. The true-length diagram shown in Fig. 10.10 could very well have been constructed by this method.

10.12

To Develop a Right Pyramid. To develop (unfold) the lateral surface of a right pyramid, it is first necessary to determine the true lengths of the edges and the true size of the base. With this information, the development can be constructed by laying out the faces in successive order with their common edges joined. If the surface is imagined to be unfolded by turning the pyramid, as shown in Fig. 10.2, each triangular face is revolved into the plane of

the paper about the edge that is common to it and the preceding face.

Since the edges of the pyramid shown in Fig. 10.11 are all equal in length, it is necessary only to find the length of the one edge $A1$ by revolving it into the position $a^F 1r$. The edges of the base, 1–2, 2–3, and so on, are parallel to the horizontal plane of projection and consequently show in their true length in the top view. With this information, the development is easily completed by constructing the four triangular surfaces.

10.13

To Develop the Surface of a Frustum of a Pyramid. To develop the lateral surface of the frustum of a pyramid (Fig. 10.12), it is necessary to determine the true lengths of edges of the complete pyramid as well as the true lengths of edges of the frustum. The desired development is obtained by first constructing the development of the complete pyramid and then laying off the true lengths of the edges of the frustum on the corresponding lines of the development.

It may be noted with interest that the true length of the edge $B3$ is equal to the length $b'3'$ on the true-length line $a^F 3'$ and that the location of point b' can be established by the shortcut method of projecting horizontally from point b^F. Point b' on $a^F 3'$ is the true revolved position of point B, because the path of point B is in a horizontal plane that projects as a line in the front view.

10.14

To Develop a Right Cone. As previously explained in Sec. 10.4, the development of a regular right circular cone is a sector of a circle. The development will have a radius equal to the slant height of the cone and an included angle at the center equal to $(r/s) \times 360°$ (Fig. 10.13). In this equation, r is the radius of the base and s is the slant height.

10.15

To Develop a Right Truncated Cone. The development of a right truncated cone must be constructed by a modified method of triangulation, in order to develop the outline of the elliptical inclined surface. This commonly used method is based on the theoretical assumption that a cone is a pyramid having an infinite number of sides. The development of the incomplete right cone shown in Fig. 10.14 is constructed on a layout of the whole

Fig. 10.11 Development of a rectangular right pyramid.

Fig. 10.12   Development of the frustum of a pyramid.

Fig. 10.13   Development of a right cone.

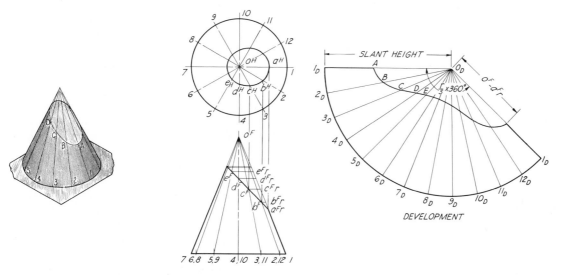

Fig. 10.14   Development of a truncated cone.

cone by a method similar to the standard method illustrated for the frustum of a pyramid in Fig. 10.12.

Elements are drawn on the surface of the cone to serve as edges of the many-sided pyramid. Either 12 or 24 are used, depending on the size of the cone. Their location is established on the developed sector by dividing the arc representing the unrolled base into the same number of equal divisions, into which the top view of the base has been divided. At this point in the procedure, it is necessary to determine the true lengths of the elements of the frustum in the same manner that the true lengths of the edges of the frustum of a pyramid were obtained in Fig. 10.12. With this information, the desired devel-

opment can be completed by setting off the true lengths on the corresponding lines of the development and joining the points thus obtained with a smooth curve.

10.16
Triangulation Method of Developing Approximately Developable Surfaces.   A nondevelopable surface may be developed approximately if the surface is assumed to be composed of a number of small developable surfaces (Fig. 10.15). The particular method ordinarily used for warped surfaces and the surfaces of oblique cones is known as the *triangulation method*. The procedure consists of completely covering the lateral surface with numerous small triangles that will lie approximately on

Fig. 10.15   Triangulation of a surface.

Fig. 10.16 Triangulation of an oblique cone.

the surface (Fig. 10.16). These triangles, when laid out in their true size with their common edges joined, produce an approximate development that is accurate enough for most practical purposes.

Although this method of triangulation is sometimes used to develop the lateral surface of a right circular cone, it is not recommended for such a purpose. The resulting development is not as accurate as it would be if constructed by one of the standard methods (Secs. 10.14 and 10.15).

10.17
To Develop an Oblique Cone Using the Triangulation Method. A development of the lateral surface of an oblique cone is constructed by a method similar to that used for an oblique pyramid. The surface is divided into a number of unequal triangles having sides that are elements on the cone and bases that are the chords of short arcs of the base.

The first step in developing an oblique cone (Fig. 10.17) is to divide the circle representing the base into a convenient number of equal parts and draw elements on the surface of the cone through the division points (1, 2, 3, 4, 5, and so on). To construct the triangles forming the development, it is necessary to know the true lengths of the elements (sides of the triangles) and chords. In the illustration, all the chords are equal. Their true lengths are shown in the top view. The true lengths of the oblique elements may be determined by one of the standard methods explained in Sec. 10.11.

Since the seam should be made along the shortest element, $A1$ will lie on the selected starting line for the development and $A7$ will be on the center line. To obtain the development, the triangles are constructed in order, starting with the triangle A–1–2 and proceeding around the cone in a clockwise direction (as shown by the arrow in the top view). The first step in constructing triangle A–1–2 is to set off the true length a^F1' along the starting line. With point A_D of the development as a center, and with a radius equal to a^F2', strike an arc; then, with point 1_D as a center, and with a radius equal to the chord 1–2, strike an arc across the first arc to locate point 2_D. The triangle $A_D2_D3_D$ and the remaining triangles are formed in exactly the same manner. When all the triangles have been laid out, the development of the whole conical surface is completed by drawing a smooth curve through the end points of the elements.

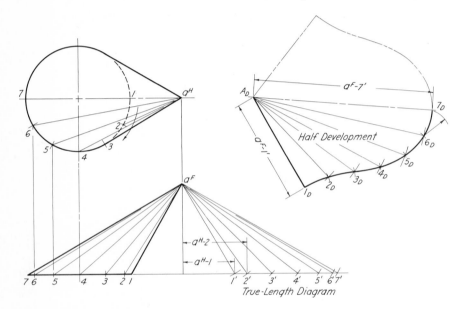

Fig. 10.17 Development of an oblique cone.

10.18
Transition Pieces. A few of the many types of transition pieces used for connecting pipes and openings of different shapes and sizes are illustrated pictorially in Fig. 10.18.

10.19
To Develop a Transition Piece Connecting Rectangular Pipes. The transition piece shown in Fig. 10.19 is designed to connect two rectangular pipes of different sizes on different axes. Since the piece is a frustum of a pyramid, it can be accurately developed by the method explained in Sec. 10.13.

Fig. 10.18 Transition pieces.

DEVELOPMENT

Fig. 10.19 Transition piece.

TOP VIEW

HALF DEVELOPMENT

FRONT VIEW

TRUE-LENGTH DIAGRAM

Fig. 10.20 Transition piece connecting two pipes.

10.20

To Develop a Transition Piece Connecting Two Circular Pipes. The transition piece shown in Fig. 10.20 connects two circular pipes on different axes. Since the piece is a frustum of an oblique cone, the surface must be triangulated, as explained in Sec. 10.17, and the development must be constructed by laying out the triangles in their true size in regular order. The general procedure is the same as that illustrated in Fig. 10.17. In this case, however, since the true size of the base is not shown in the top view, it is necessary to construct a partial auxiliary view to find the true lengths of chords between the end points of the elements.

10.21

To Develop a Transition Piece Connecting a Circular and a Square Pipe. A detailed analysis of the transition piece shown in Fig. 10.21 reveals that it is composed of four isosceles triangles whose bases form the square base of the piece and four conical surfaces that are parts of oblique cones. It is not difficult to develop this type of transition piece because, since the

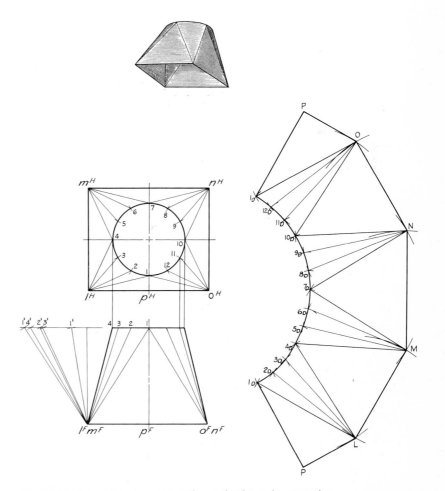

Fig. 10.21 Transition piece connecting a circular and square pipe.

193

whole surface may be "broken up" into component surfaces, the development may be constructed by developing the first and then each succeeding component surface separately (Fig. 10.15). The surfaces are developed around the piece in a clockwise direction, in such a manner that each successive surface is joined to the preceding surface at their common element. In the illustration, the triangles $1LO$, $4LM$, $7MN$, and $10NO$ are clearly shown in top view. Two of these, $1LO$ and $10NO$, are visible on the pictorial drawing. The apexes of the conical surfaces are located at the corners of the base.

Before starting the development, it is necessary to determine the true lengths of the elements by constructing a true-length diagram, as explained in Sec. 10.11. The true lengths of the edges of the lower base (LM, MN, NO, and OL) and the true lengths of the chords (1–2, 2–3, 3–4, and so on) of the short arcs of the upper base are shown in the top view. The development is constructed in the following manner: First, the triangle 1_DPL is constructed, using the length p^Hl^H taken from the top view and true lengths from the diagram. Next, using the method explained in Sec. 10.17, the conical surface, whose apex is at L, is developed in an attached position. Triangle 4_DLM is then added, and so on, until all component surfaces have been drawn.

10.22
To Develop a Transition Piece Having an Approximately Developable Surface by the Triangulation Method.

Figure 10.22 shows a half development of a transition piece that has a warped surface instead of a partially conical one, such as that discussed in Sec. 10.21. The method of constructing the development is somewhat similar, however, in that it is formed by laying out, in true size, a number of small triangles that approximate the surface. The true size of the circular intersection is shown in the top view, and the true size of the elliptical intersection is shown in the auxiliary view, which was constructed for that purpose.

The front half of the circle in the top view should be divided into the same number of equal parts as the half-auxiliary view. By joining the division points, the lateral surface may be initially divided into narrow quadrilaterals. These in turn may be subdivided into triangles by drawing diagonals, which, though theoretically curved lines, are assumed to be straight. The true lengths of the elements and the diagonals are found by constructing two separate true-length diagrams by the method illustrated in Fig. 10.12.

10.23
To Develop a Sphere.

The surface of a sphere is a double-curved surface that can

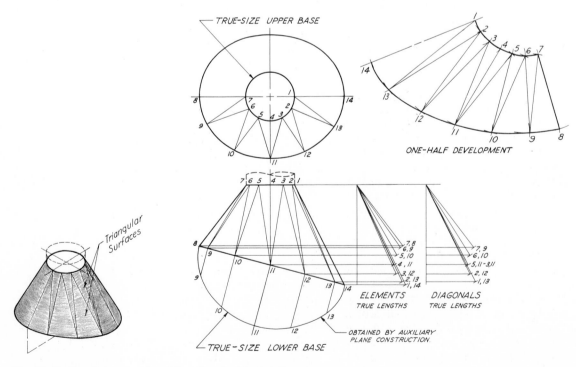

Fig. 10.22 Development of transition piece by triangulation.

be developed only by some approximate method. The standard methods commonly used are illustrated in Fig. 10.23.

In (*a*) the surface is divided into a number of equal meridian sections of cylinders. The developed surfaces of these form an approximate development of the sphere. In drawing the development it is necessary to develop the surface of only one section, for this can be used as a pattern for the developed surface of each of the others.

In (*b*) the sphere is cut by parallel planes, which divide it into a number of horizontal sections, the surfaces of which approximate the surface of the sphere. Each of these sections may be considered the frustum of a right cone whose apex is located at the intersection of the chords extended.

B □ Intersections

10.24

Lines of Intersection of Geometric Surfaces. The line of intersection of two surfaces is a line that is common to both. It may be considered the line that would contain the points in which the elements of one surface would pierce the other. Almost every line on a practical orthographic representation is a line of intersection; therefore, the following discussion may be deemed an extended study of the same subject. The methods presented in this chapter are the recognized easy procedures for finding the more complicated lines of intersection created by intersecting geometric surfaces.

In order to complete a view of a working drawing or a view necessary for developing the surfaces of intersecting geometric shapes, one frequently must find the line of intersection between surfaces. On an ordinary working drawing the line of intersection may be "faked in" through a few critical points. On a sheet-metal drawing, however, a sufficient number of points must be located to obtain an accurate line of intersection and an ultimately accurate development.

The line of intersection of two surfaces is found by determining a number of points common to both surfaces and drawing a line or lines through these points in correct order. The resulting line of intersection may be straight, curved, or straight and curved. The problem of finding such a line may be solved by one of

Fig. 10.23 **Approximate development of a sphere.**

two general methods, depending on the type of surfaces involved.

For the purpose of simplifying this discussion of intersections, it should be assumed that all problems are divided into these two general groups:

Group I. Problems involving two geometric figures, both of which are composed of plane surfaces.

Group II. Problems involving geometric figures which have either single-curved or double-curved surfaces.

For instance, the procedure for finding the line of intersection of two prisms is the same as that for finding the line of intersection of a prism and a pyramid; hence, both problems belong in the same group (Group I). Since the problem of finding the line of intersection of two cylinders and the problem of finding the line of intersection of a cylinder and a cone both involve single-curved surfaces, these two also belong in the same group (Group II).

Problems of the first group are solved by locating the points through which the edges of each of two geometric shapes pierce the other. These points are vertices of the line of intersection. Whenever one of two intersecting plane surfaces appears as a line in one view, the points through which the lines of the other surfaces penetrate it usually may be found by inspecting that view.

Problems of the second group may be solved by drawing elements on the lateral surface of one geometric shape in the region of the line of intersection. The points at which these elements intersect the surface of the other geometric shape are points that are common to both surfaces and consequently lie on their line of intersection. A curve, traced through these points with the aid of a French curve, will be a representation of the required intersection. To obtain accurate results, some of the elements must be drawn through certain critical points at which the curve changes sharply in direction. These points usually are located on contour elements. Hence, the usual practice is to space the elements equally around the surface, starting with a contour element.

10.25

Determination of a Piercing Point by Inspection (Fig. 10.24). It is easy to determine where a given line pierces a surface when the surface appears as an edge view (line) in one of the given views. For example, when the given line AB is extended as shown in (a), the F-view of the piercing point C is observed to be at c^F, where the frontal view of the line AB extended intersects the line view of the surface. With the position of c^F known, the H-view of point C can be quickly found by projecting upward to the H-view of AB extended.

In (b) the H-view (f^H) of the piercing point F is found first by extending d^He^H to intersect the edge view of the surface pierced by the line. By projecting downward, f^F is located on d^Fe^F extended.

In (c) the views of the piercing point K are found in the same manner as in (b), the only difference being that the edge view of the surface pierced by the line appears as a circle arc in the H-view instead of a straight line. It should be noted that a part of the line is invisible in the F-view because the piercing point is on the rear side of the cylinder.

The F- and H-views of the piercing point R in (d) may be found easily by projection after the P-view (r^P) of R has been once established by extending p^Pq^P to intersect the line view of the surface.

10.26

Determination of a Piercing Point Using a Line-projecting Plane. When a line pierces a given oblique plane and an edge view is not given, as in Fig. 10.25, a line-projecting plane (cutting plane) may be used to establish a line of intersection that will contain the piercing point. In the illustration, a vertical projecting plane was selected that would contain the given line RS and intersect the given plane ABC along line DE, as illustrated by the pictorial drawing.

Solution: Draw the H-view of the projecting plane through r^Hs^H to establish d^He^H, as shown in the H-view in (c). Locate d^Fe^F and draw the F-view of the

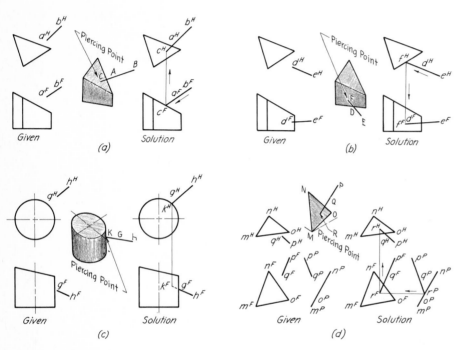

Fig. 10.24 Determination of a piercing point by inspection.

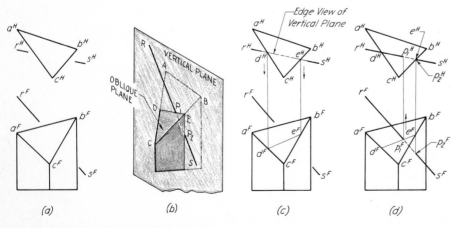

Fig. 10.25 Use of a line-projecting plane.

line of intersection. Then, complete the line view $r^F s^F$ to establish p_1^F at the point of intersection of $r^F s^F$ and $d^F e^F$. Finally, locate p_1^H on $r^H s^H$ by projecting upward from p_1^F, as shown in (d).

10.27

To Find Where a Line Pierces a Geometric Solid-cylinder–cone–sphere Using Projecting Planes (Fig. 10.26). The points where a line pierces a cylinder, cone, or sphere may be found easily through the use of a projecting plane (cutting plane) that contains the given line, as illustrated in (a), (b), and (c).

In (a) the intersections of the projecting plane and the cylinder are straight-line elements because the projecting plane used is parallel to the axis of the cylinder. The use of planes parallel to the axis permits the rapid solution of this type of problem. As shown by the pictorial drawing, the vertical projecting plane cuts elements on both the right and left sides of the cylinder. The line AB intersects the element RS at C and the other element at D. Points C and D are the piercing points.

The piercing points of a line and a cone are the points of intersection of the line and the two specific elements of the cone that lie in the projecting plane containing the line as shown in (b). The vertex of the cone and the given line fix the position of the projecting plane, the plane of the elements. In the illustration, the vertical projecting plane, taken through the line EF and the vertex of the cone T, cuts the base of the cone at U and V, the points needed to establish the F-views of the elements lying in the plane. The points of intersection of the given line EF and these elements are points G and H, the points where the line pierces the cone. If the given line had not been in a position to intersect the axis of the cone, it would have been necessary to use an oblique cutting plane through the apex.

A projecting plane that contains a line piercing a sphere will cut a circle on the surface of the sphere; therefore, points where the given line intersects the circle will be points where the line pierces the sphere. [See the pictorial drawing in (c).] In the illustrations, a vertical projecting plane was used containing the given line JK. The F-views of the piercing points M and N ($m^F n^F$) were found first at the points of intersection of the line and the circle. The H-views ($m^H n^H$) of the piercing points were found by projecting upward from m^F and n^F in the F-view.

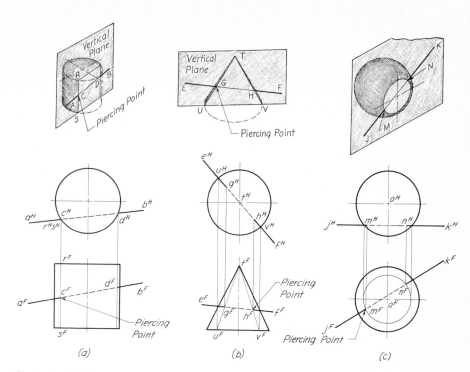

(a) (b) (c)

Fig. 10.26 To determine where a line pierces a geometric solid.

10.28

To Determine the Points Where a Line Pierces a Cone, General Case. In Sec. 10.27, the statement was made that the piercing points of a line and a cone are the points of intersection of the line and the two specific elements of the cone that lie in the projecting plane containing the line and the apex of the cone [Fig. 10.26(b)]. This pertained to a special condition for which a line-projecting plane could be used. For other cases, the following general statement applies: The piercing points of a line and any surface must lie on the lines of intersection of the given surface and a cutting plane that contains the line. Obviously, an infinite number of cutting planes could have been assumed that would have contained the line AB in Fig. 10.27, but all would have resulted in curved lines of intersection, except in the case of the one plane that was selected to pass through the apex O of the cone. As can be noted by observing the pictorial drawing, this choice gives a plane that intersects the cone along two straight-line elements.

Solution: (1) Form a cutting plane containing the line AB and the apex O by drawing a line from O to an assumed point M on line AB. Lines AB and OM define the cutting plane. (2) Extend the

Fig. 10.27 To determine the points where a line pierces a cone—general case.

elements of intersection o^F1 and o^F2 in the F-view. Points $p_1{}^F$ and $p_2{}^F$, where the F-views of these elements intersect a^Fb^F, are the F-views of the piercing points of the line AB and the given cone. (4) Project the F-views of the piercing points to the H-view to locate $p_1{}^H$ and $p_2{}^H$.

When the line and cone are both oblique an added view (auxiliary view), showing the base as an edge, may be needed to obtain a quick and accurate solution employing the steps illustrated in Fig. 10.27.

**10.29
To Find the Intersection of Two Planes, Line-projecting Plane Method.** The intersection of two oblique planes may be determined by finding where two of the lines of one plane pierce the other plane, as illustrated by the pictorial drawing in Fig. 10.28. The procedure that is illustrated employs line-projecting planes to find the piercing points of the lines XY and XZ and the oblique plane RST. Therefore, it might be said that the solution requires the determination of the piercing point of a line and an oblique plane, as explained in Sec. 10.26.

Given: The oblique planes RST and XYZ.

Solution: Since the (vertical) line-projecting plane C_1P_1 is to contain the line XY of the plane XYZ, draw the line-view representation of this projecting plane to coincide with x^Hy^H. Next, project the line of intersection AB between the line-projecting plane C_1P_1 and plane RST from the top view, where it appears as a^Hb^H in the edge-view representation of C_1P_1 to the front view. Then, since it is evident that the line AB is not parallel to XY, which lies in the projecting plane (see F-view), the line XY intersects AB. The location of this intersection at E is established first in the F-view, where the line x^Fy^F intersects a^Fb^F at e^F. The H-view of E, that is, e^H, is found by projecting upward from e^F in the F-view to the line view x^Hy^H. The other end of the line of intersection between the two given planes at F is found by using the line-projecting plane C_2P_2 and following the same procedure as for determining the location of point E.

**10.30
To Find the Intersection of a Cylinder and an Oblique Plane.** There are two distinct and separate methods shown in Fig. 10.29 for finding the line of intersection of an oblique plane and a cylinder. Both meth-

cutting plane until it intersects the plane of the base of the cone. In the pictorial illustration it should be noted that OM and ON extended establish the line of intersection QR between the cutting plane and the plane of the base of the cone. Line ON, drawn from O to an assumed point N, is an additional line in the plane. (3) Project points 1 and 2, where the line q^Hr^H intersects the curve of the base of the cone in the H-view, to the F-view, and draw the views of the

Fig. 10.28 To find the intersection line of two planes.

ods appear together on the drawing at the left. The selected line method is illustrated pictorially in (b), while the cutting-plane method is shown in (c).

In the application of the selected line method, any line of the given plane, such as line BR, is drawn in the F- and H-views (a). It can be noted by observing, in the pictorial drawing in (b), that this particular line pierces the cylinder at points P_1 and P_2 to give two points on the line of intersection. On the multiview drawing, the locations of the H-views of points P_1 and P_2 can readily be recognized as being at the points labeled $p_1{}^H$ and $p_2{}^H$, where the line view $b^H r^H$ intersects the edge view of the surface of the cylinder. The F-views $(p_1{}^F$ and $p_2{}^F)$ of points P_1 and P_2 were found by projecting downward from $p_1{}^H$ and $p_2{}^H$ to the line view $b^F r^F$. Additional points along the line of intersection, as needed, can be obtained by using other lines of the plane.

The line of intersection of this same plane and cylinder could have been almost as easily determined through the use of a series of line-projecting (cutting) planes passed parallel to the axis of the cylinder, as illustrated in (c). It should be noted that the vertical cutting plane shown cuts elements on the cylinder that intersect XY, the line of intersection of the cutting plane and the given plane, at points P_3 and P_4. Points P_3 [not visible in (c)] and P_4 are two points on the line of intersection, for they lie in both the cutting plane and the given plane and are on the surface of the cylinder. After the position of the cutting plane has been established in (a) by drawing the line representation in the H-view, x^H and y^H must be projected downward to the corresponding lines of the plane in the F-view. Line $x^F y^F$ as then drawn is the F-view of the line of intersection of the cutting plane and given plane. Finally, as the last step, the intersection elements that appear as points in the H-view at $p_3{}^H$ and $p_4{}^H$ must be drawn in the F-view. The F-views of points P_3 and $P_4(p_3{}^F$ and $p_4{}^F)$ are at the intersection of the F-views of the elements and the line $x^F y^F$.

A series of selected planes will give the points needed to complete the F-view of the intersection.

10.31

To Find the Intersection of a Cone and an Oblique Plane. When the intersecting plane is oblique, as is true in Fig. 10.30, it is usually desirable to employ the cut-

Fig. 10.29 To find the intersection of a cylinder and a plane.

ing-plane method shown in (a) rather than to resort to the use of an additional view, as in (b). Since the given cone is a right cone, any one vertical cutting plane, passing through the apex O, will simultaneously cut straight lines on the conical surface and across the given oblique plane. The pictorial illustration shows that the cutting plane XY intersects the cone along the two straight-line elements O–2 and O–8 and the plane along the line RS. Points G and H, where the elements intersect the line RS, are points along the required line of intersection because both points lie in the given plane $ABCD$ and are on the surface of the cone. In (a), the line representation of the cutting plane XY in the H-view establishes the position of the H-view $(r^H s^H)$ of the line RS. The H-views of elements O–2 and O–8 also lie in the edge view of the cutting plane XY. With this much known, the F-views of the two elements and line RS may be drawn. Points g^F and h^F are at the intersection of $r^F s^F$ and o^F–2 and o^F–8, respectively. A series of cutting planes passed similarly furnishes the points needed to complete the solution.

At times one might prefer to determine the line of intersection through the use of a constructed auxiliary view that shows the given plane as an edge. In this case, when selected elements may be seen to intersect the line view of the plane in the auxiliary view, the solution becomes quite

Fig. 10.30   To find the intersection of a cone and a plane.

simple, because all that is required is to project the point of intersection of an element and the line view of the plane to the F- and H-views of the same element. For example, the A-view (o^A–8) of the element O–8 can be seen to intersect the edge line $a^Ab^Ac^Ad^A$ at h^A, the A-view of point H. By projecting to line o^H–8, the H-view of H (h^H) may be easily established. The F-view of H (h^F) lies directly below h^H on o^F–8. Through the projected

views of other points, located similarly, smooth curves may be drawn to form the F- and H-view representations, as shown.

10.32
To Find the Intersection of a Sphere and an Oblique Plane.   Horizontal cutting planes have been used to find the line of intersection of the sphere and oblique plane shown in Fig. 10.31. Two approaches to the solution have been given on the line drawing. The horizontal cutting planes, as selected, cut circles from the sphere and straight lines from the given oblique plane. For example, the cutting plane CP_3 in the F-view cuts the horizontal line 3–3 from plane $ABCD$ and a circle from the sphere. This circle appears as an edge in the F-view and shows in its true diameter. In the H-view it will show in true shape. The line and circle intersect at two points that also have been identified by the number 3, the number assigned to the cutting plane in which these two points lie. These points now located in the H-view are projected to the CP_3 line in the F-view. The curved line through points, that have been determined by a series of planes, is a line common to both surfaces and is therefore the line of intersection.

This problem could also have been solved by using an auxiliary view showing the plane $ABCD$ as an edge. As before, the horizontal cutting planes will cut circles from the sphere and lines from the plane. However, in this case both the lines

Fig. 10.31   To find the intersection of a sphere and an oblique plane.

and the intersections show as points in the A-view. For each CP, these points must be projected from the auxiliary view to the corresponding circle in the H-view. The F-views of these points on the line of intersection may be found by projection and by using measurements taken from the A-view.

10.33

To Find the Intersection of Two Prisms. In Fig. 10.32, points A, C, and D, through which the edges of the horizontal prism pierce the vertical prism, are first found in the top view (a^H, c^H, and d^H) and are then projected downward to the corresponding edges in the front view. Point B, through which the edge of the vertical prism pierces the near face of the triangular prism, cannot be found in this manner because the side view from which it could be projected to the front view is not shown. Its location, however, can be established in the front view without even drawing a partial side view, if some scheme like the one illustrated in the pictorial drawing is used. In this scheme, the intersection line AB, whose direction is shown in the top view as line a^Hb^H, is extended on the triangular face to point X on the top edge. Point x^H is projected to the corresponding edge in the front view and a light construction line is drawn between the points a^F and x^F. Since point B is located on line AX (see pictorial) at the point where the edge of the prism pierces the line, its location in the front view is at point b^F where the edge cuts the line a^Fx^F.

10.34

To Find the Intersection of a Pyramid and a Prism (Fig. 10.33). The intersection of a right pyramid and a prism may be found by the same general method used for finding the intersection of two prisms (Sec. 10.33).

10.35

To Determine the Intersection of a Prism and a Pyramid Using Line-projecting Planes. Frequently, it becomes necessary to draw the line of intersection between two geometric shapes so positioned that the piercing points of edges cannot be found by inspection if only the principal views are to be used. In this case, one must resort to the method discussed in Sec. 10.26 to determine where a line, such as the edge line GD of the prism shown in Fig. 10.34, pierces a surface. As illustrated

by the pictorial drawing, a vertical plane passed through the edge DG of the prism, intersects the surface ABC of the pyramid along line MN that contains point D, the piercing point of DG. In (a), the H-view (m^Hn^H) of the line MN lies along d^Hg^H extended to m^H on the edge of the pyra-

Fig. 10.32 Intersecting prisms.

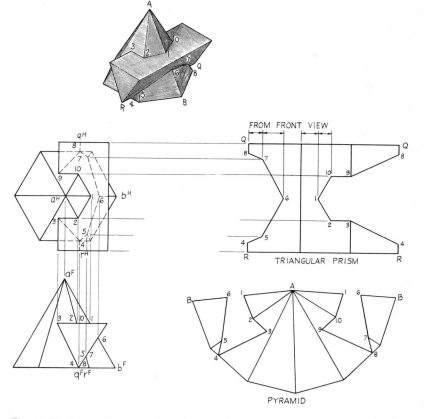

Fig. 10.33 Intersecting pyramid and prism.

(a)

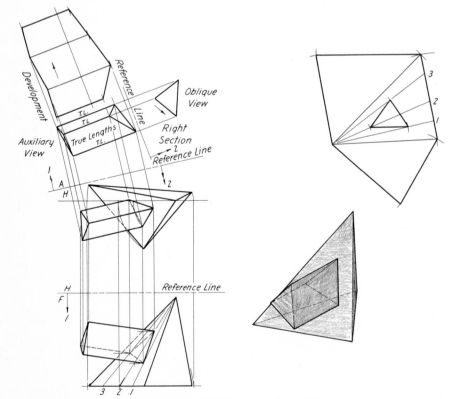

(b)

Fig. 10.34 Intersecting pyramid and prism.

mid, because the *H*-view of the cutting plane appears as an edge that coincides with $d^H g^H$. With the *F*-view of *MN* established by projecting downward from $m^H n^H$ in the *H*-view, the frontal view of the piercing point *D* is at d^F where the view of the edge line *DG* of the prism intersects $m^F n^F$. The *H*-view of *D* is found by projecting upward from d^F. The two other piercing points, at *E* and *F*, are found in the same manner using two other line-projecting planes.

10.36

To Construct a Development Using Auxiliary Views. When one of the components is oblique to the principal planes of projection, as is the prism in Fig. 10.35, the construction work needed for the development can be simplified somewhat through the use of an auxiliary view to find the true lengths of the edges and an oblique (secondary auxiliary) view to show a right section. Since the plane on which the auxiliary view is projected is a vertical one that is parallel to the edges of the prism, the distances perpendicular to the *AH* reference line are height distances. In making the construction, distances are taken from the *F*-view in the direction of the single-headed arrow for use in the auxiliary view in the direction indicated by a

similar arrow. For the oblique view, projected on an *O*-plane that is perpendicular to both the *A*-plane and the edges of the prism, distances are taken from the *AH* reference line in the direction of the arrow numbered 2 to be laid out from the reference line for the *O*-view. Since this is the second auxiliary, the arrows indicating the direction for equal distances have been given two heads.

If there is sufficient space available, the true-length measurements for the edges in the development may be projected directly from the auxiliary view showing the true lengths. The true distances between these edges, taken from the right section in the direction of the arrow, are laid off along the stretch-out line. The arrow on the development indicates the direction in which the successive faces are laid down when the prism is turned to unroll the lateral surface inside-up.

10.37

To Find the Intersection of Two Cylinders. If a series of elements are drawn on the surface of the small horizontal cylinder, as in Fig. 10.36, the points *A*, *B*, *C*, and *D* in which they intersect the vertical cylinder will be points on the line of intersection (see pictorial). These points, which are shown as a^H, b^H, c^H, and d^H in the top view, may be located in the front view by projecting them downward to the corresponding elements in the front view, where they are shown as points a^F, b^F, c^F, and d^F. The desired intersection is represented by a smooth curve drawn through these points.

10.38

To Find the Intersection of Two Cylinders Oblique to Each Other. The first step in finding the line of intersection of two cylinders that are oblique to each other (Fig. 10.37) is to draw a revolved right section of the oblique cylinder directly on the front view of that cylinder. If the circumference of the right section is then divided into a number of equal divisions and elements are drawn through the division points, the points *A*, *B*, *C*, and *D* in which the elements intersect the surface of the vertical cylinder will be points on the line of intersection (see pictorial). In the case of the illustration shown, these points are found first in the top view and then are projected downward to the corresponding elements in the front view. The line of intersection in the front view is represented by a smooth curve drawn through these points.

Fig. 10.35 To construct a development using auxiliary views.

Fig. 10.36 Intersecting cylinders.

Fig. 10.37 Intersecting cylinders.

10.39

To Find the Intersection of Two Cylinders Using Line-projecting (Cutting) Planes.
The line of intersection of the two cylinders shown in Fig. 10.37 could have been determined through the use of a series of parallel line-projecting (cutting) planes passed parallel to their axes (Fig. 10.38). The related straight-line elements cut on the cylinders by any one cutting plane, such as C, intersect on the line of intersection of the cylinders. As many line-projecting planes as are needed to obtain a smooth curve should be used and they should be placed rather close together where a curve changes sharply.

10.40

To Find the Intersection of a Cylinder and a Cone. The intersection of a cylinder and a cone may be found by assuming a number of elements on the surface of the cone. The points at which these elements cut the cylinder are on the line of intersection (see Figs. 10.39 and 10.40). In selecting the elements, it is the usual practice to divide the circumference of the base into a number of equal parts and draw elements through the division points. To obtain

needed points at locations where the intersection line will change suddenly in curvature, however, there should be additional elements.

In Fig. 10.39, the points at which the elements pierce the cylinder are first found in the top view and are then projected to the corresponding elements in the front view. A smooth curve through

Fig. 10.38 To find the intersection of two cylinders using line-projecting planes.

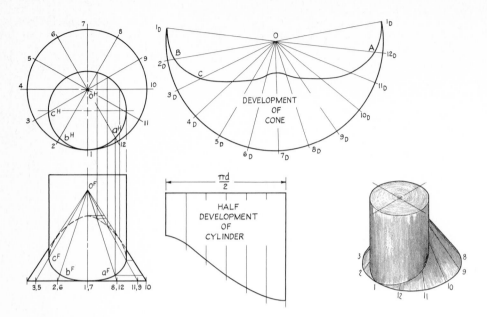

Fig. 10.39 Intersecting cylinder and cone.

from the conical surface and two straight-line elements from the cylindrical surface. Cutting plane 5, for example, cuts one element (numbered 5) from the near side of the cone and an upper and lower element from the surface of the cylinder. The intersection of these three elements, all numbered 5, establish the location of points A and B on the line of intersection.

An alternative method for finding the line of intersection of a cylinder and a right cone is illustrated in Fig. 10.41. Here horizontal cutting planes are passed through both geometric shapes in the region of their line of intersection. In each cutting plane, the circle cut on the surface of the cone will intersect elements cut on the cylinder at two points common to both surfaces (see pictorial). A curved line traced through a number of such points in different planes is a line common to both surfaces and is therefore the line of intersection.

these points forms the figure of the intersection.

To find the intersection of the cone and cylinder combination shown in Fig. 10.40, line-projecting cutting planes were passed through the vertex O parallel to the axis of the cylinder to cut intersecting elements on both geometric forms. The partial auxiliary is needed to establish these planes, because it is only in a view showing the axis of the cylinder as a point that these planes and the surface of the cylinder will show as edge views. Each cutting plane cuts one needed straight-line element

10.41
To Find the Intersection of an Oblique Cone and a Paraboloid. The method that is illustrated in Fig. 10.42 for finding the line of intersection of a cone and a paraboloid may be used effectively for the solution of many other intersection problems involving different geometric forms, such as two cones or a cone and a cylinder having parallel bases. For the arrangement as shown, a horizontal cutting plane will intersect each geometric form in a circle. When a number of horizontal cutting planes are passed through both geometric shapes in the region of their line of intersection, each plane cuts a circle on the cone that intersects a circle cut on the paraboloid at two points common to both surfaces. The curved line traced through

Fig. 10.40 To find the intersection of a cone and a cylinder.

Fig. 10.41 Intersecting cylinder and cone.

the points that have been determined by the several planes is a line common to both surfaces and is therefore the line of intersection.

10.42

To Find the Intersection of a Prism and a Cylinder.

In Fig. 10.43 it is required to find the intersection between the cylinder and two of the plane surfaces of the prism so that a pattern for the hole can be cut in the prism to match the cylinder. Although the intersection could have been secured merely by determining, through inspection, the piercing points of elements drawn arbitrarily on the cylinder, vertical line-projecting planes were used to illustrate an approach to this type of problem that is used by many people. The cutting planes are located in the view showing the right section of the cylinder. From this first step, the positions for the lines and elements cut on the prism and cylinder, respectively, can be determined by projection. Each line-projecting (cutting) plane cuts a line on a plane surface of the prism and related elements on the cylinder that intersect on the line of intersection. A sufficient number of line-projecting planes should be used to enable one to draw a smooth curved-line representation of the intersection. If more points are desired, at a location where a curve changes sharply, additional cutting planes may be added.

10.43

To Find the Intersection of a Prism and a Cone.

The complete line of intersection may be found by drawing elements on the surface of the cone (Fig. 10.44) to locate points on the intersection as explained in Sec. 10.40. To obtain an accurate curve, however, some thought must be given to the placing of these elements. For instance, although most of the elements may be equally spaced on the cone to facilitate the construction of its development, additional ones should be drawn through the critical points and in regions where the line of intersection changes sharply in curvature. The elements are drawn on the view that will reveal points on the intersection, then the determined points are projected to the corresponding elements in the other view or views. In this particular illustration a part of the line of intersection in the top view is a portion of the arc of a circle that would be cut by a horizontal plane containing the bottom surface of the prism.

If the surfaces of the prism are parallel

to the axis of the cone, as in Fig. 10.45, the line of intersection will be made up of the tips of a series of hyperbolas. The intersection may be found by passing planes that will cut circles on the surface of the cone. The points at which these cutting circles pierce the faces of the prism are points common to the lateral surfaces of both shapes and are therefore points on the required line of intersection. It should be noted that the resulting solution represents a chamfered bolthead.

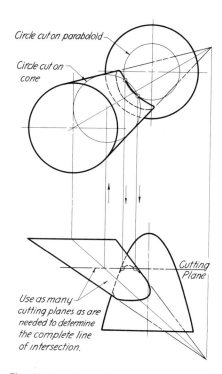

Fig. 10.42 Intersecting cone and paraboloid.

Fig. 10.43 To find the intersection of a prism and a cylinder.

Fig. 10.44 Intersecting cone and prism.

Fig. 10.45 Intersecting cone and hexagonal prism.

Problems

1. (Fig. 10.46). Develop the lateral surface of one or more of the prisms as assigned.

2. (Fig. 10.47). Develop the lateral surface of one or more of the pyramids as assigned. Make construction lines light. Show construction for finding the true lengths of the lines.

3. (Fig. 10.48). Develop the lateral surface of one or more of the cylinders as assigned. Use a hard pencil for construction lines and make them light.

4. (Fig. 10.49). Develop the lateral surface of one or more of the cones as assigned. Show all construction. Use a hard pencil for construction lines and make them light. In each case start with the shortest element and unroll, inside-up. It is suggested that 12 elements be used, in order to secure a reasonably accurate development.

5. (Fig. 10.50). Develop the lateral surface of one or more of the transition pieces as assigned. Show all construction lines in light, sharp pencil lines. Use a sufficient number of elements on the curved surfaces to ensure an accurate development.

6. (Fig. 10.51). Develop the sheet-metal connections. On pieces 3 and 4, use a sufficient number of elements to obtain a smooth curve and an accurate development.

7–8. (Figs. 10.52 and 10.53). Draw the line of intersection of the intersecting geometric shapes as assigned. Show the invisible portions of the lines of intersection as well as the visible. Consider that the interior is open.

9–10. (Figs. 10.54 and 10.55). Draw the line of intersection of the intersecting geometric shapes as assigned. It is suggested that the elements used to find points along the intersection be spaced 15° apart. Do not erase the construction lines. One shape does not pass through the other.

11. (Fig. 10.56). Draw the line of intersection of the intersecting geometric shapes as assigned. Show the invisible portions of the line of intersection as well as the visible. The interior of the combination is hollow. One shape does not pass through the other. Show construction with light, sharp lines drawn with a hard pencil.

Fig. 10.46   Prisms.

Fig. 10.47   Pyramids.

Fig. 10.48   Cylinders.

Fig. 10.55 Intersecting surfaces.

Fig. 10.56 Intersecting surfaces.

Shown is a pictorial illustration that was prepared by a NASA artist to promote the Apollo Applications Program. Illustrations of this type are used to sell an idea to administrators and political leaders. (*Courtesy National Aeronautics and Space Administration*)

pictorial presentation 11

11.1

Introduction. An orthographic drawing of two or more views describes an object accurately in form and size, but, since each of the views shows only two dimensions without any suggestion of depth, such a drawing can convey information only to those who are familiar with graphic representation. For this reason, multiview drawings are used mainly by engineers, draftsmen, contractors, and shopmen.

Frequently, however, engineers and draftsmen find they must use conventional picture drawings to convey specific information to persons who do not possess the trained imagination necessary to construct mentally an object from views. To make such drawings, several special schemes of one-plane pictorial drawing have been devised that combine the pictorial effect of perspective with the advantage of having the principal dimensions to scale. But pictorial drawings, in spite of certain advantages, have disadvantages that limit their use. A few of these are as follows:

1. Some drawings frequently have a distorted, unreal appearance that is disagreeable.

2. The time required for execution is, in many cases, greater than for an orthographic drawing.

3. They are difficult to dimension.

4. Some of the lines cannot be measured.

Even with these limitations, pictorial drawings are used extensively for technical publications, Patent Office records, piping diagrams, and furniture designs. Occasionally they are used, in one form or another, to supplement and clarify machine and structural details that would be difficult to visualize (Fig. 11.1).

11.2
Divisions of Pictorial Drawing. Single-plane pictorial drawings are classified in three general divisions: (1) axonometric projection, (2) oblique projection, and (3) perspective projection (Fig. 11.2).

Perspective methods produce the most realistic drawings, but the necessary construction is more difficult and tedious than the construction required for the conventional methods classified under the other two divisions. For this reason, engineers customarily use some form of either axonometric or oblique projection. Modified methods, which are not theoretically correct, are often used to produce desired effects.

Fig. 11.1 Pictorial illustration. (*Courtesy Lockheed Aircraft Corp.*)

AXONOMETRIC OBLIQUE PERSPECTIVE

Fig. 11.2 Axonometric, oblique, and perspective projection.

A □ Axonometric projection

11.3

Divisions of Axonometric Projection. Theoretically, axonometric projection is a form of orthographic projection. The distinguishing difference is that only one plane is used instead of two or more, and the object is turned from its customary position so that three faces are displayed (Fig. 11.3). Since an object may be placed in a countless number of positions relative to the picture plane, an infinite number of views may be drawn, which will vary in general proportions, lengths of edges, and sizes of angles. For practical reasons, a few of these possible positions have been classified in such a manner as to give the recognized divisions of axonometric projection: (1) isometric, (2) dimetric, and (3) trimetric.

Isometric projection is the simplest of these, because the principal axes make equal angles with the plane of projection and the edges are therefore foreshortened equally.

11.4

Isometric Projection. If the cube in Fig. 11.3 is revolved through an angle of 45° about an imaginary vertical axis, as shown in II, and then tilted forward until its body diagonal is perpendicular to the vertical plane, the edges will be foreshortened equally and the cube will be in the correct position to produce an isometric projection.

The three front edges, called *isometric axes*, make angles of approximately 35° 16′ with the vertical plane of projection, or picture plane. In this form of pictorial, the angles between the projections of these axes are 120°, and the projected lengths of the edges of an object, along and parallel to these axes, are approximately 81% of their true lengths. It should be observed that the 90° angles of the cube appear in the isometric projection as either 120° or 60°.

Now, if instead of turning and tilting the object in relation to a principal plane of projection, an auxiliary plane is used that will be perpendicular to the body diagonal, the view projected on the plane will be an axonometric projection. Since the auxiliary plane will be inclined to the principal planes on which the front, top, and side views would be projected, the auxiliary view, taken in a position perpendicular to the body diagonal, will be a secondary auxiliary view, as shown in Fig. 11.4.

11.5

Isometric Drawing. Objects seldom are drawn in true isometric projection, the use of an isometric scale being inconvenient and impractical. Instead, a conventional method is used in which all foreshortening is ignored, and actual true lengths are laid off along isometric axes and isometric lines. To avoid confusion and to set this method apart from true isometric projection, it is called isometric drawing.

The isometric drawing of a figure is slightly larger (approximately 22½%) than

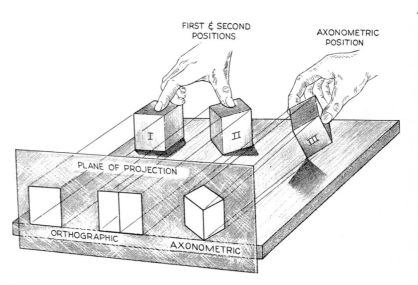

Fig. 11.3 Theory of axonometric projection.

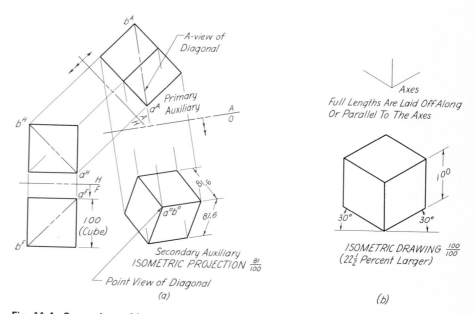

Fig. 11.4 Comparison of isometric projection and isometric drawing.

the isometric projection, but, since the proportions are the same, the increased size does not affect the pictorial value of the representation (see Fig. 11.4). The use of a regular scale makes it possible for a draftsman to produce a satisfactory drawing with a minimum expenditure of time and effort.

In isometric drawing, lines that are parallel to the isometric axes are called *isometric lines*.

11.6

To Make an Isometric Drawing of a Rectangular Object. The procedure followed in making an isometric drawing of a rectangular block is illustrated in Fig. 11.5. The three axes that establish the front

edges, as shown in (*b*), should be drawn through point *A* so that one extends vertically downward and the other two upward to the right and left at an angle of 30° from the horizontal. Then the actual lengths of the edges may be set off, as shown in (*c*) and (*d*), and the remainder of the view completed by drawing lines parallel to the axes through the corners thus located, as in (*e*) and (*f*).

Hidden lines, unless absolutely necessary for clearness, always should be omitted on a pictorial representation.

11.7

Nonisometric Lines. Those lines that are inclined and are not parallel to the isometric axes are called *nonisometric lines*. Since a line of this type does not appear in its true length and cannot be measured directly, its position and projected length must be established by locating its extremities. In Fig. 11.6, *AB* and *CD*, which represent the edges of the block, are nonisometric lines. The location of *AB* is established in the pictorial view by locating points *A* and *B*. Point *A* is on the top edge, *X* distance from the left-side surface. Point *B* is on the upper edge of the base, *Y* distance from the right-side surface. All other lines coincide with or are parallel to the axes and, therefore, may be measured off with the scale.

The pictorial representation of an irregular solid containing a number of nonisometric lines may be conveniently constructed by the box method; that is, the object may be enclosed in a rectangular box so that both isometric and nonisometric lines may be located by points of contact with its surfaces and edges (see Fig. 11.7).

A study of Figs. 11.6 and 11.7 reveals the important fact that lines that are parallel on an object are parallel in the pictorial view, and, conversely, lines that are not parallel on the object are not parallel on the view. It is often possible to eliminate much tedious construction work by the practical application of the principle of parallel lines.

11.8

Coordinate Construction Method. When an object contains a number of inclined surfaces, such as the one shown in Fig. 11.8, the use of the coordinate construction method is desirable. In this method, the end points of the edges are located in relation to an assumed isometric base line located on an isometric reference plane.

Fig. 11.5 Procedure for constructing an isometric drawing.

Fig. 11.6 Nonisometric lines.

For example, the line *RL* is used as a base line from which measurements are made along isometric lines, as shown. The distances required to locate point *A* are taken directly from the orthographic views.

Irregular curved edges are most easily drawn in isometric by the offset method, which is a modification of the coordinate construction method (Fig. 11.9). The position of the curve can be readily established by plotted points located by measuring along isometric lines.

11.9

Angles in Isometric Drawing. Since angles specified in degrees do not appear in true size on an isometric drawing, angular measurements must be converted in some manner to linear measurements that can be laid off along isometric lines. Usually, one or two measurements taken from an orthographic view may be laid off along isometric lines on the pictorial drawing to locate an inclined edge that has been specified by an angular dimension. The scale used for the orthographic view must be the same as the one being used in preparing the pictorial drawing.

In Fig. 11.10(*a*), the position of the inclined line *AB* was established on the isometric drawing by using the distance *X* taken from the front view of the orthographic drawing. When an orthographic drawing has already been prepared to a different scale than the scale being used for the pictorial representation, one can draw a partial orthographic view and take off the needed dimensions. A practical application of this idea is shown in (*b*). By making the construction of a partial view at the place where the angle is to appear on the isometric drawing, the position of the required line can be obtained graphically.

If desired, the tangent method, as explained in Sec. 3.11, may be used, as shown in (*c*). In using this method, a length equal to 10 units (any scale) is laid off along an isometric line that is to form one side of the angle. Then, a distance equal to 10 times the tangent of the angle is set off along a second isometric line that represents the second leg of the right triangle in pictorial. A line drawn through the end points of these lines will be the required line at the specified angle.

11.10

Circle and Circle Arcs in Isometric Drawing. In isometric drawing, a circle ap-

Fig. 11.7 Box construction.

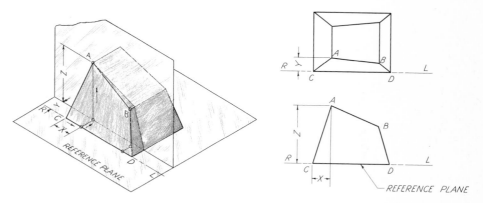

Fig. 11.8 Coordinate construction.

pears as an ellipse. The tedious construction required for plotting an ellipse accurately (Figs. 11.11 and 11.12) often is avoided by using some approximate method of drawing. The representation thus obtained is accurate enough for most work, although the true ellipse, which is slightly narrower and longer, is more pleasing in shape (Fig. 11.11). For an approximate construction, a four-center method is generally used.

Fig. 11.9 Offset construction.

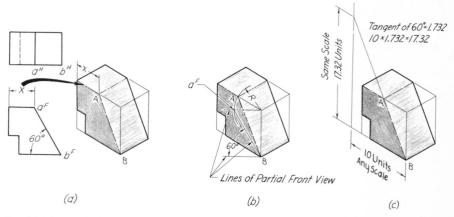

(*a*)　　　(*b*)　　　(*c*)

Fig. 11.10 Angles in isometric.

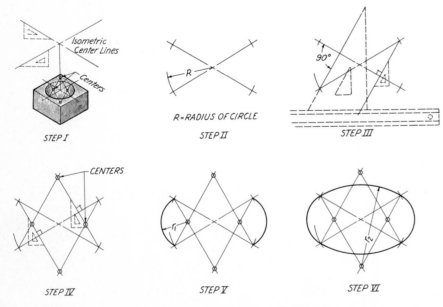

Fig. 11.11 Pictorial ellipses.

Fig. 11.12 To plot an isometric circle.

Fig. 11.13 Four-center approximation.

Fig. 11.14 Steps in drawing a four-center isometric circle (ellipse).

To draw an ellipse representing a pictorial circle, a square is conceived to be circumscribed about the circle in the orthographic projection. When transferred to the isometric plane in the pictorial view, the square becomes a rhombus (isometric square) and the circle an ellipse tangent to the rhombus at the midpoints of its sides. If the ellipse is to be drawn by the four-center method (Fig. 11.13), the points of intersection of the perpendicular bisectors of the sides of the rhombus will be centers for the four arcs forming the approximate ellipse. The two intersections that lie on the corners of the rhombus are centers for the two large arcs, while the remaining intersections are centers for the two small arcs. Furthermore, the length along the perpendicular from the center of each arc to the point at which the arc is tangent to the rhombus (midpoint) will be the radius. All construction lines required by this method may be made with a T-square and a 30° × 60° triangle.

The amount of work may be still further shortened and the accuracy of the construction improved by following the procedure shown in Fig. 11.14. The steps in this method are as follows:

Step I. Draw the isometric center lines of the required circle.

Step II. Using a radius equal to the radius of the circle, strike arcs across the isometric center lines.

Steps III and IV. Through each of these points of intersection erect a perpendicular to the other isometric center line.

Steps V and VI. Using the intersection points of the perpendiculars as centers and lengths along the perpendiculars as radii, draw the four arcs that form the ellipse (Fig. 11.15).

A circle arc will appear in pictorial representation as a segment of an ellipse. Therefore, it may be drawn by using as much of the four-center method as is required to locate the needed centers (Fig. 11.16). For example, to draw a quarter circle, it is only necessary to lay off the true radius of the arc along isometric lines drawn through the center and to draw intersecting perpendiculars through these points.

To draw isometric concentric circles by the four-center method, a set of centers must be located for each circle (Fig. 11.17).

When several circles of the same diameter occur in parallel planes, the construc-

Fig. 11.15 Isometric circles.

Fig. 11.16 Isometric circle arcs.

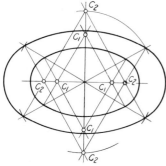

Fig. 11.17 Isometric concentric circles.

Fig. 11.18 Isometric parallel circles.

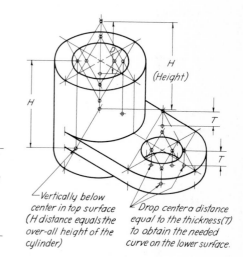

—Vertically below center in top surface (H distance equals the over-all height of the cylinder)

Drop center a distance equal to the thickness(T) to obtain the needed curve on the lower surface.

tion may be simplified. Figure 11.18 shows two views of an object and its corresponding isometric drawing. In Fig. 11.18, the centers for the ellipse representing the upper base of the large cylinder are found in the usual way, while the centers for the lower base are located by moving the centers for the upper base downward a distance equal to the height of the cylinder. By observing that portion of the object projecting to the right, it can be noted that corresponding centers lie along an isometric line parallel to the axis of the cylinder.

Circles and circle arcs in nonisometric planes may be plotted by using the offset or coordinate method. Sufficient points for establishing a curve must be located by transferring measurements from the orthographic views to isometric lines in the pictorial view. There is a rapid and easy way for drawing the cylindrical portion of the object shown in Fig. 11.19. The semicircular arc must be plotted on the rear surface as the first step. Then, after this has been done, each point is brought forward to the inclined face. The offset

Fig. 11.19 Circles in nonisometric planes.

217

Fig. 11.20 Isometric drawing of a sphere.

distances (D_1, D_2, D_3, etc.) at each level are taken from the side view in (*a*).

The pictorial representation of a sphere is the envelope of all of the great circles that could be drawn on the surface. In isometric drawing, the great circles appear as ellipses and a circle is their envelope. In practice it is necessary to draw only one ellipse, using the true radius of the sphere and the four-center method of construction. The diameter of the circle is the long diameter of the ellipse (Fig. 11.20).

11.11
Positions of Isometric Axes. It is sometimes desirable to place the principal isometric axes so that an object will be in position to reveal certain faces to a better advantage (Fig. 11.21).

The difference in direction should cause no confusion, since the angle between the axes and the procedure followed in constructing the view are the same for any position. The choice of the direction may depend on the construction of the object,

but usually this is determined by the position from which the object is ordinarily viewed.

Reversed axes (*b*) are used in architectural work to show a feature as it would be seen from a natural position below.

Sometimes long objects are drawn with the long axis horizontal, as shown in Fig. 11.22.

11.12
Isometric Sectional Views (Fig. 11.23). Generally, an isometric sectional view is used for showing the inner construction of an object when there is a complicated interior to be explained or when it is desirable to emphasize features that would not appear in a usual outside view. Sectioning in isometric drawing is based on the same principles as sectioning in orthographic drawing. Isometric planes are used for cutting an object, and the general procedure followed in constructing the representation is the same as for an exterior view.

Figure 11.23 shows an isometric half section. It is easier, in this case, to outline the outside view of the object in full and then remove a front quarter with isometric planes.

Figure 11.24 illustrates a full section in isometric. The accepted procedure for constructing this form of sectional view is to draw the cut face and then add the portion that lies behind.

Section lines should be sloped at an angle that produces the best effect, but they should never be drawn parallel to object lines. In Fig. 11.25, (*a*) illustrates the slope that is correct for most drawings, while (*b*), (*c*), and (*d*) show the poor effect produced when this phase of section lining is ignored. Ordinarily, isometric section lines are drawn at 60°.

11.13
Dimetric Projection. The view of an object that has been so placed that two of its

Fig. 11.21 Convenient positions of axes.

Fig. 11.22 Main axis horizontal—long objects.

Fig. 11.23 Isometric half section.

Fig. 11.24 Isometric full section.

axes make equal angles with the plane of projection is called a *dimetric projection.* The third axis may make either a smaller or larger angle. All of the edges along or parallel to the first two axes are foreshortened equally, while those parallel to the third axis are foreshortened a different amount. It might be said that dimetric projection, a division of axonometric projection, is like isometric projection in that the object must be placed to satisfy specific conditions. Similarly, a dimetric projection may be drawn by using the auxiliary view method. The secondary auxiliary view is the dimetric projection. The procedure is the same as for an isometric projection (Fig. 11.4), except that the line of sight is taken in the direction necessary to obtain the desired dimetric projection. Obviously, an infinite number of dimetric projections is possible.

In practical application, dimetric projection is sometimes modified so that regular scales can be used to lay off measurements to assumed ratios. This is called *dimetric drawing* [Fig. 11.26(a)].

The angles and scales may be worked out* for any ratios, such as $1:1:\frac{1}{2}$ (full size:full size:half size); $1:1:\frac{3}{4}$ (full size: full size:three-fourths size). For example, the angles for the ratios $1:1:\frac{1}{2}$ are 7° 11′ and 41° 25′. After the scales have been assumed and the angles computed, an enclosing box may be drawn in conformity to the angles and the view completed by following the general procedure used in isometric drawing, except that two scales must be used. The positions commonly used, along with the scale ratios and corresponding angles, are shown in Fig. 11.26(b). The first scale given in each ratio is for the vertical axis. Since two of the axes are foreshortened equally, while the third is foreshortened in different ratio, obviously two scales must be used. This is an effective method of representation.

11.14

Trimetric Projection (Fig. 11.28). A trimetric projection of an object is the view obtained when each of the three axes makes a different angle with the plane of projection. As might be expected, a trimetric projection may be constructed by drawing successive auxiliary views. How-

ever, since there are an unlimited number of possible lines of sight that will produce unequal foreshortening in the directions of the three axes, considerable thought must be given to the selection of a position that will show the object pictorially to the best advantage in the second auxiliary view. Making this decision is not easy. This form of pictorial representation has been used to some extent by certain aircraft companies for the preparation of production illustrations.

11.15

To Construct an Axonometric Projection (Isometric, Dimetric, or Trimetric) Directly from the Orthographic Views. A true axonometric projection can be prepared by projecting directly from two or more of the orthographic views of an object. The method, commonly called the method of intersections, was developed some years prior to 1937, the year that it was first published by L. Eckhart of the Vienna College of Engineering. Since then it has been rather widely used in the United States by those whose principal task is the preparation of pictorial drawings of complicated parts for which the orthographic views may be available. The underlying principles and basic procedure for the preparation of an isometric projection have been illustrated in Fig. 11.27.

In (*a*) the relationship of the axonometric plane to the three principal planes has been shown pictorially, and it must be understood at this point in the discussion that, since the picture plane and the axonometric plane are coincident, the axonometric triangle will be in true size. Also two other facts must be noted. First, since axonometric projection is true orthographic projection, the projection lines are perpendicular to the plane of projection. Second, when a point is projected on adjacent planes [see (*b*)] and one of the planes is revolved into the plane of the other, the two views of the point will lie on a single line that is perpendicular to the

Fig. 11.25 Section lining.

DIMETRIC DRAWING (APPROX)
(a)

DIMETRIC AXES
(b)

Fig. 11.26 Approximate dimetric drawing.

---

*Formula: $\cos \alpha = -\sqrt{2s_1{}^2s_2{}^2 - s_2{}^4}/2s_1s_2$. In this formula, α is one of the equal angles, s_1 is one of the equal scales, and s_2 is the third scale.

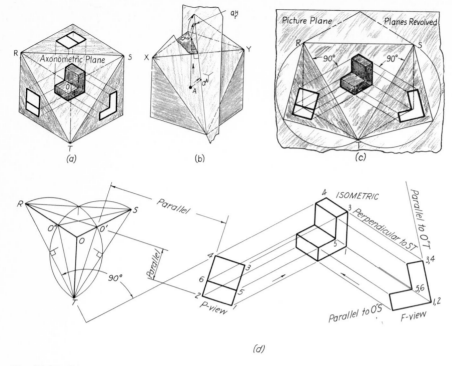

Fig. 11.27 To construct an axonometric projection from the orthographic views.

fold line—that is, the line of intersection XY of the two planes.

In (c) the principal planes have been shown revolved into the plane of the axonometric triangle, and it should now become evident to the reader how orthographic views might be used to obtain an axonometric projection once their revolved positions have been determined. By observing the drawing in (c) one can see that the projecting lines are perpendicular to the axis of rotation and that the principal edges of each view are parallel to the corresponding edges of the revolved plane.

To start the graphical diagram in (d) three mutually perpendicular lines were drawn to represent the three intersecting edges OR, OS, and OT of the cube [see (a)]. By geometry, OR will be perpendicular to ST, OS will be perpendicular to RT, and OT will be perpendicular to RS. With this diagram as now laid out, the right triangle OST may readily be revolved into the axonometric plane RST. One of the advantages of starting the diagram with the axes, as was done in the illustration, is that the forward triangle may then be revolved into the axonometric plane to determine the position of the orthographic views, instead of the rear triangle as in (c), which would require the construction of the entire box. In revolving the right tri-

angle OST, which must show in true size as $O''ST$ in the axonometric plane, it is necessary to apply the geometric principle that any two lines will form a right angle that are drawn from a point on the circumference of a circle to the end points of its diameter. Hence, point O moves along OR, which is perpendicular to ST, to a position on the semicircle that has been constructed on ST as a diameter. Similarly, O moves to O' on OS when the right triangle ORT has been revolved into the axonometric plane.

With the diagram constructed as explained, the step-by-step procedure for preparing the axonometric drawing shown is as follows:

1. Draw the F-view in some convenient location with the principal edges of the view parallel to the revolved positions of the two axes for the view, that is, with the base edge parallel to $O''S$ and the edge that is perpendicular to the base parallel to $O''T$.

2. Draw the P-view at another location with the edges of the view parallel to the revolved position of the axes for the P-view.

3. From the corners of the F-view draw projecting lines perpendicular to ST and parallel to OR.

4. From the corners of the P-view draw projecting lines perpendicular to RT and parallel to OS.

The intersection of these projecting lines with those from the F-view establishes the desired isometric projection.

When none of the angles formed by the coordinate axes of the diagram are equal, the resulting pictorial representation will be a trimetric projection. When two of the angles are equal a dimetric projection is obtained. In Fig. 11.27(d) all three of the angles were made equal because an isometric projection was deemed desirable (since engineers and draftsmen are accustomed to seeing isometric drawings and sketches in their daily work).

In preparing the trimetric projection shown in Fig. 11.28, a preliminary pictorial sketch was made to show the general appearance desired for the finished pictorial. Then the axes in the diagram were drawn very nearly parallel to the principal lines of the sketch. Otherwise, the procedure followed in constructing the pictorial representation was the same as has been explained for an isometric projection. To draw the elliptical projection of the circular arc, two points, numbered 1 and 2 in both orthographic views, were assumed along the arc. More points may be used when needed.

B □ Oblique projection

11.16

Oblique Projection. In oblique projection, the view is produced by using parallel projectors that make some angle other than 90° with the plane of projection. Generally, one face is placed parallel to the picture plane and the projection lines are taken at 45°. This gives a view that is pictorial in appearance, as it shows the front and one or more additional faces of an object. In Fig. 11.29, the orthographic and oblique projections of a cube are shown. When the angle is 45°, as in this illustration, the representation is sometimes called *cavalier projection*. It is generally known, however, as an *oblique projection* or an *oblique drawing*.

11.17

Principle of Oblique Projection. The theory of oblique projection can be explained by imagining a vertical plane of projection in front of a cube parallel to one of its faces (Fig. 11.29). When the projectors make an angle of 45° in any direction with the picture plane, the length of any oblique projection $A'B'$ of the edge AB is equal to the true length of AB. Note that the projectors could be parallel to any element of a 45° cone having its base in the plane of projection. With projectors at this particular angle (45°), the face parallel to the plane is projected in its true size and shape and the edges perpendicular to the picture plane are projected in their true length. If the projectors make a greater angle, the oblique projection will be shorter, while if the angle is less, the projection will be longer.

11.18

Oblique Drawing. This form of drawing is based on three mutually perpendicular axes along which, or parallel to which, the necessary measurements are made for constructing the representation. Oblique drawing differs from isometric drawing principally in that two axes are always perpendicular to each other, while the third (receding axis) is at some convenient angle, such as 30°, 45°, or 60° with the horizontal (Fig. 11.31). It is somewhat

Fig. 11.28 To construct a trimetric projection.

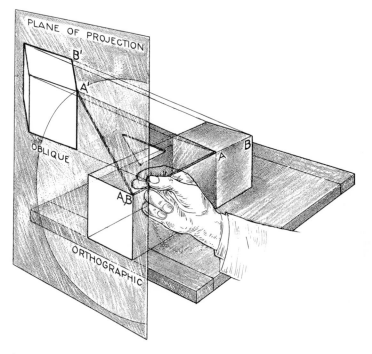

Fig. 11.29 Theory of oblique projection.

more flexible and has the following advantages over isometric drawing: (1) circular or irregular outlines on the front face show in their true shape; (2) distortion can be reduced by foreshortening along the receding axis; and (3) a greater choice is permitted in the selection of the positions of the axes. A few of the various views that can be obtained by varying the inclination of the receding axis are illustrated in Fig. 11.30. Usually, the selection of the position is governed by the character of the object.

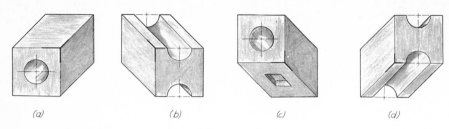

Fig. 11.30 Various positions of the receding axis.

11.19

To Make an Oblique Drawing. The procedure to be followed in constructing an oblique drawing of an adjustable guide is illustrated in Fig. 11.31. The three axes that establish the perpendicular edges in (*b*) are drawn through point *O* representing the front corner. *OA* and *OB* are perpendicular to each other and *OC* is at any desired angle (say, 30°) with the horizontal. After the width, height, and depth have been set off, the front face may be laid out in its true size and shape, as in (*c*), and the view can be completed by drawing lines parallel to the receding axis through the established corners. The circle

and semicircle are shown parallel to the picture plane in order to avoid distortion and because, from the draftsman's standpoint, it is easier to draw a circle than to construct an ellipse.

In general, the procedure for constructing an oblique drawing is the same as for an isometric drawing.

11.20

Use of a Basic Plane. If the front face of an object is in one plane, it will appear in the oblique drawing exactly the same as in orthographic drawing. Note this fact in Fig. 11.31. But the front of many objects is composed of two or more parallel planes whose relationship must be carefully established. The convenient way to accomplish this is to use one of the planes as a basic (starting) plane and work from it in the direction of the receding axis, as illustrated in Fig. 11.32. Since the front surface *A* presents the contour shape, it should be selected as the basic plane and drawn first as the front of the oblique projection, as shown in (*c*). The center *P* of the circles for surface *B* may be located easily by measuring along the axis of the hole from *O*, in plane *A*, a distance *X* equal to the distance between the planes. The measurement must be made forward along the axis from *O*, because surface *B* is in front of *A*. The centers of the arcs on the surface *C* are located in a similar manner, except that the direction of making the measurements is from the basic plane toward the back.

When an object has one or more inclined surfaces with a curved outline, it may be drawn by constructing a right section and making offset measurements, as shown in Fig. 11.36.

11.21

Rules for Placing an Object. Generally, the most irregular face, or the one containing the most circular outlines, should be placed parallel to the picture plane, in order to minimize distortion and simplify construction. By following this practice, all or most of the circles and circle arcs can be drawn with a compass, and the tedious construction that would be required to draw their elliptical representations in a receding plane is eliminated. In selecting the position of an object, two rules should be followed. The first is to place the face having the most irregular contour, or the most circular outlines, parallel to the picture plane. Note in Fig. 11.33 the advantage of following this rule.

Fig. 11.31 Procedure for constructing an oblique drawing.

Orthographic Views
(a)

(b)

(c)

(d)

(e)

Fig. 11.32 Basic plane theory of construction.

When the longest face of an object is used as the front face, the pictorial view will be distorted to a lesser degree and, therefore, will have a more realistic and pleasing appearance. Hence, the second rule is to place the longest face parallel to the picture plane. Compare the views shown in Fig. 11.34 and note the greater distortion in (*a*) over (*b*).

If these two rules clash, the first should

govern. It is more desirable to have the irregular face show its true shape than it is to lessen the distortion in the direction of the receding axis.

11.22

Angles, Circles, and Circle Arcs in Oblique. As previously stated, angles, circles, and irregular outlines on surfaces parallel to the plane of projection show in true size and shape. When located on receding faces, the construction methods used in isometric drawing may usually be applied. Figure 11.35 shows the method of drawing the elliptical representation of a circle on an oblique face. Note that the method is identical with that used for constructing isometric circles, except for the slight change in the position of the axes.

Circle arcs and circles on inclined planes must be plotted by using the offset or coordinate method (Fig. 11.36).

11.23

Reduction of Measurements in the Direction of the Receding Axis. An oblique drawing often presents a distorted appearance that is unnatural and disagreeable to the eye. In some cases the view constructed by this scheme is so misleading in appearance that it is unsatisfactory

Poor Practice
(a)

Good Practice
(b)

Fig. 11.33 Irregular contour parallel to picture plane.

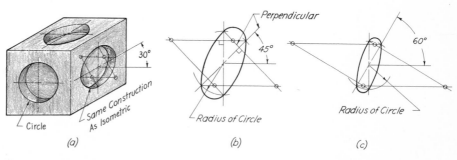

(a)

(b)

Fig. 11.34 Long axis parallel to picture plane.

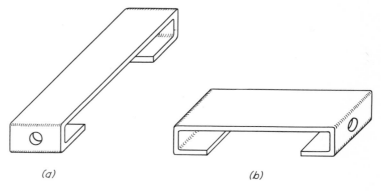

(a)

(b)

(c)

Fig. 11.35 Oblique circles.

Fig. 11.36 Curved outlines on an inclined plane.

OBLIQUE
(a)

PERSPECTIVE
(b)

Fig. 11.37 Comparison of oblique and perspective.

Oblique
(a)

(b)

Cabinet
(c)

Fig. 11.38 Foreshortening in the direction of the receding axis.

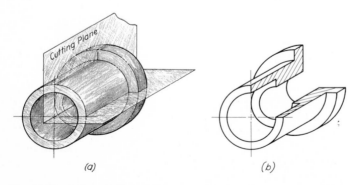

(a)

(b)

Fig. 11.39 Oblique half section.

for any practical purpose. As a matter of interest, the effect of distortion is due to the fact that the receding lines are parallel and do not appear to converge as the eye is accustomed to anticipating (Fig. 11.37).

The appearance of excessive thickness can be overcome somewhat by reducing the length of the receding lines. For practical purposes, measurements are usually reduced one-half, but any scale of reduction may be arbitrarily adopted if the view obtained will be more realistic in appearance. When the receding lines are drawn one-half their actual length, the resulting pictorial view is called a *cabinet drawing*. Figure 11.38 shows an oblique drawing (*a*) and a cabinet drawing (*c*) of the same object, for the purpose of comparison.

11.24
Oblique Sectional Views. Oblique sectional views are drawn to show the interior construction of objects. The construction procedure is the same as for an isometric sectional view, except that oblique planes are used for cutting the object. An oblique half section is illustrated in Fig. 11.39.

11.25
Pictorial Dimensioning. The dimensioning of isometric and other forms of pictorial working drawings is done in accordance with the following rules:

1. Draw extension and dimension lines (except those dimension lines applying to cylindrical features) parallel to the pictorial axes in the plane of the surface to which they apply (Fig. 11.40).

2. If possible, apply dimensions to visible surfaces.

3. Place dimensions on the object, if, by so doing, better appearance, added clearness, and easy readings result.

4. Notes may be lettered either in pictorial or as on ordinary drawings. When lettered as on ordinary drawings the difficulties encountered in forming pictorial letters are avoided (Fig. 11.40).

5. Make the figures of a dimension appear to be lying in the plane of the surface whose dimension it indicates, by using vertical figures drawn in pictorial (Fig. 11.40). (*Note:* Guide lines and slope lines are drawn parallel to the pictorial axes.)

11.26
Conventional Treatment of Pictorial Drawings. When it is desirable for an isometric

or an oblique drawing of a casting to present a somewhat more or less realistic appearance, it becomes necessary to represent the fillets and rounds on the unfinished surfaces. One method commonly used by draftsmen is shown in Figure 11.41(a). On the drawing in (b) all of the edges have been treated as if they were sharp. The conventional treatment for threads in pictorial is illustrated in (b) and (c).

Fig. 11.40 Extension and dimension lines in isometric (left); numerals, fractions, and notes in oblique (right).

C □ Perspective projection

11.27

Perspective. In perspective projection an object is shown much as the human eye or camera would see it at a particular point. Actually, it is a geometric method by which a picture can be projected on a picture plane in much the same way as in photography. Perspective drawing differs from the methods previously discussed in that the projectors or visual rays intersect at a common point known as the *station point* (Fig. 11.44).

Since the perspective shows an object as it appears instead of showing its true shape and size, it is rarely used by engineers. It is more extensively employed by architects to show the appearance of proposed buildings, by artist–draftsmen for production illustrations, and by illustrators in preparing advertising drawings.

Figure 11.1 shows a type of production illustration that has been widely used in assembly departments as an aid to those persons who find it difficult to read an orthographic assembly. This form of presentation, which may show a mechanism both exploded and assembled, has made it possible for industrial concerns to employ semitrained personnel. Figure 11.56 shows a type of industrial drawing made in perspective that has proved useful in aircraft plants. Because of the growing importance of this type of drawing, and also because engineers frequently will find perspective desirable for other purposes, its elementary principles should be discussed logically in this text. Other books on the subject, some of which are listed in the bibliography, should be studied by architectural students and those interested in a more thorough discussion of the various methods.

The fundamental concepts of perspective can be explained best if the reader will imagine himself looking through a picture

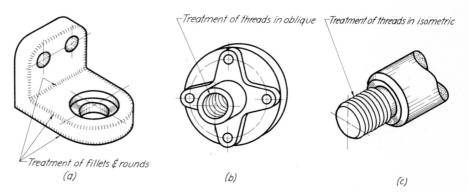

Fig. 11.41 Conventional treatment of fillets, rounds, and threads in pictorial.

plane at a formal garden with a small pool flanked by lampposts, as shown in Fig. 11.42. The point of observation, at which the rays from the eye to the objects in the scene meet, is called the *station point*, and the plane on which the view is formed by the piercing points of the visual rays is known as the *picture plane* (PP). The piercing points reproduce the scene, the size which depends on the location of the picture plane.

It should be noted that objects of the same height intercept a greater distance on the picture plane when close to it than when farther away. For example, rays from the lamppost at 2 intercept a distance 1–2 on the picture plane, while the rays from the pole at 4, which actually is the same height, intercept the lesser distance 3–4. From this fact it should be observed that the farther away an object is, the smaller it will appear, until a point is reached at which there will be no distance intercepted at all. This happens at the horizon.

Fig. 11.42 Picture plane.

Fig. 11.43 The picture (perspective).

Figure 11.43 shows the scene observed by the man in Fig. 11.42 as it would be formed on the picture plane. The posts farther from the picture plane diminish in height, as each one has a height on the picture plane equal to the distance it intercepts, (Fig. 11.42). The lines of the pool

and hedge converge to the center of vision or vanishing point, which is located directly in front of the observer on the horizon.

11.28
Perspective Nomenclature. Figure 11.44 illustrates pictorially the accepted nomenclature of perspective drawing. The *horizon line* is the line of intersection of the horizontal plane through the observation point (eye of the observer) and the picture plane. The horizontal plane is known as the *plane of the horizon*. The *ground line* is the line of intersection of the ground plane and the picture plane. The *CV point* is the center of vision of the observer. It is located directly in front of the eye in the plane of the horizon on the horizon line.

11.29
Location of Picture Plane. The picture plane is usually placed between the object and the *SP* (station point). In parallel perspective (Sec. 11.35) it may be passed through a face of the object in order to show the true size and shape of the face.

11.30
Location of the Station Point. Care must be exercised in selecting the location for the station point, for its position has much to do with the appearance of the finished perspective drawing. A poor choice of position may result in a distorted perspective that will be decidedly displeasing to the eye.

In general, the station point should be offset slightly to one side and should be located above or below the exact center of the object. However, it must be remembered that the center of vision must be near the center of interest for the viewer.

One should always think of the station point as the viewing point, and its location should be where the object can be viewed to the best advantage. It is desirable that it be at a distance from the picture plane equal to at least twice the maximum dimension (width, height, or depth) of the object, for at such a distance, or greater, the entire object can be viewed naturally, as a whole, without turning the head.

A wide angle of view is to be avoided in the interest of good picturization. It has been determined that best results are obtained when the visual rays from the station point (*SP*) to the object are kept within a cone having an angle of not more than 30° between diametrically opposing elements (see Fig. 11.45).

In locating an object in relation to the picture plane, it is advisable to place it so that both of the side faces do not make the same angle with the picture plane and thus will not be equally visible. It is common practice to choose angles of 30° and 60° for rectangular objects.

11.31

Position of the Object in Relation to the Horizon. When making a perspective of a tall object, such as a building, the horizon usually is assumed to be at a height above the ground plane equal to the standing height of a man's eye, normally about 5 ft 6 in.

A small object may be placed either above or below the horizon (eye level), depending on the view desired. If an object is above the horizon, it will be seen looking up from below, as shown in Fig. 11.46. Should the object be below the horizon line, it will be seen from above.

11.32

Lines. The following facts should be recognized concerning the perspective of lines:

1. Parallel horizontal lines vanish at a single *VP* (vanishing point). Usually the *VP* is at the point where a line parallel to the system through the *SP* pierces the *PP* (picture plane).

2. A system of horizontal lines has its *VP* on the horizon.

3. Vertical lines, since they pierce the picture plane at infinity, will appear vertical in perspective.

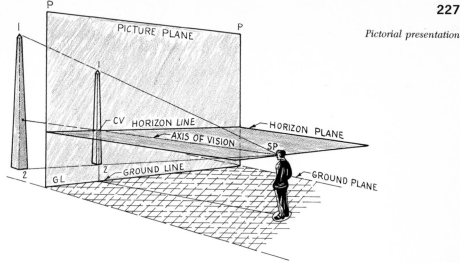

Fig. 11.44   Nomenclature.

4. When a line lies in the picture plane, it will show its true length because it will be its own perspective.

5. When a line lies behind the picture plane, its perspective will be shorter than the line.

11.33

Perspective by Multiview Projection. Adhering to the theory that a perspective drawing is formed on a picture plane by visual rays from the eye to the object, as illustrated in Fig. 11.42, a perspective can be drawn by using multiview projection (see Fig. 11.47). The multiview method may be the easiest for a student to understand but it is not often used by an expe-

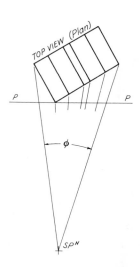

Fig. 11.45   Angle of vision.

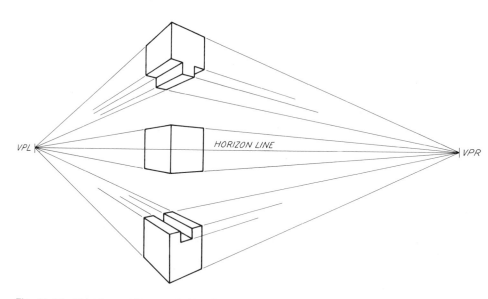

Fig. 11.46   Objects on, above, or below the horizon.

Fig. 11.47 Perspective drawing—orthographic method.

After the preliminary layout has been completed point A may be located in the perspective by the following procedure:

> **Step I.** Draw the top view ($SP^H a^H$) and the side view ($SP^P a^P$) of the visual ray from the eye to point A.
>
> **Step II.** From a' (the top view of the piercing point of the ray) draw a projection line downward.
>
> **Step III.** From a'' (the side view of the piercing point) draw a horizontal projection line to intersect the one drawn from a'.

Point A of the perspective is at this intersection.

Point B and the other points that are needed for the perspective representation are found in the same manner as point A.

11.34

Types of Perspective. In general, there are two types of perspective: *parallel perspective* and *angular perspective*. In parallel perspective, one of the principal faces is parallel to the picture plane and is its own perspective. All vertical lines are vertical, and the receding horizontal lines converge to a single vanishing point. In angular perspective, the object is placed so that the principal faces are at an angle with the picture plane. The horizontal lines converge at two vanishing points.

11.35

Parallel Perspective. Figure 11.48 shows the parallel perspective of a rectangular block. The PP line is the top view of the picture plane, SP_H is the top view of the station point, and CV is the center of vision. The receding horizontal lines vanish at CV. The front face, since it lies in the picture plane, is its own perspective and shows in its true size. The lines representing the edges back of the picture plane are found by projecting downward from the points at which the visual rays pierce the picture plane, as shown by the top views of the rays. Figure 11.49 shows a parallel perspective of a cylindrical machine part.

11.36

Angular Perspective. Figure 11.50 shows pictorially the graphical method for the preparation of a two-point perspective drawing of a cube. To visualize the true layout on the surface of a sheet of drawing paper, it is necessary to revolve mentally the horizontal plane downward into the vertical or picture plane. On completion of

Fig. 11.48 Parallel perspective.

rienced person because considerably more line work is required if the scene or object to be represented is at all complicated. The top and side views are drawn in multiview projection and the picture plane (as an edge view) and the station point are shown in each case. SP^H and its related views of the rays are in the top view, while SP^P and its projections of the rays belong to the side view. The front view of the picture plane is in the plane of the paper.

Fig. 11.49 Circles in parallel perspective.

Sec. 11.36, it is suggested that the reader turn back and endeavor to associate the development of the perspective in Fig. 11.51 with the pictorial presentation in Fig. 11.50. For a full understanding of the construction in Fig. 11.51, it is necessary to differentiate between the lines that belong to the horizontal plane and those that are on the vertical or picture plane. In addition, it must be fully realized that there is a top view for the perspective that is a line and that in this line view lie the points that must be projected downward to the perspective representation (front view).

Figure 11.51 shows an angular perspective of a block. The block has been placed so that one vertical edge lies in the picture plane. The other vertical edges are parallel to the plane, while all of the horizontal lines are inclined to it so that they vanish at the two vanishing points, VPL and VPR, respectively.

In constructing the perspective shown in this illustration, an orthographic top view was drawn in such a position that the visible vertical faces made angles of 30° and 60° with the picture plane. Next, the location of the observer was assumed and the horizon line was established. The vanishing points VPL and VPR were found by drawing a 30° line and a 60° line through the SP. Since these lines are parallel to the two systems of receding horizontal lines, each will establish a required vanishing point at its intersection with the picture plane. The vertical line located in the picture plane, which is its own perspective, was selected as a measuring line on which to project vertical measurements from the orthographic front view. The lines shown from these division points along this line to the vanishing points (VPL–VPR) established the direction of the receding horizontal edge lines in the perspective. The positions of the back edges were determined by projecting downward from the points at which the projectors from the station point (SP) to the corners of the object pierced the picture plane, as shown by the top view of the object and projectors.

11.37
Use of Measuring Lines. Whenever the vertical front edge of an object lies in the picture plane, it can be laid off full length in the perspective, because theoretically, it will be in true length in the picture formed on the plane by the visual rays (see Fig. 11.51). Should the near vertical edge lie

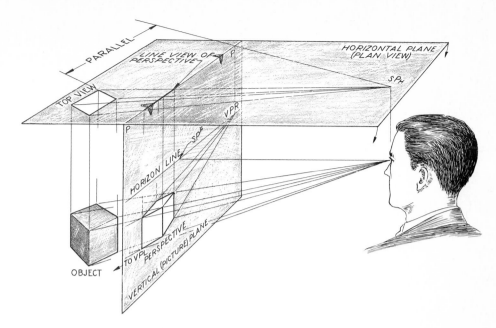

Fig. 11.50 Angular perspective.

behind the picture plane, as is the case with the edge line AB in Fig. 11.52, the use of a measuring line becomes desirable. The measuring line $a'b'$ is the vertical edge AB moved forward to the picture plane, where it will appear in its true height. Some prefer to think of the vertical side as being extended to the picture

Fig. 11.51 Angular perspective.

Fig. 11.52 Use of measuring lines.

the picture plane (point X) is projected downward to the front-view picturization. Points A and B must fall on the vanishing lines from a' and b' to VPR respectively.

A measuring line may be used to establish the "picture height" of any feature of an object. For example, in Fig. 11.52, the vertical measuring line through c' was used as needed to locate point C and the top line of the object in the perspective.

11.38

Vanishing Points for Inclined Lines. In general, the perspective of any inclined straight line can be found by locating the end points of the line perspectively. By this method, an end point may be located by drawing the perspective representations of any two horizontal lines intersecting at the end point. Where several parallel inclined lines vanish to the same vanishing point, it may prove to be worthwhile to locate the VP for the system in order to conserve time and achieve a higher degree of accuracy.

Just as in the case of other lines, the vanishing point for any inclined line may be established by finding the piercing point in the picture plane (PP) of a line through SP parallel to the given inclined line.

In Fig. 11.53 the inclined lines are AB and CD. If one is to understand the construction shown for locating vanishing point VPI, he must recognize that vertical planes vanish in vertical lines and that a line in a vertical plane will vanish at a point on the vanishing line of the plane. The preceding statements being true, the vanishing point VPI for line CD must lie at some point on a vertical line through VPR, since CD and CE lie in the same vertical plane. VPI is the point at which a line drawn parallel (in space) to CD through SP pierces the picture plane.

On the drawing, the distance D that VPI is above VPR may be found easily through the construction of a right triangle with $SP^H Q$ as a base. Angle β is the slope angle for line CD (see side view). The needed distance D is the line QR, or the short leg of the triangle.

11.39

Use of Measuring Points. Frequently, it is desirable to use measuring points in preparing a perspective drawing, because their use permits the laying off of a series of direct measurements that can be transferred quickly and accurately to the perspective. Measuring points are, in real-

plane so that the true height of the side is revealed. The length and position of AB is established in the perspective picture by first drawing vanishing lines from a' and b' to VPR; then the top view of the edge in

Fig. 11.53 Vanishing point for inclined lines.

ity, special vanishing points that are used to establish distances in perspective along perspective lines.

To understand the measuring point method, one must realize that, in theory, a vertical face is rotated into the picture plane so that direct measurements may be made along the horizontal ground line to the same scale as the top view. In Fig. 11.54 the vertical face containing line AB was revolved into the picture plane to position ab_r. The vanishing point of the line bb_r is MPR. The position of MPR was established by first drawing a line from SP parallel to bb_r to a piercing point in the picture plane and then projecting to the horizon line.

Another method for determining the location of measuring points is included in this same illustration. By this second method, the location of a measuring point can be readily found by swinging an arc with the plan view of the vanishing point as a center and using a radius equal to the distance from this center to the SP. For example, the location of MPR on the horizon was determined by swinging an arc from Y using the distance from Y to SP as a radius. The point of intersection of this arc and picture plane (in plan) projected to the horizon established the position of MPR for use in preparing the perspective drawing shown. MPL was similarly located.

The scaled lengths (D_1, D_2, and D_3) laid off along ab_r, when vanished to MPR, established the desired lengths along the line AB, which represents the perspective of the lower edge of the object. Height distances were laid off along the measuring line through the vertical edge lying in the picture plane at a. With width, height, and depth distances established, the perspective drawing was completed by following the procedure given in Sec. 11.36.

11.40

Circles in Perspective. If a circle is on a surface that is inclined to the picture plane (PP), its perspective will resemble an ellipse. It is the usual practice to construct the representation within an enclosing square by finding selected points along the curve in the perspective, as shown in Fig. 11.55(a). Any points might be used, but it is recommended that points be located on 30° and 60° lines.

In (a), the perspective representation of the circle was found by using visual rays and parallel horizontal lines in combination. In starting the construction, the po-

Fig. 11.54 Use of measuring points.

sitions of several selected points, located on the circumference and lying on horizontal lines in the plane of the circle, were established in both views. After these lines had been drawn in the perspective in the usual manner, the locations of the points along them were determined through the use of visual rays, as shown. Specifically,

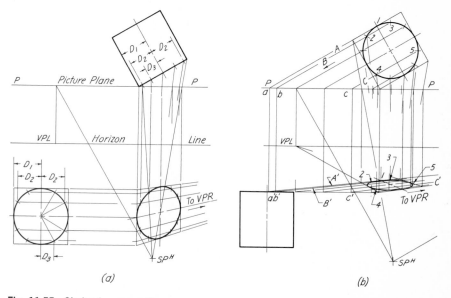

Fig. 11.55 Circles in perspective.

the position of a point in the perspective was found by projecting downward from the piercing point of the ray from the point and the picture plane (see line view) to the perspective view of the line on which the point must lie. In (*b*), the same method was applied to construct the perspective view of a circle in a horizontal plane. It should be noted in this case that the horizontal lines, as established in the top view, were extended to cut the picture plane so that the true height of these lines at the plane could be used in the perspective view for locating the end points of the perspective representations (at the left).

D □ Industrial illustrations

11.41

Technical Illustration Drawings. The design and manufacturing procedures of present-day mass production require various types of pictorial illustrations to communicate ideas and concepts to large numbers of persons who are all working toward a common objective. These illustration drawings are of value at all stages of a project from the design phase, where they may be only pencil-shaded freehand sketches, through all of the stages of production that we may consider to include not only the assembly but final installa-

tion of the systems as well. They are used in operation and maintenance manuals to make complex and difficult tasks understandable to those persons who may be unable to interpret conventional drawings (Fig. 11.56). Pictorial illustrations range from simple types of line drawings (Figs. 11.57 and 11.58) that have already been discussed to artists' renderings that have the realism of photographs. An artist's rendering, such as the one shown in Fig. 11.60, is usually prepared to reinforce an oral or written report that is to be presented before a decision-making group that consists to some extent of persons who would otherwise be unable to understand construction details. Drawings of this type are depended on to sell a project (Fig. 11.61).

Illustration drawings are used for many purposes in every field of engineering technology (Fig. 11.59). They appear in advertising literature, operation and service manuals, patent applications (Fig. 11.64), and textbooks (Fig. 11.71). Illustration drawings may be working drawings, assembly drawings, piping and wiring diagrams, and architectural and engineering renderings that are almost true-to-life. Typical examples of pictorial drawings that were prepared to facilitate assembly are shown in Figs. 11.1 and 16.11. Figure 22.1 shows an electronic diagram that appeared in a service manual.

Fig. 11.56 Production illustration prepared in perspective. (*Courtesy Craftint Mfg. Co.*)

**EXPERIMENTAL
CATALYTIC CONVERTER**

Fig. 11.57 **This pictorial representation of a catalytic converter appeared in a technical report prepared by General Motors.** Pictorial drawings such as this one are often needed in final design reports. Some engineers and technologists think that the utilization of a complex combination of components along with a catalytic converter in the exhaust line will prove to be the most effective method of treating hydrocarbons and carbon monoxide. (*Courtesy General Motors Corporation*)

Fig. 11.58 **Items other than the catalytic converter that are a part of the total package of emission control are shown in the pictorial representation of the advanced emission control system under development at General Motors.** This presentation was extracted from a technical report prepared by GM engineers. (*Courtesy General Motors Corporation*)

**ADVANCED EMISSION CONTROL SYSTEM
(JANUARY 1972)**

QUICK HEAT MANIFOLD

IMPROVED CARBURETION AND CHOKE ALTITUDE AND TEMPERATURE COMPENSATION

OXIDIZING CATALYTIC CONVERTER

AIR INJECTION PUMP

EXHAUST GAS RECIRCULATION

MODIFIED SPARK TIMING

PCV

DOMED TANK VAPOR SEPARATOR

ELECTRONIC IGNITION

CARBON CANISTER

Fig. 11.59 **Industrial illustration.** (*Courtesy Ford Motor Company*)

1 CAP — GAS GENERATOR PROTECTIVE
2 BODY — GAS GENERATOR
3 PIPE PLUG SOCKET
4 PLUNGER — FILL VALVE
5 TANK — PRESSURE (66 CU IN)
6 SPRING — FILL VALVE
7 RETAINER — FILL VALVE SPRING
8 HOUSING — GAS GENERATOR
9 NOZZLE — GENERATOR
10 POWDER CHARGE
11 DISC — RUPTURE
12 SPACER — GAS GENERATOR
13 GUARD — ORIFICE PLUG

14 RETAINER — ACTUATOR
15 ORIFICE PLUG
16 ACTUATOR ASM
17 DIFFUSER ASM
18 KNEE BAG
19 VALVE — KNEE BAG
20 PLATE — KNEE BAG RETAINER
21 MAIN BAG
22 BOARD — GAS GENERATOR CIRCUIT
23 PRESSURE SWITCH
24 GASKET — CAP PROTECTIVE
25 INITIATOR ASM

Fig. 11.60 Components of the Allied Chemical Corporation air-bag modules to be used in the air-bag/seat-belt restraint system now under development by the Ford Motor Company. The system has a main bag (cushion) and a small knee bag. (*Courtesy Ford Motor Company*)

11.42

Design Illustrations. Design illustrations are prepared to clarify conventional engineering design and production drawings and written specifications. These are used for the communication of ideas and concepts concerning the details of compli-cated designs. Properly prepared, drawings of this type reveal the relationship of the components of a system so clearly that the principles of operation of a unit can be understood by almost everyone, even by persons who may be relatively unfamiliar with graphic methods. It is common practice to prepare a series of such drawings to clarify complex details of construction, to indicate the function of closely related parts, and reveal structural features. A pictorial of this type is shown in Fig. 11.62. It was used to supplement a technical paper presented at a recent Society of Automotive Engineers Congress held in Detroit.

11.43

Shading Methods. A pictorial illustration can be improved and given a more realistic appearance by shading to produce the effect of surface texture. To this shades and shadows may be added to give additional realism. The use of pencil shading is most common. Ink shading (see Fig. 11.71) produces clean illustrations of high quality that are well suited for reproductions in texts and brochures. The techniques of surface shading by means of ink lines will be discussed in Sec. 11.44.

Some of the other more basic methods of representing surface textures under light and shade involve the use of Ross-board, Craftint paper, Zip-a-tone overlay film, and the airbrush.

Most of the shaded pictorial drawings in

Fig. 11.61 Artist's concept of a rocket control system. (*Courtesy Aeronutronic Division, Philco-Ford*)

Fig. 11.62 New combustion process developed by the Ford Motor Company, cutaway view. (*Courtesy Ford Motor Company*)

Fig. 11.63 Pictorial representation of the principal elements of the inertial guidance system in a missile vehicle. (*Courtesy General Motors Corporation*)

this text were drawn on Rossboard. This popular drawing paper, with its rough plaster-type surface, is available in many textured patterns. Surface shading is done with a very soft pencil.

Craftint papers, single-tone or double-tone, have the pattern in the paper. Drawings are prepared on these papers in the usual manner and regular black water-proof drawing ink is used for the finished lines. Solid black areas are filled in with ink before the areas where shading is desired are brushed with a developer that brings out the surface pattern. These papers are available in many shading patterns. The drawing shown in Fig. 11.56 was prepared on Craftint paper.

Zip-a-tone clear cellulose overlay screens with printed shading patterns of dots and lines provide an easy method of surface shading suitable for high-quality printed reproductions (Fig. 11.63). The screen, backed with a clear adhesive, is applied as a sheet or partial sheet to the areas to be shaded. Unwanted portions are removed with a special cutting needle or razor blade, before the screen is rubbed down firmly to complete the bond between screen and paper.

A high degree of realism can be achieved using a small spray gun that in the language of the artist is known as an *airbrush.* This delicate instrument sprays a fine mist of diluted ink over the surface of the drawing to produce variations of

tone. A capable artist can produce a representation that has the realism of a photograph. The pictorial illustration in Fig. 11.61 is an excellent example of airbrush rendering by a commercial illustrator. The airbrush is sometimes used to retouch photographs to improve their appearance for reproduction.

11.44
Surface Shading by Means of Lines (Fig. 11.64). Line shading is a conventional method of representing, by ruled lines, the varying degrees of illumination on the surfaces of an object. It is a means of giving clearer definition to the shapes of objects and a finished appearance to certain types of drawings. In practice, line shading is used on Patent Office drawings, display drawings, and on some illustrations prepared for publications. It is never used on ordinary drawings, and for this reason few draftsmen ever gain the experience necessary to enable them to employ it effectively.

In shading surfaces, the bright areas are left white and the dark areas are represented by parallel shade lines (Figs. 11.65 and 11.66). Varying degrees of shade may be represented in one of the following ways:

1. By varying the weight of the lines while keeping the spacing uniform, as in Fig. 11.65.

Fig. 11.64 Patent drawing with surface shading.

Fig. 11.65 Surface shading on a prism.

Fig. 11.66 Surface shading on a cylinder.

Fig. 11.67 Pyramid.

Fig. 11.68 Cone.

Fig. 11.69 Surface shading on a sphere.

2. By using uniform straight lines and varying the spacing.

3. By varying both the weight of the lines and the spacing, as in Fig. 11.66.

The rays of light are assumed to be parallel and coming from the left, over the shoulder of the draftsman (Fig. 11.65). In accordance with this, two of the visible faces of the hexagonal prism, shown in Fig. 11.65, would be illuminated, while the remaining visible inclined face would be dark. It should be noted that the general principle of shading is modified in the case of flat inclined surfaces, which, theoretically, would be uniformly lighted. Such surfaces are shaded in accordance with a conventional scheme, the governing rule of which may be stated as follows: *The portion of an illuminated inclined surface nearest the eye is the lightest, while the portion of a shaded inclined surface nearest the eye is the darkest.* In the application of this rule, the shading on an inclined illuminated face will increase in density as the face recedes, while on an unilluminated inclined surface it will decrease in density as the face recedes, as shown in Fig. 11.65.

A cylinder would be shaded as shown in Fig. 11.66. The lightest area on the surface is at the *brilliant line*, where light strikes it and is reflected directly to the eye; the darkest is along the *shade line*, where the light rays are tangent to the surface. The brilliant line passes through the point at which the bisector for the angle formed by a light ray to the center and the visual ray from the center line to the eye would pierce the external surface. In shading a cylinder, the density of the shading is increased in both directions, from the bright area along the brilliant line

to the contour element on the left and the shade line on the right. The density of the shading is slightly decreased on the shaded portion beyond the shade line. Since one would expect this area to be even darker, some draftsmen extend the dark portion to the contour element. On very small cylinders, the bright side (left side) usually is not shaded.

A pyramid may be shaded as shown in Fig. 11.67.

The surface of a cone will be the darkest at the element where the light rays are tangent to the surface (Fig. 11.68).

A sphere is shaded by drawing concentric circles having either the geometric center of the view or the "brilliant point" as a center (Fig. 11.69). The darkest portion of the surface is along the shade line, where the parallel rays of light are tangent to the sphere and the lightest portion is around the *brilliant point*. The *brilliant point* is where the bisector of the angle between a light ray to the center and the visual ray from the center to the eye pierces the external surface. The construction necessary to determine the dark line and the brilliant point is shown in Fig. 11.69(*b*).

Although good line shading requires much practice and some artistic sense, a skillful draftsman should not avoid shading the surfaces of an object simply because he never before has attempted to do so. After careful study, he should be able to produce fairly satisfactory results. Often the shading of a view makes it possible to eliminate another view that otherwise would be necessary.

Figures 11.70 and 11.71 show applications of line shading on technical illustrations.

Fig. 11.70 Line-shaded drawing. (*From Doyle, Tool Engineering, Prentice-Hall, Inc.*)

Fig. 11.71 Line-shaded pictorial drawing. (*Courtesy Socony-Vacuum Oil Co.*)

Problems

The student will find that a preliminary sketch will facilitate the preparation of isometric and oblique drawings of the problems of this chapter. On such a sketch he may plan the procedure of construction. Since many technologists and designers frequently find it necessary to prepare pictorial sketches during discussions with untrained persons who cannot read orthographic views, it is recommended that some problems be sketched freehand on either plain or pictorial grid paper (see Fig. 6.34). Additional problems may be found at the end of Chapter 6.

1–9. (Figs. 11.72–11.80). Prepare instrumental isometric drawings or freehand sketches of the objects as assigned.

10–14. (Figs. 11.76–11.80). Prepare instrumental oblique drawings or freehand sketches of the objects as assigned.

15. (Fig. 11.81). Make an isometric drawing of the differential spider.

Fig. 11.72

Fig. 11.75

Fig. 11.78

Fig. 11.73

Fig. 11.76

Fig. 11.79

Fig. 11.74

Fig. 11.77

Fig. 11.80

Fig. 11.81 Differential spider.

Fig. 11.82 Stepladder.

16. (Fig. 11.82). Make an isometric drawing of the stepladder. Select a suitable scale.

17. (Fig. 11.83). Make an isometric drawing of the sawhorse. Select a suitable scale.

18. (Fig. 11.84). Make an oblique drawing of the locomotive driver nut.

19. (Fig. 11.85). Make an oblique drawing of the adjustment cone.

20. (Fig. 11.86). Make an oblique drawing of the fork.

21. (Fig. 11.87). Make an oblique drawing of the feeder guide.

22. (Fig. 11.88). Make an isometric drawing of the hinge bracket.

23. (Fig. 11.89). Make an isometric drawing of the alignment bracket.

24. (Fig. 11.90). Make an isometric drawing of the stop block.

25. (Figs. 11.85–11.87). Make a parallel perspective drawing as assigned.

26. (Figs. 11.88–11.90). Make an angular perspective drawing as assigned.

27. (Fig. 11.91). Make a pictorial drawing (oblique, isometric, or perspective) of the slotted bell crank.

28. (Fig. 11.92). Make a pictorial drawing (oblique, isometric, or perspective) of the control guide.

Fig. 11.83 Sawhorse.

Fig. 11.84 Locomotive driver nut.

Dimensions are in millimeters

Fig. 11.85 Adjustment cone.

Fig. 11.86 Fork.

Fig. 11.87 Feeder guide.

Fig. 11.88 Hinge bracket.

Fig. 11.89   Alignment bracket.

Fig. 11.90   Stop block.

(Do not show fillets and rounds)

Fig. 11.91   Slotted bell crank.

Fig. 11.92   Control guide.

PART 3

DESIGN

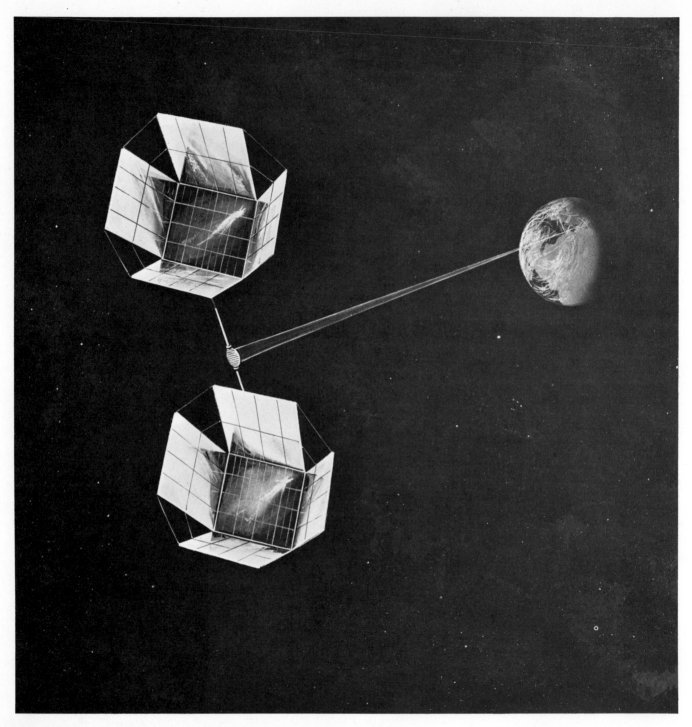

"Four major U.S. corporations are now working together to investigate the feasibility of large scale satellite systems that would harness the abundant energy of the sun for use on earth. . . .

"A satellite solar power station of the size presently under consideration would be capable of producing 10,000 megawatts which is sufficient energy to meet the base load demands of a city such as New York in the year of 2000. A network of satellites such as these would have the potential to meet a significant portion of future energy needs. More importantly, it would be virtually pollution-free and would not consume irreplaceable energy sources on earth." (*Courtesy Grumman Corporation—one of four companies taking part in the investigation*)

the design process and graphics 12

A □ The Design process

12.1
Design. In the dictionary, design is defined as follows: (1) to form or conceive in the mind, (2) to contrive a plan, (3) to plan and fashion the form of a system (structure), and (4) to prepare the preliminary sketches and/or plans for a system that is to be produced. Engineering design is a decision-making process used for the development of engineering systems for which there is human need (Fig. 12.1). To design is to conceive, to innovate, to create. One may design an entirely new system or modify and rearrange existing things in a new way for improved usefulness or performance. Engineering design begins with the recognition of a social or economic need (Fig. 12.2). The need must first be translated into an acceptable idea by conceptualization and decision making. Then the idea must be tested against the physical laws of nature before one can be certain that it is workable. This requires that the designer have a full knowledge of the fundamental physical laws of the basic sciences, a working knowledge of the engineering sciences, and the ability to communicate ideas both graphically (Fig. 12.21) and orally. The designer should be well grounded in economics, have some knowledge of engineering materials and be familiar to a limited extent with manufacturing methods. In addition, some knowledge of both marketing and advertising will prove worthwhile, since usually what is produced must be distributed at a profit. Proficiency in designing can be attained only through total involvement,

243

FUEL NOZZLE

COMBUSTION CHAMBER

DISPLACER PISTON

COOLED EXHAUST
OUTLET

HOT EXHAUST

COMBUSTION AIR
INLET

REGENERATOR

CYLINDER

POWER PISTON

Fig. 12.1 Sterling engine.
The Sterling engine is an interesting engine that can be considered
as an alternative for use in the automobile. It is considerably
different from the Wankel. The cross-sectional view of a single-
cylinder engine shows its complexity. The engine has rather low
hydrocarbon and carbon monoxide emissions. It tends to be rather
heavy and expensive to produce at this time. (*Courtesy General
Motors Corporation*)

since it is only through practice that the designer acquires the art of continually providing new and novel ideas. In developing the design, the engineer or engineering technologist must apply his knowledge of engineering and material sciences while taking into account related human factors, reliability, visual appearance, manufacturing methods, and sale price. It may therefore be said that the ability to design is both an art and a science.

A creative person will almost never follow a set pattern of action in developing an idea. To do so would tend to structure his thinking and might limit the creation of possible solutions. The design process calls for unrestrained creative ingenuity and continual decision making by a free-wheeling mind. However, the total development of an idea, from recognition of a need to the final product, does appear to proceed loosely in stages that are recognized by authors and educators (Fig. 12.2).

Creative thinking usually begins when a design team, headed by a project leader, has been given an assignment to develop something that will satisfy a particular need. The need may have been suggested by a salesman, a housewife, or even an engineer from another company now using a product or a machine produced by the design team's own company. Most often the directive will come down from top management, as was the case with the development of the electric knife some years ago. Although it is always pleasant for an individual to think about the careers of famous inventors of the past and dream of the fame and fortune that might await the development and marketing of one of his ideas, the fact is that almost all new and improved products, from food choppers to aircraft engines, represent a team effort.

12.2
Design Synthesis. The process of combining constituent elements in a new or altered arrangement to achieve a unified entity is known as design synthesis. It is a process that involves reasoning from assumed propositions and known principles to arrive at comparatively new design solutions to recognized problems. The synthesis of systems for simple combinations as well as for complicated assemblies requires creative ability of the very highest order. The synthesis of both parts and systems usually requires successive trials to create new arrangements of old components and new features.

Proof of this point is the Land camera, considered by many people to be an entirely new product, although in reality the camera represents a combination of features and principles common to existing cameras to which Mr. Land added several new ideas of his own, including a new type of film and film pack that for the first time made it possible to develop film and pictures within the camera itself. Land's design activity no doubt started with a recognition of the need and desire that people had to take pictures that could be seen almost immediately. The early automobile is another example that bears out the fact that old established products have features that are used as a starting basis for a new product. For example, at the turn of the century the automobile looked like a horseless carriage; the horse was taken to the barn and a motor was added in its place.

12.3

Design of Systems and Products. In general most design problems may be classified as being either a systems design or a product design, even though it may be quite difficult in many instances to recognize a problem as belonging entirely to one classification or the other. This is due to the fact that there often will be an overlap of identifying characteristics.

A systems design problem involves the interaction of numerous components that together form an operating unit. A complex system such as the climatic-control system for an automobile (heating and cooling), an automatic movie projector, a stadium, an office building, or even a parking lot represents a composition of several component systems that together form the complete composite system (Fig. 12.3). Some of the component systems of an office building, aside from the structural system itself, are: the electrical system, the plumbing system (including sewers), the heating and cooling systems, the elevator system, and the parking facilities system. All of these component systems, when combined, will meet the needs of a total system; but usually the total system design will involve more than just a technological approach. In the design of composite systems for use by the public, as in the design of a highway, an office building, or a stadium, the designer must adhere to the availability of funds. He must build into his design the safety features that are required by law and, at the same time, give due consideration to human factors, trends, and even present-day social problems. In many cases, a designer will find that there are political and special-interest groups that limit his freedom of decision making. Under such conditions he must be willing to compromise in the best interests of all concerned. This means that a successful designer must have experience in dealing with people. Social studies courses taken along with and in addition to the scientific courses offered in engineering colleges and technical schools will prove to be helpful.

Product design is concerned with the design of some appliances, systems components, and other similar unit-type items for which there appears to be a market. Such a product may be an electric lock, a power car-jack, a lawn sprinkler, a food grinder, a toy, a piece of specialized furniture, an electrical component, a valve, or other item that can be readily marketed as a commercial unit. Such products are designed to perform a specific function and to satisfy a particular need. Total product design includes not only the design of an item, but the testing, manufacture, and distribution of the product as well. Design does not end with the solution of a problem through creative thinking. The design phase for a product to be sold on the open market ends only when the item has received wide acceptance by the public. Of course, if the item being designed is not for the consumer market, the designer will have less reason to be concerned with details of marketing. This would be true in the case of components for systems and for items to be used in products produced in finished form by other companies.

After initially recognizing a need or desire for a new product, the creative individual enters the next step in the design procedure—the research and exploration phase. During this phase, the possible ways and the feasibility of fulfilling the need are investigated (Fig. 12.4). That is, the question is raised at this point as to whether the contemplated product can be marketed at a competitive price or at a price the public will be willing to pay. Except in the case of the public sector, where a system is being designed under the direction of a government unit (Fig. 12.5),

```
┌──────────────────────────┐
│   PROBLEM RECOGNITION     │
└──────────────────────────┘
            ⇓
┌──────────────────────────┐
│   PRELIMINARY IDEAS       │
└──────────────────────────┘
            ⇓
┌──────────────────────────┐
│   CONCEPTUALIZATION       │
│   AND REFINEMENT          │
└──────────────────────────┘
            ⇓
┌──────────────────────────┐
│   ANALYSIS                │
└──────────────────────────┘
            ⇓
┌──────────────────────────┐
│   DESIGN DESCRIPTION      │
└──────────────────────────┘
            ⇓
┌──────────────────────────┐
│   DECISION                │
└──────────────────────────┘
            ⇓
┌──────────────────────────┐
│   IMPLEMENTATION          │
│   (Working Drawings)      │
└──────────────────────────┘
```

Fig. 12.2 The design process.

Fig. 12.3 Xerox 400 Telecopier transceiver.
This transceiver marks the beginning of a new era in facsimile transmission. It sends and receives, over regular telephones, a letter-size document or drawing in approximately 4 minutes. (*Courtesy Business Products Group—Xerox Corporation*)

Fig. 12.4 Minigap System.
Creative thinking produces new ways of life for man. The Minigap System links cars together into a caravan that is led by a specially built leader vehicle. Computers inside the cars take over control of brakes, accelerator, and steering to free the motorist from the task of driving as long as he remains hooked-up. Caravans can mix in the traffic flow with other vehicles. (*Courtesy Ford Motor Company*)

profit is the name of the game. A search of literature and patent records will often show that there have been previous developments with the possibility of infringements on the design and patents of other companies. The discovery that others already may have legal protection for one or more of the possible solutions will tend to control the paths that innovative thinking may take in seeking an acceptable design solution. Although freewheeling innovative thinking is the basic element of successful design, the experienced engineer never fails to consider known restraints when he is brainstorming possible solutions. In addition to avoiding patent infringements, the designer must be fully aware of possible changes in desires and needs and give

Fig. 12.5 NEMO on the ocean floor.
NEMO is a 168-cm-diameter sphere of acrylic plastic. The sphere is constructed from 12 identical curved pentagons. The capsule, with 63.5-mm walls, is bonded with acrylic adhesive. One of the first uses of NEMO will be as a diver control center at points of Seabee underwater construction sites. In operation, the manned observatory is lowered into the sea from a Navy support vessel with the crewmen flooding the ballast tanks for the descent. NEMO operates independently, controlled by an array of push buttons (41) linked to a solid-state circuitry control system. NEMO has been designed for a normal stay underwater of 8 hours. Under emergency conditions it may remain submerged for as long as 24 hours. (*Official Navy photograph. Courtesy* The Military Engineer Magazine)

full consideration to the production processes. Also not to be overlooked in the decision-making stage are the visual design (styling aspects) and the price range within which the product must be sold.

The ultimate success of a design is judged on the basis of its acceptability in the marketplace and on how well it satisfies the needs of a particular culture (Fig. 12.9). In these latter decades of the twentieth century, we have added to these judgments a requirement that what has been created must not damage the physical environment around us, an environment which has been rapidly deteriorating because of some past technological advances. New requirements are being prepared and approved in rapid order that set standards to reduce if not eliminate pollution. These restrictions were almost unknown to the creative designers of the first half of this century. Today, it has become almost mandatory for a designer selecting a material or an energy source to ask himself if the material can be recycled and whether any residue discharged into the atmosphere is damaging to life.

12.4
Innovative Design—Individuals and Groups.
In the past there have been a number of creative individuals, working almost alone, who have created products that have advanced our culture, products for which people still have a great desire. Examples of such products are the printing press, the steam engine, the gasoline engine, the automobile, the telephone, the phonograph, the motion picture camera and projector, the radio, the television, and many more. These individuals were keen observers of their culture and they possessed inquisitive attitudes that led to experimentation. Some called them dreamers. If so, their eyes were wide open, they worked long hours, and they persisted in making a new try after each failure. Their fame and their fortune, although rightly due them, resulted from the fact that each brought forth a new product or system at a time when it would be readily accepted and when it could be more or less mass-produced. These people possessed a good sense of timing and they were not afraid of failure or ridicule. They belonged in the age in which they lived.

At present, it is the practice of large industrial organizations to use group procedures in order to stimulate the imagina-

tions of the individual members of the group and thereby benefit from their combined thinking about a specific problem (Fig. 12.6). Within the group, the innovative idea of one individual stimulates another individual to present an alternative suggestion. Each idea forms the basis for still other ideas until a great number have been listed that hopefully will lead to a workable solution to the problem at hand. This group attack on a problem produces a long list of ideas that would be difficult for a single individual to assemble in so short a time (Fig. 12.7). A single person attempting to solve the problem alone would find it necessary to conduct extensive searches of current literature and patent records and to hold numerous discussions with knowledgeable experts in seeking the guidance needed for making necessary decisions.

The most widely used of the group-related procedures is known as *brainstorming*. In applying this procedure to seek the solution to a design problem, a group of optimal size meets in a room

Fig. 12.6 Passive restraint system for rear-seat passengers now being developed by Firestone is designed to provide accident protection in one-thirtieth of a second. It is shown in a stop-action sequence: (1) The rear passenger section with safety blanket stored in a compartment built into the back of the front seat. (2) The blanket about to be deployed. (3) The blanket in place for restraint. (4) This shows how the blanket may be released so that the passengers can leave the car. (*Courtesy The Firestone Rubber Company*)

Fig. 12.7 Artist's concept of space shuttle system.
Shown is the complete system, including the airplane-like orbiter, large external propellant tank for the orbiter's three liquid hydrogen/liquid oxygen main engines, flanked by duel solid-rocket boosters.
(*Courtesy National Aeronautics and Space Administration*)

where they will be relatively free of interruptions and distractions. The people selected should be knowledgeable and there should be no one assigned to the group who may take a strong negative attitude toward the design problem to be considered. Usually, group size ranges from six to fifteen persons. If the group consists of less than six people, the back-and-forth interchange of ideas is reduced and the length of the list of ideas, which will be the basis for eventual decision making, is shorter. On the other hand, should the group be larger than fifteen, some individuals who may be capable of making excellent suggestions will have little chance to talk. Also, very large groups tend to be dominated by a few individuals.

During the brainstorming phase of design, no appraisals or judgments should be made nor should criticisms or ridicule of any nature be permitted. The group leader, if he is a capable person, will encourage positive thinking and stimulating comments. He will discourage those who may want to dominate the discussion.

The ideas and suggestions of the members of the group should be listed on a chalk-board and, no matter how long the list, none should be omitted. All ideas should be welcomed and recorded. After the meeting, all the ideas presented may be typed and reproduced for the information of the group and for reference at later discussions.

As the list of suggestions lengthens, several possible solution patterns usually emerge (Fig. 12.8). These in turn lead to still more suggestions for other possible combinations and for improvements to likely solutions to the design problem.

One or two lengthy brainstorming sessions can result in several acceptable design solutions that represent the combined suggestions of the individuals in the group. However, making a group evaluation of the ideas that have evolved from brainstorming is another matter, since no one member feels completely responsible for the results. The major weakness of any group procedure, such as brainstorming, is that individual motivation is dampened to some extent. However, since group procedures are productive and have become widely used in industry, there have been studies made that will hopefully lead to changes that will minimize this recognized weakness and improve group effectiveness. Means must be found to raise the motivation of each participant to the highest possible level.

12.5
Creation of New Products. A product or a system may be said to have been produced either through evolutionary change or by what appears to be pure innovation. The word *appears* has been used appropriately, since few, if any, products are ever entirely new in every respect. Most products that appear to have drawn heavily on innovation usually combine both old and new ideas in a new and more workable arrangement. A product of evolutionary change, however, develops rather slowly, over a long period of time, and with slight improvements being made only now and then. Such a product may be reliable and virtually free of design and production errors, but the small amount of design work involved, done at infrequent intervals, will never really challenge a creative person.

In today's competitive world, when products are produced for world-wide consumption, evolutionary change is hardly sufficient to ensure either the economic well-being or even the survival of those companies that seem to be willing to let well enough alone. Rapid technological changes coupled with new scientific discoveries have increased the emphasis on the importance of new and marketable products that can gain a greater share of the total market than is possible with the product the company may now be pro-

self contained
crane

Bulk plastic tanks
re-supply by air-drop

Basic vehicle-
including feed
blowers for
trucks

extruding plant

refrigerator unit
for cooling pipe

power
drive

control station
for extruder

1. cab 2. sleeping area 3. antenna 4. crane
5. power plant 6. Bulk plastic in tanks 7. feed
blower 8. shielded coupling 9. heating element
10. extruder 11. refrigerator plant 12. control
station 13. power rollers 14. extruder pipe

Fig. 12.8 Design of a self-contained pipelayer that could be in use in 1985. It is intended to facilitate irrigation of large tracts of desert land. The equipment as designed is capable of transporting sufficient bulk plastic to lay approximately 2 miles of plastic pipe from each pair of storage tanks. Tanks are to be discarded when empty and replaced by air drop. (*Courtesy Donald Desky Associates, Inc., and Charles Bruning Company*)

moting. In meeting this need for new and marketable ideas, the designer will find that his innovative ability and his experience and knowledge are being taxed to the limit and, since he may in one sense be stepping into the unknown, he will be taking some risks.

The unusual characteristics that seem to be a part of the general makeup of every outstanding designer are: (1) the ability to recognize a problem, (2) the ability to take a questioning approach toward all possible solutions, (3) the possession of an active curiosity about a problem at hand, (4) the innate willingness to take responsibility for what he has done or may do, (5) the ability to make needed decisions and to defend his deci-

sions in writing and orally, and (6) the possession of intellectual integrity.

William Lear, one of the most prominent designers of the last three decades, has spent his entire working life discovering needs and then finding ways to fulfill them. In the case of the development of the eight-track stereo tape, he was working from economic considerations that required more repertoire on tape without adding more tape. This meant either running the tape slower or adding more tracks. The practical answer from Lear's viewpoint was to add more tracks. In addition to the Learoscope, an automatic direction-finder for use on airplanes, Lear is responsible for the development of the car radio, the automatic pilot, the Lear jet plane, and more than 150 other inventions. A few years ago, he began development work on the problem of steam-powered automobile engines. Lear's inventing has been done when he was surrounded by many people with considerable know-how. His work involved the gathering of a maze of information and ideas from which he could pick out the salient facts and discard the unimportant ones, while always keeping the goal in mind and solving the problem at the least possible cost.

12.6
Background for Innovative Designing. Designing should be done by people who have a diversified background and who are not entirely unfamiliar with the problem at hand. As an example, even though the design of a product may be thought of as being in the field of mechanical engineering, a designer with a knowledge of electrical applications and controls will find it a distinct advantage since many of our present-day products use electric current as an energy source. In this case, even though an electrical engineer may be a member of the design group, the mechanical engineer and others should have at least some understanding of his suggestions. This added knowledge will enable them to modify their thinking about a product that is largely a mechanical device. Examples of these products include electric locks, electric food choppers, and electric typewriters.

The background required will vary considerably depending upon the field in which the individual works. For example, a person who may be designing small household appliances would probably never need more than the knowledge acquired from his basic engineering courses, while a designer in the aerospace field would need a background based upon advanced study in chemistry, physics, and mathematics. On the other hand, there are respected and competent designers in industry who have had as little as two years of technical education. With this limited training and several years of on-the-job experience, these men and women have become able to design fairly complex machines.

Because of the increasing complexity of engineering, the rapid development of new materials, and the accumulation of new knowledge at an almost unbelievable rate, it has become absolutely necessary for engineering design to become a team effort in some fields. Under such conditions, the design effort becomes the responsibility of highly qualified specialists. A project requiring designers of varied specialized backgrounds might need, for example, people with experience in mechanical, electrical, and structural design and persons with considerable knowledge of materials and chemical processes (Fig. 12.9). If it is decided that styling is important, then one or more stylists must be added to the team. A complete design group for a major project could include pure scientists, metallurgists, craftsmen, and stylists in addition to the designers.

Finally, graphics must not be overlooked when one considers the background needed to become a successful designer. Anyone who hopes to enter the field of design, other than as a specialist or craftsman, must have a thorough training in this area. He must have a working knowledge of all of the forms of graphical expression that are presented in this text and at the same time be capable of expressing himself well both orally and in written form in his preliminary and final reports. The methods used for the preparation of oral and written reports are discussed in Sec. 12.24.

12.7
History and Background of Human Engineering. Human engineering is a relatively new area in the field of design; it has gained the recognition that it rightly deserves only within the last few decades. In order to simplify matters we might define human engineering as adapting design to the needs of man; that is, we engineer our designs to suit human behavior, human motor activities, and human physical and mental characteristics. The applications of human engineering apply not only to man–machine systems and consumer products but to work methods and to work environments as well. In the early years of its development, human engineering was concerned mainly with the working environment and with the comfort and general welfare of all human beings. As time passed, designers came to realize that more than just safety and comfort should be considered, and that almost all the designs with which they were concerned were in some way related to man's general physical characteristics, his behavior, and his attitudes. At this point, designers began to include these added human factors in their approach to the solution of a design problem so as to

Fig. 12.9 Jet engine.
The ultimate goal of every design must be the production of a product or system useful to man such as the jet engine shown being assembled. (*Courtesy Pratt & Whitney Aircraft Division of United Aircraft Corporation*)

secure the most satisfying and efficient man–machine relationship possible.

It is in the area of human engineering that an engineering designer must adhere to the input of specialists who may be involved in a wide range of disciplines. These disciplines may be industrial engineering, industrial psychology, medicine, physiology, climatology, and statistics. Stimulated to a great extent by the space program, scientists from all of these disciplines have, together and separately, become deeply involved in basic research and laboratory experimentation, which has led to a continuous input of new man–machine information into design. Although in the past human engineering has been associated mainly with industrial engineering, industrial psychology, and industrial design, designers in all fields of engineering must now not only be knowledgeable about the principles of human engineering but they must be capable of utilizing these principles and related information whenever they are developing a product that involves human relationship.

Typical body dimensions, representing an average-size person, are used when a product is being designed for general use.

The measurements of typical adult males and females, as determined from studies made by Henry Dryfuss, may be used for most designs requiring close adaptation to human physical characteristics. Since many designs involve both foot and hand movements these average body dimensions, as tabulated, include arms, hands, legs, and feet along with other parts of the human body. Data relating to body proportions and dimensions are known as anthropometric data. Information relating to body proportions may be obtained from *The Measure of Man*, Whitney Library of Design.

The first known serious studies of the human body were made by Leonardo da Vinci. To record his studies for his own use and for the use of others, he made some of the finest and most accurate detail sketches and drawings of the human body known to man. These drawings show even the intricate details of muscle formation. His work, done in defiance of the laws of his time, is still used today in several textbooks. Da Vinci's studies of the human body mark the beginning of the science of biomechanics. See Fig. 12.10.

From the time of Leonardo da Vinci

Fig. 12.10 Human proportions and body dimensions as illustrated by Leonardo da Vinci (1452–1519).

until early in the twentieth century, very little work was done toward the development of this science. This lack of interest in people and their relationship with equipment and tasks to be performed was due largely to the fact that from the beginning of the Industrial Revolution (which many people say began with invention of the steam engine patented by James Watt in 1769) until the early 1900s, the interest of designers was centered mainly on the creation of new products and the raising of production efficiency to the high levels needed to compete in world markets. Until about 1911, revolutionary change was the order of the day and there was very little time available for consideration of the human anatomical, physiological, behavior, and attitude factors that have now become the basis of our present man–machine–task systems. Even though our computer-programmed numerically controlled machine tools permit us to do

almost any task with only minimal human intervention, man is still needed in most of our man–machine–task systems and he must be taken largely as he has been created. In our designs we must not overlook even the possibility of boredom, for should man become sufficiently bored he might just "pull the plug" and everything would stop.

12.8

Human Engineering in Design. There are a number of factors in human engineering, other than human anthropometric measurements, that must be dealt with in design. These factors include motor activities (Fig. 12.11) and body orientation; the five human senses (sight, hearing, touch, smell, and sometimes even taste); atmospheric environment, temperature, humidity, and light; and, finally, accelerative forces if they are exceptional and are likely to cause undue physical discomfort.

THE PHYSIOLOGIC-MECHANISTIC SYMBIOSIS

Fig. 12.11 Boston Arm.
The range of movement of the Boston Arm is demonstrated in the upper photograph. The arm, developed as a joint project of Liberty Mutual Insurance Companies, Harvard Medical School, Massachusetts General Hospital, and the Massachusetts Institute of Technology, acts as does a normal arm through thought-impulses transmitted from the brain to existing arm muscles. The design of a product for the handicapped can provide much satisfaction to the designer. (*Courtesy Liberty Mutual*)

During the first half of this century most of the research done in human engineering, aside from the anthropometric studies already mentioned, was directed toward: work areas and the position of controls; physical effort and fatigue; and the speed and accuracy to be expected in the performance of particular tasks. Finally, when it became evident that this was not enough, industrial engineers and industrial psychologists turned their attention to the more complex activities of the average human being. These new studies dealt primarily with receiving information through sight and sound, the making of decisions in response to stimuli, and, finally, the performance in direct response to these decisions.

The study of body motion deals primarily with the effective range of operation of parts of the body, usually the arms and legs, and the amount of body force a human being may reasonably be expected to exert in the performance of an assigned task. For example, many time and motion studies have been made of the range of operation of persons performing given tasks while seated at assigned work areas. At the same time, in many of these studies attention has been given to the location and the amount of force required to operate levers and controls in relation to the size and strength of the operator.

Vision is an important factor in all designs where visual gauges or colored lights on control panels are a part of a control system that involves manual operation or, in the case of numerically controlled machines, monitoring for manual intervention at specified times. Control panels for such equipment must be designed to be within the visual range of the average person, and distinctive colors must be selected and used for the colored lights. The colors selected must be easily recognizable and must be capable of quickly attracting the operator's attention. Not to be overlooked is the fact that vision studies have produced much new information for use in highway design. From these same studies have come new ideas for our highway warning and information signs along with suggestions for their placement.

When a designer must consider the working environment as a part of the total design, he should realize that environment includes (1) temperature, (2) humidity, (3) lighting, (4) color schemes, and (5) sound. These are a few of the factors that also deserve full consideration in the design of a large industrial plant, a particular work area in a plant, the cockpit of an airplane, or the cabin of a space vehicle.

The overall environment and the design of the working areas and the living quarters of the undersea laboratory shown in Fig. 12.12 were based on human requirements. The design of any undersea craft or laboratory involves problems that are similar in many ways to those encountered in the design of space vehicles, in that an artificial living environment must be created and maintained for extended periods of time. This requires a self-contained atmosphere. The members of the crew also must have ample space in which to work and live under climatic conditions that duplicate those on land. Crew members must be able to perform their tasks under normal lighting conditions and they must be able to see to the outside. Aside from the design of features and components that are related directly to the performance of the research assignments, the overall development of the undersea craft can be said to be based on the physical needs and the psychological attitudes and reactions of human beings.

Human engineering is applied to a wide range of consumer items. Automobiles (Fig. 12.13), refrigerators, furniture, office equipment, lawn mowers, hand tools, and like items have long been designed with human factors in mind.

The automotive industry, which has been designing and redesigning cars for more than 75 years with human factors in mind, is now producing cars in accordance with stricter government safety regulations after being accused by a vocal few of having produced American cars for comfort, power, style, and speed with less than full attention being given to the safety of the occupants. Even though the charge in this case is probably unfounded designers must realize that this situation can happen whenever engineers ignore, even to limited extent, some of their professional responsibilities and give in to the whims of stylists and to the suggestions of administrators and sales managers before having made a thorough study of each and every proposal that has been presented.

Under pressure from consumer groups and organizations interested in safety and in protecting our natural environment, there has been a rising tide of governmental laws and regulations. Many of these new laws and directives in a sense provide controls in the field of human

Fig. 12.12 Artist's conception of the Atlantis undersea habitat.
The project, which was to be designed and developed as a joint project of the
University of Miami and the Space Division of Chrysler Corporation, has been
abandoned at this time. The project goal was to explore the continental shelf
along our coasts. (*Courtesy Space Division Chrysler Corporation*)

engineering. In the case of environmental
pollution, automobile companies, acting
to meet regulations set forth by the Environmental Protection Agency, have developed several new devices to reduce
undesirable pollutants at a scheduled
percentage rate to produce an almost pollution-free car by 1980. In the interest of
safety, some type of passive protection
system, such as the air-bag device shown
in Fig. 12.13, must be added by a set date.

These government laws and regulations
have come into existence because of a
growing interest on the part of the general
public in human engineering. Designers in
the years ahead must be fully cognizant of
all such regulations and must be willing to
abide by them or seek to have them
changed should they appear to be unrea-
sonable or impractical.

12.9

Visual Design. Visual design includes the
use of line, form, proportion, texture, and
color to produce the eye-pleasing appearance needed to bring about the acceptance of a consumer item. Without this acceptance there would be no profit, and

AIR-BAG/SEAT-BELT RESTRAINT SYSTEM
WITH STARTER INTERLOCK

**Fig. 12.13 Major elements of an air-bag/seat-
belt restraint system now under development
by the Ford Motor Company.**
The air-bag assembly and the seat-belt
starter-interlock components are identified. It
should be noted that there are separate signal lights for the seat-belt and air-bag systems. (*Courtesy Ford Motor Company*)

even though the item might otherwise have been carefully engineered, it would soon disappear from the marketplace. The sketches shown in Fig. 12.14, for the dash panel of an automobile, tastefully combine these visual elements into an attractive design. An illusion of depth has been obtained by means of pencil shading.

Since artistic styling is now recognized as being one of the most important factors in sales, many engineers have come to accept the role of stylists in the development of a consumer product, particularly when they are employed by a company that is small and cannot afford to employ one or more trained stylists. Large companies, such as the Ford Motor Company (Fig. 12.15), General Motors, and the Boeing Company (Fig. 12.16), have styling divisions. Medium-size companies often turn to nationally known organizations to get needed help; the styling is then done

under contract agreement. Many books have been written about aesthetics that have proved to be helpful to engineers (see the Bibliography). At present, design engineers, who have been trained largely to solve technical problems, are reading more about art in design and they are considering the eye-appeal and the overall appearance of products as part of their engineering interests.

12.10
Constructive Criticism. There are people who seem to find it easier to criticize than to mix praise with alternative suggestions. In any group meeting a critic should show respect for good ideas and be able to offer constructive suggestions. If there is to be feedback, which will hopefully lead to the introduction of more ideas, the discussion must be free of any harsh criticism. Harsh criticism may cause a sensitive person to assume a defensive position or to withdraw almost entirely from participation in group action. It is the responsibility of the project leader to prevent this from happening.

12.11
Recognition of a Need. A design project usually begins with the recognition of a need and with the willingness of a company to enter the market with a new product. At other times an idea may be initiated and developed by an individual who either seeks economic benefits for himself or who seeks to solve some social or environmental problem. In either case, the identification of the need in itself represents a high order of creative thinking and the search for a solution to the need requires considerable self-confidence and inner courage. As can be easily observed from reviewing the achievements of our distinguished inventors of the past, those who are closely attuned to life around them become aware of needs or less-than-ideal situations that are worthy of their attention.

It is important that a proposed design activity have clear and definite objectives that will justify the money and effort to be expended in product design and development. The statement defining the objectives should identify the need and state the function the product is to perform in satisfying this need. The identification of the need may be based on the designer's personal observations, suggestions from salesmen in the field, opinion surveys, or on new scientific concepts. The identifica-

Fig. 12.14 Design sketches for a dashboard panel. (*Courtesy General Motors Corporation*)

Fig. 12.15 Stylist at work. (*Courtesy Ford Motor Company*)

tion of the design problems involved in creating the needed product comes later in the design process.

12.12

Formal Proposal. The statement covering the recognition of need can be used as a basis for a formal proposal that may be either a few short paragraphs or several typewritten pages. A complete proposal may include supporting data in addition to the description of the plan of action that is to be taken to solve the problem as identified (Fig. 12.17). The report should have the same general form as other technical reports and might include a listing of requirements and possible limitations as then recognized. In preparing the report one should keep in mind that the proposal, when approved, gives the broad general parameters of an agreement under which the project will be developed to its conclusion.

12.13

Phases of the Design Process. Many people in the past have prepared outlines of the steps that can be followed in the process of design (Fig. 12.18). They have prepared these outlines in order to give some semblance of order to the total design process from the point of recognition of the need to the point of marketing the product. One must recognize, however, that there are actually many combinations of steps in the overall procedure, with no single listing either the best or the one and only combination. The design procedure required in many cases can be very complex and successful designers have found different ways to achieve their goals. However, the phases of design, as recognized by this author, have been listed here in sequential order to provide some degree of direction to the student who is making his first attempt to design a product under a contrived classroom situation. More experienced persons may find it desirable to alter this outline to make it more suitable to their own method of designing.

The basic phases in the design process can be thought of as being (1) identification of need, (2) task definition (goal), (3) task specifications, (4) ideation, (5) conceptualization, (6) analysis, (7) experimental testing (Fig. 12.26), (8) design (solution) description, and (9) design for production. Phase 9 (implementation) is not usually a primary concern of a designer. Also, the designer may or may not

Fig. 12.16 Several styling design sketches prepared for the U.S. supersonic transport (SST). Visual appeal is important to the success of the total design project. (*Courtesy The Boeing Company*)

give some thought to manufacture, distribution, and consumption of the product. Consideration of these factors may be thought of as being the tenth, eleventh, and twelfth stages of total design.

12.14

Task Definition—Definition of Goals. Briefly stated, task definition is the expression of a commitment to produce either a product or a system that will satisfy the need as identified. By means of broad statements, the product and the goals of the project are identified. The statements, as written, must be clear and concise to avoid at least some of the difficulties (often encountered in design) that can be traced directly back to poorly defined goals.

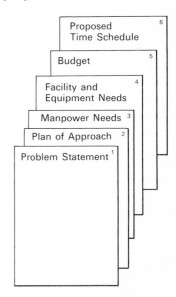

Fig. 12.17 Elements of a proposal.

Even though it is probable that the person who has initiated the project has already gathered some pertinent information and has some preconceived ideas, it is desirable that this material not be included (Fig. 12.19). It is better to present the goals in terms of objectives and then allow the designers to pursue the project in their own way, as free of restrictions as possible. The task definition should be included in the proposal.

12.15

Task Specifications. This is a listing of parameters and data that will serve to control the design. This stage will ordinarily be preceded by some preliminary research to collect information related to the goal as defined. In preparing the task specifications, the designer or design group lists all the pertinent data that can

be gathered from research reports, trade journals, patent records, catalogues, and other sources that possess information relating to the project at hand. Included in this listing should be the parameters that will tend to control the design. Other factors that deserve consideration, such as materials to be used, maintenance, and cost may also be noted.

12.16

Ideation. The ideation phase of design has been discussed for group procedure situations in Sec. 12.4. It is recommended that the reader review this discussion to refresh his memory regarding brainstorming procedures. It is well to remember that often a lasting solution to a problem has resulted from a creative idea selected from a number of alternative ones (Fig. 12.20). The mathematical likelihood of finding an optimum solution is greatly enhanced as the list of possible alternatives grows longer. Truly great creations are possible when one, acting either alone or in a group, lets his imagination soar with little restraint. If an engineer on a design assignment can set aside his engineering know-how and blind himself, at least temporarily, to traditional approaches he is in the right mood to accept almost any challenge. This is the mood and the mental approach to great discoveries. Some call it ideation; others have called it *imagineering.* If we can learn to open up our engineering minds to new approaches to our technical problems as well as to our existing problems of air pollution, industrial waste, transportation, and even unemployment, there is no limit to what we can accomplish.

This open-minded *imagineering* approach to problems must be tempered with a sense of professional responsibility. It is no longer acceptable to solve an immediate problem using a solution that in years to come can endanger the environment. Imagineering is engineering for the total well-being of mankind and all other living things.

12.17

Conceptualization. Conceptualization follows the preliminary idea (ideation) stage when all the rough sketches (Figs. 12.21–12.23) and notes have been assembled and reviewed to determine the one or more apparent solutions that seem to be worthy of further consideration. In evaluating alternative solutions, consideration must be given to any restrictions that

Fig. 12.18 Basic phases of total design.

Fig. 12.19 Initial steps of the design process.

have been placed on the final design. It is at this stage that the preliminary sketches should be restudied to see that all worthwhile ideas are being included and that none have been inadvertently overlooked. At no time during this phase should the designer become so set in his thinking that he does not feel free to develop still another and almost entirely new and different concept, if necessary. He should realize that it is more sensible to alter or even abandon a concept at this stage than later, when considerable money and time will have been invested in the project. The conceptualization stage of design is that stage where alternative solutions are developed and evaluated in the form of concepts. Considerable research may be necessary and task specifications must be continually reviewed. As activity progresses, many idea sketches are made as alternative approaches are worked out; these approaches are evaluated for the best possible chance of product success. It is not necessary at this stage of the design procedure for any of the alternative solutions to be worked out in any great detail.

12.18

Selection of Optimum Concept. As the design of a product or system progresses, a point is reached in the procedure when it becomes necessary to select the best design concept to be presented to the administrators in the form of a proposal. In making this final selection, a more-or-less complete design evaluation is made for each of the alternative concepts under

Fig. 12.20 Idea sketches for a small portable safety alarm that will give warning for both fire and an attempted break-in. (Dimension values are in millimeters)

Fig. 12.21 Portion of a design sketch. (*Courtesy General Electric Company*)

Fig. 12.22 Idea sketch for a quick-acting machine clamp.

Fig. 12.23 Idea sketch showing remote control system for a motor boat.

consideration. These evaluations may reveal ways that costs can be reduced and value improved; means of simplifying the design to reduce costs may also become apparent.

12.19
Design Analysis. After a design concept has been chosen as the best possible solution to the problem at hand, it must be subjected to a design analysis; that is, it must be tested against physical laws and evaluated in terms of certain design factors that are almost certain to be present (Fig. 12.24).

Total analysis of a proposed design will include a review of the engineering principles involved and a study of the materials to be used. In addition, there should also be an evaluation of such design considerations as (1) the environmental conditions under which the device will operate, (2) human factors, (3) possible production methods and production problems, (4) assembly methods, (5) maintenance requirements, (6) cost, and (7) styling and market appeal. If the design is based on newly discovered scientific principles, some research may be in order before a final decision can be reached.

It is at this stage in the design procedure that physics, chemistry, and the engineering sciences are utilized most fully. In making the usual design analysis, the engineer or engineering technologist must depend on the formal training that he has received in school, and although considerable mathematics may be needed in making most of the necessary calculations, he will find it convenient at times to resort to graphical methods. Over the years, graphical methods have proved to be most helpful in evaluating and developing a design. For example, descriptive geometry methods can be employed for making spatial analyses and critical information can be obtained by scaling accurate drawings.

If the design analysis should prove that the design as proposed is inadequate and does not meet requirements, the designer may then either make certain modifications or incorporate into his designs some new concept that might well be a modification of an earlier idea that was abandoned along the way.

12.20
Experimental Testing. The experimental testing phase of the design process ranges from the testing of a single piece of software or hardware to verify its workability, durability, and operational characteristics through the construction and testing of a full-size prototype of the complete physical system.

A component of a product should be tested in such a way that the designer can predict its durability and performance under the conditions that will be encountered in its actual use. Needed tests may be performed using standard test apparatus or with special devices that have been produced for a particular test.

There are three types of models that may be constructed for the purpose of testing and evaluating a product. These are: (1) the mockup, (2) the scale model, and (3) the prototype.

12.21
Mockups. A mockup is a full-size *dummy* constructed primarily to show the size, shape, component relationships, and styling of the finished design. At this point the designer can see his conception begin to take shape for the first time. Automobile manufacturers customarily produce mockups to evaluate proposed changes in the styling of automobile bodies for new models (Fig. 12.25). Needed modifications in size and body configuration can be determined by studying the mockup and analyzing its overall appearance. Since interior styling is important also, interiors are modeled in clay to reveal the aesthetic appearance the stylist had in mind. In the automobile industry, mockups are made to secure early approval of management for model change. A mockup is more meaningful than a sketch to those whose

Fig. 12.24 Analysis phase of design.

support and approval is needed. One must realize, however, that numerous sketches and artistic renderings (Fig. 12.16) are made before any work is started on a mockup. These sketches are used as guides. Mockups may be made of clay, wood, plastic, and so forth.

12.22

Preliminary and Scale Models. Models may be made at almost any stage in the design process to assist the designer in evaluating and analyzing his design. Models are made to strengthen three-dimensional visualization, to check the motion and clearance of parts, and to make necessary tests to clear up questions that have arisen in the designer's own mind or in the mind of a colleague.

The designer may prepare a preliminary model to understand more fully what the shape of a component should be, how well it may be expected to operate, and how it might be fabricated most economically. In some cases, the model might be so simple that it could be made of paper, wood, or clay.

Scale and test models may be constructed either for analysis and evaluation or for the purpose of presenting for approval the design as developed in a more or less refined stage. Scale models may be made of balsa wood, plastic, aluminum, wire, steel, or any other material that can be used to a good advantage. The designer should select a scale that will make the model large enough to permit the movement of parts should a demonstration of movement be desirable.

12.23

Prototypes. A prototype is the most expensive form of model that can be constructed for experimental purposes. Yet, since it will yield valuable information that is difficult to obtain in any other way, its cost is usually justified. Since a prototype is a full-size working model of a physical system that has been built in accordance with final specifications, it represents the final step of the experimental stage (Fig. 12.26). In it the designers and stylists see their ideas come to life. From a prototype the designers can gain information needed for mass-production procedures that are to come later. Much can be learned at this point about workability, durability, production techniques, assembly procedures, and, most important of all, performance under actual operating conditions. Since prototype testing offers

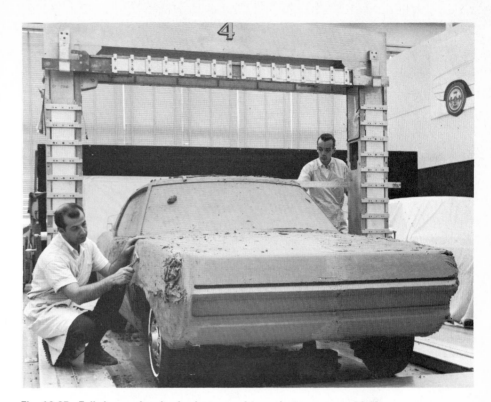

Fig. 12.25 Full-size mockup in clay is prepared to evaluate a proposed body configuration for a forthcoming model. (*Courtesy Ford Motor Company*)

the last chance for modification of the design, possible changes to improve the design should not be overlooked, nor should a designer ever be reluctant to make a desirable change.

Since a prototype is a one-of-a-kind working model, it is made by hand using

Fig. 12.26 Full-size prototype of the Pioneer Satellite A-B-C-D series.
A prototype provides the best means for an evaluation of a complex design. (*Courtesy TRW Systems Group*)

general purpose machine tools. Although it might be best to use the same materials that will be used for the mass-produced product, this is not always done. Materials that are easier to fabricate by hand are often substituted for hard-to-work materials.

In the development of a design, designers deal first with the mockup; next they work out specific problems relating to single features with preliminary models; then they evaluate the design with scale models; and, finally, they may, if desirable, test the whole concept using a prototype. This order in the use of models maintains a desirable relationship between concept and analysis and represents a logical procedure for the total design process.

12.24

Design (Solution) Description—Final Report. In the solution description the designer describes his design on paper, to communicate his thinking to others. Although the purpose of the final report may be to sell the idea to upper management, it may also be used to instruct the production division on how to construct the product. It usually will contain specific information relating to the product or system. In some cases a process will be described in considerable detail.

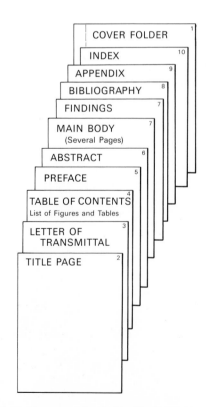

Fig. 12.27 Contents of a design (final) report.

A complete design description, prepared as the main part of a formal report, should include: (1) a detailed description of the device or system, (2) a statement of how the device or system satisfies the need, (3) an explanation of how the device operates, (4) a full set of layout drawings, sketches, and graphs, (5) pictorial renderings, if needed, (6) a list of parts, (7) a breakdown of costs, and (8) special instructions to ensure that the intent of the designer is followed in the production stage (Fig. 12.27). After the design description is accepted and approved, there remain only the commercial stages of the total design process, namely, implementation, manufacture, distribution (sales), and consumption.

12.25

Implementation—Design for Production. Implementation is that phase of the total design process when working drawings are prepared for the men in the shops who must fabricate the nonstandard parts and assemble the product. A complete set of working drawings, both detail and assembly, are needed to perimit the manufacture of a product.

Even though in theory the production design phase directly follows the preliminary design phase and the acceptance of the design recommended in the final report, in actual practice there is often no clear dividing line between these two phases. This is because detailed drawings of some components may have been started and to some extent completed well back in the preliminary design stage, along with one or more design layout assemblies (see Fig. 12.28). Furthermore, it is not unusual for detail design drawings to be made for two or more likely solutions in the design-for-production stage and the final selection of the one best solution delayed until the information derived from these detail design drawings and related assemblies can be used in making the final decision. Such a delay in decision making tends to cause the preliminary design and the production design phases to appear to melt into one another. However, there is still a division line even though the designers themselves may not recognize this fact.

12.26

Manufacture (Fig. 12.29). From the time the task specifications are written and through all the stages up to the manufacture, the designer works closely with a

PC.NO	NAME
1	BASE
2	FLYWHEEL
3	GEAR
4	CONNECTING ROD
5	LEVER
6	CAM
7	CAM FOLLOWER
8	SHAFT
9	CROSSHEAD
10	PAWL
11	CROSSHEAD SHOULDER SC.
12	PAWL SHOULDER SC.
13	FLYWHEEL SHOULDER SC.
14	LEVER PIN
15	PAWL PIN
16	PAWL SPRING PIN
17	$\frac{1}{2}$-NC. STD. HEX. NUT
18	#2×1$\frac{1}{4}$ STD TAPER PIN
19	#304 WDRF. KEY
20	#404 WDRF. KEY
21	PAWL SPRING

16 PITCH-30 TOOTH GEAR

*#6 PRESS FIT IN LEVER
#3 RUNNING FIT IN BEARING*

*REAM IN ASSEMBLY FOR
#2×1$\frac{1}{4}$ TAPER PIN*

12 STATIONS AS SHOWN EQUALLY SPACED

*INDEXING MECHANISM
PRESS DIAL TABLE*

SECTION A-A

Fig. 12.28 Design layout drawing.

production engineer who is familiar with the available shop facilities, production methods, inspection procedures, quality control, and the assembly line. If this is done, the problems encountered in the manufacturing stage will be few in number.

12.27
Distribution. Since a designer usually has little expertise in this area, task problems relating to distribution are passed along to marketing specialists. These specialists have the knowledge and the supporting staff required to decide on the proper release date and to set a competitive price based upon market testing and cost and profit studies. Included among these specialists will be experienced advertising men who prepare the needed advertising and promotional literature. However, these specialists do consult the designer frequently during this stage since he knows more about the product than anyone else. The designer is the one person

who can be depended upon to supply needed technical data and information concerning the capabilities and limitations of the product. Furthermore, the sales promotion people expect ideas from the designer that will lead to a wide and favorable distribution.

12.28
Consumption. There should hopefully be a consumer feedback from this last stage of the design process that will prove useful to the designer when it becomes necessary to alter and improve the product at the time of the next model change. Most of this feedback information will come from the sales force, from distributors, and from service departments. Some of this information will be in the form of complaints made by irate users, but much of it will be received as constructive suggestions. These constructive suggestions take the designer back to the need stage for the next model, and a new round of the design cycle starts all over again.

Fig. 12.29 Detail drawing, showing the use of dual-dimensioning.
The reader should note the procedures used when both inch measurements and millimeters are
shown. Some industrial drawings are now being dimensioned in this manner. See Chapter 13.
(*Courtesy Ford Motor Company*)

The consumption stage represents the goal sought by the designer and this is where the design has its ultimate test. At this stage the design will be pronounced a success or a failure by the users and consumers who are and always will be the final judges.

B □ Patents and patent office drawings

12.29

Patent. A patent, when granted to an inventor, excludes others from manufacturing, using, or selling the device or system covered by the patent anywhere in the country for a period of seventeen years. In the patent document, in which the invention is fully described, the rights and privileges of the inventor are set forth and defined. Upon the issuance of the patent, the inventor has the right to either manufacture and sell the product himself with a

protected market or he can assign his rights to others and charge them for the manufacture and sale or use of his invention. After the seventeen-year period has expired, the inventor no longer has any protection and the invention becomes public property and may be produced, sold, and used by anyone for the good of all.

To ensure full protection of the patent laws, patented products must be marked *Patent* with the patent number following. Even though an invention is not legally protected until a patent actually has been issued, a product often bears a statement that reads: *Patent Pending.*

In accordance with patent law "any person who has invented any new and useful process, machine, manufacture or composition of matter, or any new and useful improvement thereof, may obtain a patent," subject, however, to restrictions, conditions, and requirements imposed by patent law. To be patentable an invention must be new and original and be uniquely different; it must perform a useful function; and finally the invention must not have been previously described in any publication anywhere nor have been sold or in general use in the United States before the applicant made the invention. It should be noted that an idea in itself is not patentable, since it is a requirement that a specific design and design description of a device accompany an application for a patent.

12.30

Patent Attorney. When an inventor has decided that he has a patentable device, he should engage a patent attorney who will help prepare the necessary application for a patent. The inventor should depend upon this attorney to advise him as to whether or not the product or process that he has conceived may infringe upon the rights of others. Since most patent attorneys are also graduate engineers, the inventor will find that he has allied himself with a person who can guide his application through the searching, investigating, and processing that takes place before a patent is granted. Generally, an application remains pending for as long as four or more years before a patent is finally granted. In this period the pending patent application may well be amended to include engineering changes that have been made in the device or system.

If after some investigation, the attorney–engineer thinks that the device is novel and therefore patentable, an application for a patent should be prepared and filed. A patent application includes a formal portion consisting of the petition, a power of attorney, and an oath or declaration. This is followed by a description of the invention, called the specifications, and a list of claims relating to it. If the device can be illustrated, one or more drawings should be included (Fig. 12.30).

In selecting a patent attorney, the inventor should remember that this man can be of service to him over many years going well beyond the time when the patent is issued. The attorney or his firm may be retained to prepare all agreements covering the sale and leasing of patent rights, to assist in negotiations relating to these rights, and if the need should arise, to handle charges of alleged infringements.

Many of the large corporations have patent divisions that operate as a section of the home office legal staff. An additional office to deal with patent matters may also be maintained in Washington,

D.C. In this case, both offices, staffed with patent attorneys, will have all activities coordinated by a director of patents.

12.31

Role of the Inventor in Obtaining a Patent. Even though an inventor must rely almost entirely upon a patent attorney to take the final steps to secure a patent, the inventor plays a vital role up to the time of filing the application. Very often what he has done or not done determines whether his rights to his invention can be safeguarded by his attorney.

Fig. 12.30 Patent drawing, showing the sequence of events from launch to landing.

Since court decisions in patent suits usually depend upon the inventor's ability to prove that certain design events happened on a specific date, it is always advisable to keep the design records in a hardcover permanently bound notebook so that there will be no question as to whether or not pages have been added (as could be charged if a loose leaf notebook had been used). A well-kept patent notebook is a complete file of information covering a design. It can serve as the basis of project reports; it can spare the designer and his company the unnecessary expense of repeating portions of the experimental work; and, finally, it furnishes indisputable proof of dates of conception and development.

To prepare a legally effective patent notebook the designer should:

1. Use a bound notebook having printed page numbers.

2. Make entries directly using either a black ink pen or an indelible pencil.

3. Keep entries in a chronological order. Do not add retroactive entries.

4. Add references to sources of information.

5. Describe all procedures, equipment, and instruments used for development work.

6. Insert photos of instrument and equipment setups. Add other photos showing models, mockups, and so forth.

7. Sign and date all photos.

8. Have a qualified witness sign and date every completed page. The witness cannot be a co-inventor nor can he have a financial interest in the development of the device. The witness must be a person who is capable of understanding the construction and operation of the device or system. He must also be experienced in reading drawings and understand the specifications.

9. Have the witness write the words "Witnessed and Understood" and then write and date his signature.

10. Have two or more witnesses sign and date every page after reading it.

11. Avoid having blank spaces on any page.

12. Have all new entries witnessed at least once a week.

13. Have the notebook evaluated periodically by a patent attorney. Follow his suggestions.

It takes a considerable time and effort to adhere to these thirteen recommendations for the preparation of an effective patent notebook. However, if an inventor has once become entangled in an infringement lawsuit he realizes that a little extra effort pays off.

12.32

Patent Drawings. A person who has invented a new machine or device, or an improvement for an existing machine, and who applies for a patent is required by law to submit a drawing showing every important feature of his invention. When the invention is an improvement, the drawing must contain one or more views of the invention alone, and a separate view showing it attached to the portion of the machine for which it is intended.

Patent drawings must be carefully prepared in accordance with the strict rules of the U.S. Patent Office. These rules are published in a pamphlet entitled *Rules of Practice in the United States Patent Office*, which may be obtained, without charge, by writing to the Commissioner of Patents, Washington, D.C.

In the case of a machine or mechanical device, the complete application for a patent will consist of a petition, a "specification" (written description), and a drawing. An applicant should employ a patent attorney, preferably one who is connected with or regularly employs competent draftsmen capable of producing well-executed drawings that conform with all of the rules. Ordinary draftsmen lack the skill and experience necessary to produce such drawings. Three sheets of drawings for a patent for a radiation protection system are shown in Figs. 12.31–12.33.

Several U.S. Patent Office publications that are available to the general public have been listed in the Bibliography.

12.33

Rules. The following rules are quoted verbatim from the pamphlet, **Guide for Patent Draftsmen.***

> **81. Drawings required.** The applicant for patent is required by statute to furnish a drawing of his invention whenever the nature of the case admits of it; this

*This publication may be obtained from the U.S. Government Printing Office at a minimal cost (30 cents).

United States Patent Office

3,465,153

Patented Sept. 2, 1969

1

3,465,153
RADIATION PROTECTION SYSTEM AND
APPARATUS
Willard F. Libby, Los Angeles, Calif., assignor to
McDonnell Douglas Corporation, Santa Monica,
Calif., a corporation of Maryland
Filed Aug. 14, 1964, Ser. No. 389,734
Int. Cl. G21f *1/12, 3/02, 7/00*
U.S. Cl. 250—108   **16 Claims**

My present invention relates generally to astronautics, the science of space flight, and more particularly to a system and apparatus and method for the protection of astronauts from the hazards of suddenly encountered radiation fields of extreme intensity in space.

Manned space flights have now been successfully achieved by both the United States and the Soviet Union. Such flights will be followed by manned space probes including lunar and interplanetary missions for the manned exploration of the Moon, Mars and Venus. The manned space program of the United States is directed towards manned exploration first of the moon and then initially only of the two planets Mars and Venus of the solar system since all of its other planets appear to be barren and lifeless. These and other probes will, of course, eventually lead to interstellar journeys over vast distances to other stellar systems for the purpose of conducting explorations aimed at discovering new worlds which are susceptible to colonization by the human race

2

by using a greater number of stages are offset by the additional complexity involved, and it is very difficult to increase the propellant-weight ratio much beyond a certain value in the present concepts of vehicle systems.

The only feasible alternative remaining is to reduce the inert weight which is not useful for propulsion in the vehicle system so that a greater payload weight can be obtained without the need to increase initial launch weight of the vehicle system. This is an important consideration since any unnecessary inert weight in the various stages of the vehicle system imposes a heavy, additional demand on required engine thrust which is functionally related to launch weight. In a large, three stage booster system to be used on a lunar flight, for example, any change in weight of the final stage will be reflected in a similar change in the total launch weight multiplied, however, by a growth factor which may easily number in the hundreds.

In undertaking manned, space exploration missions, the astronauts may be exposed to radiation fields of high intensity in space. Biological damage is done by the ionization produced by radiation and high energy charged particles which pass through the tissues of the astronauts. A lethal action arises when the radiation dosage is excessive such that changes in living cells result in their death when they attempt division. Of course, extremely high radiation dose rates which may be lethal to an astronaut after a relatively short exposure period are

Fig. 12.31 Portion of the first of six sheets of specifications (petition) for the radiation protection system illustrated in Figs. 12.32 and 12.33. Copies of patents may be purchased from the Patent Office (50 cents each).

Fig. 12.32 Patent drawing showing a radiation protection system. See also Fig. 12.33.

Fig. 12.33 Drawings showing the details that are related to the claims.

drawing must be filed with the application. Illustrations facilitating an understanding of the invention (for example flow sheets in cases of processes, and diagrammatic views) may also be furnished in the same manner as drawings, and may be required by the Office when considered necessary or desirable.

35 U. S. C. 113. Drawings. *When the nature of the case admits, the applicant shall furnish a drawing.*

No names or other identification will be permitted within the "sight" of the drawing, and applicants are expected to use the space above and between the hole locations to identify each sheet of drawings. This identification may consist of the attorney's name and docket number or the inventor's name and case number and may include the sheet number and the total number of sheets filed (for example, "sheet 2 of 4").

83. Content of drawing. (a) The drawing must show every feature of the invention specified in the claims. However, conventional features disclosed in the description and claims, where their detailed illustration is not essential for a proper understanding of the invention, should be illustrated in the drawing in the form of a graphical drawing symbol or a labeled representation (e.g. a labeled rectangular box).

(b) When the invention consists of an improvement on an old machine the drawing must when possible exhibit, in one or more views, the improved portion itself, disconnected from the old structure, and also in another view, so much only of the old structure as will suffice to show the connection of the invention therewith.

84. Standards for drawings.

(*a*) *Paper and Ink.* Drawings must be made upon pure white paper of a thickness corresponding to two-ply or three-ply bristolboard. The surface of the paper must be calendered and smooth and of a quality which will permit erasure and correction with India ink. India ink, or its equivalent in quality, must be used for pen drawings to secure perfectly black solid lines. The use of white pigment to cover lines is not acceptable.

(*b*) *Size of Sheet and Margins.* The size of a sheet on which a drawing is made must be exactly 8½ by 14 inches. One of the shorter sides of the sheet is regarded as its top. The drawing must include a top margin of 2 inches and bottom and side margins of one-quarter inch from the edges, thereby leaving a "sight" precisely 8 to 11¾ inches. Margin border lines are not permitted. All work must be included within the

"sight." The sheets may be provided with two ¼-inch-diameter holes having their centerlines spaced eleven-sixteenths inch below the top edge and 2¾ inches apart, said holes being equally spaced from the respective side edges.

(*c*) *Character of Lines.* All drawings must be made with drafting instruments or by a process which will give them satisfactory reproduction characteristics. Every line and letter must be absolutely black and permanent; the weight of all lines and letters must be heavy enough to permit adequate reproduction. This direction applies to all lines however fine, to shading, and to lines representing cut surfaces in sectional views. All lines must be clean, sharp, and solid, and fine or crowded lines should be avoided. Solid black should not be used for sectional or surface shading. Freehand work should be avoided wherever it is possible to do so.

(*d*) *Hatching and Shading.* (1) Hatching should be made by oblique parallel lines, which may be not less than about one-twentieth inch apart.

(2) Heavy lines on the shade side of objects should be used except where they tend to thicken the work and obscure reference characters. The light should come from the upper left-hand corner at an angle of 45°. Surface delineations should be shown by proper shading, which should be open.

(*e*) *Scale.* The scale to which a drawing is made ought to be large enough to show the mechanism without crowding when the drawing is reduced in reproduction, and views of portions of the mechanism on a larger scale should be used when necessary to show details clearly; two or more sheets should be used if one does not give sufficient room to accomplish this end, but the number of sheets should not be more than is necessary.

(*f*) *Reference Characters.* The different views should be consecutively numbered figures. Reference numerals (and letters, but numerals are preferred) must be plain, legible and carefully formed, and not be encircled. They should, if possible, measure at least one-eighth of an inch in height so that they may bear reduction to one twenty-fourth of an inch; and they may be slightly larger when there is sufficient room. They must not be so placed in the close and complex parts of the drawing as to interfere with a thorough comprehension of the same, and therefore should rarely cross or mingle with the lines. When necessarily grouped around a certain part, they should be placed at a little distance, at the closest point where there is available space, and connected by lines with the parts to which they refer. They should not be placed upon hatched or shaded surfaces but when necessary, a blank space may be left in the hatching or shading where the character occurs so that it shall appear perfectly distinct and separate from the work. The same part of an invention appearing in more than one view of the drawing must always be designated by the same character, and the same character must never be used to designate different parts.

(g) *Symbols, Legends.* Graphical drawing symbols and other labeled representations may be used for conventional elements when appropriate, subject to approval by the Office. The elements for which such symbols and labeled representations are used must be adequately identified in the specification. While descriptive matter on drawings is not permitted, suitable legends may be used, or may be required in proper cases, as in diagrammatic views and flow sheets or to show materials or where labeled representations are employed to illustrate conventional elements. Arrows may be required, in proper cases, to show direction of movement. The lettering should be as large as, or larger than, the reference characters.

(i) *Views.* The drawing must contain as many figures as may be necessary to show the invention; the figures should be consecutively numbered if possible, in the order in which they appear. The figures may be plan, elevation, section, or perspective views, and detail views of portions or elements, on a larger scale if necessary, may also be used. Exploded views, with the separated parts of the same figure embraced by a bracket, to show the relationship or order of assembly of various parts are permissible. When necessary a view of a large machine or device in its entirety may be broken and extended over several sheets if there is no loss in facility of understanding the view (the different parts should be identified by the same figure number but followed by the letters, **a, b, c,** etc., for each part). The plane upon which a sectional view is taken should be indicated on the general view by a broken line, the ends of which should be designated by numerals corresponding to the figure number of the sectional view and have arrows applied to indicate the direction in which the view is taken. A moved position may be shown by a broken line superimposed upon a suitable figure if this can be done without crowding, otherwise a separate figure must be used for this purpose. Modified forms of construction can only be shown in separate figures. Views should not be connected by projection lines nor should center lines be used.

(j) *Arrangement of Views.* All views on the same sheet must stand in the same direction and should if possible, stand so that they can be read with the sheet held in an upright position. If views longer than the width of the sheet are necessary for the clearest illustration of the invention, the sheet may be turned on its side so that the two-inch margin is on the right-hand side. One figure must not be placed upon another or within the outline of another.

(k) *Figure for Official Gazette.* The drawing should, as far as possible, be so planned that one of the views will be suitable for publication in the Official Gazette as the illustration of the invention.

(l) *Extraneous Matter.* An inventor's, agent's, or attorney's name, signature, stamp, or address, or other extraneous matter, will not be permitted upon the face of a drawing, within or without the margin, except that identifying indicia (attorney's docket number, inventor's name, number of sheets, etc.) should be placed within three-fourths inch of the top edge and between the hole locations defined in paragraph (b) of this rule. Authorized security markings may be placed on the drawings provided they be outside the illustrations and are removed when the material is declassified.

(m) *Transmission of Drawings.* Drawings transmitted to the Office should be sent flat, protected by a sheet of heavy binder's board, or may be rolled for transmission in a suitable mailing tube; but must never be folded. If received creased or mutilated, new drawings will be required.

85. Informal drawings. The requirements of rule 84 relating to drawings will be strictly enforced. A drawing not executed in conformity thereto, if suitable for reproduction, may be admitted, but in such case the drawing must be corrected or a new one furnished, as required. The necessary corrections or mounting will be made by the Office upon applicant's request or permission and at his expense.

86. Draftsman to make drawings. (a) Applicants are advised to employ competent draftsmen to make their drawings.

(b) The Office may furnish the drawings at the applicant's expense as promptly as its draftsmen can make them, for applicants who cannot otherwise conveniently procure them.

88. Use of old drawings. If the drawings of a new application are to be identical with the drawings of a previous application of the applicant on file in the Office, or with part of such drawings, the old drawings or any sheets thereof may be used if the prior application is, or is about to be, abandoned, or if the sheets to be used are cancelled in the prior application. The new application must be accompanied by a letter requesting the transfer of the drawings, which should be completely identified.

123. Amendments to the drawing. (a) No change in the drawing may be made except by permission of the Office. Permissible changes in the construction shown in any drawing may be made only by the Office. A sketch in permanent ink showing proposed changes, to become part of the record, must be filed. The paper requesting amendments to the drawing should be separate from other papers.

(b) Substitute drawings will not ordinarily be admitted in any case unless required by the Office.

Design patents

152. Drawing. The design must be represented by a drawing made in conformity with the rules laid down for drawings of mechanical inventions and must contain a sufficient number of views to constitute a complete disclosure of the appearance of the article. Appropriate surface shading must be used to show the character or contour of the surfaces represented.

Letters and figures of reference must be carefully formed. Several types of lettering and figure marks are shown in Fig. 12.34, however, the draftsman may use any style of lettering that he may choose.

Place heavy lines on the shade side of objects, assuming that the light is coming from the upper left-hand corner at an angle of 45° (Fig. 12.34). Make these heavy lines the same weight throughout the various views on the drawings.

Descriptive matter is not permitted on patent drawings. Legends may be applied when necessary but only plain black lettering should be used.

The different views should be consecutively numbered.

Surface delineations should be shown by proper shading (Fig. 12.35). The amount of shading necessary depends on the size of the diameter of the shaft, etc. Note that a single heavy line on the shadow side is sufficient shading for small pipes, rods, and shafts. When more than one shade line is used on cylindrical surfaces, the shading is blended from the second line. Note that the outer line is a light line. This rule on shading applies to spherical as well as cylindrical surfaces (Fig. 12.35).

Make all lines clear and sharp so that they will reproduce properly.

India ink, or its equivalent in quality, must be used for pen drawings to secure perfectly black solid lines.

Heavy shade lines on perspective views are placed on the edges closest to the eye (Fig. 12.36). The rule of the light coming from the upper left-hand corner at a 45° angle does not apply to perspective views. If a very light line is placed either side of the heavy line, a more finished appearance of the article is obtained. The addition of horizontal ground line as shown in the lower perspective view of the block emphasizes support for the same.

The appearance of a mirror or shiny surfaces can be illustrated by the oblique shading shown on the two views in Fig. 12.36.

Reference characters should be placed at a little distance from the parts to which they refer (Fig. 12.37). They should be connected with these parts by a short lead line, never by a long lead line. When necessary blank spaces must be left on shaded and hatched areas for applying the numerals.

Use wood graining sparingly on parts of wood in section. Excessive wood graining is objectionable as it blurs the view and is very confusing (Fig. 12.37).

Various methods of shading are shown; however, the conventional surface shading should be used until the draftsman has obtained enough experience to attempt the more involved types of shading (Fig. 12.37).

Inclined surfaces are distinguished

Fig. 12.34 Examples of line work and lettering for patent drawings.

from flat surfaces by using the shading shown on the illustration in the upper left in Fig. 12.38. You will note that the outer line is always a light line. This gives a slanting effect to the surface as the heavy line is placed on the edge of the upper plane. The surface shading is blended from this heavy line giving the desired appearance.

The other figures in the illustration show various methods of shading. Flat, shiny surfaces may be shown as illustrated in the circular figure.

The scale to which a drawing is made ought to be large enough to show the mechanism without crowding.

Several methods of illustrating threads are shown in Fig. 12.39. The conventional thread may be shown on small bolts and openings. The detail method should be used on large pipes and threaded portions. Solid black shading as shown is very effective in illustrating the threads but care should be used in applying same.

Convex and concave surfaces are defined by the shading shown in the illustrations of the ball and socket (Fig. 12.39).

Plan the views properly so that one figure is not placed upon another or within the outline of another.

The conventional method of illustrating bevel gears is clearly shown on the two figures in Fig. 12.40. Particular care must be given to the correct spacing between the gear teeth and also to the weight of the shade lines used. Both must be correctly shown to obtain the desired effect.

Two types of bearings are also shown in Fig. 12.40. The roller bearing is clearly disclosed by the use of the conventional cylindrical shading. The fanciful black shading shown on the ball bearing is very effective in bringing out the idea of an object being shiny as well as round.

The use of white pigment to cover lines is not acceptable.

**SURFACE SHADING FOR
VARIOUS SIZES OF PIPES & SHAFTS**

SURFACE SHADING FOR SPHERICAL OBJECTS

Fig. 12.35 Surface shading for shafts and spherical objects.

Fig. 12.36 Shading patent drawings.

SURFACE SHADING ON BEVEL EDGES

Fig. 12.38 Shading surfaces and beveled edges.

NUMERALS MUST BE PLACED
AS CLOSE AS POSSIBLE TO THE
PART TO WHICH THEY REFER

SHADING FOR ROUND
HANDLES, ETC.

Fig. 12.37 Numbering parts and shading round
handles and cylindrical surfaces.

THREADS-CONVENTIONAL METHOD

Fig. 12.39 Treatment of threads, fasteners, and
small components.

BEVEL GEARS

NOTE—TEETH OF
EACH GEAR
HAVE THE
SAME
SLANT

BALL BEARING

INNER
RACE

OUTER
RACE

NOTE—ALL TEETH CONVERGE IN A CENTRAL
POINT.—BROKEN LINES ARE FOR INSTRUCTION
PURPOSES AND ARE NOT TO BE PLACED ON DRAWINGS

TOP PLAN VIEW

ROLLER BEARING

Fig. 12.40 Treatment of gears and bearings on patent drawings.

Design projects

Note to Instructor: Design projects may be included within traditional courses covering engineering drawing. Either two or three weeks of a semester may be set aside for design projects, or the design tasks may be spread out over a much longer time to parallel the work on the regularly assigned classroom drawing problems. In developing the design of a simple product, a student working in a group can acquire much of the basic knowledge deemed essential.

The following problems are offered as suggestions to stimulate creativity and give some additional experience in both pictorial and multiview sketching. Students who have an inclination to design useful mechanisms should be encouraged to select a problem for themselves, for the creative mind works best when directed to a task in which it already has some interest. However, the young beginner should confine his activities to ideas for simple mechanisms that do not require extensive training in machine design and the engineering sciences.

It is suggested that the group leader for the design project prepare a design event and activities schedule using a form that is similar to the one shown in Fig. 12.41. A carefully prepared record should be kept of the progress of the design work.

1. Prepare design sketches (both pictorial and multiview) for an open-end wrench to fit the head of a bolt having a body diameter of 24 mm. Give dimensions on the multiview sketch and specify the material. See Table 13 in the Appendix.

2. Prepare design sketches (both pictorial and multiview) for a wrench having a head with four or more fixed openings to fit the heads of bolts having nominal diameters of $\frac{5}{8}$, $\frac{3}{4}$, $\frac{7}{8}$, and 1 in. Dimension the multiview sketch and specify the material.

DESIGN SCHEDULE & PROGRESS REPORT

Project:_____

Design Team:_____

Leader:_____

Member:_____

_____

_____

Work Periods:_____

Due Dates

Investigation Report:_____

Proposal:_____

Preliminary Report:_____

Final Report:_____

ASSIGNMENT		EST. HRS	START DATE	PERCENT COMPLETE				ACTUAL HOURS
	Members			25%	50%	75%	100%	
1								
2								
3								
4								
5								

Fig. 12.41 Design schedule and progress record.

3. Prepare a series of design sketches for a bumper hitch (curved bumper) for attaching a light two-wheel trailer to a passenger automobile. Weight of trailer is 135 kg.

4. Prepare sketches for a hanger bracket to support a 24 mm control rod. The bracket must be attached to a vertical surface to which the control rod is parallel. The distance between the vertical surface and the center line of the control rod is 100 mm.

5. Prepare sketches for a mechanism to be attached to a two-wheel hand truck to make it easy to move the truck up and down stairs with a heavy load.

6. Prepare sketches for a quick-acting clamp that can be used to hold steel plates in position for making a lap weld.

7. Prepare a pictorial sketch of a bracket that will support an instrument panel at an angle of 45° with a vertical bulkhead to which the bracket will be attached. The bracket should be designed to permit the panel to be raised or lowered a height distance of 100 mm as desired.

8. Prepare sketches for an adjustable pipe support for a $1\frac{1}{2}$-in. pipe that is to carry a chemical mixture in a factory manufacturing paint. The pipe is overhead and is to be supported at 10-ft intervals where the adjustable supports can be attached to the lower chords of the roof trusses. The lower chord of a roof truss is formed by two angles $2\frac{1}{2} \times 2\frac{1}{2} \times \frac{5}{16}$ that are separated by $\frac{3}{8}$-in.-thick washers.

9. Prepare sketches and working drawings for an easy-to-operate fast-release glider hitch. It is suggested that the student talk with several members of a local glider club to determine the requirements for the hitch and what improvements can be made for the hitch that is being used. Follow the stages of design listed in Sec. 12.13.

10. Prepare design sketches for a camera mounting that may be quickly attached and removed from any selected surface on an automobile, boat, or other type of moving vehicle. The device is to sell for not more than $9.95. Standard parts are to be used if possible. Follow through the several stages of design listed in Sec. 12.13 as required by the instructor. Make either a complete set of working drawings or drawings of selected parts.

11. Design a tire pump that will be more efficient and easier to operate than the ones now on the market.

12. Design an electric door lock that can be opened by pushing buttons in a special pattern.

13. Design a thermostatically controlled, electrically heated sidewalk upon which snow and ice will not accumulate.

14. Design an automatic automobile theft alarm.

15. Design a weather-controlled window-closing system.

16. Design a child's toy.

17. Design an automatic pet food dispenser.

18. *Class Projects.* Each student in the class is to prepare a description (on a single sheet of paper) of an innovational idea of a needed design that he deems to be suitable for a group project. From among the ideas collected several of the most worthy will be assigned to the class for development, one idea to each group. Each group of students is to be considered as a project group and shall be headed by one student member who may be thought of as being the project engineer. The instructor will assume the role of coach for all of the groups. About midway in the total time period assigned for the development of the design, each group is to submit a written preliminary design report that shall be accompanied by sketches. A final design report (see Sec. 12.24) will be due when the project has been completed. Each report will be judged on the following: (1) evidence of good group organization, (2) quality of technical work, (3) function and appearance, (4) economic analysis, (5) manufacturing methods and requirements, and (6) over-all effectiveness of communication (written and/or oral). (*Note:* The instructor will decide whether or not finished shop drawings are to be made for all of the parts. Information needed for the preparation of finished drawings may be found in Chapters 13–16.)

PART 4

GRAPHICS FOR DESIGN AND COMMUNICATION

Metric drawing. *(Courtesy International Harvester Company)*

dimensions, notes, limits, and geometric tolerances 13

A □ Fundamentals and techniques

13.1
Introduction. A detail drawing, in addition to giving the shape of a part, must furnish information such as the distances between surfaces, locations of holes, kind of finish, type of material, number required, and so forth. The expression of this information on a drawing by the use of lines, symbols, figures, and notes is known as *dimensioning*.

Intelligent dimensioning requires engineering judgment and a thorough knowledge of the practices of patternmaking, forging, and machining.

13.2
Theory of Dimensioning. Any part may be dimensioned easily and systematiclally by dividing it into simple geometric solids. Even complicated parts, when analyzed, are usually found to be composed principally of cylinders and prisms and, frequently, frustums of pyramids and cones

(Fig. 13.1). The dimensioning of an object may be accomplished by dimensioning each elemental form to indicate its size and relative location from a center line, base line, or finished surface. A machine drawing requires two types of dimensions: *size dimensions* and *location dimensions*.

13.3
Size Dimensions (Fig. 13.2). Size dimensions give the size of a piece, component part, hole, or slot.

Figure 13.1 should be carefully analyzed, as the placement of dimensions

279

Prism   Cylinder   Pyramid   Cone   Sphere

Fig. 13.1 Dimensioning geometric shapes.

Fig. 13.2 Size dimensions.

Fig. 13.3 Location dimensions.

shown is applicable to the elemental parts of almost every piece.

The rule for placing the three principal dimensions (width, height, and depth) on the drawing of a prism or modification of a prism is as follows: *Give two dimensions on the principal view and one dimension on one of the other views.*

The circular cylinder, which appears as a boss or shaft, requires only *the diameter and length, both of which are shown preferably on the rectangular view.* It is better practice to dimension a hole (negative cylinder) by giving the diameter and operation as a note on the contour view with a leader to the circle (Figs. 13.1 and 13.54).

Cones are dimensioned by giving *the diameter of the base and the altitude on the same view.* A taper is one example of a conical shape found on machine parts (Fig. 13.49).

Pyramids, which frequently form a part of a structure, are dimensioned by giving *two dimensions on the view showing the shape of the base.*

A sphere requires only the diameter.

13.4
Location Dimensions. Location dimensions fix the relationship of the component parts (projections, holes, slots, and other significant forms) of a piece or structure (Fig. 13.3). Particular care must be exercised in the selection and placing of location dimensions because on them depends the accuracy of the operations in making a piece and the proper mating of the piece with other parts. To select location dimensions intelligently, one must first determine the contact surfaces, finished surfaces, and center lines of the elementary geometric forms and, with the accuracy demanded and the method of production in mind, decide from what other surface or center line each should be located. Mating location dimensions must be given from the same center line or finished surface on both pieces.

Location dimensions may be from center to center, surface to center, or surface to surface (Fig. 13.4).

13.5
Procedure in Dimensioning. The theory of dimensioning may be applied in six steps, as follows:

1. Mentally divide the object into its component geometrical shapes.

2. Place the size dimensions on each form.

3. Select the locating center lines and surfaces after giving careful consideration to mating parts and to the processes of manufacture.

4. Place the location dimensions so that each geometrical form is located from a center line or finished surface.

5. Add the overall dimensions.

6. Complete the dimensioning by adding the necessary notes.

13.6
Placing Dimensions. Dimensions must be placed where they will be most easily understood—in the locations where the reader will expect to find them. They generally are attached to the view that shows the contour of the features to which they apply, and a majority of them usually will appear on the principal view (Fig. 13.14). Except in cases where special convenience and ease in reading are desired, or when a dimension would be so far from the form to which it referred that it might be misinterpreted, dimensions should be placed outside a view. They should appear directly on a view only when clarity demands.

All extension and dimension lines should be drawn before the arrowheads

L_1 – Center to Center
L_2 – Surface to Center
L_3 – Surface to Surface

Fig. 13.4 Types of location dimensions.

have been filled in or the dimensions, notes, and titles have been lettered. Placing dimension lines not less than 15 mm (.62 in.) from the view and at least 10 mm (.38 in.) from each other will provide spacing ample to satisfy the one rule to which there is no exception: *Never crowd dimensions*. If the location of a dimension forces a poor location on other dimensions, its shifting may allow all to be placed more advantageously without sacrificing clarity. Important location dimensions should be given where they will be conspicuous, even if a size dimension must be moved.

The person dimensioning a drawing must make certain that every feature has been completely dimensioned and that no dimension has been repeated in a second view.

13.7

Dimensioning Practices. A generally recognized system of lines, symbols, figures, and notes is used to indicate size and location. Figure 13.5 illustrates dimensioning terms and notation.

A *dimension line* is a lightweight line that is terminated at each end by an arrowhead. A numerical value, given along the dimension line, specifies the number of units for the measurement that is indicated (Fig. 13.6). When the numerals are in a single line, the dimension line is broken near the center, as shown in (*a*) and (*b*). Under no circumstances should the line pass through the numerals.

Extension lines are light, continuous lines extending from a view to indicate the extent of a measurement given by a dimension line that is located outside a view. They start 1.5 mm (0.6 in.) from the view and extend 3 mm (.12 in.) beyond the dimension line (Fig. 13.5).

Arrowheads are drawn for each dimension line, before the figures are lettered. They are made with the same pen or pencil used for the lettering. The size of an arrowhead, although it may vary with the size of a drawing, should be uniform on any one drawing. To have the proper proportions, the length of an arrowhead must be approximately three times its spread. This length for average work is usually 3 mm (.12 in.). Figure 13.7 shows enlarged drawings of arrowheads of correct proportions. Although many draftsmen draw an arrowhead with one stroke, the beginner will get better results by using two slightly concave strokes drawn toward the point (*a*) or, as shown in (*b*), one

Fig. 13.5 Terms and dimensioning notation. (Dimension values are in millimeters.)

Fig. 13.6 Dimension line.

Fig. 13.7 Formation of arrowheads.

stroke drawn to the point and one away from it.

A *leader* or *pointer* is a light, continuous line (terminated by an arrowhead) that extends from a note to the feature of a piece to which the note applies (Fig. 13.5). It should be made with a straightedge and should not be curved or made freehand.

A leader pointing to a curve should be radial, and the first 3 mm (.12 in.) of it should be in line with the note (Fig. 13.5).

Finish marks indicate the particular surfaces of a rough casting or forging that are to be machined or "finished." They are placed in all views, touching the visible or invisible lines that are the edge views of surfaces to be machined (Fig. 13.8). When a surface is to be finished by the removal of material and it is not necessary to indicate surface quality, a bar is added to the check-mark (✔) portion of the texture symbol at the top of the short leg. Also see Sec. 13.38.

It is not necessary to show finish marks on holes. They are also omitted, and a title note, "finish all over," is substituted, if the piece is to be completely machined. Finish marks are not required when limit dimensions are used.

Dimension figures should be lettered either horizontally or vertically with the whole numbers equal in height to the capital letters in the notes and guidelines and slope lines must be used. The numerals must be legible; otherwise, they might be misinterpreted in the shop and cause errors, which would be embarrassing to the draftsman.

13.8

Fractional Dimensioning. For ordinary work, where accuracy is relatively unimportant, shopmen work to nominal dimensions given as common fractions of an inch, such as $\frac{1}{2}$, $\frac{1}{4}$, $\frac{1}{8}$, $\frac{1}{16}$, $\frac{1}{32}$, and $\frac{1}{64}$.

Use of ✔ finish symbol

Fig. 13.8 Finish marks.

When dimensions are given in this way, many large corporations specify the required accuracy through a note on the drawing that reads as follows: *Permissible variations on common fraction dimensions to machined surfaces to be ±.010 unless otherwise specified.* It should be understood that the allowable variations will differ among manufacturing concerns because of the varying degree of accuracy required for different types of work.

13.9

Decimal System. At present, the fractional system, which dominated the whole of American industry until the Ford Motor Company adopted the decimal-inch some 40 years ago, is still used in those fields that are not under the direct influence of the automotive and aircraft companies. How long the inch system of fractions and decimals will last cannot be foretold at this time. In the long run this matter may be of little importance, since it may be only a few years before the millimeters of the metric system completely replace the old English inch system. In the United States more and more multinational companies are making metric drawings (Figs. 5.1 and 16.3).

However, this need not be of great concern to the student, for he should be primarily interested in the selection and placement of dimensions. At a later time he will find it easy to use either the fractional or decimal inch or metric systems as required in his field of employment. To assist one to use any system, a standard conversion table has been provided in the Appendix (Table 3).

In Fig. 13.9 a drawing is shown that illustrates decimal-inch dimensioning.

The following practices are recommended for decimal-inch dimensioning.

(*a*) Two-place decimals should be used for dimensions where tolerance limits of ±.01 or more can be allowed (Fig. 13.9).

(*b*) Decimals to three or more places should be used for tolerance limits less than ±.010 [see Fig. 13.6(*b*)].

(*c*) In the case of a two-place decimal, the second decimal place should preferably be an even digit such as .02, .04, and .06 so that when it is divided by 2, the result will remain a two-place decimal. Odd two-place decimals may be used where necessary for design reasons.

(*d*) Common fractions may be used to

indicate standard nominal sizes for materials and for features produced by standard tools as in the case of drilled holes, threads, keyways, and so forth.

(*e*) When desired, decimal equivalents of nominal commercial sizes may be used for materials and for features such as drilled holes, threads, and so forth that are produced by standard tools.

13.10

Use of the Metric System (Fig. 13.10). Since the metric system has worldwide acceptance and has been legalized for use in this country (approved by act of Congress in 1866), engineers and draftsmen in the United States will find it advantageous to be able to interpret and quite possibly prepare detail drawings using metric dimensioning. At present, there are persons who argue strongly that American industry must change over to this system if the United States is to retain leadership in world trade. In any case, metric dimensioning is appearing more frequently than in the past on drawings made in the United States. The unit of measurement is the millimeter (mm) and, since this fact is generally known by all persons using the system, no indicating marks are needed.

Under the metric system, drawings are prepared to scales based on divisions of 10, such as 1 to 2, 1 to 5, 1 to 10, and so forth. A millimeter is one-thousandth part of a meter, which our government has established as being 39.37 in. in length.

Most of the illustrations in this chapter have been dimensioned using millimeters.

B □ General dimensioning practices

13.11

Selection and Placement of Dimensions and Notes. The reasonable application of the selected dimensioning practices that follow should enable a student to dimension acceptably. The practices in boldface type should never be violated. In fact, these have been so definitely established by practice that they might be called rules.

When applying the dimensioning practices that follow, the student should realize that each company has its own drafting standards and thus the practices set forth in this chapter may not be exactly the same as the practices followed in all corporations throughout the United

DET. 15 RIVET SET
1– TYPE "CC" STEEL
STK. 1⅜ WT. .84 LBS.
HEAT IN CYANIDE TO 1525°F
QUENCH IN OIL
DRAW AT 350°F
ROCKWELL 54–59C

Fig. 13.9 Decimal-inch dimensioning.
(*Courtesy Ford Motor Company*)

States. However, the practices given are in general agreement with recommendations set forth in ANSI standards.

1. Place dimensions using either of two recognized methods—aligned or unidirectional.

a. *Aligned method.* Place the numerals for the dimension values so that they are readable from the bottom and right side of the drawing. An aligned expression is placed along and in the direction of the dimension line (Fig. 13.11). Make the values for oblique dimensions readable from the directions shown in Fig. 13.31.

b. *Unidirectional method.* Place the numerals for the dimension values so that they can be read from the bottom of the drawing (see Fig. 13.12). The fraction bar for a common fraction should be parallel to the bottom of the drawing.

2. Place dimensions outside a view, unless they will be more easily and quickly understood if shown on the view (Figs. 13.12 and 13.13).

3. Place dimensions between views unless the rules, such as the contour rule, the rule against crowding, and so forth, prevent their being so placed.

4. Do not use an object line or a center line as a dimension line.

5. Locate dimension lines so that they will not cross extension lines.

6. If possible, avoid crossing two dimension lines.

7. A center line may be extended to serve as an extension line (Fig. 13.14).

Fig. 13.10 Metric dimensioning.
(Dimensions are in millimeters.)

Fig. 13.11 Reading dimensions aligned.

All dimension values read from the bottom

**Fig. 13.12 Reading dimensions—
unidirectional.**

Some dimensions may be
placed on a view for added
clearness and ease of reading

Keep area around numerical
values free of section lines

Fig. 13.13 Dimensions on the view.

Fig. 13.16 Dimensioning an angle bracket.
(Dimension values are in millimeters.)

8. Keep parallel dimensions equally spaced [usually 10 mm (.38 in.) apart] and the figures staggered (Fig. 13.15).

9. Always give locating dimensions to the centers of circles that represent holes, cylindrical projections, or bosses (Figs. 13.5 and 13.14).

10. If possible, attach the location dimensions for holes to the view on which they appear as circles (Fig. 13.16).

11. Group related dimensions on the view showing the contour of a feature (Fig. 13.14).

12. Arrange a series of dimensions in a continuous line (Fig. 13.17).

13. Dimension from a finished surface, center line, or base line that can be readily established (Figs. 13.38 and 13.50).

14. Stagger the figures in a series of parallel dimension lines to allow sufficient space for the figures and to prevent confusion (Fig. 13.15).

15. Place longer dimensions outside shorter ones so that extension lines will not cross dimension lines.

16. Give three overall dimensions lo-

Fig. 13.14 Contour principle of dimensioning.

Fig. 13.17 Consecutive dimensions.

cated outside any other dimensions (unless the piece has cylindrical ends—see 42 and Fig. 13.42).

17. When an overall is given, one intermediate distance should be omitted unless noted (REF) as being given for reference (Figs. 13.18 and 13.19).

18. Do not repeat a dimension. One of the duplicated dimensions may be missed if a change is made. Give only those dimensions that are necessary to produce or inspect the part.

19. Make decimal points of a sufficient size so that dimensions cannot be misread.

20. When dimension figures appear on a sectional view, show them in a small uncrosshatched portion so that they may be easily read. This may be accomplished by doing the section lining after the dimensioning has been completed (Fig. 13.20).

21. When an arc is used as a dimension line for an angular measurement, use the vertex of the angle as the center [Fig. 13.21(*a*)]. It is usually undesirable to terminate the dimension line for an angle at lines that represent surfaces. It

Fig. 13.15 Parallel dimensions.

is better practice to use an extension line [Fig. 13.21(*b*)].

22. Place the figures of angular dimensions so they will read from the bottom of a drawing, except in the case of large angles (Fig. 13.22).

23. Dimension an arc by giving its radius preceded by the abbreviation R, and indicate the center with a small cross. [Locate the center by dimensions. (Fig. 13.23).]

24. TRUE R precedes the radius value, where the radius is dimensioned in a view that does not show the true shape of the arc (Fig. 13.24).

25. Show the diameter of a circle, never the radius. If it is not clear that the dimension is a diameter, the figures should be preceded by the diameter symbol ϕ (Figs. 13.25 and 13.26). Often this will allow the elimination of one view.

26. When dimensioning a portion of a sphere with a radius the term SPHER R is added (Fig. 13.27).

27. Letter all notes horizontally (Fig. 13.50).

28. Make dimensioning complete, so that it will not be necessary for a workman to add or subtract to obtain a desired dimension or to scale the drawing.

29. Give the diameter of a circular hole, never the radius, because all hole-forming tools are specified by diameter. If the hole does not go through the piece, the depth may be given as a note (Fig. 13.28).

30. Never crowd dimensions into small spaces. Use the practical methods suggested in Fig. 13.29.

31. Avoid placing inclined dimensions in the shaded areas shown in Fig. 13.30. Place them so that they may be conven-

Fig. 13.18 Omit unnecessary dimensions.

Fig. 13.19 Mark dimensions REF, if given for reference.

Fig. 13.20 Dimension figures on a section view.

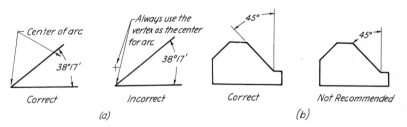

Fig. 13.21 Dimensioning an angle.

Fig. 13.22 Angular dimensions.

Fig. 13.23 Dimensioning a circular arc. (Dimension values are in millimeters.)

Fig. 13.24 Dimensioning a circular arc— true R.

Fig. 13.25 Dimensioning a cylindrical piece.

Note:

*Although it is better practice to use a minimum of two views,
a cylindrical part may be completely described in one view (no
end view) by using the ⌀ symbol with the value of the diameter.*

Fig. 13.26 Dimensioning machined cylinders.

Fig. 13.27 Dimensioning a piece with a spherical end.

Fig. 13.28 Dimensioning holes.

Fig. 13.29 Dimensioning in limited spaces.

iently read from the right side of the drawing. If this is not desirable, make the figures read from the left in the direction of the dimension line [Fig. 13.31 (*b*)]. The unidirectional method is shown in (*a*).

32. Omit superfluous dimensions. Do not supply dimensional information for the same feature in two different ways.

33. Give dimensions up to 72 in. in inches, except on structural and archi-

tectural drawings (Fig. 13.32). Omit the inch marks when all dimensions are in inches.

34. Show dimensions in feet and inches as illustrated in Fig. 13.33. Note that the use of the hyphen in (*a*) and (*b*) and the cipher in (*b*) eliminates any chance of uncertainty and misinterpretation.

35. If feasible, design a piece and its elemental parts to such dimensions as .10, .40 and .50 in. or 4, 10, 15, and

Fig. 13.30 Areas to avoid.

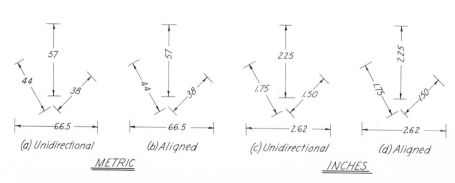

Fig. 13.31 Reading horizontal, vertical, and oblique dimensions.

20 mm. Except for critical dimensions and hole sizes, dimension values in millimeters should be given in a full number of millimeters, preferably to an even number (12, 16, 20, 40, etc.) that, when divided by 2, will remain a full number. Metric drill sizes are given in Appendix Tables 23 and 49.

36. Dimension a chamfer by giving the angle and length as shown in Fig. 13.34. For a 45° angle only, it is permissible to give the needed information as a note. The word "chamfer" may be omitted in the note form.

37. Equally spaced holes in a circular flange may be dimensioned by giving the dimeter of the bolt circle, across the circular center line, and the size and number of holes, in a note (Fig. 13.35).

38. When holes are unequally spaced on a circular center line, give the angles as illustrated in Fig. 13.36.

39. Holes that must be accurately located should have their location established by the coordinate method. Holes arranged in a circle may be located as shown in Fig. 13.37 rather than through the use of angular measurements. Figure 13.38 shows the application of the coordinate method to the location of holes arranged in a general rectangular form. The method with all dimensions referred to datum lines is sometimes called *base-line dimensioning*.

40. Dimension a curved line by giving offsets or radii.

a. A noncircular curve may be dimensioned by the coordinate method illustrated in Fig. 13.39. Offset measurements are given from datum lines.

b. A curved line, which is composed of circular arcs, should be dimensioned by giving the radii and locations of either the centers or points of tangency (Figs. 13.40 and 13.41).

41. Show an offset dimension line for an arc having an inaccessible center (Fig. 13.41). Locate with true dimensions the point placed in a convenient location that represents the true center.

42. Dimension, as required by the method of production, a piece with cylindrical ends. Give the diameters and center-to-center distance (Fig. 13.42). No overall is required.

Fig. 13.32 Dimension values.

Fig. 13.33 Feet and inches.

Fig. 13.34 Dimensioning a chamfer.

Fig. 13.35 Equally spaced holes.

Fig. 13.36 Locating holes on a circle by polar coordinates.

Fig. 13.37 Accurate location dimensioning of holes.

Fig. 13.38 Location dimensioning of holes.

43. The method to be used for dimensioning a piece with rounded ends is determined by the degree of accuracy required and the method of production (Figs. 13.43–13.45).

a. It has been customary to give the radii and center-to-center distance for parts and contours that would be laid out and/or machined using centers and radii. A link (Fig. 13.43) or a pad with a slot is dimensioned in this manner to satisfy the requirements of the pattern-

maker and machinist. An overall dimension is not needed.

b. Overall dimensions are recommended for parts having rounded ends when considerable accuracy is required. The radius is indicated but is not dimensioned when the ends are fully rounded. In Fig. 13.44, the center-to-center hole distance has been given because the hole location is critical.

c. Slots are dimensioned by giving length and width dimensions. They are located by dimensions given to their longitudinal center line and to either one end or a center line (Fig. 13.45).

d. When the location of a hole is more critical than the location of a radius from the same center, the radius and the hole should be dimensioned separately (Fig. 13.46).

44. A keyway on a shaft or hub should be dimensioned as shown in Fig. 13.47.

45. When knurls are to provide a rough surface for better grip, it is necessary to specify the pitch and kind of knurl, as shown in Fig. 13.48(*a*) and (*c*). When specifying knurling for a press fit, it is best practice to give the diameter before knurling with a tolerance and include the minimum diameter after knurling in the note that gives the pitch and type of knurl, as shown in (*b*).

46. Dimension special tapers as illustrated in Fig. 13.49. Standard tapers

Fig. 13.39 Dimensioning curves by offsets.

Fig. 13.40 Dimensioning curves consisting of circular arcs.

Fig. 13.41 Dimensioning curves by radii.

Fig. 13.42 Dimensioning a part with cylindrical ends—link.

Fig. 13.43 Link with rounded ends.

Fig. 13.44 Dimensioning a part with rounded ends.

Fig. 13.45 Dimensioning a slot.

Fig. 13.46 Dimensioning a piece with a hole and rounded end.

Fig. 13.47 Dimensioning keyways and keyslots.

require one diameter, the length, and a note specifying the taper by number. The usual practice is to give the diameter at the large end.

Special tapers may be dimensioned in different ways, as illustrated in Fig. 13.49. As can be observed, the following dimensions may be used in different combinations to specify the size and form of tapered surfaces of conical tapers.

1. The diameter at an end of the tapered feature.

2. The length of the tapered portion of the piece.

3. The diameter at a selected cross-sectional plane. The position of this plane, which may or may not be within the length of the tapered portion of the piece, is shown with a basic dimension.

4. A dimension locating a cross-sectional plane at which a basic diameter is specified.

5. The rate of taper (note).

6. The included angle.

The dimensions and notes given in (*b*) and (*e*) are considered adequate for tapers that engage one another permanently or intermittently.

Flat tapers may be dimensioned as shown in (*c*) or (*f*).

47. A half section may be dimensioned through the use of hidden lines on the external portion of the view (Fig. 13.50).

48. The fact that a dimension is out of scale may be indicated by a straight line placed underneath the dimension value (Fig. 13.51).

49. In sheet-metal work, mold lines are used in dimensioning instead of the centers of the arcs (see Fig. 13.52). A mold line (construction line) is the line at the intersection of the plane surfaces adjoining a bend.

13.12

Dimensions from Datum. When it is necessary to locate the holes and surfaces of a part with a considerable degree of accu-

Fig. 13.48 Dimensioning knurls.

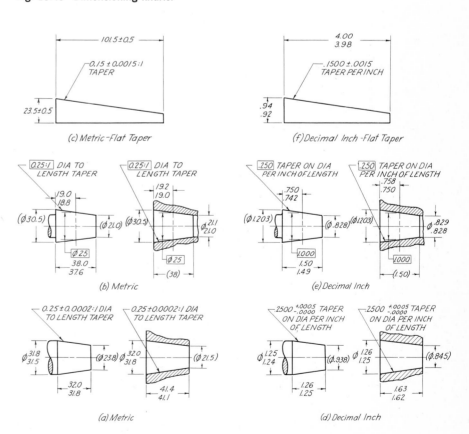

Fig. 13.49 Dimensioning tapers.

racy, it is the usual practice to specify their position by dimensions given from a datum (Fig. 13.53) in order to avoid cumulative tolerances. By this method the different features of a part are located with respect to carefully selected datums and not with respect to each other (Fig. 13.38). Lines and surfaces that are se-

lected to serve as datums must be easily recognizable and accessible during production. Corresponding datum points, lines, or surfaces must be used as datums on mating parts.

13.13

Notes (Fig. 13.54). The use of properly composed notes often adds clarity to the presentation of dimensional information involving specific operations. Notes also are used to convey supplementary instructions about the kind of material, kind of fit, degree of finish, and so forth. Brevity in form is desirable for notes of general information or specific instruction. In the case of threaded parts one should use the terminology recommended in Chapter 14.

13.14

Drawing Symbols. The symbols shown in Fig. 13.55 are used with dimensions and notes. A symbol must be shown before its associated linear dimension value in a dimension line or in place with a note as shown by the examples. The symbols have been shown enlarged at the left in the illustration and their interpretations directly follow.

13.15

Dual Dimensioning. A dual-dimensioning procedure has been adopted recently by several of our large industrial organizations; the procedure calls for showing both U.S. inch units and the SI (Système International) metric units of measurements on the same drawing. See Fig. 13.56.

Dimension values, given in both inch and SI units, must insure the interchangeability of parts. The interchangeability of a part is determined by the number of decimal places retained in rounding off a converted dimension and by the

Fig. 13.50 Dimensioning a half section.

Fig. 13.51 Dimension out of scale.

Fig. 13.52 Profile dimensioning.

Fig. 13.53 Dimensions from datum lines.

Fig. 13.54 Shop notes.

Symbol	Interpretation	Examples
$\varnothing$	*Diameter.* Formerly given as DIA or D.	$\varnothing 10 \pm 0.1$
R	*Radius*	R12
□	*Square Section.* Formerly shown as SQ or SQUARE.	$\square 20 \pm 0.5$
T	*Depth.* Formerly spelled out as DEPTH or DEEP.	T10 MAX
$\vee$	*Countersink.* Formerly noted as CSK.	$\varnothing 6 \pm 0.1$ $\vee 82°\varnothing 10 \pm 0.1$
⊔	*Counterbore or Spotface.* Formerly spelled out as CBORE or SF.	$\varnothing 6 \pm 0.1$ $\sqcup \varnothing 10 \pm 0.1$ T 3 ± 0.1
I	*Thickness or Thick.* Used to indicate thickness of a part or section. Formerly spelled out as THICK.	I 5 ± 0.1
△	*Break sharp corners or chamfer edges.* Formerly spelled out as BREAK or CHAMFER.	△ 0.5 MAX △45°×5
X	*Places or times.* Formerly spelled out as 2 HOLES, 2 PLACES, etc.	(2X) $\varnothing 8 \pm 0.2$
( )	*Parentheses.* Used around quantity callouts and around multiple operations for repetitive features.	(2) (4X) $\varnothing 12 \pm 0.2$

Fig. 13.55 General drawing symbols.

SELF-PIERCING CLINCH NUT

DIMENSIONS [ ] ARE IN INCHES

Fig. 13.56 Drawing showing dual-dimensioning. (*Courtesy Ford Motor Company*)

Fig. 13.57 Position method applied to **dual-dimensioning.**

Fig. 13.58 Bracket method applied to **dual-dimensioning.**

degree that the tolerance limits of the conversion have been permitted to violate the limits of the original dimension. There are several reliable sources for information on conversion principles. These are: (1) Rules for Conversion and Rounding—ASTM E380 Standard Metric Practice Guide and (2) Conversion of Toleranced Linear Dimensions—SAE 1390 Dual Dimensioning Standard.

Each drawing should have noted on it how the dimension values given in inch units and those given in SI units may be identified. For linear dimensions, the SI unit is the millimeter. Two methods have been commonly employed on drawings to distinguish between values given in these different units. These methods are: (1) the position method and (2) the bracket method.

When the position method is used, the millimeter dimension may be placed either above the inch dimension as shown in Fig. 13.57, (*a*) or to the left of the inch dimension with a slash line separating the values (*d*). Another method of display is similar except that the inch dimension value is placed above or to the left of the metric value.

In using the bracket method, the inch dimension and millimeter dimension values may also be displayed in two ways but only one method should be used on a single drawing. In the first method, the millimeter dimension value is enclosed in square brackets as shown in Fig. 13.58. The bracketed millimeter dimension value may be placed either above or below or to the right or left of the inch value. However, the practice that is adopted must be followed for the entire drawing. The second and alternate method is similar except that the inch values (instead of the milli-

meter dimensions) are enclosed in brackets.

Some of the drawings that are made during this decade will be dual-dimensioned since it is expected that U.S. industry will be changing over to the metric system in order to remain competitive in world trade.

C □ Limit dimensioning and cylindrical fits— cylindrical fits-inches

13.16

Limit Dimensions. Present-day competitive manufacturing requires quantity production and interchangeability for many closely mating parts. The production of each of these mating parts to an exact decimal dimension, although theoretically possible, is economically unfeasible, since the cost of a part rapidly increases as an absolute correct size is approached. For this reason, the commercial draftsman specifies an allowable error (tolerance) between decimal limits (Fig. 13.59). The determination of these limits depends on the accuracy and clearance required for the moving parts to function satisfactorily in the machine. Although manufacturing experience is often used to determine the proper limits for the parts of a mechanism, it is better and safer practice to adhere to the fits recommended by the American National Standards Institute in ANSI B4.1–1967. This standard applies to fits between plain cylindrical parts. Recommendations are made for preferred sizes, allowances, tolerances, and fits for use where applicable.

There are many factors that a designer

Fig. 13.59 Limit dimensioning for the production of interchangeable parts.

must take into consideration when selecting fits for a particular application. These factors might be the bearing load, speed, lubrication, materials, and length of engagement. Frequently temperature and humidity must be taken into account. Considerable practical experience is necessary to make a selection of fits or to make the subsequent adjustments that might be needed to satisfy critical functional requirements. In addition, manufacturing economy must never be overlooked.

Those interested in the selection of fits should consult texts on machine design and technical publications, for coverage of this phase of the dimensioning of cylindrical parts is not within the scope of this book. However, since it is desirable to be able to determine limits of size following the selection of a fit, attention in this section will be directed to the use of Table 48 in the Appendix. Whenever the fit to be used for a particular application has not been specified in the instructions for a problem or has not been given on the drawing, the student should consult his instructor after a tentative choice has been made based on the brief descriptions of fits as given in this section.

To compute limit dimensions it is necessary to understand the following associated terms.

Nominal size. Nominal size is the approximate size used for the purpose of identification. In Fig. 13.60, the nominal size of the hole and shaft is ½, or .50 in.

Basic size. Basic size is the theoretical size from which the limits of size are determined by the application of allowances and tolerances. Values to be used for plus-and-minus tolerancing are given in Tables 48A, B, C, D, and E in the Appendix.

Allowance. An allowance is the intentional difference between the maximum material limits (clearance or maximum interference) of mating parts. For clearance fits the allowance is positive. For interference fits it will be negative. In Fig. 13.60, the allowance is positive and is equal to the difference between the smallest hole and the largest shaft (.5000 − .4988 = .0012).

Tolerance. Tolerance is the permissible variation of a size that has been prescribed for a part. It is the difference between the specified limits. In Fig. 13.60, the tolerance for the shaft is .0010, the difference between the upper limit (.4988) and the lower limit (.4978).

Limits of size. The limits of size are the extreme maximum and minimum sizes specified by a toleranced dimension. In Fig. 13.60, the limits of size for the hole are .5016 and .5000.

Fit. Fit is the term commonly used to signify the tightness or looseness that may result from the application of a specific combination of allowances and tolerances in the design of mating parts. The fit may be either a clearance fit, an interference fit, or a transition fit.

Clearance fit. A clearance fit results in limits of size that assures clearance between assembled mating parts. In Fig. 13.60, the RC6 fit provides clearance under maximum material conditions. The allowance or minimum clearance is .0012.

Interference fit. An interference fit has limits of size that result in an interference between two mating parts. In the case of a hole and shaft, the shaft will be larger than the hole, to give an actual interference of metal that will result in either a force or a press fit. This has the effect of producing an almost permanent assembly for two assembled parts.

Transition fit. A transition fit has limits of size that can lead to either clearance or interference. A shaft may be either larger or smaller than the hole in a mating part.

Basic hole system. For the widely used basic hole system, the design size of the hole (smallest size) is the basic size. The allowance (if any) is applied to the shaft. In Fig. 13.60, the basic hole size is .5000, the smallest diameter. The allowance of .0012 can be subtracted from this diameter to obtain the diameter of the largest shaft (.4988). In design, the basic hole system is used whenever conditions permit because holes are usually formed and checked using standard tools and gauges while cold-rolled shafting can be much more easily machined to any specified size.

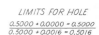

HOLE   *SHAFT*

NOMINAL SIZE

(a)

LIMITS FOR HOLE

$$\frac{0.5000 + 0.0000 = 0.5000}{0.5000 + 0.0016 = 0.5016}$$

LIMITS FOR SHAFT

$$\frac{0.5000 - 0.0012 = 0.4988}{0.5000 - 0.0022 = 0.4978}$$

COMPUTATIONS

(b)

HOLE   *SHAFT*

LIMITS

(c)

Fig. 13.60 Computation of limits (basic hole system).

Hole	$\dfrac{.5000 + .0012}{.5016 + .0012} = \dfrac{.5012}{.5028}$
Shaft	$\dfrac{.4988 + .0012}{.4978 + .0012} = \dfrac{.5000}{.4990}$

**Fig. 13.61 Computation
of limits (basic shaft
system).**

Basic shaft system. For the basic shaft system the design size (maximum diameter) of the shaft is the basic size. The allowance, if any, is applied to the hole. In Fig. 13.61, the basic shaft size is .5000. The basic shaft system should be used only when several parts having different fits are to be mounted on standard-size cold-finished shafting of fixed diameter. For each part to be assembled on the shaft the allowance can be applied to the basic shaft size to obtain the smallest diameter of the hole.

Tables 48A, B, C, D, and E in the Appendix cover three general types of fits: running fits, locational fits, and force fits. For educational purposes standard fits may be designated by means of letter symbols, as follows:

RC—Running or Sliding Clearance Fit

LC—Locational Clearance Fit

LT—Transition Clearance or Interference Fit

LN—Locational Interference Fit

FN—Force or Shrink Fit

It should be understood that these letters are not to appear on working drawings. Only the limits for sizes are shown.

When a number is added to these letter symbols a complete fit is represented. For example, FN4 specifies, symbolically, a class 4 force fit for which the limits of size for mating parts may be determined from use of Table 48E. The minimum and maximum limits of clearance or interference for a particular application may be read directly from this table.

Classes of fits as given in these tables are as follows:

Running and Sliding Fits—Classes RC1–RC9

Locational Clearance Fits—Classes LC1–LC11

Locational Transition Fits—Classes LT1–LT6

Locational Interference Fits—Classes LN1–LN3

Force and Shrink Fits—Classes FN1–FN5

Running and sliding fits. Running and sliding fits (RC1-RC9) provide a similar running performance, with suitable lubrication allowance, throughout the range of sizes. The clearances for the RC1 and RC2 classes, which are used mainly for slide fits, increase more slowly with diameter than the other classes, so that accurate location can be maintained even at the expense of free relative motion.

Locational fits. These fits are intended for mating parts requiring some degree of accuracy of location. Locational fits provide for parts requiring rigidity and accurate alignment (interference fits—LN) as well as fits for mating parts where some freedom of location is permissible (locational clearance—LC). Locational fits are divided into three groups: locational clearance fits—LC, locational transition fits—LT, and locational interference fits—LN. Locational transition fits offer a compromise between LC and LN fits and are used where either a small amount of clearance or interference can be permitted.

Force fits. Force fits and shrink fits constitute a special type of interference fit that is normally characterized by maintenance of constant bore pressures throughout the range of sizes. Force fits may be light drive fits—FN1, medium drive fits—FN2, heavy drive fits—FN3, and force fits—FN4 and FN5. Medium drive fits are suitable for ordinary steel parts. FN4 and FN5 force fits are for parts that can be highly stressed. Normally, FN2 fits are used for shrink fits on light sections, while FN3 is applicable for shrink fits in medium sections.

13.17

Computation of Limits of Size for Cylindrical Parts. To obtain the correct fit between two engaging parts, compute limit dimensions that modify the nominal size of both. Numerical values of the modifications necessary to obtain the proper allowance and tolerances for various diameters for all fits mentioned previously are given in Tables 48A, B, C, D, and E in the Appendix.

The two systems in common use for computing limit dimensions are (1) the basic hole system, and (2) the basic shaft system. The same ANSI tables may be used conveniently for both systems.

13.18

Basic Hole System. Because most limit dimensions are computed on the basic hole system, the illustrated example shown in Fig. 13.60 involves the use of this system. If, as is the usual case, the nominal size is known, all that is necessary to determine the limits is to convert the nominal size to the basic hole size and apply the figures given under "standard limits," adding or subtracting (according to their signs) to or from the basic size to obtain the limits for both the hole and the shaft.

Example

Suppose that a $\frac{1}{2}$-in. shaft is to have a class RC6 fit in a $\frac{1}{2}$-in. hole [Fig. 13.60(a)]. The nominal size of the hole is $\frac{1}{2}$ in. The basic hole size is the exact theoretical size .5000.

From Table 48A it is found that the hole may vary between +.0000 and +.0016, and the shaft between −.0012 and −.0022. As can be readily observed, these values result in a variation (tolerance) of .0016 between the upper and lower limits of the hole, while the variation (tolerance) for the shaft will be .0010. The allowance (minimum clearance) is .0012, as given in the table.

The limits of the hole are

$$\frac{.5000 + .0000}{.5000 + .0016} = \frac{.5000}{.5016}$$

The limits on the shaft are

$$\frac{.5000 - .0012}{.5000 - .0022} = \frac{.4988}{.4978}$$

It is recommended that the maximum limit always be placed directly above the minimum limit where dimensions are associated with dimension lines (ANSI Y14.5–1973). However, when both limits are given in a single line in association with a leader or note, the minimum limit is to be given first.

13.19

Basic Shaft System. When a number of parts requiring different fits but having the same nominal size must be mounted on a shaft, the basic shaft system is used because it is much easier to adjust the limits for the holes than to machine a shaft of one nominal diameter to a number of different sets of limits required by different fits.

For basic shaft fits the maximum size of the shaft is basic. The limits of clearance or interference are the same as those shown in Tables 48A, B, C, D, and E for the corresponding fits. The symbols for basic shaft fits are identical with those used for the standard fits with a letter S added. For example, LC4S specifies a locational clearance fit, class 4, as determined on a basic shaft basis.

Basic shaft system— clearance fits

To determine the needed limits, increase each of the limits obtained, using the basic hole system, by the value given for the upper shaft limit. For example, if the same supposition (nominal size and fit requirement) is made as for the preceding illustration, the limits shown in Fig. 13.61 can be most easily obtained by adding .0012 to each of the limits shown for the hole and shaft in Fig. 13.60.

Basic shaft system— interference and transition fits

To determine the needed limits, subtract the value shown for the upper shaft limit from the basic hole limits.

Example

The basic shaft limits are to be determined using an FN2 fit and the same nominal diameter as for the two previous illustrations.

$$\text{Hole:} \quad \frac{.5000 - .0016}{.5007 - .0016} = \frac{.4984}{.4991}$$

Shaft:

$$\frac{.5016 - .0016}{.5012 - .0016} = \frac{.5000 \text{ (basic size)}}{.4996}$$

In brief, it can be stated that the limits for hole and shaft, as given in Tables 48A–E, are increased for clearance fits or decreased for transition or interference fits by the value shown for the upper shaft limit, which is the amount required to change the maximum shaft to basic size.

Limits and fits— SI metric system

13.20

SI Metric—Limits and Fits. Representing approximately 60 nations, the ISO (International Organization for Standardization) has developed a universally recognized system of limits and fits for component parts. The related symbologies used ensure ready recognition of nominal size and tolerance.

The components of a toleranced dimension are given in the following order: (1) diameter symbol, if appropriate, (2) the basic size, and (3) the tolerance symbol (Fig. 13.62). Upper-case letters are used for hole tolerances and lower-case letters

Fig. 13.62 Shaft with a toleranced dimension.

Upper Deviation = +0.021
Lower Deviation = 0

Limits of Size $\phi 30 ^{+0.021}_{0}$

OR $\phi 30 + 0.021$

Hole
(a)

Upper Deviation = -0.007
Lower Deviation = -0.020

Limits of Size $\phi 30 ^{-0.007}_{-0.020}$

OR $\phi 29.993 - 0.013$

Shaft
(b)

Fig. 13.63 Computation of limits for hole and mating shaft.

are used for shaft tolerances (Fig. 13.63).

Even though it is common practice to present a tolerance as given in Fig. 13.62(*a*), the upper and lower deviations may also be placed in parentheses as in (*b*). Still a third alternative is to present toleranced dimensions as shown in (*c*) in order to avoid further calculations.

Fig. 13.63 shows the limits for a H7/g6 mating hole and shaft. The basic size is 30 mm. The upper and lower deviation values were obtained from Tables 20 A and B in the Appendix. The deviations for the hole are 0 and +0.021 mm. The shaft will be less than basic size, since the deviations are −0.007 mm and −0.020 mm. H7/g6 produces a sliding fit that is not intended to run freely but to locate accurately and move and turn easily.

Fits used by designers range from the interference fits (negative allowance) through the transition fits and clearance fits (positive allowance). A transition fit has size limits that can lead to either interference or clearance. That is, a shaft may be either larger or smaller than the hole in the mating part.

In engineering design, fits between mating parts can be achieved with a limited selection of holes and shafts with most practical requirements satisfied. The surprisingly small selection that is listed covers a range of fits from a loose running fit (H11/c11) to a shrink fit (H7/s6). The type of fit for each combination (hole/shaft) has been identified. The first five fits are clearance fits, the next two are transition fits, and the final two fits are interference fits.

13.21

Tolerances. Necessary tolerances may be expressed by general notes printed on a drawing form or they may be given with definite values for specific dimensions as in the case shown in Fig. 13.68, where the part has been dimensioned using millimeters. When expressed in the form of a printed note, the wording might be as follows: ALLOWABLE VARIATION ON ALL FRACTIONAL DIMENSIONS IS ±.010 UNLESS OTHERWISE SPECIFIED. A general note for tolerance on decimal dimensions might read: ALLOWABLE VARIATION ON DECIMAL DIMENSIONS IS ±.001. This general note would apply to all decimal dimensions where limits were not given.

The general notes on tolerances should be allowed to apply to all dimensions where it is not necessary to use specific tolerances.

13.22

Tolerancing Methods. A tolerance applied directly to a dimension may be expressed by giving limits (maximum and minimum values) or by means of plus-and-minus tolerancing. When limit dimensioning is used, the high limit is given above the low limit, as: $.750 \atop .742$. When given in a single line, the low limit precedes the high limit, with a dash separating the values, as: .742–.750.

When the metric system is used, the same practices are followed. It should be noted that when the high and low limits are placed one above the other, they are not separated by a line.

Examples

$31.8 \atop 31.5$ mm   31.5–31.8 mm

13.23

Plus-and-Minus Tolerancing. When plus-and-minus tolerancing is used, the speci-

Selected ISO fits (hole basis)		
Hole	Shaft	Type of fit
H11	c11	Loose running fit
H9	d9	Free running fit
H8	f7	Close running fit
H7	g6	Sliding fit-intended to move freely
H7	h6	Locational clearance fit
H7	k6	Locational transition fit
H7	n6	Locational transition fit
H7	p6	Locational interference fit-press fit
H7	s6	Medium drive fit or shrink fit

fied size is given first, followed by a plus-and-minus expression of the tolerance as shown in Fig. 13.64.

Unilateral plus-and-minus tolerancing is illustrated in (a) for both the metric and English systems of measurement. It should be noted for metric dimensioning that the nil value is shown by a zero with no plus-or-minus symbol being given. When the tolerance is a plus value, it is placed above the zero. When the tolerance has a minus value, it is placed below the zero.

Bilateral tolerancing is illustrated in (b). For equal plus-and-minus values, the mean value is preceded by a ± symbol. The plus value is shown above when unequal plus-and-minus tolerancing is used. Both tolerance values should be given using the same number of decimal places.

13.24

Cumulative Tolerances. An undesirable condition may result when either the location of a surface or an overall dimension is affected by more than one tolerance dimension. When this condition exists, as illustrated in Fig. 13.65(a), the tolerances are said to be cumulative. In (a) surface B is located from surface A and surface C is related in turn to surface B. With the tolerances being additive, the tolerance on C is the sum of the separate tolerances (±.002). In respect to A, the position of C may vary from 1.998 to 2.002. This tolerance, as illustrated by the shaded rectangle, is .004 in. When consecutive dimensioning is used, one dimension should always be omitted to avoid serious inconsistency. The distance omitted should be the one requiring the least accuracy. To avoid the inconsistency of cumulative tolerances it is the preferred practice to locate the surfaces from a datum plane, as shown in (b), so that each surface is affected by only one dimension. The use of a datum plane makes it possible to take full advantage of permissible variations in size and still satisfy all requirements for the proper functioning of the part.

13.25

Specification of Angular Tolerances. Angular tolerances may be expressed in degrees, minutes, or seconds (see Fig. 13.66). If desired, an angle may be given in degrees and decimal parts of a degree with the tolerance in decimal parts of a degree.

D □ Geometric dimensioning and tolerancing

13.26

Geometric Tolerancing. Geometric tolerances specify the maximum variation that can be allowed in form or position from true geometry. Actually, a geometrical tolerance is either the width or diameter of a tolerance zone within which a surface or axis of a hole or cylinder can lie with the resulting part satisfying the necessary standards of accuracy for proper functioning and interchangeability. Whenever tolerances of form are not specified on a drawing for a part, it is understood that the part as produced will be acceptable regardless of form variations. Expressions of tolerances of form control straightness, flatness, parallelism, squareness, concentricity, roundness, angular displacement, and so forth.

13.27

Symbols for Tolerances of Position and Form. The characteristic symbols shown in Fig. 13.67 have been adopted for use in lieu of notes to express positional and form tolerances. In general, these symbols are the same as those given in Mil. Std. 8C–1962 for use by the armed services. Figure 13.68 shows typical feature control symbols applied to a drawing. After making a careful study of this drawing, the reader is urged to relate and compare the symbolic callouts used with callouts and their illustrated significance as given in Figs. 13.69–13.76 inclusive.

Use of Symbols

Geometric characteristic symbols. These symbols denoting geometric characteristics are shown in Fig. 13.67. Figures 13.68–13.76 show their use.

Datum identifying symbol. The datum identifying symbol consists of a frame (box) containing the datum reference letter. It may be associated with the datum feature in one of the ways shown in Figs. 13.68–13.76.

Basic dimension symbol. In order that a basic dimension may be identified, it is enclosed in a frame as shown in Fig. 13.68.

Supplementary symbols—M, MC, RFS, and diameter. The symbols Ⓜ and Ⓢ are used to designate "maximum material condition" and "regardless of feature size" (Fig. 13.67). In notes, the ab-

(a) Unilateral Tolerancing

(b) Bilateral Tolerancing

Fig. 13.64   Plus-and-minus tolerancing.

Cumulative Tolerances
(a)

Non-cumulative Tolerances
(b)

Fig. 13.65   Cumulative tolerances.

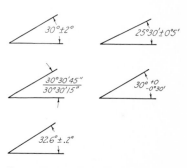

Fig. 3.66   Angular tolerances.

297

CHARACTERISTIC SYMBOLS			
Form Tolerances	Single features	Straightness	—
		Flatness	▱
		Contour of any surface	⌒
		Roundness–Circularity	○
		Cylindricity	⌭
	Related Features	Perpendicularity-Squareness	⊥
		Parallelism	//
		Angularity	∠
		Runout	↗
Positional Tolerances		True Position	⊕
		Concentricity	◎
		Symmetry	≡
SUPPLEMENTARY SYMBOLS			
MMC	Maximum Material Conditions		Ⓜ
RFS	Regardless of Feature Size		Ⓢ
	Diameter		⌀

Examples:

—	0 0.04		○	0.05

┌Symbol

⊥	0.15	A		⌭	0.04

┌Datum Reference

┌Tolerance

//	0.05	A		∠	0.04	A

┌Diameter Symbol

⊕	⌀0.12 Ⓜ	A	D Ⓜ	B

Fig. 13.67 Geometric characteristic symbols.*

*Compiled from ANSI Y14.5–1973.

breviations MMC and RFS are used.

The symbol φ is used to designate a diameter (Figs. 13.67–13.68). On a drawing it is used to replace DIA or D. It always precedes the associated dimension on the drawing and the specified tolerance in a feature control symbol.

Combined symbols. Individual symbols (characteristic and supplementary), datum reference letters, and the needed tolerance may be combined in one frame, as shown in the lower portion of Fig. 13.67, to express a callout for a tolerance symbolically.

In using the universally recognized symbols given in Fig. 13.67, a position or form tolerance is expressed by means of a feature control symbol consisting of a frame that contains the appropriate geometric characteristic symbol followed by the allowable tolerance. It should be noted in Fig. 13.68 that a vertical line separates the symbol from the tolerance. Where applicable, the tolerance should be proceded by the diameter symbol and followed by the symbol Ⓜ or Ⓢ. See Fig. 13.68.

Where a tolerance of position of form is related to a datum, this relationship shall be indicated in the feature control symbol by placing the datum reference letter(s) following the tolerance. As can be noted in the several examples shown in Fig. 13.67, vertical lines separate these entries. Each datum reference letter entered is supplemented by the symbol for MMC or RFS when applicable (Fig. 13.68).

Placement of feature control symbols. A symbol is related to the feature to which

it applies by one of the four methods listed.

(*a*) By adding the symbol to a note (Fig. 13.68).

(*b*) By a leader from the feature to the symbol (Fig. 13.68).

(*c*) By attaching the symbol frame to an extension line from the feature (Fig. 13.68).

(*d*) By attaching a side or end of the symbol frame to a dimension line (Fig. 13.68).

The symbols shown in Fig. 13.67 provide the best means for specifying geometric characteristics and tolerances. Notes have been extensively used in the past but notes can be inconsistent and require more space. In addition, these symbols are understood internationally and surmount language barriers. This has been the principal reason for their quick adoption by multinational companies for use on drawings.

13.28
Tolerance Specification for Straightness. Straightness is specified by symbol as shown in Fig. 13.69. The given example illustrates straightness control of individual longitudinal surface elements of a cylindrical part. The symbol meaning is set forth in (*b*). In (*a*) the tolerance zone has been given as 0.05 mm, while in (*c*) it is .002 in. These values are equivalent.

13.29
Tolerance Specification for Flatness. The callout used to control flatness specifies that all points of the actual surface must lie between two parallel planes that are a distance apart equal to the specified tolerance (see Fig. 13.70). The symbol in (*a*) is interpreted to read: The entire actual surface shall be flat within 0.1 (mm) total tolerance zone. When necessary, the expressions MUST NOT BE CONCAVE or MUST NOT BE CONVEX may be added below the feature control symbol.

13.30
Tolerance Specification for Perpendicularity. Perpendicularity is the condition of a surface, median plane, or axis that is at a 90° angle to a datum plane or axis. In Fig. 13.71, the perpendicularity tolerance specifies a cylindrical tolerance zone perpendicular to a datum plane *A* within which the axis of the considered feature (hole) must lie. From the feature control

Fig. 13.68 Use of symbols in specifying positional and form tolerances.

symbol we learn that when the feature is at MMC, the maximum perpendicularity tolerance is 0.10 (mm). The φ symbol was used to indicate a cylindrical tolerance zone.

The example in Fig. 13.72 illustrates perpendicularity tolerance applied to a surface. The symbol in (a) when interpreted specifies that the surface indicated must be perpendicular to datum plane A within an 0.1 (mm)-wide tolerance zone.

13.31
Tolerance Specification for Parallelism. Parallelism is the condition of a surface or axis equidistant at all points from a datum plane or axis. For the example in Fig. 13.73, the datum is considered as a plane established by the high points of surface A. All points of the other surface must lie between two planes that are parallel to the datum. In (a) the two planes are 0.05 (mm) apart.

13.32
Tolerance Specification for Concentricity (Fig. 13.74). When the cylindrical or conical features of a part must be basically concentric, it is the practice to specify permissible eccentricity in terms of the maximum permissible deviation from concentricity. Concentricity tolerance, as illustrated in Fig. 13.74(b), is the diameter of the cylindrical tolerance zone within which the axis of the feature must lie. The axis of this tolerance zone must coincide with the axis of the datum feature that has been indicated.

13.33
Tolerance Specification for Angularity. The feature control symbol that is commonly used to specify the tolerance for control of angularity is shown in Fig. 13.75. Angularity is the condition of a surface or axis that is at some specified angle (other than 90°) from a datum plane or axis. The symbol may be interpreted as: This surface must be at a 45° angle in relation to datum plane A within a 0.05 (mm)-wide tolerance zone.

13.34
Profile Tolerancing. Where a uniform amount of variation may be permitted along a profile, a zone tolerance may be specified. The zone is indicated at a conspicuous location by two phantom lines drawn parallel to the profile when the zone is bilateral, that is, symmetrical about the contour line, as shown in Fig. 13.76. Only

Significance:
Any longitudinal element of the cylindrical surface must lie between two parallel lines where the two lines and the nominal axis of the part share a common plane

Fig. 13.69 Specification for straightness.

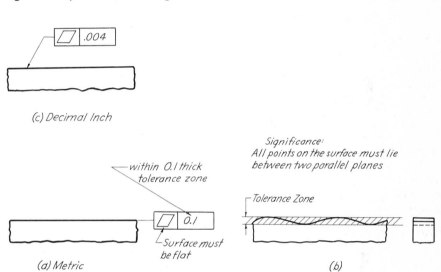

Significance:
All points on the surface must lie between two parallel planes

Fig. 13.70 Specification for flatness.

Significance:
Regardless of actual size, the axis of the hole must lie within a cylindrical zone which is perpendicular to the datum plane A

Fig. 13.71 Specification of perpendicularity.

Fig. 13.72 Specification for perpendicularity.

(c) Decimal Inch

(a) Metric

Significance:
The actual surface must lie between
two parallel planes which are
perpendicular to the datum plane A

Tolerance Zone

Datum Plane

(b)

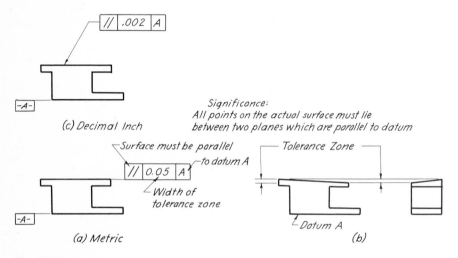

Fig. 13.73 Specification for parallelism.

(c) Decimal Inch

(a) Metric

Significance:
All points on the actual surface must lie
between two planes which are parallel to datum

Surface must be parallel
to datum A

Width of
tolerance zone

Tolerance Zone

Datum A

(b)

Fig. 13.74 Specification for concentricity.

(c) Decimal Inch

(a) Metric

Feature shall be concentric

Datum

within Ø 0.05 RFS

Significance:
The feature axis must lie within a
cylindrical zone having a diameter
equal to the concentricity tolerance.
The axis of the cylindrical zone
coincides with the datum axis.

Tolerance Zone

(b)

one phantom line is needed for a unilateral zone that may lie on either side of the true profile. As can be observed from the illustration, the finished surface must lie within the specified tolerance zone. The variation to be permitted and the extent of the tolerance zone must be specified. On the drawing, the applicable feature control symbol should appear with the view where the surface is represented in profile.

13.35

True-position Dimensioning. In the past, it has been the usual practice to locate points by means of rectangular dimensions given with tolerances. A point located in this manner will lie within a square tolerance zone when the positioning dimensions are at right angles to each other, as in Fig. 13.77. Where features are located by radial and angular dimensions with tolerances, wedge-shaped tolerance zones result.

In making a comparison of coordinate tolerancing and true-position tolerancing of circular features, it can be noted in the case of coordinate tolerancing, as illustrated in Fig. 13.77, that the actual position of the feature can be anywhere within the 0.1 square and that the maximum allowable variation from the desired position occurs along the diagonal of the square. With this allowable variation along the diagonal being 1.4 times the specified tolerance, the diameter of the cylinder for true-position tolerancing of the same feature could be 1.4 times the tolerance that would be used in coordinate tolerancing without any increase in the maximum allowable variation. True-position tolerancing increases the permissible tolerance in all directions, without detrimental effect on the location of the feature. True-position dimensioning takes into full account the relations that must be maintained for the interchangeable assembly of mating parts and it permits the design intent to be expressed more simply and precisely. Furthermore, the true-position approach to dimensioning corresponds to the control furnished by position and receiver gauges with round pins. Such gauges are commonly used for the inspection of patterns of holes in parts that are being mass produced.

Positional tolerancing can be used for specific features of a machine part. When the part contains a number of features arranged in groups, positional tolerances can be used to relate each of the groups to one another as necessary and to tolerance the position of the features within a group independently of the features of the other groups.

The term *true position* denotes the theoretically exact position for a feature. In practice, the basic (exact) location is given with untoleranced dimensions that are excluded from the general tolerances, usually specified near the title block. This is done by enclosing each of the true-position-locating dimensions in a rectangular box to identify them. See Fig. 13.79.

When the alignment between mating parts depends on some functional surface, this surface is selected as a datum for dimensioning, and the datum is identified.

The requirement of true-position dimensioning for a cylindrical feature is illustrated in Fig. 13.78(*a*). It must be understood that the axis of the hole at all points must lie within the specified cylindrical tolerance zone having its center located at true position. This cylindrical tolerance zone also defines the limits within which variations in the squareness of the axis of the hole in relation to the flat surface must be confined.

For noncircular features, such as slots and tabs, the positional tolerance is usually applied only to surfaces related to the center plane of the feature. In applying true-position dimensioning to such features, it will be found that the principal difference is in the geometric form of the tolerance zone within which the center plane of the feature must be contained. See Fig. 13.78(*b*). The center plane of the tolerance zone must be located at true position. It should be noted that this tolerance zone also defines the limits within which variations in the squareness of the

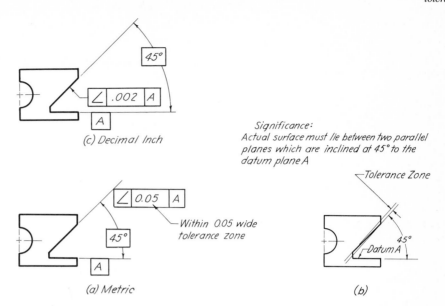

Fig. 13.75 Specification for angularity.

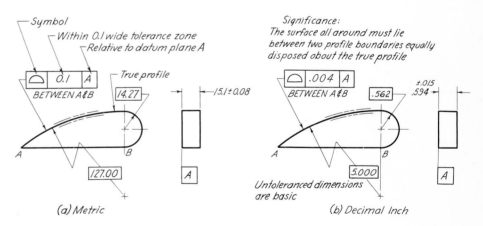

Fig. 13.76 Profile (zone) tolerancing.

Fig. 13.77 Comparison between coordinate tolerancing and true-position tolerancing.

Fig. 13.78 Meaning of true-position dimensioning.

Fig. 13.79 Position tolerancing and use of feature control symbol.

(a) Metric

(b) Decimal Inch

Fig. 13.80 Hole pattern with datum surfaces implied.

(a) Metric

(b) Decimal Inch

Fig. 13.81 Positional tolerancing with datum references.

center plane of the slot must be contained.

13.36

True-position Dimensioning—Application of the MMC Principle. The least desirable situation exists for the assembly of mating parts when both parts are at their maximum metal condition (designated MMC). The expression *maximum metal condition* by itself, as applied to an internal feature of a finished part, means that the internal feature (hole, slot, etc.) is at its minimum allowable size. In the case of external features (shafts, lugs, tabs, etc.), a maximum material condition exists when these features are at their maximum allowable sizes. MMC occurs for mating parts, say, for a hole and a shaft, when the shaft is at its maximum size and the hole is at its smallest size. Thus, at MMC,

there is least clearance between these parts. In general, tolerance of position and the MMC of mating features are considered together in relationship to each other. This leads to the situation where the specified limits of location frequently may be exceeded and acceptable parts produced when the mating features are away from their maximum material limits of size. With this latter condition permissible, it becomes desirable to indicate the fact that specified limits of location need be observed only under MMC.

True-position tolerancing on the MMC basis is both practical and economical for the mass production of interchangeable parts. However, the MMC basis should not be applied where it would be inconsistent with functional requirements.

In those relatively few cases where the more economical MMC basis is not applicable and the positional tolerance must be stated without reference to MMC, the more restrictive "regardless of feature size" (RFS) basis is specified. This is accomplished by adding the abbreviation RFS (or S, its symbolic equivalent), to the true-position callout.

Additional information concerning the meaning of MMC as related to positional tolerances may be found in ANSI Y14.5–1973.

The 0.14 position tolerance, illustrated in Fig. 13.77, would usually be based on the MMC size of the hole. Then as the hole size deviates from the MMC size, the position of the hole is permitted to shift off its true position somewhat beyond the specified tolerance zone to offer a realistic bonus to the extent of that departure. In Fig. 13.79, the position tolerance zone would show considerable enlargement as the hole size departs from MMC (17.5 mm) to its high limit size of 17.6 mm. Thus, position tolerancing provides greater production tolerances while meeting fully all design requirements.

Datum planes (surfaces), as the basis for position relationships, may be either specified on a drawing or implied. They have not been specified for the part in Fig. 13.80. In this case, the dimension (19 ± 0.1 mm) from the left edge and a like dimension from the lower edge establish the relationship of the hole pattern to these edges as implied datum surfaces. Datums must be specified on a drawing when feature interrelationships, involving form and position tolerances, must be directly and accurately controlled. See Fig. 13.81.

A circular pattern of six holes located by position tolerancing with respect to the center hole is shown in Fig. 13.81. In (a) the symbol specifies the six holes are to be located at position within ϕ 0.05 at 9.5 MMC size of the holes with respect to datum A and datum hole B at 24 MMC.

Tabs and slots are dimensioned as shown in Fig. 13.82. For such features the positional tolerance is applied only to surfaces related to the center plane of the feature as shown in Fig. 13.78(b). The feature control symbol reads: these features (3 tabs) must be at position within a 0.5 (total)-wide zone with the features at MMC and with respect to datum A at MMC.

Fig. 13.82 Positional tolerancing of tabs.

E □ Designation of surface texture

13.37

Surface Quality. The improvement in machining methods within recent years coupled with a strong demand for increased life for machined parts has caused engineers to give more attention to the quality of the surface finish. Not only the service life but also the proper functioning of the part as well may depend on obtaining the needed smoothness quality for contact surfaces.

On an engineering drawing a surface may be represented by line if shown in profile or it may appear as a bounded area in a related view. Machined and ground surfaces, however, do not have the perfect smoothness represented on a drawing. Actually a surface has three dimensions—length, breadth, and curvature (waviness)—as illustrated in Fig. 13.83(a). In addition, there will be innumerable peaks and valleys of differing lengths, widths, and heights. An exaggerated profile of surface roughness is shown in (b). Combined waviness and roughness is illustrated in (c).

The following terms must be understood before the surface symbol shown in Fig. 13.88 can be properly applied:

Surface texture. This term refers to repetitive or random deviations from the nominal surface, which form the pattern of the surface. Included are roughness, waviness, lay, and flaws (Fig. 13.84).

Roughness. Roughness is the relatively finely spaced surface irregularities that are produced by the cutting action of tool edges and abrasive grains on surfaces that are machined.

Fig. 13.83 Surface definitions illustrated.

Fig. 13.84 Surface texture definitions.

Roughness height. Roughness height is the average (arithmetical) deviation from the mean line of the profile. It is expressed in micrometers (microinches) (Fig. 13.84). See Table 1.

Table 1 Preferred roughness average values (R_a)

μm °	$\mu in.$ †	μm	$\mu in.$
0.025	1	1.60	63
0.050	2	3.2	125
0.10	4	6.3	250
0.20	8	12.5	500
0.40	16	25	1,000
0.80	32		

* Micrometers—μm.
† Microinches—$\mu in.$

Roughness width. Roughness width is the distance between successive peaks or ridges, which constitute the predominant pattern of roughness. Roughness width is measured in millimeters (inches) (Fig. 13.84).

Roughness width cutoff. This term indicates the greatest spacing of repetitive surface irregularities to be included in the measurement of average roughness height. It is measured in inches or millimeters (Fig. 13.84).

Waviness. Waviness is the surface undulations that are of much greater magnitude than the roughness irregularities. Waviness may result from machine or work deflections, vibrations, warping, strains, or similar causes. See Table 2.

Waviness height. Waviness height is the peak-to-valley distance (Fig. 13.84). It is rated in inches or millimeters.

Waviness width. Waviness width (rated in inches or millimeters) is the spacing of successive wave valleys or wave peaks (Fig. 13.84).

Flaws. Flaws are irregularities, such as cracks, checks, blowholes, scratches, and so forth, that occur at one place or at relatively infrequent or widely varying intervals on the surface (Fig. 13.84).

Lay. Lay is the predominant direction of the tool marks of the surface pattern (Fig. 13.89).

Microinch. A *microinch* is one-millionth (.000001) of an inch (μin.). A micrometer is one-millionth of a meter (μm).

13.38
Designation of Surface Characteristics (Fig. 13.88).

A surface whose finish is to be specified should be marked with the finish mark having the general form of a check mark ($\sqrt{}$) so that the point of the symbol shall be on the line representing the surface, on the extension line, or on a leader pointing to the surface. Good practice dictates that the long leg and the extension shall be to the right as the drawing is read. Figure 13.85 illustrates the specification of roughness, waviness, and lay by listing rating values on the symbol.

Surface texture symbols used to control surface irregularities are shown in Fig. 13.86. The recommended proportions for the basic symbol are given in (*a*). When the removal of material by machining is necessary, a horizontal bar is added at the top of the short leg (*b*). In (*c*) the added value indicates the amount of stock to be removed. This value may be given either in millimeters or in inches. When material removal is prohibited, a small circle is added in the vee (*d*). This circle indicates that the surface is to be produced without removal of material by some production process such as die casting, powder metallurgy, injection molding, forging, or extruded shape. The surface texture symbol in (*e*) is used when surface characteristics are to be specified above or below the horizontal line or to the right of the symbol as illustrated in Fig. 13.88.

Surface texture values are applied to the surface texture symbol as illustrated in Fig. 13.87. Roughness average rating (maximum value) is placed above the vee to the left of the long leg (*a*). See Table 1. Maximum waviness height rating is the first rating placed above the horizontal line (*c*). See Table 2. It is followed by the maximum waviness spacing rating. Lay designation is indicated by positioning the proper lay symbol to the right of the long leg as shown in (*f*) and (*h*). When it is necessary to specify maximum roughness spacing, the value is placed at the right of

Table 2 Preferred series maximum waviness height values

mm*	in.†	mm	in.	mm	in.
0.0005	0.00002	0.008	0.0003	0.12	0.005
0.0008	0.00003	0.012	0.0005	0.20	0.008
0.0012	0.00005	0.020	0.0008	0.25	0.010
0.0020	0.00008	0.025	0.001	0.38	0.015
0.0025	0.0001	0.05	0.002	0.50	0.020
0.005	0.0002	0.08	0.003	0.80	0.030

* Millimeters.
† Inches.

UNLESS OTHERWISE SPECIFIED
ALL SURFACES 3.2

*(b) Specification of roughness,
waviness, and lay by giving values
on symbol. Roughness is given in
micrometers.*

Fig. 13.85 Application of surface texture symbols to a drawing of a machine part.

the lay symbol (*h*). The value for roughness sampling length (cutoff rating) is placed below the horizontal line as shown in (*g*).

If the nature of the preferred lay is to be shown, it will be indicated by the addition of a combination of lines, as shown in Fig. 13.88. Parallel or perpendicular lines indicate that the dominant lines on the surface are parallel or perpendicular to the boundary line of the surface in contact with the symbol (Fig. 13.89).

The chart in Fig. 13.90 shows the expected surface roughness in microinches and micrometers for surfaces produced by common production methods.

The surface-quality symbol, which is used only when it is desirable to specify surface smoothness, should not be confused with a finish mark (⩗), which indicates the removal of material. A surface-quality symbol might be used for a surface on a die casting, forging, or extruded shape where the surface is to have a natural finish and no material is to be removed.

Surface finish should be specified only by experienced persons because the function of many parts does not depend on the smoothness quality of a surface or surfaces. In addition, surface quality need not be necessarily indicated for many parts that are produced to close dimensional tolerances because a satisfactory surface finish may result from the required machining processes. It should be remembered that the cost of producing a part will generally become progressively greater as the specification of surface finish becomes more exacting.

Fig. 13.86 **Surface texture symbols.**

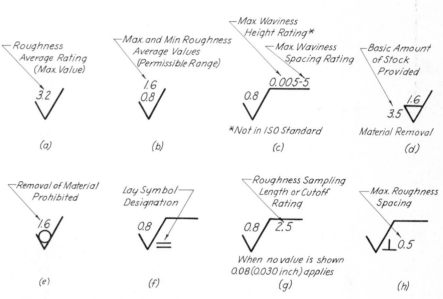

Fig. 13.87 **Application of surface texture values to symbol.**

(a) Metric

(b) Decimal Inch

Fig. 13.88 **Surface texture symbol.** Values for surface control applied—metric and decimal inch.

Fig. 13.89 Lay notations.

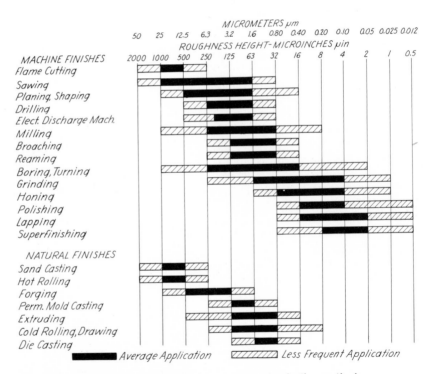

Fig. 13.90 Surface finishes expected from common production methods (micrometers μm and microinches μin.).

Problems

The following problems offer the student the opportunity to apply the rules of dimensioning given in this chapter. If it is desirable, either millimeters or decimals of an inch may be used in place of fractions. Use Table 3 in the Appendix.

1,2. (Figs. 13.91 and 13.92). Reproduce the given views of an assigned part. Determine the dimensions by transferring them from the drawing to one of the open-divided scales by means of the dividers.

Fig. 13.91 Dimensioning problems.

Fig. 13.92 Dimensioning problems.

3–9. (Figs. 13.93–13.99). Make a fully dimensioned multiview sketch or drawing of an assigned part. Draw all necessary views. Give a detail title with suitable notes concerning material, number required, etc. These parts have been selected from different fields of industry—automotive, aeronautical, chemical, electrical, etc.

Fig. 13.93 Control pedal—airplane control system.

Fig. 13.94 Guide bracket.

Fig. 13.95 Cover—mixing machine.

Fig. 13.96 Elevator bracket.

Fig. 13.97 Valve seat.

10. (Fig. 13.100). Make a fully dimensioned multiview sketch or drawing of the rocker arm. Show a detail section taken through the ribs. *Supplementary information:* (1) The distance from the center of the shaft to the center of the hole for the pin in 4.00 in. The distance from the shaft to the threaded hole is 4.50 in. (2) The nominal diameter of the hole for the shaft is 1.875 in. The hole in the rocker arm is to be reamed for a definite fit. Consult your instructor. The diameter of the pin is .969 in. (3) The diameter of the threaded boss is 2.00 in. (4)

The diameter of the roller is 2.25 in., and its length is 1.46 in. Total clearance between the roller and finished faces is to be .03 in. (5) The inside faces of the arms are to be milled in toward the hub far enough to accommodate the roller. (6) The rib is .62 in. thick. (7) The lock nut has $1\frac{1}{4}$–12 UNF thread. (8) Fillets and rounds .12 in. R except where otherwise noted.

11. (Fig. 13.101). Make a fully dimensioned drawing of an assigned part of the shaft support.

Fig. 13.98 Yoke.

Fig. 13.99 Torch holder—welding.

Fig. 13.100   Rocker arm—marine engine.

PART NO.	NAME	MATERIAL	NO.REQ.
1	BRACKET	C.I.	1
2	COVER	C.I.	1
3	BUSHING	SAE 1030	1
4	FLANGE	SAE 1030	1

Fig. 13.101   Shaft support.

TANK RETURN PORT

MANIFOLD

TRANSMISSION MAIN
PRESSURE PORT

UNLOADING VALVE

ADJUSTING
SCREW

PUMP PRESSURE
PORT

ACCUMULATOR
PORT

STEERING VALVE
PORT

CHECK VALVE

This manifold and unloading control valve is part of the hydraulic steering and braking circuits. It maintains the hydraulic system at a constant pressure. Units of this type require many threaded parts and the use of springs, bearings, and fasteners. (*Courtesy General Motors Corporation*)

threads and standard machine elements

14

A □ Screw threads

14.1

Introduction. In the commercial field, where the practical application of engineering drawing takes the form of working drawings, knowledge of screw threads and fasteners is important. There is always the necessity for assembling parts either with permanent fastenings such as rivets, or with bolts, screws, and so forth, which may be removed easily.

Engineers, detailers, and draftsmen must be completely familiar with the common types of threads and fastenings, as well as with their use and the correct methods of representation, because of the frequency of their occurrence in structures and machines. Information concerning special types of fasteners may be obtained from manufacturers' catalogues.

Young technologists in training should study Fig. 14.1 to acquaint themselves with the terms commonly associated with screw threads.

14.2

Threads. The principal uses of threads are (1) for fastening, (2) for adjusting, and (3) for transmitting power. To satisfy most of the requirements of the engineering profession, the various forms of threads shown in Fig. 14.2 are used.

Fig. 14.1 Screw-thread nomenclature.

The Unified thread has essentially the same basic profile as the ISO (International Organization for Standardization) thread form shown in Fig. 14.5. It is the thread form in general use in the United States. This will not be the case in the future for, as our large multinational corporations step up their changeover from the English inch system to the metric system for the design of new products, the Unified form will be gradually replaced as an acceptable thread form by the ISO basic form. This adoption of the ISO form, however, will not be accomplished without some minor modification of some

ISO thread characteristics to satisfy American industry.

The sharp V is used to some extent where adjustment and holding power are essential.

For the transmission of power and motion, the modified square, Acme, and Brown and Sharpe worm threads have been adopted. The modified square thread, which is now rarely used, transmits power parallel to its axis. A further modification of the square thread is the stronger Acme, which is easier to cut and more readily disengages split nuts (as lead screws on lathes). The Brown and Sharpe worm thread, with similar proportions but with longer teeth, is used for transmitting power to a worm wheel.

The knuckle thread, commonly found on incandescent lamps, plugs, and so on, can be cast or rolled.

The Whitworth and buttress threads are not often encountered by the average engineer. The former, which fulfills the same purpose as the American Standard thread, is used in England but is also frequently found in this country. The buttress or breech-block thread, which is designed to take pressure in one direction, is used for breech mechanisms of large guns and for airplane propeller hubs. The thread form has not been standardized and appears in different modified forms.

The Dardelet thread is self-locking in assembly.

14.3

Multiple-start Threads. Whenever a quick advance is desired, as on fountain pens, valves, and so on, two or more threads are cut side by side. Two threads form a double (2-START) thread; three, a triple (3-START) thread; and so on. A thread that is not otherwise designated is understood to be a single-start thread.

Fig. 14.3 shows heavy strings wound around a rod for the purpose of demonstrating single-start and double (2-START) threads. The center line of the single string, representing the single thread, assumes the form of a helix. In the case of the double (2-START) thread, it should be noted that there are two strings side by side that are shaded differently for clarity. On a double (2-START) thread, each thread starts diametrically opposite the other one.

In drawing a single or an odd-number multiple-start thread, a crest is always diametrically opposite a root; in a double or other even-number multiple-start

Fig. 14.2 Screw threads.

thread, a crest is opposite a crest and a root opposite a root.

14.4

Right-hand and Left-hand Threads. A right-hand thread advances into a threaded hole when turned clockwise; a left-hand thread advances when turned counter-clockwise. They can be easily distinguished by the thread slant. A right-hand thread on a horizontal shank always slants upward to the left (\) and a left-hand, upward to the right (/). A thread is always considered to be right-hand if it is not otherwise specified. A left-hand thread is always marked LH on a drawing.

14.5

Pitch. The pitch of a thread is the distance from any point on a thread to the corresponding point on the adjacent thread, measured parallel to the axis, as shown in Fig. 14.1.

14.6

Lead. The lead of a screw may be defined as the distance advanced parallel to the axis when the screw is turned one revolution (Fig. 14.4). For a single thread, the lead is equal to the pitch; for a double (2-START) thread, the lead is twice the pitch; for a triple (3-START) thread, the lead is three times the pitch, and so on.

14.7

ISO Metric Screw Threads. Some thirty years ago (1946) an ISO Committee was organized to investigate the possibility for the establishment of a single international system for screw threads. The futility of the idea soon became evident and it was agreed that two systems would be accepted, a new SI Metric Standard System and the Unified Inch System then already developed for use in the United States, Great Britain, and Canada. Fortunately for all concerned, the two thread standards are somewhat alike, with both using essentially the same basic thread profile. See Fig. 14.5. The principal differences between these two recognized standards are in the basic sizes, the magnitude and application of allowances and tolerances, and the method for designating and specifying threads.

When it became evident at the beginning of the 1970s that the United States was about to start a slow conversion to the metric system, the American National Standards Institute realized that the opportunity was at hand to make needed

Fig. 14.3 Demonstration of a helix as related to a single-start and double (2-START) screw thread.

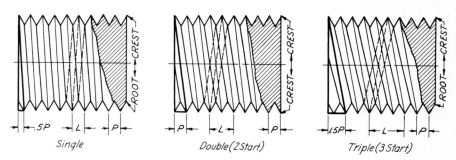

Fig. 14.4 Single-start and multiple-start threads.

major changes in the many forms of fasteners in use. Performance could be improved and redesign would offer the possibility for more optimum use of materials and for the limitation of products to the fewest number of styles, types, and sizes consistent with present-day requirements of American industry.

In 1971 a special committee was created by ANSI to study and prepare an optimum metric fastener system for use in the United States. This committee was

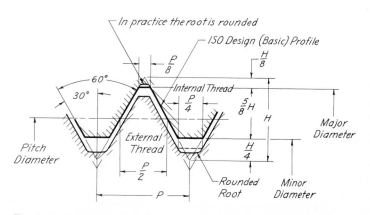

Fig. 14.5 ISO metric internal and external thread design profiles.

asked to develop a total system of mechanical fasteners that would improve fastener performance through product redesign. All dimensions and properties of the new system were to be given in metric (SI) units. As one of its first tasks the committee made a thorough study of existing screw-thread systems. This project led to the preparation and release of OMFS Recommendations 2, 3, 4, and 6 for threads as an extension of ISO Metric Screw Threads. More information on these threads will be found in Sec. 14.10. The diameter–pitch combinations of the ISO Metric Screw Thread Standard Series may be found in Table 10 in the Appendix.

14.8

Designation of ISO Metric Screw Threads (Fig. 14.6). In general, ISO metric threads will be specified in this text using only the basic designation. The basic designation consists of the letter M followed by the nominal size (basic major diameter in millimeters) and then the pitch in millimeters (See Tables 10 and 11 in the Appendix), the nominal size and pitch being separated by the sign $\times$.

Examples:

M8 $\times$ 1 (ISO designation)
M20 $\times$ 1.5

For coarse series threads the indication of the pitch may be omitted. However, it is recommended in a number of company engineering standards that this option be disregarded. Therefore, the student should follow the recommendations of his instructor and specify, say, a 10-mm coarse thread as either M10 or M10 $\times$ 1.5, whichever is suggested.

The complete designation of an ISO metric thread includes the basic designation previously discussed, followed by an identification for the tolerance class. The tolerance-class designation includes the symbol for the pitch-diameter tolerance followed by the symbol for the crest-diameter tolerance. As can be observed from the examples that follow, each of these symbols consists of a number indicating the tolerance grade followed by a letter indicating the tolerance position. The tolerance-class designation is separated from the basic designation by a dash.

Example

When the pitch-diameter and crest-diameter tolerance symbols are identical, the symbol is only given once. It is not repeated.

Example

M6 $\times$ 1 — 6g

Pitch-diameter and crest-diameter tolerance symbols (equal)

For the principal thread elements (pitch diameter and crest diameter) the ISO standard provides a number of tolerance grades that reflect various magnitudes. Basically, three Metric Tolerance Grades are recommended by ISO. These are grades 4, 6, and 8. Grade 6 is commonly used for general-purpose threads with normal lengths of engagement. As might be expected, grade 6 is the closest ISO grade to our Unified 2A and 2B fits.

Under this system, the numbers of the tolerance grades reflect the size of the tolerance. Tolerances below grade 6 are smaller and are specified for fine-quality requirements or short lengths of engage-

Fig. 14.6 Detail drawing with ISO (metric) thread specification.

ment. Tolerances above grade 6 are larger and are therefore recommended for coarse quality or long lengths of engagement.

Tolerance positions establish the maximum material limits of the pitch and crest diameters for both internal and external threads. The series of tolerance-position symbols as established by ISO to reflect varying amounts of allowance are as follows:

For external threads (bolts, etc.):
Lowercase e—large allowance
Lowercase g—small allowance
Lowercase h—no allowance
For internal threads (nuts, etc.):
Uppercase G—small allowance
Uppercase H—no allowance

A desired fit between mating threads may be specified by giving the internal thread tolerance-class designation followed immediately by the external thread tolerance class. The two designations are separated by a slash (/), as shown by the example.

Example

M6 × 1–6H/6g (or, since this is a coarse thread, as M6–6H/6g)

14.9
ISO Metric Screw Thread Standard Series (Table 10). For use at this time a series of diameter-pitch combinations for nominal diameters from 1.6 through 36 has been adopted at General Motors for threads considered to be necessary for product design purposes. As ANSI continues its work and new recommendations are forthcoming, this GM interim standard will be replaced by a complete standard for ANSI Metric Screw Threads.

A detail drawing of a plug having a tapered metric thread is shown in Fig. 14.7. It should be noted that the taper has been specified as 1:16.

14.10
Metric Threads for Production of Screws, Bolts, and Nuts. Early in January of 1974 the ANSI special committee that had been assigned the task of developing an optimum metric fastener system issued OMFS-7 which recommended a metric screw thread series for screws, bolts, and nuts. The basic profile of the recommended thread was identical to ISO 68 except for a slight reduction made to the thread height to permit a larger root radius for the external thread. Later that

Fig. 14.7 Part with a tapered metric thread. (*Courtesy General Motors Corporation— Standards Section*)

same month, the Standards and Technical Practices Committee of the Industrial Fasteners Institute (IFI), seeing few if any manufacturing problems in producing these threads in conformance with boundary profiles, endorsed the OMFS-7 recommendations. The committee then authorized the publication of IFI-500 which was reviewed and finally endorsed by AN Standards Committee B1 in October of that same year. Following its endorsement, Committee B1, which has responsibility for the screw threads standards, authorized the preparation of standard B1.18. The standard for fastener threads recognizes only one series of diameter-pitch combinations as listed here.

M1.6 × 0.35	M20 × 2.5
M2 × 0.4	M24 × 3
M2.5 × 0.45	M30 × 3.5
M3 × 0.5	M36 × 4
M3.5 × 0.6	M42 × 4.5
M4 × 0.7	M48 × 5
M5 × 0.8	M56 × 5.5
M6 × 1	M64 × 6
M8 × 1.25	M72 × 6
M10 × 1.5	M80 × 6
M12 × 1.75	M90 × 6
M14 × 2	M100 × 6
M16 × 2	

These 25 thread sizes, adopted for metric fasteners, covers the size range of the inch series from No. 0 (.060) through 4 inches.

14.11

Detailed Screw-thread Representation.
The true representation of screw threads
by helical curves, requiring unnecessary
time and laborious drafting, is rarely used.
The detailed representation, closely approximating the actual appearance, is
preferred in commercial practice, for it is
much easier to represent the helices with
slanting lines and the truncated roots and
crests with sharp V's (Fig. 14.8). Since
detailed rendering is also time consuming,
its use is justified only in those few cases
where appearance and permanency are
important factors and when it is neces-

sary to avoid the possibility that confusion
might result from the use of one of the
symbolic methods. The preparation of a
detailed representation is a task that belongs primarily to a draftsman, the engineer being concerned only with specifying
that this form be used.

The steps in drawing a metric or Unified
thread are shown in Figs. 14.9 and 14.10.

The stages in drawing the detailed representation of modified square and Acme
threads are shown in Figs. 14.11, 14.12,
and 14.13. All lines of the finished square
thread are made the same weight. The
root lines of the Acme thread may be
made heavier than the other lines.

14.12

**American National Standard Conventional
Thread Symbols (Fig. 14.14).** To save
valuable time and expense in the preparation of drawings, the American National
Standards Institute has adopted the
"schematic" and "simplified" series of
thread symbols to represent threads having a diameter of 25 mm (1 in.) or less.

The root of the thread for the simplified
representation of an external thread is
shown by invisible lines drawn parallel to
the axis [Fig. 14.15(a)].

The schematic representation consists
of alternate long and short lines perpen-

Fig. 14.8 Detailed representation.

Fig. 14.9 Detailed representation of Unified and American and sharp V-threads (external).

**Fig. 14.10 Detailed representation of
Unified and American and sharp V-threads
(internal).**

Fig. 14.11 Detailed representation of square threads (modified).

Fig. 14.12 Detailed representation of
square threads (internal).

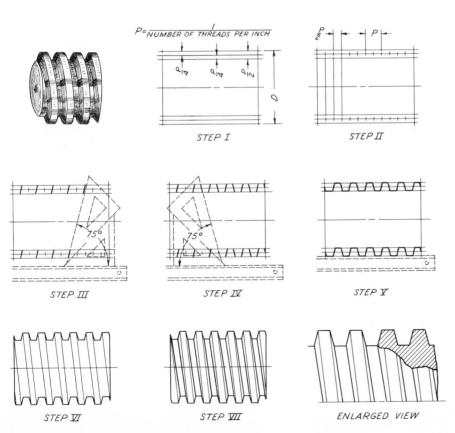

Fig. 14.13 Detailed representation of Acme thread.

Fig. 14.14 External thread representation.

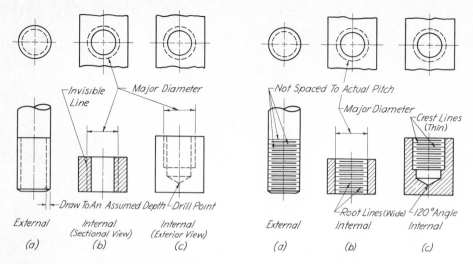

Invisible Line — *Major Diameter*

Draw To An Assumed Depth — *Drill Point*

External (a) Internal (Sectional View) (b) Internal (Exterior View) (c)

Fig. 14.15 Simplified representation.

Not Spaced To Actual Pitch

Major Diameter — *Crest Lines (Thin)*

Root Lines (Wide) — *120° Angle*

External (a) Internal (b) Internal (c)

Fig. 14.16 Schematic representation.

ACTUAL OR APPROXIMATE TO A CONVENIENT SCALE DIVISION

CREST LINES FINISHED WEIGHT *FINISHED WEIGHT* *ROOT LINES*

LIGHT *30°* *HEAVY*

Fig. 14.17 Drawing conventional threads—schematic representation.

(a) ELEVATION (b) SECTION TAPPED THROUGH (c) SECTION

Thread Depth Drill Depth

(d) (e) TAP DRILL SHOWN (f)

Depth

(g) (h) BOTTOM TAPPED (i)

Fig. 14.18 Representation of internal threads.

dicular to the axis. Although these lines, representing the crests and roots of the thread, are not spaced to actual pitch, their spacing should indicate noticeable differences in the number of threads per inch of different threads on the same working drawing or group of drawings (Fig. 14.16). The root lines are made heavier than the crest lines (Fig. 14.17).

Before a hole can be tapped (threaded), it must be drilled to permit the tap to enter. See Tables 10 and 25 in the Appendix for tap drill sizes for standard threads. Since the last of the thread cut is not well formed or usable, the hole must be shown drilled and tapped deeper than the screw will enter [Fig. 14.18(d), (e), and (f)]. To show the threaded portion extending to the bottom of the drilled hole indicates the use of a bottoming tap to cut full threads at the bottom. This is an extra and expensive operation not justified except in cases where the depth of the hole and the distance the screw must enter are limited [see Fig. 14.18(g), (h), and (i)].

Figure 14.19 shows a simplified method of representation for square threads.

14.13
Threads in Section. The detailed representation of threads in section, which is used for large diameters only, is shown in Fig. 14.10. Since the far side of an internal thread in section is visible, the crest and root lines incline in the opposite direction to those of an external thread having the same specifications.

The schematic and simplified representations for threads of small diameter are shown in Figs. 14.14 and 14.18.

A sectioned assembly drawing is shown in Fig. 14.20. When assembled pieces are both sectioned, the detailed representation is used, and the thread form is drawn. Schematic representation could have been used for the thread on the cap screw instead of the simplified representation shown (Fig. 13.101).

14.14
American–British Unified Thread. The Unified Thread Standard came into existence after the representatives of the United States, Great Britain, and Canada signed a unification agreement on Nov. 18, 1948 in Washington, D.C. This accord, which made possible the interchangeability of threads for these countries, created a new thread form (Fig. 14.21) that is a compromise between our own American National Standard design and the British

Whitworth. The external thread of the new form has a rounded root and may have either a flat or rounded crest. The Unified thread is a general-purpose thread for screws, bolts, nuts, and other threaded parts (ANSI B1.1–1960). See Sec. 14.2.

14.15

Unified and American Screw-thread Series. The Unified and American screw thread series, as given in ANSI B1.1–1960 consists of six series and a selection of special threads that cover special combinations of diameter and pitch. Each series differs from the other by the number of threads per inch for a specific diameter (see Tables 25 and 26 in the Appendix).

The coarse-thread series (UNC and NC) is designated UNC for sizes above $\frac{1}{4}$ in. in diameter. This series is recommended for general industrial use.

The fine-thread series (UNF and NF), designated UNF for sizes above $\frac{1}{4}$ in., was prepared for use when a fine thread is required and for general use in the automotive and aircraft fields.

The extra-fine-thread series (UNEF and NEF) is used for automotive and aircraft work when a maximum number of threads is required for a given length.

The 8-thread series (8N) is a uniform pitch series for large diameters. It is sometimes used in place of the coarse-thread series for diameters greater than 1 in. This series was originally intended for high-pressure joints.

The 12-thread series (12UN or 12N) is a uniform pitch series intended for use with large diameters requiring threads of medium-fine pitch. This series is used as a continuation of the fine-thread series for diameters greater than $1\frac{1}{2}$ in.

The 16-thread series (16UN or 16N) is a uniform pitch series for large diameters requiring a fine-pitch thread. This series is used as a continuation of the extra-fine-thread series for diameters greater than 2 in.

14.16

Unified and American Screw-thread Classes. Classes of thread are determined by the amounts of tolerance and allowance specified. Under the new unified system, classes 1A, 2A, and 3A apply only to external threads; classes 1B, 2B, and 3B apply to internal threads. Classes 2 and 3 from the former American Standard (ASA) have been retained without change in the Unified and American Thread Standard for use in the United States only, but they are not among the unified classes even though the thread forms are identical. These classes are used with the American thread series (NC, NF, and N series), which covers sizes from size 0 (.060) to 6 in.

Class 1A and class 1B replace class 1 of the old American Standard.

Class 2A and class 2B were adopted as the recognized standards for screws, bolts, and nuts.

Class 3A and class 3B invoke new classes of tolerance. These classes along with class 2A and class 2B should eventually replace class 2 and class 3 now retained from the American (National) Standard. Class 2 and class 3 are defined in the former standard ANSI B1.1–1935 as follows:

Class 2 fit. Represents a high quality of commercial thread product and is recommended for the great bulk of interchangeable screw-thread work.

Class 3 fit. Represents an exceptionally high quality of commercially threaded product and is recommended only in cases where the high cost of precision tools and continual checking is warranted.

14.17

Identification Symbols for Unified Screw Threads. Threads are specified under the unified system by giving the diameter, number of threads per inch, initial letters (UNC, UNF, etc.), and class of thread (1A,

Fig. 14.19 Simplified representation of a square thread.

Fig. 14.20 Threads in section.

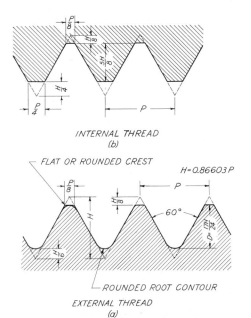

INTERNAL THREAD
(b)

FLAT OR ROUNDED CREST

$H = 0.86603P$

60°

ROUNDED ROOT CONTOUR

EXTERNAL THREAD
(a)

Fig. 14.21 American–British Unified thread.

2A, and 3A; or 1B, 2B, and 3B) (see Fig. 14.22).

Unified and American National threads are specified on drawings, in specifications, and in stock lists by thread information given as shown in Fig. 14.23. A multiple-start thread is designated by specifying in sequence the nominal size, pitch, and lead.

14.18
Thread Dimensioning. In general, the thread length dimension shown on a drawing should be the length of the complete (full-form) threads. The incomplete threads should be beyond this dimensioned length.

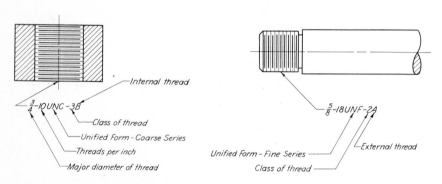

Fig. 14.22 Unified thread identification symbols.

Fig. 14.23 Thread identification symbols.

14.19
Square Threads. Square threads can be completely specified by a note. The nominal diameter is given first, followed by the number of threads per inch and the type of thread (see Fig. 14.23).

14.20
Acme and Stubb Acme Threads. Acme threads have been standardized in a preferred (single) series of diameter–pitch combinations by the American National Standards Institute (B1.5–1952). The standard provides two types of Acme threads, the general-purpose and centralizing. The three classes of general-purpose threads (2G, 3G, and 4G) have clearances on all diameters for free movement. The five classes of centralizing threads (2C, 3C, 4C, 5C, and 6C) have a limited clearance at the major diameters of the external and internal threads to ensure alignment of mating parts and to prevent wedging on the flanks of the thread.

Acme threads are specified by giving the diameter, number of threads per inch, type of thread, and class (Fig. 14.23). The letter A or B is added to indicate an external or an internal thread.

Examples

$1\frac{3}{8}$–4 ACME–2GA or 1.375–4 ACME–2GA

$1\frac{3}{4}$–4 ACME–2GB–LH

$\frac{3}{4}$–6 ACME–4CB or .750–6 ACME–4CB

$2\frac{3}{4}$–3 ACME–3GA–2-START (2-START indicates a double thread)

14.21
Buttress Threads. The buttress thread is designed for exceptionally high stress along the axis of the thread in one direction only. No pitch-diameter series has been recommended, because of the need for special design of most components. The ANSI B1.9–1953 standard covers three classes of buttress threads. These are 1A, 2A, and 3A for external threads and 1B, 2B, and 3B for internal threads.

Buttress threads are specified by giving, in order, the diameter, threads per inch, type of thread (National Buttress), and class.

Examples

$\frac{5}{8}$–20 BUTTRESS–2A

10–10 BUTTRESS–3B (2-START) (Optionally, these threads may be designated by giving the pitch followed by the letter P instead of the number of threads per inch.)

.625–.05P BUTTRESS–2A

B □ Fasteners

14.22

Metric Fasteners. In the United States, designers and detailers have held back from using metric fasteners while waiting for the development of American National Standards that would represent a guarantee of finality both here and abroad. They have been unwilling to adopt an interim metric system for fasteners that would eventually be phased out for an ANSI–ISO approved system having universal acceptance. At the very least, they did not see any reason for making a change as long as fasteners of the old inch system were readily available.

Realizing that American industry was faced with the certainty of a changeover to metric, the only question being how soon, the Industrial Fasteners Institute and other similar organizations began working to develop metric threaded fastener standards that could be submitted to the ANSI B18 committee as first drafts. The IFI standards for fasteners were finalized only after many rounds of negotations between standards engineers and fastener manufacturers in the United States and in foreign countries using the metric system. Optimal sizing for all types of fasteners was ultimately approved under a single system of strength grades, thread characteristics, clearances, finishes, and so forth. These industrial standards have been accepted by committee B18 with minor changes and adjustments as tentative standards. The minor differences that no doubt will still exist between these tentative standards and ISO standards or draft recommendations must then be resolved by further negotation before the final standards bearing the approval of both ANSI and ISO will become available for worldwide use. These new standards will assure international interchangeability as well as more accurate sizing and will result in fewer sizes and fewer styles. Fewer sizes and fewer styles in turn will lead to reduced inventories and to bulk pricing advantages. Metric fasteners are standard in one metric thread series (see Sec. 14.10). The thread pitches are between those of the coarse-thread and fine-thread series of the present Unified (inch) threads.

It is lamentable that at the time this book was placed in production approved and published ANSI standards for the common fasteners were not available.

Therefore, it became necessary to prepare Tables 12–24 in the Appendix by using data gathered from a number of sources such as ISO and IFI standards and from the catalogs of distributors of metric components.

Except for metric socket screws, all standard fastener products have one tolerance grade, 6g for externally threaded parts (bolts, machine screws, etc.) and 6H for nuts. The 6H/6g metric threads closely match our present 2A/2B threads for fasteners in that there is an allowance on the external thread and no allowance on the internal thread. See Sec. 14.8. Socket screws have 4g6g threads.

Except for machine and tapping screws, metric screws and bolts have standard thread lengths based on D, the nominal diameter. These thread lengths are:

2D + 6 mm for fastener lengths up to 125 mm
2D + 12 mm for lengths over 125 mm up to 200 mm
2D + 25 mm for lengths over 200 mm

If the length of the fastener is shorter than the standard thread length for its diameter, it should normally be threaded for its full length.

14.23

Standard Bolts and Nuts (Fig. 14.24). Commercial producers of bolts and nuts manufacture their products in accordance with standard specifications given in approved standards. Up to the present most bolts and nuts have been produced in accordance with the recommendations set forth in the American National Standard (ANSI B18.2.2–1965) entitled Square and Hexagon Bolts and Nuts. See Table 29 in the Appendix. Although bolts and nuts will continue to be produced to the specifications of this standard for a time, designers will find that fastener manufacturers will gradually shift their standard production to metric specifications given in standards that will have been approved and published by the early years of the 1980's. Tables 13 and 14 in the Appendix give the dimensions for metric hex bolts and nuts—styles 1 and 2.

The present American Standard covers the specifications for three series of bolts and nuts:

1. *Regular series.* The regular series was adopted for general use.

2. *Heavy series.* Heavy bolt heads and nuts are designed to satisfy the special

Fig. 14.24 Bolts and nuts.

commercial need for greater bearing surface.

3. *Light-series nuts.* Light nuts are used under conditions requiring a substantial savings in weight and material. They are usually supplied with a fine thread.

The amount of machining is the basis for further classification of hexagonal bolts and nuts in both the regular and heavy series as unfinished and semifinished.

Square-head bolts and nuts are standardized as unfinished only.

Unfinished heads and nuts are not washer-faced, nor are they machined on any surface.

Semifinished bolt heads and nuts are machined or treated on the bearing surface to provide a washer face for bolt heads and either a washer face or a circular bearing surface for nuts. Nuts, not washer-faced, have the circular bearing surface formed by chamfering the edges.

Bolts and nuts are *always* drawn across corners in all views. This recognized commercial practice, which violates the principles of true projection, prevents confusion of square and hexagonal forms on drawings.

The chamfer angle on the tops of heads and nuts is 30° on hexagons and 25° on squares, but both are drawn at 30° on bolts greater than 1 in. in diameter.

Specification of American Standard hex bolts

American Standard bolts are specified in parts lists and elsewhere by giving the diameter, number of threads per inch, series, class of thread, length, finish, and type of head.

Examples

$\frac{1}{2}$–13 UNC–2A $\times$ 1$\frac{3}{4}$
SEMIFIN HEX HD BOLT

Frequently it is advantageous and practical to abbreviate the specification thus:

$\frac{1}{2}$ $\times$ 1$\frac{3}{4}$ UNC SEMIFIN HEX HD BOLT

Although American Standard bolt lengths have not been standardized in construction practice, because of the varied requirements in engineering design, length increments for length under the head to the end of a hexagonal bolt can be considered as $\frac{1}{8}$ in. for bolts $\frac{1}{4}$–$\frac{3}{4}$ in. in length, $\frac{1}{4}$ in. for bolts $\frac{3}{4}$–3 in. in length, and $\frac{1}{2}$ in. for bolts 3–6 in. in length. Length increments for square-head bolts are $\frac{1}{8}$ in. for bolts $\frac{1}{4}$–$\frac{3}{4}$ in. in length, and $\frac{1}{4}$ in. for bolts $\frac{3}{4}$–4$\frac{3}{4}$ in. in length.

The minimum thread length for bolts up to and including 6 in. in length will be twice the diameter plus $\frac{1}{4}$ in. For lengths over 6 in. the minimum thread length will be twice the diameter of the bolt plus $\frac{1}{2}$ in. (ANSI B18.2–1965).

Specification of metric hex bolts

Hex bolts are specified in parts lists and elsewhere by giving in sequence the nominal size, thread pitch, nominal length, material, product name, and, if required, the protective coating.

Examples

M10 $\times$ 1.5 $\times$ 100 HEX BOLT
M24 $\times$ 3 $\times$ 50, STAINLESS STEEL HEX BOLT

14.24
To Draw Bolt Heads and Nuts. Using the dimension taken from the tables, draw the

lines representing the top and contact surfaces of the head or nut and the diameter of the bolt. Lay out a hexagon about an inscribed chamfer circle having a diameter equal to the distance across the flats (Fig. 14.25) and project the necessary lines to block in the view. Draw in the arcs after finding the centers, as shown in Fig. 14.25.

A square-head bolt or nut may be drawn by following the steps indicated in Fig. 14.26.

The engineer and experienced draftsman wisely resort to some form of template for drawing the views of a bolt head or nut (see Fig. 2.11). To draw the views as shown in Figs. 14.25 and 14.26 consumes valuable time needlessly.

14.25

Studs. Studs, or stud bolts, which are threaded on both ends as shown in Fig. 14.27, are used where bolts would be impractical and for parts that must be removed frequently (cylinder heads, steam chest covers, pumps, and so on). They are first screwed permanently into the tapped holes in one part before the removable member with its corresponding clearance holes is placed in position. Nuts are used on the projecting ends to hold the parts together.

Since most studs are not standard they must be produced from specifications given on a detail drawing. In dimensioning a stud, the length of thread must be given for both the stud end and nut end along with an overall dimension. The thread information is given by note.

In a bill of material, studs may be specified as follows:

Unified inch studs

Examples

½–13 UNC–2A × 2¾ STUD

It is good practice to abbreviate the specification thus:

½ × 2¾ STUD

Metric studs

Example

DOUBLE-END STUD, M10 × 1.5 ×100, STEEL, CADMIUM PLATED

14.26

Cap Screws (Fig. 14.28). American Standard cap screws are similar to machine screws. They are available in four stand-

ard heads, usually in finished form. When parts are assembled, the cap screws pass through clear holes in one member and screw into threaded holes in the other (Fig. 14.29). American Standard hexagonal cap screws have a washer face $\frac{1}{64}$ in. thick with a diameter equal to the distance across flats. All cap screws 25 mm (1 in.) or less in length are threaded very nearly to the head.

American Standard cap screws are specified by giving the diameter, number

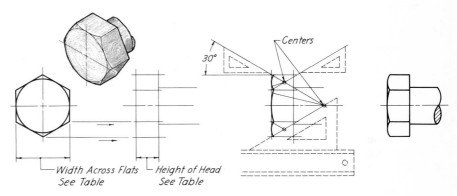

Fig. 14.25 Steps in drawing a hexagonal bolt head.

Fig. 14.26 Steps in drawing a square bolt head.

Fig. 14.27 Stud bolt.

Fig. 14.28 Hexagonal-head cap screw.

SOCKET   SOCKET   FILLISTER
HEAD   HEAD   HEAD
(Hexagonal)   (Fluted)   (Slotted)

HEXAGONAL   ROUND   FLAT
HEAD   HEAD   HEAD
              (Slotted)   (Slotted)

Fig. 14.29 American Standard cap screws.

Fig. 14.30 Use of a machine screw.

of threads per inch, series, class of thread, length, and type of head.

American Standard cap screws

Examples

$\frac{5}{8}$–11 UNC–2A $\times$ 2 FIL HD CAP SC

It is good practice to abbreviate the specification thus:

$\frac{5}{8} \times 2$ UNC FIL HD CAP SC

Metric cap screws

Examples

M6 $\times$ 1 $\times$ 40 HEX CAP SCREW
M14 $\times$ 2 $\times$ 80 HEX CAP SC, CADMIUM PLATED

14.27

Machine Screws. American Standard machine screws, which fulfill the same purpose as cap screws, are used chiefly for small work having thin sections (Fig. 14.30). Under the approved American National Standard they range from No. 0 (.060 in. diam.) to $\frac{3}{4}$ in. (.750 in. diam.) and are available in either the American Standard Coarse or Fine-Threaded Series. The four forms of heads shown in Fig. 14.31 have been standardized.

Metric machine screws may have either slotted or cross recess drives with flat countersunk heads, oval countersunk heads, and pan heads. In addition, hex head and hex washer head machine screws are available to the designer. They may be purchased in ten sizes in a range from M2 through M12. Table 16 in the Appendix gives the head dimensions for four types.

Machine screws M3 and smaller have a thread length of 25 mm. Larger diameters have 38 mm of thread length.

American Standard machine screws

To specify American Standard machine screws, give the diameter, threads per inch, thread series, class of thread, length, and type of head.

Examples

No. 12–24 NC–3 $\times$ $\frac{3}{4}$ FIL HD MACH SC

It is good practice to abbreviate by omitting the thread series and class of fit.

No. 12–24 $\times$ $\frac{3}{4}$ FIL HD MACH SC

Metric machine screws

Metric machine screws are specified in parts lists and elsewhere by giving in sequence the nominal size, thread pitch, nominal length, product name (including type of head), material, and protective finish, if needed.

Examples

M10 $\times$ 1.5 $\times$ 40 SLOTTED FLAT HEAD MACHINE SCREW

M12 $\times$ 1.75 $\times$ 80 HEX HEAD MACHINE SCREW

14.28
Commercial Lengths: Studs, Cap Screws, Machine Screws. Unless a fastening of any of these types carries a constant and appreciable fatigue stress, the usual practice is to have it enter a distance related to its nominal diameter (Fig. 14.32). If the depth of the hole is not limited, it should be drilled to a depth of 1 diameter beyond the end of the fastener to permit tapping to a distance of $\frac{1}{2}$ diameter below the fastener.

The length of the fastening should be determined to the nearest commercial length that will allow it to fulfill minimum conditions. In the case of a stud, care should be taken that the length allows for a full engagement of the nut. Commercial lengths for fasteners increase by the following increments:

Standard length increments

Inch increments

For fastener lengths:

$\frac{1}{4}$–1 in. = $\frac{1}{8}$ in.

For fastener lengths:

1–4 in. = $\frac{1}{4}$ in.

Millimeter increments

Standard lengths—short fasteners:

8 mm, 12 mm, 14 mm, 16 mm, and 20 mm

For fastener lengths:

20 mm–100 mm = 5 mm increments
110 mm–200 mm = 10 mm increments

It is generally recognized that the shortest practical length for any metric fastener should not be less than 1.5 times its diameter. For example, an M8 metric fastener should be 12 mm or longer and an M24 fastener not less than 35 mm in length.

For fastenings and other general-purpose applications, the engagement length should be equal to the nominal diameter (*D*) of the thread when both components are of steel. For steel external threads in cast iron, brass, or bronze, the engagement length should be 1.5*D*. When assembled into aluminum, zinc, or plastic, the engagement should be 2*D*.

14.29

Set Screws. Set screws are used principally to prevent rotary motion between two parts, such as that which tends to occur in the case of a rotating member mounted on a shaft. A set screw is screwed through one part until the point presses firmly against the other part (Fig. 14.33).

The several forms of safety heads shown in Fig. 14.34 are available in combination with any of the points. Headless set screws comply with safety codes and should be used on all revolving parts. The many serious injuries that have been caused by the projecting heads of square-head set screws have led to legislation prohibiting their use in some states [Fig. 14.33(*c*)]. Tables 18 and 32 in the Appendix give dimensions for metric and American Standard set screws.

American Standard set screws

American Standard set screws are specified by giving the diameter, number of threads per inch, series, class of thread, length, type of head, and type of point.

Examples

¼–20 UNC–2A × ½
SLOTTED CONE PT SET SC

The preferred abbreviated form gives the diameter, number of threads per inch, length, type of head, and type of point.

¼–20 × ½ HEX SOCKET CONE PT SET SC

Metric set screws

Examples

M6×1×12 SLOTTED CONE PT SET SC
M8×1.25×14 HEX SOCKET
FLAT PT SET SC

14.30

Keys. Keys are used in the assembling of machine parts to secure them against relative motion, generally rotary, as is the case between shafts, cranks, wheels, and

so on. When the relative forces are not great, a round key, saddle key, or flat key is used (Fig. 14.35). For heavier duty, rectangular keys are more suitable (Fig. 14.36).

The square key (Fig. 14.37) and the Pratt and Whitney key (Fig. 14.38) are the two keys most frequently used in machine design. A plain milling cutter is used to cut the keyway for the square key, and an end mill is used for the Pratt and Whitney keyway. Both keys fit tightly in the shaft and in the part mounted on it.

The gib-head key (Fig. 14.39) is designed so that the head remains far enough from the hub to allow a drift pin to be driven to remove the key. The hub side of the key is tapered ⅛ in. per ft to ensure a fit tight enough to prevent both axial and rotary motion. For this type of key, the keyway must be cut to one end of the shaft.

*Flat
Head*   *Round
Head*   *Fillister
Head*   *Oval
Head*

Fig. 14.31 Types of machine screws.

*DRILLED
(a)*

A= Minimum distance fastener
   must enter threaded hole
B= Standard length of fastener
D= Major diameter of fastener
C+E= ½D (Minimum)

*TAPPED
(b)*

*FASTENER IN POSITION
(c)*

A (MINIMUM)
STEEL – A= D
CAST IRON, BRASS, BRONZE – A = 1½D
ALUMINUM – A=2D

Fig. 14.32 Threaded hole and fastener.

(a)   *(b)*   *(c)*

Fig. 14.33 Use of set screws.

Slotted *Hexagonal Socket* *Fluted Socket*

Cone Point *Flat Point* *Oval Point* *Cup Point* *Full Dog Point* *Half Dog Point*

Fig. 14.34 Set screws.

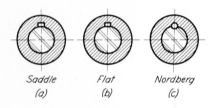

Saddle *Flat* *Nordberg*
(a) *(b)* *(c)*

Fig. 14.35 Light-duty keys.

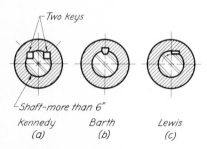

Two keys

Shaft-more than 6″

Kennedy *Barth* *Lewis*
(a) *(b)* *(c)*

Fig. 14.36 Heavy-duty keys.

14.31
Woodruff Keys. A Woodruff key is a flat segmental disc with either a flat or a round bottom (Fig. 14.40). It is always specified by a number, the last two digits of which indicate the nominal diameter in eighths of an inch, while the digits preceding the last two give the nominal width in thirty-seconds of an inch.

A practical rule for selecting a Woodruff key for a given shaft is as follows: Choose a standard key that has a width approximately equal to one-fourth of the diameter of the shaft and a radius nearly equal (plus or minus) to the radius of the shaft. Table 40 in the Appendix gives the dimensions for American Standard Woodruff keys. Table 24 lists dimensions for metric keys.

When Woodruff keys are drawn, it should be remembered that the center of the arc is placed above the top of the key at a distance shown in column E in the table.

14.32
Taper Pins. A taper pin is commonly used for fastening collars and pulleys to shafts,

as illustrated in Fig. 14.41. The hole for the pin is drilled and reamed with the parts assembled. When a taper pin is to be used, the drawing callout should read as follows:

DRILL AND REAM FOR #4 TAPER PIN WITH PC #6 IN POSITION

Drill sizes and exact dimensions for taper pins are given in Table 42 in the Appendix.

14.33
Locking Devices. A few of the many types of locking devices that prevent nuts from becoming loose under vibration are shown in Figs. 14.42 and 14.43.

Figure 14.42 shows six forms of patented spring washers. The ones shown in (D), (E), and (F) have internal and external teeth.

In common use is the castellated nut with a spring cotter pin that passes through the shaft and the slots in the top [Fig. 14.43(a)]. This type is used extensively in automotive and aeronautical work.

Figure 14.43(b) shows a regular nut that is prevented from loosening by an American National Standard jam nut.

In Fig. 14.43(c) the use of two jam nuts is illustrated.

A regular nut with a spring-lock washer is shown in Fig. 14.43(d). The reaction provided by the lock washer tends to prevent the nut from turning.

A regular nut with a spring cotter pin through the shaft, to prevent the nut from backing off, is shown in Fig. 14.43(e).

Special devices for locking nuts are illustrated in Fig. 14.43(f) and (g). A set screw may be held in position with a jam nut, as in (h).

SQUARE KEY

Fig. 14.37 Square key.

P&W KEY

Fig. 14.38 Pratt and Whitney key.

GIB HEAD KEY

Fig. 14.39 Gib-head key.

Fig. 14.40 Woodruff key.

Fig. 14.41 Use of a taper pin.

14.34

Areo Thread. The Areo-thread screw-thread system allows the use of high-strength cap screws and studs in light soft metals, such as aluminum and magnesium, through the use of a phosphor bronze or stainless steel coilspring lining in the tapped hole, as shown in Fig. 14.44. This coil (screw bushing) is formed to fit a modified American Standard thread. Special tools are needed for inserting the coil in the tapped hole.

14.35

Miscellaneous Bolts, Screws, and Nuts. Other types of bolts and screws that have been adopted for commercial use are illustrated in Fig. 14.45.

Wood screws have threads proportioned for the holding strength of wood. They are available with different forms of heads (flat, round, and oval).

Some of the fastenings shown in Fig. 14.45 have been standardized by the American National Standards Institute.

14.36

Phillips Head. The Phillips head, shown in Fig. 14.46 for a wood screw, is one of various types of recessed heads. Although special drivers are usually employed for installation, an ordinary screwdriver can be used. Machine screws, cap screws, and many special types of fasteners are available with Phillips heads.

14.37

Rivets. Rivets are permanent fasteners used chiefly for connecting members in such structures as buildings and bridges and for assembling steel sheets and plates for tanks, boilers, and ships. They are cylindrical rods of wrought iron or soft steel, with one head formed when manu-

Fig. 14.42 Special lock washers.

Fig. 14.43 Locking schemes.

Fig. 14.44 Aero thread.

Fig. 14.45 Miscellaneous bolts, screws, and nuts.

Fig. 14.46 Phillips head screw.

Fig. 14.47 Riveting procedure.

factured. A head is formed on the other end after the rivet has been put in place through the drilled or punched holes of the mating parts. A hole for a rivet is generally drilled, punched, or punched and reamed $\frac{1}{16}$ in. larger than the diameter of the shank of the rivet [Fig. 14.47(*a*)]. Figure 14.47(*b*) illustrates a rivet in position. Small rivets, less than $\frac{1}{2}$ in. in diameter, may be driven cold, but the larger sizes are driven hot. For specialized types of engineering work, rivets are manufactured of chrome–iron, aluminum, brass, copper, and so on. Standard dimensions for small rivets are given in Table 37 in the Appendix.

The type of rivets and their treatment are indicated on drawings by the American National Standard conventional symbols shown in Fig. 25.12.

14.38

Riveted Joints. Joints on boilers, tanks, and so on, are classified as either lap joints or butt joints (Fig. 14.48). Lap joints are generally used for seams around the circumference. Butt joints are used for longitudinal seams, except on small tanks where the pressure is to be less than 100 lb per sq in.

C □ Springs

14.39

Springs. In production work, a spring is largely a matter of mathematical calculation rather than drawing, and it is usually purchased from a spring manufacturer, with the understanding that it will fulfill specified conditions. For experimental work, and when only one is needed, it may be formed by winding oil-tempered spring wire or music wire around a cylindrical bar. As it is wound, the wire follows the helical path of the screw thread. For this reason the steps in the layout of the representation for a spring are similar to the screw thread. Pitch distances are marked off, and the coils are given a slope of one-half of the pitch. Figure 14.49(*a*) shows a partial layout of a tension spring. Other types of ends are shown in (*b*). A compression spring layout, with various types of ends, is illustrated in Fig. 14.50. Single-line symbols for the representation of springs are shown in Fig. 14.51.

When making a detail working drawing of a spring, it should be shown to its free length. On either an assembly or detail drawing, a fairly accurate representation, neatly drawn, will satisfy all requirements.

It is common practice in industry to rely on a printed spring drawing, accompanied by a filled-in printed form, to convey the necessary information for the production of a needed spring by a reliable manufacturer. The best procedure is to give the spring characteristics along with a list of necessary dimensions and then depend on an experienced spring designer, at the plant where the spring will be produced, to finalize the design.

The method of representing and dimensioning a compression spring is shown in Fig. 14.52. The spring is represented by a rectangle with diagonals (printed or drawn). Pertinent information is added as dimensions and notes. Either an ID (inside diameter) or an OD (outside diameter) dimension is given, depending on how the spring is to be used.

Fig. 14.48 Forms of riveted joints.

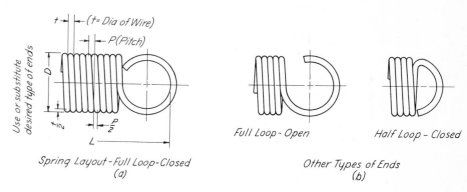

Fig. 14.49 Tension springs.

A method of representing and dimensioning an extension spring is shown in Fig. 14.53. Although drawings of this type may be printed forms, it is often necessary to prepare a drawing showing the ends and a few coils, since extension springs may have any one of a wide variety of types of ends. Needed information

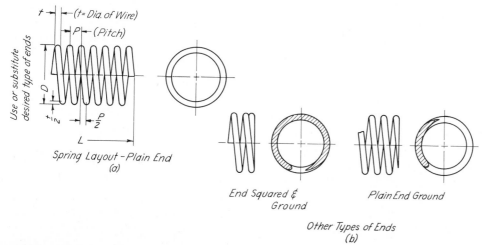

Fig. 14.50 Compression springs.

Fig. 14.51 Single-line representation of springs.*

*ANSI Z14.1–1946.

is presented in tabular form, as shown, with or without a complete spring design. When no printed form is available, a drawing similar to the one shown in Fig. 14.53 must be prepared.

A torsion spring offers resistance to a torque load. The extended ends form the torque arms, which are rotated about the axis of the spring. One method of representing and dimensioning torsion springs is shown in Fig. 14.54. A printed form may be used when there is sufficient uniformity in product requirements to warrant the preparation of a printed form or several printed forms. When a printed form is not available, a drawing similar to the one shown must be prepared.

The term *flat spring* includes all springs made of a strip material. One method of representing and dimensioning flat springs is shown in Fig. 14.55.

D □ Bearings

14.40

Bearings. A draftsman ordinarily is never called on to make a detail drawing of a ball or roller bearing, because bearings of these two types are precision-made units that are purchased from reliable manufacturers. All draftsmen working on machine drawings, however, should be familiar with the various types commonly used and should be able to represent them correctly on an assembly drawing. An engineer will find it necessary to determine shaft-mounting fits and housing-mounting fits from a manufacturer's handbook, in order to place the correct limits on shafts and housings. Figure 14.56 shows two types of ball bearings. A roller bearing is shown in Fig. 14.57.

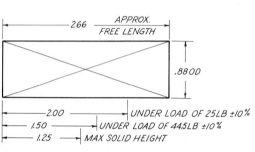

Fig. 14.52 Compression spring drawing.

Fig. 14.53 Extension spring drawing.

Fig. 14.54 Torsion spring representation and dimensioning.

Fig. 14.55 Flat spring drawing.

Ball bearings may be designed for loads either perpendicular or parallel to the shaft. In the former, they are known as radial bearings and in the latter, as thrust bearings. Other types, designated by various names, are made to take both radial and thrust loads, either light or heavy. In most designs, bearings are forced to take both radial and thrust loads. Ball bearings are designated by a letter and code number, the last number of which represents the bearing bore. They may be extra light, light, medium, or heavy and still have the same bore number. That is, bearings of different capacities and different outer diameters can fit shafts having the same nominal size. Figure 14.58 shows typical mountings in a single mechanism.

Tables 21 and 22 in the Appendix give all the necessary information on the bearings used in the problems in this text.

Roller bearings are designed for both radial and thrust loads. The bearing consists of tapered rollers that roll between an inner and an outer race. The rollers are enclosed in a retainer (cage) that keeps them properly spaced.

SINGLE ROW *(a)*   DOUBLE ROW *(b)*

Fig. 14.56 Ball bearings.

Fig. 14.57 Roller bearing.

Fig. 14.58 Ball bearings. (*Courtesy New Departure Bearing Co.*)

Problems

Excellent practice in drawing (or sketching) the representations of threads, threaded fasteners, keys, and rivets is provided by the problems of this chapter.

Use Table 3 in the Appendix if it should be desirable to convert inches and fractions of an inch to millimeters. A metric scale has been given for Prob. 1.

1. Draw or sketch the three layouts shown in Fig. 14.59 to full size, using the given scale to determine the measurements. On layout ① complete the drawing to show a suitable fastener on center line AA. On layout ② show a M12 ($\frac{1}{2}$ in.) hexagonal-head cap screw on center line BB. On layout ③ show a $\frac{3}{8}$-in. button-head rivet on center lines CC and a No. 608 Woodruff key (or equivalent metric key) on center line (CL) DD. Use the schematic symbol for the representation of threads.

2. (Fig. 14.60). Reproduce the views of the assembly of the alignment bearing. On CLs A show $\frac{1}{4}$-in. button-head rivets (four required). On CL B show a $\frac{5}{16} \times \frac{1}{2}$ American Standard square-head set screw. Do not dimension the views.

3. (Fig. 14.61). Reproduce the views of the assembly of the impeller drive. On CLs A show $\frac{1}{4}$–20 UNC $\times \frac{1}{2}$ round-head machine screws and regular lock washers. On CL B show a No. 406 Woodruff key. On CL C show a standard No. 2 $\times 1\frac{1}{2}$ taper pin.

4. (Fig. 14.62). Reproduce the views of the assembly of the bearing head. On CLs A show M12 $\times$ 1.75 studs with regular lock washers and regular semifinished hexagonal nuts (four required). On CLs B show M10 $\times$ 1.5 $\times$ 35 hexagonal-head cap screws (two required). On CL C drill through and tap $\frac{1}{8}$-in. pipe thread.

5. (Fig. 14.63). Reproduce the views of the assembly of the air cylinder. On CL AA at the left end of the shaft show an M24 $\times$ 3 semifinished hexagonal nut. At the right end show a hole tapped M20 $\times$ 2.5 $\times$ 40 deep. Between the piston and the (right) end plate draw a spring 75 mm OD, five full coils, 6.5 mm wire. On CLs B show M10 $\times$ 1.5 $\times$ 30 hexagonal-head cap screws. On CLs C draw M6 $\times$ 1 $\times$ 20 flathead machine screws with heads to the left. On CL D show a $\frac{1}{4}$-in. standard pipe thread. On CLs E show M12 $\times$ 1.75 $\times$ 45 semifinished hexagonal-head bolts. Use semifinished hexagonal nuts. Show visible fasteners on the end view.

Fig. 14.59

Fig. 14.60   Alignment bearing.

Fig. 14.61 Impeller drive.

Fig. 14.62 Bearing head.

Fig. 14.63 Air cylinder.

① FACE (Lathe)

CHAMFER (Lathe) ③

⑥ (a) DRILL, (b) COUNTERSINK
⑦ 4 HOLES (Drill Press)

⑧ (a) DRILL, (b) TAP (thread)
⑨ (Drill Press)

FACE (surface) ⑤

BORE (hole) ② (Lathe)

CHAMFER (Lathe) ④

Ø7.1 – ⌵82° Ø13
4 HOLES

R22

B.C. 92

58

29

31.82 / 31.76 — BORE

R7

Ø5.00 – 14 DEEP
M6×1 – 4 HOLES
9.5 MIN. DEPTH OF FULL THD.

15

R1.5

72

1.5 ×45° CHAMFER

1.5

60

114

76

3.2

3.2

8

32

FILLETS & ROUNDS R3
EXCEPT AS NOTED

INLET COVER
C.I.   1 REQ'D.

Detail drawing of a machined casting.

shop processes, shop terms, and tool drawing

15

15.1

Shop Processes. An engineering draftsman must be thoroughly familiar with the fundamental shop processes before he is qualified to prepare drawings that will fulfill the requirements of the production shops. In preparing working drawings, he must consider each and every individual process involved in the production of a piece and then specify the processes in terms that the shopman will understand (see the facing page). All too frequently, drawings that specify impractical methods and impossible operations are sent to the shops. Most of these impractical specifications are the result of a lack of knowledge, on the draftsman's part, of what can or cannot be done by skilled craftsmen using modern machines and tools.

Although an accurate knowledge of shop processes can be acquired only through actual experience in the various shops, it is possible for an apprentice draftsman to obtain a working knowledge of the fundamental operations through study and observation. This chapter presents and explains the principal operations in the pattern shop, foundry, forge shop, and machine shop.

15.2

Die Casting. Die casting is an inexpensive method for producing certain types of machine parts, particularly those needing no great strength, in mass production. The castings are made by forcing molten metal or molten alloy into a cavity between metal dies in a die-casting machine. Parts thus produced usually require little or no finishing.

15.3

Sand Casting. Castings are formed by pouring molten metal into a mold or cavity. In sand molding, the molten metal assumes the shape of the cavity that has been formed in a sand mold by ramming a prepared moist sand around a pattern and then removing the pattern. Although a casting shrinks somewhat in cooling, the metal hardens in the exact shape of the pattern used (Fig. 15.1).

A sand mold consists of at least two sections. The upper section, called the *cope,* and the lower section, called the *drag,* together form a box-shaped structure called a *flask.*

When large holes [20 mm (.80 in.) and over] or interior passageways and openings are needed in a casting, dry sand cores are placed in the cavity. Cores exclude the metal from the space they occupy and thus form desired openings. Large holes are cored to avoid an unnecessary boring operation. A dry sand core is formed by ramming a mixture of sand and a binding material into a core box that has been made in the pattern shop. To make a finished core rigid, the coremaker places it in a core oven, where it is baked until it is hard.

The molder when making a mold inserts in the sand a sprue stick that he removes after the cope has been rammed. This resulting hole, known as the *sprue,* con-

ducts the molten metal to the *gate,* which is a passageway cut to the cavity. The adjacent hole, called the *riser,* provides an outlet for excess metal.

15.4

Pattern Shop. The pattern shop prepares patterns of all pieces that the foundry is to cast. Although special pattern drawings are frequently submitted, the patternmaker ordinarily uses a drawing of the finished piece that the draftsman has prepared for both the pattern shop and the machine shop (see the chapter-facing page). The finish marks on such a drawing are just as important to the patternmaker as to the machinist, for he must allow, on each surface to be finished, extra metal, the amount of which depends on the method of machining and the size of the casting. In general, this amount varies from 1.5 mm (.06 in.), on very small castings, to as much as 20 mm (.80 in.), on large castings.

It is not necessary for the draftsman to specify on his drawing the amount to be allowed for shrinkage, for the patternmaker has available a "shrink rule," which is sufficiently oversize to take care of the shrinkage.

A pattern usually is first constructed of light, strong wood, such as white pine or mahogany, which, if only a few castings are required, may be used in making sand molds. In quantity production, however, where a pattern must be used repeatedly, the wooden one will not hold up, so a metal pattern (aluminum, brass, and so on) is made from it and is used in its place.

Every pattern must be constructed in such a way that it can be withdrawn from each section of the sand mold. If the pattern consists of two halves (split), the plane of separation should be so located that it will coincide with the plane of separation of the cope and the drag (Fig. 15.1). Each portion of the pattern must be slightly tapered, so that it can be withdrawn without leaving a damaged cavity. The line of intersection, where the dividing plane cuts the pattern, is called the *parting line.* Although this line is rarely shown on a drawing, the draftsman should make certain that the design will allow the patternmaker to establish it. Ordinarily, it is not necessary to specify the slight taper, known as *draft,* on each side of the parting line, for the patternmaker assumes such responsibility when constructing the pattern.

Fig. 15.1 Sand mold.

A "filled-in" interior angle on a casting is called a *fillet*, to distinguish it from a rounded exterior angle, which is known as a *round* (Fig. 5.46). Sharp interior angles are avoided for two reasons: They are difficult to cast; and they are likely to be potential points of failure because the crystals of the cooling metal arrange themselves at a sharp corner in a weak pattern. Fillets are formed by nailing quarter rounds of wood or strips of leather into the sharp angles or by filling the angles with wax.

15.5

Foundry. Although a draftsman is not directly concerned with the foundry, since the patternmaker takes his drawing and prepares the pattern and core boxes for the molder, it is most important that he be familiar with the operations in making a sand mold and a casting. Otherwise, he will find it difficult to prepare an economical design, the cost of which depends on how simple it is to mold and cast.

15.6

Forge Shop. Many machine parts, especially those that must have strength and yet be light, are forged into shape, the heated metal being forced into dies with a drop hammer. Drop forging, since heated metal is made to conform to the shape of a cavity, might be considered a form of casting. However, because dies are difficult to make and are expensive, this method of production is used principally to make parts having an irregular shape that would be costly to machine and could not be made from casting material. Forgings are made of a high-grade steel. Dies are made by expert craftsmen who are known simply as diemakers.

Generally, special drawings, giving only the dimensions needed, are made for the forge shop.

15.7

Machine Shop. In general, the draftsman is more concerned with machine-shop processes than with the processes in other shops, for all castings and forgings that have been prepared in accordance with his drawings must receive their final machining in the machine shop (see chapter-facing page). Since all machining operations must be considered in the design and then properly specified, a draftsman must be thoroughly familiar with the limitations as well as the possibilities of such common machines as the lathe, drill press, boring

machine, shaper, planer, milling machine, and grinder (Fig. 15.2).

15.8

Standard Stock Forms. Many types of metal shapes, along with other materials that are used in the shops for making parts for structures, are purchased from manufacturers in stock sizes. They are made available from the stock department, where rough stock, such as rods, bars, plates, sheet metal, and so on, is cut into sizes desired by the machine shop.

15.9

Lathe. Many common operations, such as turning, facing, boring, reaming, knurling, threading, and so on, may be performed with this widely used machine. In general, however, it is used principally for machining (roughing-out) cylindrical surfaces to be finished on a grinding machine. Removing metal from the exterior surfaces of cylindrical objects is known as *turning* and is accomplished by a sharp cutting tool that removes a thin layer of metal each time it travels the length of a cylindrical surface on the revolving work (Fig. 15.3). The piece, which is supported in the machine between two aligned centers, known as the *dead center* and the *live center*, is caused to rotate about an axis by power transmitted through a lathe dog, chuck, or faceplate. The work revolves against the cutting tool, held in a tool post, as the tool moves parallel to the longitudinal axis of the piece being turned.

Fig. 15.2 Shop operations required to produce a part from cold-rolled stock. The numbers indicate a possible order in which the operations could be performed in the shops.

Fig. 15.3 Lathe operation—turning.

Cutting an interior surface is known as *boring* (Fig. 15.4). A note is not necessary on a drawing to indicate that a surface is to be turned on a lathe.

When a hole is reamed, it is finished very accurately with a fluted reamer of the exact required diameter. If the operation is performed on a lathe, the work revolves as the nonrotating reamer is fed into the hole by turning the handwheel on the tail stock (see Fig. 15.5).

Screw threads may be cut on a lathe by a cutting tool that has been ground to the shape required for the desired thread. The thread is cut as the tool travels parallel to the axis of the revolving work at a fixed speed (Fig. 15.6).

Knurling is the process of roughening or embossing a cylindrical surface. This is accomplished by means of a knurling tool containing knurl rollers that press into the work as the rollers are fed across the surface (Fig. 15.7).

15.10

Drill Press. A drill press is a necessary piece of equipment in any shop because, although it is used principally for drilling, as the name implies (Fig. 15.8), other operations, such as reaming, counterboring, countersinking, and so on, may be performed on it by merely using the proper type of cutting tool. The cutting tool is held in position in a chuck at the end of a vertical spindle that is made to revolve, through power from a motor, at a particular speed suitable for the type of metal being drilled. The most flexible drill press, especially for large work, is the radial type, which is so designed that the spindle is mounted on a movable arm that can be revolved into any desired position for drilling. With this machine, holes may be drilled at various angles and locations without shifting the work, which may be either clamped to the horizontal table or held in a drill vise or drill jig. The ordinary type of drill press without a movable arm is usually found in most shops along with the radial type. A multispindle drill is used for drilling a number of holes at the same time.

Figure 15.9 shows a setup on a drill press for performing the operation of counterboring. A counterbore is used to enlarge a hole to a depth that will allow the head of a fastener, such as a fillister-head cap screw, to be brought to the level of the surface of the piece through which it passes. A counterbore has a piloted end having approximately the same diameter as the drilled hole.

Figure 15.10 shows a setup for the op-

Fig. 15.4 Boring on a lathe.

Fig. 15.5 Reaming on a lathe.

eration of countersinking. A countersink is used to form a tapering depression that will fit the head of a flathead machine screw or cap screw and allow it to be brought to the level of the surface of the piece through which it passes.

A plug tap is used to cut threads in a drilled hole (Fig. 15.11).

A spotfacer is used to finish a round spot that will provide a good seat for the head of a screw or bolt on the unfinished surface of a casting. Figure 15.12 shows various cutting tools commonly used for forming holes and cutting threads.

Fig. 15.6 Cutting threads on a lathe.

Fig. 15.7 Knurling.

Fig. 15.8 Drill press.

Fig. 15.9 Counterboring on a drill press.

Fig. 15.10 Countersinking on a drill press.

Fig. 15.11 Tapping on a drill press.

Fig. 15.12 Various cutting tools.

PLUG TAP

BOTTOMING TAP

STRAIGHT REAMER

TAPER REAMER

COUNTERSINK   COUNTERSINK

COUNTERBORE

TWIST DRILL

THREAD DIE

15.11

Hand Reaming. A hole may be finished to an accurate size by hand reaming, as shown in Fig. 15.13. The reamer in this illustration is of a special type known as a *line reamer.*

15.12

Boring (Fig. 15.14). Boring is the operation of enlarging a circular hole for accuracy in roundness or straightness and may be accomplished on a lathe, drill press, milling machine, or boring mill. When the hole is small and of considerable length, the operation may be performed on a lathe. If the hole is large, the

work is usually done on a boring mill, of which there are two types—the vertical and the horizontal. On a vertical boring machine, the work is fastened on a horizontal revolving table, and the cutting tool or tools, which are stationary, advance vertically into it as the table revolves. On a horizontal boring machine, the tool revolves and the work is stationary.

15.13

Milling Machine. A milling machine is used for finishing plane surfaces and for milling gear teeth, slots, keyways, and so on. In finishing a plane surface, a rotating circular cutter removes the metal for a desired cut as the work, fastened to a moving horizontal bed, is automatically fed against it. Several types of milling cutters are shown in Fig. 15.16.

Figure 15.15 shows a setup for milling gear teeth in a gear blank. Note the form of this particular type of cutter.

15.14

Shaper (Fig. 15.17). A shaper is used for finishing small plane surfaces and for cutting slots and grooves. In action, a fast-moving reciprocating ram carries a tool across the surface of the work, which is fastened to an adjustable horizontal table. The tool cuts only on the forward stroke.

15.15

Planer. The planer is a machine particularly designed for cutting down and finishing large flat surfaces. The work is fastened to a long horizontal table that moves back and forth under the cutting

Fig. 15.13 Hand reaming.

Fig. 15.14 Boring on a boring mill.

Fig. 15.15 Cutting gear teeth on a milling machine.

tool. In action, the tool cuts as the table moves the surface against it. Unlike the cutter on the shaper, it is stationary except for a slight movement laterally for successive cuts.

15.16
Grinding Machine. A grinding machine has a rotating grinding wheel that, ordinarily, is either an emery wheel (fine or coarse) or some type of high-speed wheel made of carborundum (Fig. 15.18). Grinding consists of bringing the surface to be ground into contact with the wheel. Although grinding machines are often used for "roughing" and for grinding down projections and surfaces on castings, their principal use, as far as a draftsman is concerned, is for the final finishing operation that will bring the cylindrical surface of a piece of work down to accurate dimensions. Very fine surface finishes with tolerances as close as 0.025 or 0.050 micrometer (1 or 2 microinches) may be obtained by grinding. Grinding wheels of special form are used for grinding threads when close fits are desired and for forming other shapes.

When a flat surface is to be brought to a super finish, a surface grinder is used. The work, clamped to either a reciprocating or rotating worktable, passes under a rapidly rotating abrasive wheel, as illustrated in Fig. 15.18(a).

In grinding the external surface of a shaft, as illustrated in (d), the work, mounted between centers, rotates slowly while in contact with the rapidly rotating grinding wheel moving along the work as indicated by the arrows. The grinding wheel is mounted so that it may be moved up to the work or away from it. Normally,

Fig. 15.16 Milling cutters and operations.

the depth of cut is about 0.05 mm (.002 in.) per pass. However, finishing cuts may be made as fine as 0.003 mm (.0001 in.) per pass to bring the finished size within the dimensional limits given on a detail drawing.

As might be expected with centerless grinding, the work is not held between

Grinding a Flat Surface
(a)

Internal Grinding
(b)

Grinding a Curved Outline
(c)

Principle of Cylindrical Grinding
(d)

Principle of Centerless Grinding
(e)

Fig. 15.18 Grinding surfaces.

Fig. 15.17 Shaper.

Fig. 15.19 Broach for forming a rectangular hole.

Fig. 15.20 Cutting a keyway.

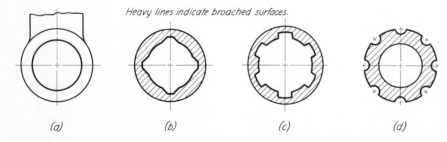

Fig. 15.21 Some typical broached contours.

centers, as in the case of cylindrical grinding. Rather, it may be said to be fed along against the grinding wheel by a regulating wheel while being supported on a rest blade. See Fig. 15.18(*e*). It should be noted that, with both the grinding wheel and regulating wheel rotating in a clockwise direction, the work is forced to rotate counterclockwise. Pressure from the grinding wheel forces the work against both the work rest and the regulating wheel. A centerless grinder produces work that is round and of constant diameter.

Internal grinding for fine finish and close tolerances is illustrated in (*b*). Form grinding is shown in (*c*).

15.17
Broaching. A broach is a tool used to cut keyways and to form square, rectangular, hexagonal, or irregular-shaped holes. It is a hard, tempered cutting tool with serrated cutting edges that enlarge a drilled, punched, or cored hole to a required shape. In operation, each tooth removes a chip of material from the surface except the last few teeth, which are to size (Fig. 15.19). A broach produces a fine finish with the work accurately sized. This is accomplished with a single pass. Broaches, whether used to finish an internal (hole) or an external surface, are either pushed or pulled. A special broaching machine is used for pulling broaches. Some form of press, hydraulic or otherwise, is required for push broaches.

The cutting of a keyway using a slotter (broach) is illustrated in Fig. 15.20. A guide bushing has been inserted in the hole in the piece to hold the tool in position.

Some typical broached contours are shown in Fig. 15.21.

15.18
Superfinishing: Polishing, Honing, and Lapping. Polishing consists of bringing a ground surface into contact with a revolving disc of leather or cloth, thus producing a lustrous smoothness that would be impossible to obtain by using even the finest grinding wheel. The operation is specified on a drawing by a note: "polish" or "grind and polish."

Honing and lapping are methods of producing superfinishes, after grinding, through the use of abrasives. For honing cylindrical bores, fine-grained abrasive sticks which are available in many styles and grades, are used to minimize scratches and to finish the bore to precision limits.

Lapping is a final stock-removing operation that is performed by rubbing the surfaces of parts to be lapped over a lap of soft metal to which has been applied a powdered abrasive mixed with a lubricant.

15.19

Jigs and Fixtures. Often, when an operation must be performed many times in making a part in quantities on a general machine, one of two devices, a *jig* or a *fixture*, may be used to facilitate production and ensure accuracy without making repeated measurements. In general, it might be said that jigs and fixtures are devices for holding work while various machining operations are being performed. Such devices play an important role in the present quantity production of interchangeable parts, for their use greatly reduces the amount of labor required in performing accurate machining operations. However, the advantage that may be gained in the form of lower labor cost in the assembly of units having parts that are interchangeable may be just as important.

Although there are some people who use the terms denoting these devices interchangeably, there is a definite difference between a jig and a fixture that is quite generally recognized. Specifically, a jig is a work-holding device that is capable of controlling the path of the cutting tool. Since it is not fastened to the machine, a jig may be moved around so that several holes may be drilled, tapped, or reamed in their proper locations as established by the position of the drill bushings. Although it is usual for the work to be held by the jig, there are cases where the jig may be clamped on the piece. A well-designed jig should permit the work to be quickly inserted and removed.

A fixture, as the name implies, is fixed or fastened to the table of the machine to locate and hold the work securely in a definite position. Either the cutting tool is moved into position for the operation or the table is moved under the tool. A fixture does not guide the cutting tool. Fixtures are used for such operations as milling, honing, broaching, grinding, and welding.

The three factors that must be considered and evaluated when determining whether or not the cost of making a jig or a fixture will be justified are (1) the number of parts to be produced, (2) the saving in labor cost, and (3) the possibility that

the device may be needed for making replacement parts at a later time. Using simple arithmetic, one can compute and compare the cost of producing one unit in the conventional manner against making the part using a jig or fixture. Knowing the cost of the jig or fixture, the number of units to be produced, and the savings per unit, it will not be difficult to arrive at a decision. Generally, the use of the device should result in a definite saving of time and labor.

More could be given here relating to the economic problems surrounding the use and design of jigs and fixtures but it would be beyond the scope of this chapter.

15.20

Jig Borers. Jig borers are provided with accurate lead screws and special measuring devices that make possible the accurate location and boring of holes. Their use greatly reduces the cost of making jigs since the usual layout work with scriber and punch is not required. It is a common practice in many shops to use a jig borer or its equivalent to machine pieces in small lots rather than go to the expense of making jigs.

15.21

Types of Jigs. There are a number of types of jigs and fixtures that have been developed and are suitable for a wide range of work. The drawing for a simple jig for drilling a jacket tube is shown in Fig. 15.22. The work, shown in phantom line, is usually drawn in red, while the jig itself is drawn in black. In preparing the drawing, the visibility of the lines of the jig itself are as they would be if the work were not in place. This treatment of visibility along with showing the piece to be drilled in red is universal practice.

15.22

Fixtures. Fixtures are designed for varied uses. They may be classified as vise, boring, milling machine, tapping, lathe, grinding, broaching, honing, welding, or checking fixtures. Most readers should be familiar with the principle of the vise. Checking fixtures, as the name implies, are used to check accuracy of production. Figure 15.23 shows a special milling machine fixture that is designed to hold a cover while a keyway is being cut.

15.23

Tool Drawings. When the product drawings for a structure or mechanism have

NO.	REQ	NAME	SPECIFICATIONS
1	1	BASE	SAE 1020-CARB.HDN.
2	1	VEE BLOCK	SAE 1020-CARB.HDN.
3	1	BUSHING PLATE	C.R.S
4	1	SLIDING BLOCK	SAE 1020-CARB.HDN.
5	2	X-L-O BUSHING	#0-11 DR.SIZE #29
6	1	X-L-O BUSHING	#X-14 DR.SIZE $\frac{11}{16}$
7	1	SCREW	$\frac{5}{16}$-18UNC-C.R.S.
8	1	STOP BLOCK	SAE 1020-CARB.HDN.
	1	LINER BUSHING	X-L-O #0-41
	2	X-L-O LOCK SC.	#1
	1	X-L-O LOCK SC.	#2
	1	HAND KNOB	STD.1$\frac{1}{2}$ DIA.
	1	PIN	$\frac{3}{8}$ DIA ×
	1	HEX. NUT	$\frac{5}{8}$-11UNC-2B
	1	T-BOLT	$\frac{3}{8}$-16UNC
	1	HEX. NUT	$\frac{3}{8}$-16UNC-2B
	1	PLAIN WASHER	$\frac{3}{8}$ I.D.
	2	SOC.HD. SC.	$\frac{5}{16}$-18UNC
	2	SOC.HD. SC.	$\frac{3}{8}$-16UNC
	2	DOWEL	$\frac{5}{16}$×1$\frac{1}{2}$
	2	DOWEL	$\frac{3}{8}$×1$\frac{1}{2}$

Fig. 15.22 Jig for drilling a jacket tube. (*Courtesy Ross Gear Division of TRW, Inc.*)

NO.	REQ	NAME	SPECIFICATIONS
1	1	BASE	SAE 1020 CARB.HDN.
2	2	KEY	C.R.S.
3	1	CLAMP STRAP	C.R.S.
4	1	STOP	C.R.S.
5	1	LOCATING PIN	DR. ROD. HDN.
6	1	PLUG	TOOL STL.
	1	AETNA BEARING	#E-1
	1	STUD	$\frac{1}{2}$-13UNC×2$\frac{3}{8}$
	1	HEX. NUT	$\frac{1}{2}$-13UNC
	1	HEX.SOC.HD.CAP SC.	$\frac{5}{16}$-18UNC×1
	2	HEX.SOC.HD.CAP SC.	$\frac{1}{4}$-20UNC×$\frac{1}{2}$
	1	FLAT PT. SET SC.	$\frac{1}{2}$-13UNC×$\frac{3}{4}$
	1	FLAT PT. SET SC.	$\frac{5}{16}$-18UNC×$\frac{5}{8}$

Fig. 15.23 Milling machine fixture for cutting a keyway (upper cover). (*Courtesy Ross Gear
Division of TRW, Inc.*)

been completed by the engineering department, they are released to the production division, where a tool routing sheet and a worksheet are prepared by a process engineer. A routing sheet lists needed manufacturing operations in proper sequence and specifies which tools and machines are to be used. It also gives information for checking production accuracy and may specify the type of inspection instruments. The worksheets contain detailed information concerning the manufacturing processes.

In some cases, clearance specifications between mating parts may be given when critical situations are known to exist. These sheets furnish information needed by the tool designer and tool draftsman and ensure that current machine-shop and production practices will be followed.

The drawings for making the holding devices and special tools needed for the various operations are known as *tool drawings*, and they are prepared by the tool engineering branch of the production division. Along with the routing sheet and worksheet, the tool designer and tool draftsman must have catalogs at hand that furnish information about standard parts and devices that may be purchased ready-made in standard arrangements.

In the planning stage of the development of a jig or fixture, freehand sketches may be used by the designer to integrate his thoughts and record information on standard parts. Such sketches assist the designer in preparing the accurate design layouts that are needed, particularly for complicated situations and where different tool draftsmen will be assigned the task of preparing the working drawings.

The tool draftsman starts with an outline of the workpiece traced in red using phantom lines. Then he draws the outline assembly around the workpiece using black lines (Fig. 15.22). Where a black line of the assembly drawing coincides with one of the red lines of the workpiece, the black line is drawn over the phantom red line. In other words, on tool drawings the black lines are given precedence over the red ones. The practice of showing the workpiece in red and the outline of the jig or fixture in black permits easy reading of the drawing by enabling the reader to distinguish quickly between the workpiece and the jig or fixture holding it. An assembly layout may show locating dimensions and toleranced dimensions for critical conditions.

After the design assembly layout has

been completed and approved, prints are made for the use of the tool draftsman, who is to make the detail drawings. The draftsman or group of draftsmen under a squad leader prepares drawings for all nonstandard and altered standard parts, placing several detail drawings on each detail sheet, as space will permit without overcrowding. Each detail drawing is identified by the same part number shown for it on the assembly layout. In addition to showing the part number in a small circle, it is customary to letter, just below the drawing, the part name, part material, and number required for one unit (Fig. 15.23). In the case of relatively simple jigs and fixtures, the parts are dimensioned, where possible, directly on the layout so that the layout becomes a working drawing. Those parts that cannot be dimensioned conveniently on the layout are detailed in an open space adjacent to the layout (Fig. 15.23). Of course, additional sheets may be used if necessary.

15.24

Dies and Die Drawings. Since standard die-sets may be found in manufacturing catalogs in different styles and sizes, it is the usual practice of tool designers to analyze the workpiece carefully and then select a suitable die-set on which the required punches and dies can be mounted.

A standard die-set has three principal parts, namely, the punch holder (upper shoe), die holder (lower shoe), and guideposts. Other parts of a standard combination are the shank and guidepost bushings. The shank provides for fastening the punch holder to the ram of the press. The guideposts ensure accurate alignment of the punch with the die under all conditions (Fig. 15.24).

Specific information on die design may be found in the several books listed in the bibliography. It is recommended that one consult the handy reference book *Practical Design of Manufacturing Tools, Dies, and Fixtures*, that was prepared under the direction of the Society of Manufacturing Engineers.

The fundamental press-work operations involving the use of punches and dies are (1) plain blanking, (2) piercing, (3) bending or forming, (4) drawing, (5) coining. and (6) assembling. Die assembly drawings usually show three or four views of the assembled die and the same general procedure is followed in their preparation as for jig and fixture drawings except for a modification of the theoretical arrange-

Fig. 15.24 Typical die set.

ment of the views as shown in Fig. 15.25. In this illustration, the front view (front elevation) shows both of the main members (punch and die) in working position. The plan view of the lower shoe (die holder) is in the position of a top view directly above the front view. Aligned horizontally with this latter view is the plan view of the upper shoe (punch holder) shown as it would be seen looking upward from below. This is done so that the diemaker has a plan of the actual layout from which to work and to conserve space. The front view, always in section, shows the die as it would be seen from the front of the press and the views of the upper and lower shoes show all parts attached as they would be when in operation. Small part-section views and auxiliary views are used when needed. If a side view is necessary, it should be drawn in section. The assembly drawing should give general information and show the important dimensions needed by the diemaker, such as the shut height of the die, shrank diameter, the type and size of press, and the press stroke. On the die assembly drawing, each component part is numbered as shown in Fig. 15.26.

When it is necessary to detail a designed die, it is the usual practice to place the details on sheets separate from the assembly drawing.

A material list should be given on the die assembly drawing. This should show the number of each part and specify the rough material size.

15.25
Automatic Machines. In large industrial concerns, most mechanical parts are made on either semiautomatic or fully automatic machines by semiskilled operators. A discussion of even a few of these, however, is beyond the scope of a general drawing text in which each subject is limited to a few pages. Since most specialized mass-production machines, having mechanisms that control the movement of cutting tools, operate on the same general principles as the general-purpose machines, a young engineering draftsman should be able to determine their limitations and capabilities through observation, if he has a general knowledge of such machines as the lathe, shaper, drill press, milling machine, and so on. No prospective designer or draftsman should ever

Fig. 15.25 Arrangement of views for a punch and die drawing.

Fig. 15.26 Die design layout assembly drawing. (*Courtesy Ross Gear Division of TRW, Inc.*)

forgo an opportunity to observe special production machines for he must have a thorough understanding of all shop machines and methods, if his drawings are to be satisfactory for the shops.

In use at the present time are production machines that follow directions given on punched tapes. The use of tapes and computers by industry does not mean that the draftsmen, technologists, and engineers will have less work to do and that there will be fewer of them. It does mean, however, that the men assigned to both areas must keep themselves informed and up-to-date on late developments. The draftsman should learn how to dimension the drawing of a part to meet the requirement for programming the machine or machines that will perform the shop operations. He must have a more thorough understanding of basic fundamentals than in the past and at the same time be willing to accept change. Numerically controlled machine tools are discussed in Chapter 18.

The fact must be recognized that automation is here and the only question the technically trained man of today can ask himself is whether or not he, as an individual, can adjust to new knowledge and different requirements.

15.26

Manufacturing Processes and the Detail Drawing. In preparing the detail drawing that is needed for the production of a part, the draftsman must give considerable thought to the manufacturing processes that will be required to make the conception a reality. Machined surfaces must be indicated and the dimensions selected and placed with the manufacturing methods in mind. Shop notes, as they may be needed, must be prepared in accordance with the recommendations given in the appropriate ANSI standards.

The detail working drawing of the inlet cover shown on this chapter's facing page has been prepared for use by both the pattern shop and the machine shop since these two shops will be involved in the production of the part. It should and does contain all of the dimensions and notes needed, first, to make the pattern for the casting and, second, to machine the cast part to obtain the finished inlet cover ready for use.

The pictorial drawing gives an interpretation, in shop terms, of the machining operations required by the detail drawing. The sequence of these operations is indicated by the numbers enclosed within circles.

15.27

Measuring Tools. Figures 15.27, 15.28, and 15.29 show a few of the measuring tools commonly available in shops. When great accuracy is not required, calipers are used (Fig. 15.28). The outside calipers are suited for taking external measurements, as, for example, from a shaft. They are adjusted to fit the piece, and then the setting is applied to a rule to make a reading. The inside calipers have outturned toes, which fit them for taking internal measurements, as, for example, in measuring either a cylindrical or a rectangular hole. When extreme accuracy is required, some form of micrometer calipers may be used (Fig. 15.29).

Quick Reading with Decimal Equivalents

Fig. 15.27 Steel rule.

DIVIDERS   OUTSIDE CALIPERS   INSIDE CALIPERS

Fig. 15.28 Calipers.

Fig. 15.29 Micrometers.

Shop processes, shop terms, and tool drawings

Glossary
of common shop terms

Anneal (v). To heat a piece of metal to a particular temperature and then allow it to cool slowly for the purpose of removing internal stresses.

Bore (v). To enlarge a hole using a boring bar in order to make it smooth, round, and coaxial. Boring is usually done on a lathe or boring mill.

Boss (n). A circular projection, which is raised above a principal surface of a casting or forging.

Braze (v). To join two pieces of metal by the use of hard solder. The solder is usually a copper–zinc alloy.

Broach (v). To machine a hole to a desired shape, usually other than round. The cutting tool, known as a broach, is pushed or pulled through the rough-finished hole. It has transverse cutting edges.

Burnish (v). To smooth or apply a brilliant finish.

Bushing (n). A removable cylindrical sleeve, which is used to provide a bearing surface.

Carburize (v). To harden the surface of a piece of low-grade steel by heating in a carbonizing material to increase the carbon content and then quenching.

Case-harden (v). To harden a surface as described above or through the use of potassium cyanide.

Chamfer (v). To bevel an external edge or corner.

Chase (v). To cut screw threads on a lathe using a chaser, a tool shaped to the profile of a thread.

Chill (v). To cool the surface of a casting suddenly so that the surface will be white and hard.

Chip (v). To cut away or remove surface defects with a chisel.

Collar (n). A cylindrical part fitted on a shaft to prevent a sliding movement.

Color-harden (v). (See Case-harden.) A piece is color-hardened mainly for the sake of appearance.

Core (v). To form a hole or hollow cavity in a casting through the use of a core.

Counterbore (v). To enlarge the end of a cylindrical hole to a certain depth, as is often done to accommodate the head of a fillister-head screw. (n) The name of the tool used to produce the enlargement.

Countersink (v). To form a conical enlargement at the end of a cylindrical hole to accommodate the head of a screw or rivet. (n) The name of the tool used to form a conical-shaped enlargement.

Crown (n). The angular or curved contour of the outer surface of a part, such as on a pulley.

Die (n). A metal block used for forming or stamping operations. A thread-cutting tool for producing external threads.

Die Casting (n). A casting that has been produced by forcing a molten alloy having an aluminum, copper, zinc, tin, or lead base into a metal mold composed of two halves.

Die Stamping (n). A piece that has been cut or formed from sheet metal through the use of a die.

Draw (v). To form metal, which may be either cold or hot, by a distorting or stretching process. To temper steel by gradual or intermittent quenching.

Drill (v). To form a cylindrical hole in metal. (n) A revolving cutting tool designed for cutting at the point.

Drop Forging (n). A piece formed while hot between two dies under a drop hammer.

Face (v). To machine on a lathe a flat face, which is perpendicular to the axis of rotation of the piece.

Feather (n). A rectangular sliding key, which permits a pulley to move along the shaft parallel to its axis.

File (v). To shape, finish, or trim with a fine-toothed metal cutting tool, which is used with the hands.

Fillet (n). A rounded filling, which increases the strength at the junction of two surfaces that form an internal angle.

Fit (n). The tightness of adjustment between the contacting surfaces of mating parts.

Flange (n). The top and bottom member of a beam. A projecting rim added on the end of a pipe or fitting for making a connection.

Forge (v). To shape hot metals by hammering, using a hand-hammer or machine.

Galvanize (v). To coat steel or iron by immersion in a bath of zinc.

Graduate (v). To mark off or divide a scale into intervals.

Grind (v). To finish a surface through the action of a revolving abrasive wheel.

Kerf (n). A groove or channel cut by a saw or some other tool.

Key (n). A piece used between a shaft and a hub to prevent the movement of one relative to the other.

Keyway or Keyseat (n). A longitudinal groove cut in a shaft or a hub to receive a key. A key rests in a keyseat and slides in a keyway.

Knurl (*v*). To roughen a cylindrical surface to produce a better grip for the fingers.

Lap (*v*). To finish or polish with a piece of soft metal, wood, or leather impregnated with an abrasive.

Lug (*n*). A projection or ear, which has been cast or forged as a portion of a piece to provide a support or to allow the attachment of another part.

Malleable Casting (*n*). A casting that has been annealed to toughen it.

Mill (*v*). To machine a piece on a milling machine by means of a rotating toothed cutter.

Neck (*v*). To cut a circumferential groove around a shaft.

Pack-harden (*v*). To case-carburize and harden.

Pad (*n*). A low, projecting surface, usually rectangular.

Peen (*v*). To stretch or bend over metal using the peen end (ball end) of a hammer.

Pickle (*v*). To remove scale and rust from a casting or forging by immersing it in an acid bath.

Plane (*v*). To machine a flat surface on a planer, a machine having a fixed tool and a reciprocating bed.

Polish (*v*). To make a surface smooth and lustrous through the use of a fine abrasive.

Punch (*v*). To perforate a thin piece of metal by shearing out a circular wad with a nonrotating tool under pressure.

Ream (*v*). To finish a hole to an exact size using a rotating fluted cutting tool known as a reamer.

Rib (*n*). A thin component of a part that acts as a brace or support.

Rivet (*n*). A headed shank, which more or less permanently unites two pieces. (*v*) To fasten steel plates with rivets.

Round (*n*). A rounded external corner on a casting.

Sandblast (*v*). To clean the surface of castings or forgings by means of sand forced from a nozzle at a high velocity.

Shear (*v*). To cut off sheet or bar metal through the shearing action of two blades.

Shim (*n*). A thin metal plate, which is inserted between two surfaces for the purpose of adjustment.

Spline (*n*). A keyway, usually for a feather key. (See Feather.)

Spotface (*v*). To finish a round spot on the rough surface of a casting at a drilled hole for the purpose of providing a smooth seat for a bolt or screw head.

Spot Weld (*v*). To weld two overlapping metal sheets in spots by means of the heat of resistance to an electric current between a pair of electrodes.

Steel Casting (*n*). A casting made of cast iron to which scrap steel has been added.

Swage (*v*). To form metal with a "swage block," a tool so constructed that through hammering or pressure the work may be made to take a desired shape.

Sweat (*v*). To solder together by clamping the pieces in contact with soft solder between and then heating.

Tack Weld (*n*). A weld of short intermittent sections.

Tap (*v*). To cut an internal thread, by hand or with power, by screwing into the hole a fluted tapered tool having thread-cutting edges.

Taper (*v*). To make gradually smaller toward one end. (*n*) Gradual diminution of diameter or thickness of an elongated object.

Taper Pin (*n*). A tapered pin used for fastening hubs or collars to shafts.

Temper (*v*). To reduce the hardness of a piece of hardened steel through reheating and sudden quenching.

Template (*n*). A pattern cut to a desired shape, which is used in layout work to establish shearing lines, to locate holes, etc.

Tumble (*v*). To clean and smooth castings and forgings through contact in a revolving barrel. To further the results, small pieces of scrap are added.

Turn (*v*). To turn-down or machine a cylindrical surface on a lathe.

Undercut (*n*). A recessed cut.

Upset (*v*). To increase the diameter or form a shoulder on a piece during forging.

Weld (*v*). To join two pieces of metal by pressure or hammering after heating to the fusion point.

A PRODUCTION DRAWING **Commercial example of a detail drawing.** It should be noted that the drawing gives complete information for the production of the part and describes with adequate dimensions the part's size. (*Courtesy International Business Machines Corporation*)

A3 620-0306-0   MECHANICAL FORMAT   ASTRO CLOTH   N650T

production drawings: preparation and duplication

16

A □ Production (shop) drawings

16.1

Communication Drawings. These varied types of engineering drawings, ranging from design drawings to exploded pictorial drawings, have one thing in common, and that is that they are prepared to convey needed ideas and facts to others. Since all serve the same purpose they may be classed together as "communication drawings," this term being almost all-inclusive.

In this chapter we will be concerned mainly with the types of drawings that are prepared by draftsmen under an engineer's supervision and that are to serve as communications to others beyond the engineering department. The preparation of idea sketches, both multiview and pictorial, has been discussed in detail in Chapters 6 and 12. Charts and graphs, which may also be thought of as communication drawings, are presented in Chapter 19. Charts and graphs are used by engineers to supplement written reports and technical papers.

16.2

Sketches and Design Drawings. The first stage in the development of an idea for a structure or machine is to prepare freehand sketches and to make the calculations required to determine the feasibility of the design. From these sketches the designer prepares a layout, on which an accurate analysis of the design is worked out. It is usually drawn full-size and is executed with instruments in pencil (Fig. 12.28). The layout should be complete enough to allow a survey of the location of parts (to avoid interference), the accessibility for maintenance, the requirements for lubrication, and the method of assembly. See Chapter 12.

Usually, only center distances and certain fixed dimensions are given. The general dimensioning, as well as the determination of material and degree of finish of individual parts, is left for the draftsman, who makes the detail drawings while using the layout drawing as a guide.

Design layouts require both empirical and scientific design. Empirical design involves the use of charts, formulas, tables, and so forth, which have been derived from experimental studies and scientific computations. Scientific design, which requires a broad knowledge of the allied fields, such as mechanics, metalurgy, and mathematics, is used when a new machine is designed to operate under special specified conditions for which data are not available in any handbook.

16.3

Classes of Machine Drawings. There are two recognized classes of machine drawings: detail drawings and assembly drawings.

16.4

Set of Working Drawings. A complete set of working drawings for a machine consists of detail sheets giving all necessary shop information for the production of individual pieces and an assembly drawing showing the location of each piece in the finished machine. In addition, the set may inclue drawings showing a foundation plan, piping diagram, oiling diagram, and so on.

16.5

Detail Drawing. A detail drawing should give complete information for the manufacture of a part, describing with adequate dimensions the part's size. Finished surfaces should be indicated and all necessary shop operations shown. The title should give the material of which the part is to be made and should state the number of the parts that are required for the production of an assembled unit of which the part is a member. Commercial examples of detail drawings are shown in Figs. 16.1–16.4.

Since a machinist will ordinarily make one part at a time, it is advisable to detail each piece, regardless of its size, on a separate individual sheet. In some shops,

however, custom dictates that related parts be grouped on the same sheet, particularly when the parts form a unit in themselves. Other concerns sometimes group small parts of the same material together thus: castings on one sheet, forgings on another, special fasteners on still another, and so on.

16.6

Making a Detail Drawing. With a design layout or original sketches as a guide, the procedure for making a detail drawing is as follows:

1. Select the views, remembering that, aside from the view showing the characteristic shape of the object, there should be as many additional views as are necessary to complete the shape description. These may be sectional views that reveal a complicated interior construction, or auxiliary views of surfaces not fully described in any of the principal views.

2. Decide on a scale that will allow, without crowding, a balanced arrangement of all necessary views and the location of dimensions and notes. Although very small parts should be drawn double-size or larger, to show detail and to allow for dimensions, a full-size scale should be used when possible. In general, the same scale should be used for pieces of the same size.

3. Draw the main center lines and block in the general outline of the views with light, sharp 6H pencil lines.

4. Draw main circles and arcs in finished weight.

5. Starting with the characteristic view, work back and forth from view to view until the shape of the object is completed. Lines whose definite location and length are known may be drawn in their finished weight.

6. Put in fillets and rounds.

7. Complete the views by darkening the object lines.

8. Draw extension and dimension lines.

9. Add arrowheads, dimensions, and notes.

10. Complete the title.

11. Check the entire drawing carefully.

16.7

One-view Drawings. Many parts, such as shafts, bolts, studs, and washers, may

Fig. 16.1 Detail drawing. (*Courtesy John Deere and Company*)

require only one properly dimensioned view. In the case of each of these parts, a note can imply the complete shape of the piece without sacrificing clearness. Most engineering departments, however, deem it better practice to show two views.

16.8

Detail Titles. Every detail drawing must give information not conveyed by the notes and dimensions, such as the name of the part, part number, material, number required, and so on. The method of recording and the location of this information on the drawing varies somewhat in different drafting rooms. It may be lettered either in the record strip or directly below the views (Figs. 16.1 and 16.2).

If all surfaces on a part are machined, finish marks are omitted and a title note, "FINISH ALL OVER," is added to the detail title.

16.9

Title Blocks and Record Strips. The purpose of a title or record strip is to present in an orderly manner the name of the machine, name of the manufacturer, date, scale, drawing number, and other drafting-room information.

Every commercial drafting room has developed its own standard title forms, whose features depend on the processes of manufacture, the peculiarities of the plant organization, and the established

customs of particular types of manufacturing. In large organizations, the blank form, along with the borderline, is printed on standard sizes of tracing paper and/or Mylar.

A record strip is a form of title extending almost the entire distance across the bottom of the sheet. In addition to the usual title information, it may contain a

Fig. 16.2 Working drawing.

section for recording revisions, changes, and so on, with the dates on which they were adopted (Fig. 16.3).

16.10
Contents of the Title (Figs. 16.1 and 16.2). The title on a machine drawing generally contains the following information:

1. Name of the part.

2. Name of the machine or structure. (This is given in the main title and is usually followed by one of two words: *details* or *assembly*.) See Fig. 16.7.

3. Name and location of the manufacturing firm.

4. Name and address of the purchasing firm, if the structure has been designed for a particular company.

5. Scale.

6. Date. (Often spaces are provided for the date of completion of each operation in the preparation of the drawing. If only one date is given, it is usually the date of completion of the drawing.)

7. Initials or name of the draftsman who made the pencil drawing.

8. Initials of the checker.

9. Initials or signature of the chief draftsman, chief engineer, or another in authority who approved the drawing.

10. Initials of the tracer (if drawing has been traced).

11. Drawing number. This generally serves as a filing number and may furnish information in code form. Letters and numbers may be so combined to indicate departments, plants, model, type, order number, filing number, and so on. The drawing number is sometime repeated in the upper-left-hand corner (in an upside-down position), so that the drawing may be quickly identified if it should become reversed in the file.

Some titles furnish information such as material, part number, pattern number,

Fig. 16.3   Detail drawing. (*Courtesy John Deere and Company*)

finish, treatment, estimated weight, superseded drawing number, and so on.

16.11
Corrections and Alterations. Alterations on working drawings are made either by cancellation or by erasure. Cancellations are indicated by parallel inclined lines drawn through the views, lines, notes, or dimensions to be changed.

Superseding dimensions should be placed above or near the original ones. If alterations are made by erasure, the changed dimensions are often underlined.

All changes on a completed or approved drawing should be recorded in a revision record that may be located either adjacent to the title block (Fig. 16.3) or at one corner of the drawing (Fig. 16.4). This note should contain the identification symbol, date, authorization number, character of the revision, and the initials of the draftsman and checker who made the change. The identification symbol is a numeral or letter placed in a small circle near the alteration on the body of the drawing.

If the changes are made by complete erasure, record prints should be made for the file before the original is altered. Many companies make record prints whenever changes are extensive.

Since revisions on completed drawings are usually necessitated by unsatisfactory methods of production or by a customer's request, they should never be made by a draftsman unless an order has been issued with the approval of the chief engineer's office.

16.12
Pattern-shop Drawings. Sometimes special pattern-shop drawings, giving information needed for making a pattern, are required for large and complicated castings. If the patternmaker receives a drawing that shows finished dimensions, he provides for the draft necessary to draw the pattern and for the extra metal for machining. He allows for shrinkage by making the pattern oversize. When, however, the draft and allowances for finish are determined by the engineering department, no finish marks appear on the drawing. The allowances are included in the dimensions.

16.13
Forge-shop Drawings. If a forging is to be machined, separate detail drawings usually are made for the forge and machine

shops. A forging drawing gives all the nominal dimensions required by the forge shop for a completed rough forging.

16.14
Machine-shop Drawings. Rough castings and forgings are sent to the machine shop to be finished. See Fig. 16.4. Since the machinist is not interested in the dimensions and information for the previous stages, a machine-shop drawing frequently gives only the information necessary for machining.

16.15
Assembly Drawings. A drawing that shows the parts of a machine or machine unit assembled in their relative working positions is an assembly drawing. There are several types of such drawings: design assembly drawings, working assembly drawings, unit assembly drawings, installation diagrams, and so on, each of which will be described separately (Figs. 16.5–16.7 and 16.10–16.12).

16.16
Working Assembly Drawings. A working assembly drawing, showing each piece completely dimensioned, is sometimes made for a simple mechanism or unit of related parts. No additional detail drawings of parts are required.

16.17
Subassembly (Unit) Drawings. A unit assembly is an assembly drawing of a group of related parts that form a unit in a more complicated machine. Such a drawing would be made for the tail stock of a lathe, the clutch of an automobile, or the carburetor of an airplane. A set of assembly drawings thus takes the place of a complete assembly of a complex machine (Fig. 16.6).

16.18
Bill of Material or Parts List. A bill of material is a list of parts placed on an assembly drawing just above the title block, or, in the case of quantity production, on a separate sheet. The bill contains the part (item or key) number, descriptive name, material, quantity (number) required, and so on, of each piece. Additional information, such as stock size, pattern number (castings), and so forth, is sometimes listed.

Suggested dimensions for ruling are shown in Fig. 16.9. For 3.5 mm ($\frac{1}{8}$ in.) letters, the lines should never be spaced closer than 8 mm ($\frac{5}{16}$ in.). Fractions, if used, are made slightly less than full height and are centered between the lines.

When listing standard parts in a bill of material, the general practice is to omit the name of the materials and to use abbreviated descriptive titles. A pattern number may be composed of the commercial job number followed by the assigned number one, two, three, and so on. It is suggested that parts be listed in the following order: (1) castings, (2) forgings, (3) parts made from bar stock, and (4) standard parts.

Sometimes bills of material are first typed on thin paper and then blueprinted. The form may be ruled or printed (Fig. 16.8).

16.19

Title. The title strip on an assembly drawing usually is the same as that used on a detail drawing. It will be noted, when lettering in the block, that the title of the drawing is generally composed of the name of the machine followed by the word *assembly* (Figs. 16.5 and 16.7).

16.20

Making the Assembly Drawing. The final assembly may be traced from the design assembly drawing, but more often it is redrawn to a smaller scale on a separate sheet. Since the redrawing, being done from both the design and detail drawings, furnishes a check that frequently reveals errors, the assembly always should be drawn before the details are accepted as finished and the blueprints are made. The assembly of a simple machine or unit is sometimes shown on the same sheet with the details.

Accepted practices to be observed on assemblies are as follows:

1. *Sectioning.* Parts should be sectioned using the American Standard symbols shown in Fig. 7.31. The practices of sectioning apply to assemblies.

2. *Views.* The main view, which is usually in full section, should show to the best advantage nearly all the individual parts and their locations. Additional views are shown only when they add necessary information that should be conveyed by the drawing.

3. *Hidden lines.* Hidden lines should be omitted from an assembly drawing, for they tend merely to overload it and create confusion. Complete shape description is unnecessary, since parts are either standard or are shown on detail drawings.

4. *Dimensions.* Overall dimensions and center-to-center distances indicating the relationship of parts in the machine as a whole are sometimes given. Detail dimensions are omitted, except on working-assembly drawings.

5. *Identification of parts.* Parts in a machine or structure are identified on

Fig. 16.4 Detail drawing. (*Courtesy Warner & Swasey Co.*)

Fig. 16.5 An assembly drawing.

the assembly drawing by numbers that are used on the details and in the bill of material (Fig. 16.5). These should be made at least 5 mm ($\frac{3}{16}$ in.) high and enclosed in a 10 mm ($\frac{3}{8}$ in.) circle. The centers of the circles are located not less than 20 mm ($\frac{3}{4}$ in.) from the nearest line of the drawing. Leaders, terminated by arrowheads touching the parts, are drawn radial with a straightedge. The numbers, in order to be centered in the circles, should be made first and the circles drawn around them. An alternative method used in commercial practice is to letter the name and descriptive information for each part and draw a leader pointing to it in the main view.

16.21
Checking Drawings. Checking, the final

Fig. 16.6 **Tachometer mounting bracket (unit) assembly.** (*Courtesy Studebaker Corp.*)

Fig. 16.7 **Assembly drawing.** (*Courtesy Pierce Governor Co., Inc.*)

	PIERCE GOVERNOR ASSEMBLY GC-3998		
	PARTS LIST		
Key No.	Part Name	Part No.	Quantity
1	Governor Body	G-9042-16	1
2	Governor Flange	G-9138-3	1
3	Sems Fastener	X-1784	4
4	Gasket	X-1425	1
5	Hex. Nut	X-977	1
6	Hex. Head Screw	X-890-4	1
7	Welch Plug	X-2019	1
8	Shoulder Stud	G-9799	1
9	Washer	X-2307	2
10	Snap Ring	X-1923	1
11	Stop Bracket	G-9556	1
12	Governor Spring	SN-1304	1
13	Thrust Bearing	X-1336-A	1
14	Throttle Lever Assembly	A-6325	1
15	Thrust Sleeve	G-10813	1
16	Snap Ring—Internal	X-1921	1
17	Oil Cup	X-2053	1
18	Spacer	G-12614-1	1
19	Governor Pulley	G-10908-1	1
20	Oil Lever Check	X-2054	1
21	Hex. Nut	X-1011-1	2
22	Adj. Screw Eye	G-12306	1
23	Roll Pin	X-2620	1
24	Roll Pin	X-2602	1
25	Oil Lever Tag	X-1945	1
26	Spring Adj. Lever	G-5715	1
27	Bushing	X-2721	1
28	Yoke	G-9838	1
29	Sems Fastener	X-1687	2
30	Bumper Screw	G-5113-1	1
31	Hex. Nut	X-246-4	1
32	Bumper Spring	SN-1481	1
33	Spacer	G-11886-1	1
34	Laminated Weight Assembly	A-2446	4
35	Weight Pin	G-14007	4
36	"E" Retaining Ring	X-2996	4
37	Ball Bearing	X-310	1
38	Name Plate	X-581	1
39	Escutcheon Pin	X-455	2
40	Oil Seal	X-652	1
41	Rocker Shaft Oil Seal	A-6118	1
42	External Snap Ring	X-1904	2
43	Ball Bearing	X-328	2
44	Rocker Shaft	G-11698	1
45	Spider and Shaft Assembly	A-6637	1
46	Washer	X-2026-4	1
47	Elastic Stop Nut	X-1845	1

Fig. 16.8 Parts list—governor assembly (Fig. 16.7).

9	13×24 PLAIN WASHER	1	
8	M10×12 SLOTTED DOG PT. SET SC.	1	
7	M5×20 FLAT HD. MACH. SC.	6	
6	BALL	2	C.R.S.
5	HANDLE	1	C.R.S.
4	VISE SCREW	1	C.R.S.
3	JAW PLATE	2	C.R.S.
2	JAW PATT. NO. 19742-2	1	C.I.
1	BASE PATT. NO. 19742-1	1	C.I.
ITEM	NAME	NO. PER UNIT	MATERIAL

LIMITS, UNLESS OTHERWISE NOTED:
FRACTIONAL ±⅟₆₄   DECIMAL ±.010   ANGULAR ±½°

TITLE OF UNIT

VISE ASSEMBLY

SCALE *FULL SIZE*
APPROVED *WJL*

LAFAYETTE, INDIANA

TRACED BY *J.H.D.*   CHECKED BY *J.H.P.*   DATE *12-10-79*
DRAWN BY *DOE, JOHN H.*   CODE *WJL-E-15*   DRAWING NO. *19742*

Fig. 16.9 Bill of material.

assurance that the machine is correctly designed, should be done by a person (checker or squad foreman) who has not prepared the drawings but who is thoroughly familiar with the principles of the design. He must have a broad knowledge of shop practices and assembly methods. In commercial drafting rooms, the most experienced persons are assigned to this type of work. The assembly drawing is checked against the detail drawings and corrections are indicated with either a soft or colored pencil. The checker should:

1. Survey the machine as a whole from the standpoint of operation, ease of assembly, and accessibility for repair work. He should consider the type, strength, and suitability of the materials.

2. Check each part with the parts adjacent to it, to make certain that proper clearances are maintained. (To determine whether or not all positions are free of interference, it may be necessary to lay out the extreme travel of moving parts to an enlarged scale.)

3. Study all drawings to see that each piece has been illustrated correctly and that all necessary views, types of views, treatments of views, and scales have been shown.

4. Check dimensions by scaling; calculate and check size and location dimensions that affect mating parts; determine the suitability of dimensions from the standpoint of the various departments' needs, such as pattern, forge, machine, assembly shop, and so on; examine views for proper dimensioning, and mark unnecessary, repeated, or omitted dimensions.

5. Check tolerances, making sure the computations are correct and that proper fits have been used, so that there will be no unnecessary production costs.

6. See that finishes and such operations as drilling, reaming, boring, tapping, and grinding are properly specified.

7. Check specifications for material.

8. Examine notes for correctness and location.

9. See that stock sizes have been used for standard parts, such as bolts, screws, keys, and so on. (Stock sizes may be determined from catalogs.)

10. Add any additional explanatory notes that should supply necessary information.

11. Check the bill of material to see that each part is completely and correctly specified.

12. Check items in the title block.

13. Make a final survey of the drawing in its entirety, making certain there is either a check or correction for each dimension, note, and specification.

16.22

Installation Assembly Drawings. An installation drawing gives useful information for putting a machine or structure together. The names of parts, order of assembling parts, location dimensions, and special instructions for operating may also be shown.

16.23

Outline Assembly Drawings. Outline assembly drawings are most frequently made for illustrative purposes in catalogs. Usually they show merely overall and principal dimensions (Fig. 16.10). Their appearance may be improved by the use of line shading.

16.24

Exploded Pictorial Assembly Drawings for Parts Lists and Instruction Manuals. Exploded pictorial assembly drawings are used frequently in the parts lists sections of company catalogs and in instruction manuals. Drawings of this type are easily understood by those with very little experience in reading multiview drawings. Figure 16.11 shows a commercial example of an exploded pictorial assembly drawing.

16.25

Diagram Assembly Drawings. Diagram drawings may be grouped into two general classes: (1) those composed of single lines and conventional symbols, such as piping diagrams, wiring diagrams, and so on (Fig. 16.12); and (2) those drawn in regular projection, such as an erection drawing, which may be shown in either orthographic or pictorial projection.

Fig. 16.10 Outline assembly drawing. (*Courtesy General Electric Company*)

Fig. 16.11 Exploded pictorial assembly drawing. (*Courtesy Lockheed Aircraft Corporation*)

Piping diagrams give the size of pipe, location of fittings, and so on. To draw an assembly of a piping system in true orthographic projection would add no information and merely entail needless work.

A large portion of electrical drawing is composed of diagrammatic sketches using conventional electrical symbols (Fig. 16.13). Electrical engineers therefore need to know the American National (ANSI) Standard wiring symbols given in the Appendix.

16.26

Chemical Engineering Drawings. In general, the chemical engineer is concerned with plant layouts and equipment design. He must be well informed on the types of machinery used in grinding, drying, mixing, evaporation, sedimentation, and distillation, and must be able to design or select conveying machinery.

It is obvious that the determining of the sequence of operations, selecting of machinery, arranging of piping, and so on, must be done by a trained chemical engineer who can speak the basic language of the mechanical, electrical, or civil engineer with whom he must cooperate. To be able to do this, he must have a thorough

Fig. 16.12 Diagram assembly drawing. (*Courtesy Instruments Magazine*)

knowledge of the principles of engineering drawing.

Plant layout drawings, the satisfactory development of which requires numerous preliminary sketches (layouts, scale dia-

Fig. 16.13 Schematic drawing. (*Courtesy General Motors Corporation*)

grams, flow sheets, and so on), show the location of machines, equipment, and the like. Often, if the machinery and apparatus are used in the manufacturing of chemicals and are of a specialized nature, a chemical engineer is called on to do the designing. It even may be necessary for him to build experimental apparatus.

16.27
Electrical Engineering Drawings. Electrical engineering drawings are of two types: machine drawings and diagrammatic assemblies (Fig. 16.13). Working drawings, which are made for electrical machinery, involve all of the principles and conventions of the working drawings of the mechanical engineer. Diagrammatic drawings have been discussed in Sec. 16.25. The practices to be followed in preparing drawings of electrical systems are presented in Chapter 22.

16.28
Civil Engineering Drawings. The civil engineer is concerned with a broad field of construction and with civic planning. The drawings prepared for civil engineers may be in the nature of maps for city, state, and nationwide planning for streets, water systems, sewerage systems, airports, highways, railroads, harbor and waterways; or they may be design, fabrication, and erection drawings for concrete and steel structures, as in the case of buildings and bridges. Information needed for the preparation of drawings of these types may be found in Chapters 25 and 26.

B □ Reproduction and duplication of engineering drawings

16.29
Usually it is necessary to duplicate a set of drawings of a machine or structure, one or more copies being made for the office and extra copies for interested persons connected with the home organization or an outside cooperating firm. Sometimes many sets are required. Under modern production methods, the various parts of a machine or structure may be produced in numerous departments and plants that are located miles apart. Each of these departments must have exact copies of the original drawings. To satisfy this demand, several economical processes have been devised.

The various processes may be grouped in accordance with the general similarity in methods. The mechanical processes form a group that includes mimeographing, hectographing, and printing; the photochemical processes, using reflected light, include photography and photocopying; the photochemical processes, requiring transmitted light, include blueprinting, Ozalid printing, Van Dyke printing, and so on. A few processes, such as photolithography, are combinations of methods.

16.30
Blueprints. A blueprint may be considered a photographic copy of an original drawing in that the process is similar to that of photography. A piece of sensitized paper is exposed to light transmitted through a negative (tracing) and then is developed in water to bring out the image. The negative is a transparent sheet of paper or transparentized cloth on which the image of the original drawing has been traced with opaque lines. The printing paper is a white paper that has been coated with a solution of ammonia citrate of iron and ferrocyanide of potassium. When the paper is exposed through the tracing, a chemical action takes place wherever the light is able to reach the sensitized surface. In the first bath of clear water, the coating is washed away from the parts of the surface protected by these lines, exposing the original surface of the paper against a developed blue background. The contrast between the blue background and white lines may be intensified by dipping the print in an oxidizing solution of potassium bichromate in water, after which the print must be rinsed thoroughly in clean water to avoid stain marks.

Notations and changes can be made on a blueprint by lettering and drawing with an alkaline solution strong enough to bleach out the blue compound. Although a solution of sodium hydroxide (caustic soda) is best for this purpose, ordinary sodium bicarbonate (baking soda) in water may be used.

Since a good print with clear white lines against a uniform brilliant blue background can be obtained only by proper timing of the exposure, it is well to make one or two trial (part) prints on small scraps of blueprint paper before making the final print. If the background is a pale blue, the trial print was underexposed and the time of exposure should be increased;

if the trial print has a scorched appearance with indistinct lines against a very dark blue background, the print was overexposed and the time of exposure should be decreased.

16.31

Typed Material. In preparing a typewritten original for printing, a "soft record" ribbon should be used and the sheet should be "backed" with a piece of carbon paper so that the typed characters will be on two sides. Troublesome smearing can be somewhat reduced by spraying the sheet (both sides) with "fixative."

16.32

Blueprint Machine. For a large quantity of prints some form of continuous machine is desirable. On such a machine the paper unwinds continuously from a roll, as the tracings are "fed in" with the inked sides up. The exposure is made as the tracing passes upward in contact with a glass cylinder containing a bank of arc lights. Some machines are equipped with a washer and drier, thus allowing the prints to be made in one operation (Fig. 16.14), but these are so expensive that only firms making a great number of prints daily find them a worthwhile investment. Small firms have found it much cheaper to have their prints made by commercial printing concerns, of which there are at least one or two in every fairly large industrial city.

16.33

Brown Prints. A negative brown print is made on so-called *Van Dyke paper,* a thin printing paper with a yellowish-green sensitized surface that turns to a light bronze on exposure and a deep brown when washed in water. To intensify a developed brown print and make it permanent, it is "fixed" in a solution of sodium hyposulphite and again washed in water. If the print is to be used for making positive prints, it should be made by exposing with the tracing reversed so that the inked lines are toward the Van Dyke paper. A reversed negative will print faster and will produce sharper lines if it has been transparentized by rubbing the back with a specially prepared oil (white Vaseline in benzene).

A positive brown print, having brown lines on white background, is made on brown-print paper by using a brown-print negative as a substitute for the original tracing. Prints showing a change may be made without altering the original tracing if the part to be changed is "blocked out" on the negative print and drawn in on the positive print.

Fig. 16.14 Continuous printing machine with washer and drier. (*Courtesy C. F. Pease Co.*)

16.34

Blue-line Prints. A positive blue-line print, having blue lines on a white background, may be made on regular blueprint paper by exposing with a brown-print negative in the place of the original tracing. Such prints may also be made from an original tracing, if a paper coated with a solution of gum arabic, ammonia citrate of iron, and chloride of iron is used. When the print is developed in solution of ferrocyanide of potassium, the coated surface turns blue where it has been protected from light and remains white where it has been exposed. The print must be washed thoroughly in water when it is taken from the developer.

An advantage offered by blue-line prints is that alterations may be more easily indicated and read on a white background. Furthermore, if the prints are made on blueprint paper using a negative print, the original tracing is protected against wear.

16.35

Ozalid Prints. Ozalid prints, depending on the type of Ozalid paper used, show black, blue, or maroon lines on a white background. The prints are exposed in the same manner as for other contact processes. They are developed in controlled dry ammonia vapor and, since no liquid solution is used with this process, a finished print will be an undistorted exact-scale reproduction of the original tracing.

Ozalid printer–developer units, such as the one shown in Fig. 16.15, produce finished prints in two simple steps without any auxiliary apparatus. No washing, fixing, rinsing, or drying is required.

An Ozalid duplicate tracing, which is known as an *intermediate*, may be made on either Ozalid sepia-line papers (single- and double-coated), Ozalid sepia-line cloth, or ozalid foil. Since each of these types of intermediate materials is translucent, an intermediate on any one type may be used in place of the original tracing to produce drawing room and shop prints. The use of intermediates makes it possible to file away the original drawing where it will not be damaged or smudged and will be safe from fire. The intermediate can be used for all reference work and may be altered when design changes become necessary. Changes may be made and obsolete detail removed with a corrector fluid and the new lines, dimensions, and notes may then be added. A second intermediate may then be made from the first one. The use of changed intermediates makes retracing unnecessary when rather extensive alterations in design become desirable. Also, the valuable original can remain untouched in the file for record purposes. As a matter of interest, an intermediate will often make a much better print than an original tracing.

16.36

Black-and-White (Diazo) Prints (BW). Positive black-line prints may be made direct from an original tracing by using a specially prepared black-print paper. When developed, the sensitized surface turns black where it has been protected and remains white where exposed. Only two steps are required to obtain a fully developed print, exposure and development. This method of printing is similar to that used for making blueprints, except that the print is developed by a solution in a special developing unit on top of the printer (Fig. 16.16). Black-and-white prints are more desirable than blueprints, because of their greater legibility and the fact that they can be made in only two steps, whereas blueprints require five.

Colored-line (red, brown, or blue) prints having a white background may be made

Fig. 16.15 Ozalid printing machine. (*Courtesy General Analine and Film Corp.*)

making an intermediate print on which unwanted lines, figures, and notes can be removed and the new material added in ink or pencil. Prints made directly from the altered intermediate bear all of the added changes without showing the unwanted material. Drafting time has been saved for use on the other tasks, and the original tracing remains untouched for record purposes. This method of alteration may be used to delete portions of an original that might be so confidential that it should not be reproduced.

16.41
Reproduction of Worn or Mutilated Tracings. Worn or mutilated tracings that have reached a stage where it is difficult to make readable prints may be reproduced photographically rather than by hand tracing at a great saving in time and money. If desired, the scale may be changed. By taking advantage of the special types of film now available for this purpose, the intensity of the pencil lines can be increased to give a reproduction with clear and sharp lines. Stains and discolorations may be eliminated by using a proper filter on the camera or by opaquing.

16.42
Xerox Prints. Xerox prints of small documents are black-and-white positive prints that can be made in considerable number, if needed, at very little expense. The process employs a selenium-coated electrostatically charged plate that may be reused. After the image of the original has been projected onto the plate, a negatively charged powder, which will adhere to the positively charged areas (lines) of the image, is applied over the plate. Then, by means of a positive electrical charge, the powder is transferred to the paper and permanently fixed (baked) onto the surface as the last step in the process.

16.43
Long-distance Xerography (LDX). Long-distance xerography networks, utilizing microwave, coaxial cable, or special telephone line, provide a practical means for transmitting and receiving electronically any form of communication, typed or drawn, across a plant area or across the continent. The high-speed, high-capacity facsimile transmission and receiving equipment consists of an LDX scanner and LDX printer. These units are shown in Fig. 16.18. The document scanner on the

sending end converts images to video transmission signals. At the receiving end the printer electronically restores the transmitted images and produces a black-on-white copy of the original document, sketch, or drawing. The system is easy to use. The original is simply inserted into the feed slot of the scanner. It is returned to the operator in seconds. The scanner senses colors without exposure adjustment. Also see Fig. 12.3.

16.44
Microfilming Drawings. The practice of microfilming drawings is increasing rapidly, as many companies are giving more

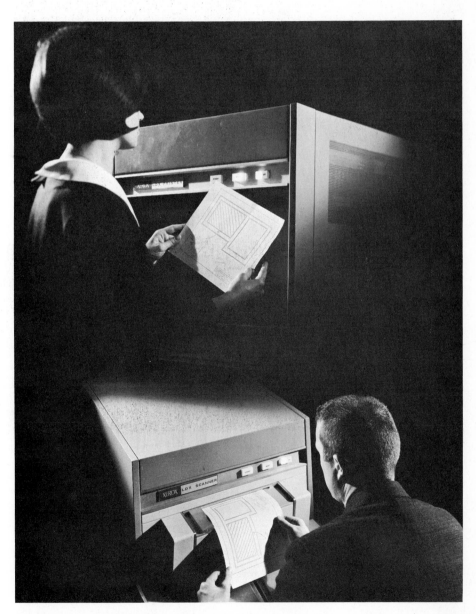

Fig. 16.18 Transmitting drawings and sketches electronically. (*Courtesy Xerox Corporation*)

attention to their reproduction methods and are being forced to save filing space. Some companies have found that a microfilming program will reduce their storage space requirements by as much as 75%.

Reduction in filing space is made possible largely because obsolete drawings may be removed from the files and destroyed after they have been microfilmed.

Of equal importance is the protection offered when microfilm prints can be filed away in fireproof vaults. Since lost original tracings may be quickly replaced by first making enlarged printing intermediates, it can be said that microfilming offers a form of insurance both against careless loss through handling and from fire.

The three sizes of film in general use today are the 16 mm, 35 mm, and 70 mm. Although the most common size is the 35 mm, the 16 mm is often used for small drawings. One large corporation uses 70 mm for tracings as large as 95 cm × 130 cm. Tracings, which may be as large as 140 cm × 600 cm, are microfilmed in sections, and related negatives are filed together so that prints can be

made at a later time should the need arise.

Microfilm negatives may be cut and filed individually in envelopes marked with the same number given the tracing. Negatives are made for subsequent revisions, thus permitting the destruction of unnecessary and outdated tracings.

A film may be mounted in an engineering document card, commonly called an *aperture card.* The card shown in Fig. 16.19 is a form of standard data card that has been keypunched for mechanized equipment capable of sorting, filing, and retrieving cards. If necessary, a keypunched and imaged duplicate aperture card can be easily produced for distribution.

An engineer may obtain any needed information concerning the progressive development of a part or mechanism by viewing the negative on a microfilm reader or by using a portable projector (Fig. 16.19). When a full-size print, a reduced-scale print, or an intermediate is needed, either can be quickly made by enlargement on paper or cloth by using the nega-

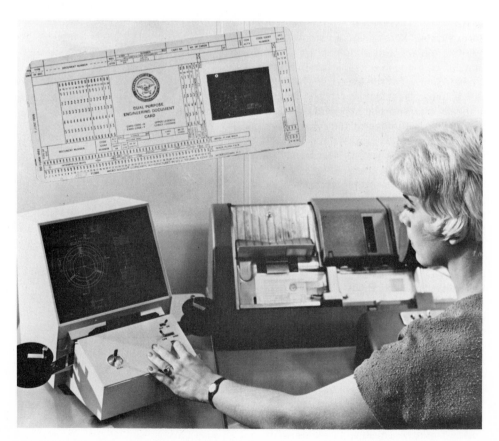

Fig. 16.19 Microfilm reader and aperture card. (*Courtesy Keuffel & Esser Co.*)

tive, mounted on an aperture card, in a microfilm printer designed for printing from microfilm.

16.45

Combining a Line Drawing with a Photograph of a Part or Mechanism. The use of photography in the preparation of assembly drawings of simple unit mechanisms has become increasingly popular because a photograph is readily understood by a relatively inexperienced person. By referring back and forth between the line drawing showing the assembly and the picture, it is possible for one to gain a full understanding of the mechanism in much less time than would be possible without the photograph.

The first step in the preparation of a photographic assembly drawing is the making of a halftone enlargement of the piece of equipment using an orthotype film. Then a positive intermediate is made from the original line drawing, and a window is cut into it at an open spot to accept the halftone positive. From this layout, consisting of a line drawing, a bill of material, a title block, and a halftone, a positive film intermediate is made that can be used to make needed prints on a direct-process machine.

At times, it has been found desirable to present a photographic representation on a line drawing showing a single part.

16.46

Use of Glass Cloth. Glass cloth, which has been a rather recent development, is a translucent drawing material that is dimensionally stable. It is similar to vellum and can be rolled for ease in handling, shipping, or storing. It has an excellent drawing surface for either pencil or ink.

Because drawings made on glass cloth will remain unvaryingly accurate, its use has opened the door to shortcut methods, which save both layout time and floor space. In the aircraft industry, it has made possible a more accurate production of templates, dies, and form blocks at a distinct saving in time and money, because its use eliminates shop layouts by hand, camera photography, and other expensive and complicated reproduction methods that have been used in the past. The use of glass cloth for the reproduction of layouts eliminates the necessity for a detailed checking that a hand-scribed layout requires.

In practice, the original full-scale drawing on glass cloth and the photosensitized template or tool stock are exposed together in a large contact printer. Development is simple, in that only a moist sponging is required with a developer. All work may be done under daylight, but not in sunlit areas.

The only drawing that need be prepared by the drafting room for use through final production is the original drawing on glass cloth.

16.47

Reproductions from Layouts on Metal Sheets. A reproduction may be made of a layout prepared on metal plate or other opaque material by the reflex contact method. The original, if prepared on metal, is drawn directly in pencil or ink on a working surface that has been prepared by spraying it with a white-matte, loft board surfacer. The special vinyl base film that is used for the intermediate is unique, in that the light that passes through it from the back during exposure strikes the prepared surface of the metal plate and is then reflected to the emulsion. For this reason one must be certain that the emulsion side of the film has been placed in contact with the front side of the plate when preparing for the exposure. After exposure, the film is removed and immersed in a developer for a few minutes and then fixed in hypo and washed.

The resulting negative may be used for making positive prints on cloth or paper or for making positive prints on metal, as desired. Prints on metal may be needed as copies for subcontractors or for use as templates if the original layouts have been prepared full-size.

16.48

Industrial Practice in Automobile-body Design. In automobile-body design, full-sized drawings of each year's models are made on the painted surface of a large sheet of aluminum. The drawings are carefully laid out with instruments directly on the painted surface; then the lines are scratched in. A detail drawing, such as one showing the details of a doorpost on a particular model, is made by tracing directly from the aluminum sheet. Aluminum is used because a body-design drawing for a model must be prepared 2 or 3 years in advance of the production date and the medium on which it is prepared must be strong enough to withstand constant wear and rough treatment. Furthermore, a drawing on an aluminum sheet is not affected by weather changes.

Problems

The four general types of problems presented in this chapter have been designed to furnish practice in the preparation of working drawings or sketches. The first type is composed of dimensioned pictorial drawings of individual pieces taken from a variety of mechanisms. The student should prepare complete working detail drawings of these pieces as they may be assigned by the instructor. It should be recognized that dimensions are not necessarily placed the same on orthographic views as they are on pictorial drawings. To make it possible for the student to apply the principles presented in Chapter 13, no special effort has been made to place dimensions in accordance with the rules of good practice.

The second type of problem is that which shows in pictorial all the parts of a unit mechanism. This gives the student an opportunity to prepare a complete set of working drawings of a simple unit. It is suggested that the detail drawings be prepared before the assembly is drawn.

The third and fourth types provide practice in both reading and preparing drawings, the third requiring the preparation of detail drawings from given assembly drawings, the fourth requiring the making of assembly drawings from the details.

Note: The inch values given on the drawings of problems may be converted to millimeters by using Appendix Tables 3 and 4. It is common practice under the metric system to give a dimension that is not critical or a size dimension for a hole or a shaft in full millimeters. For drill sizes see Tables 23 and 49. Metric threads may be found in Tables 10 and 11. See Table 6 for metric tolerance equivalents.

1–13. (Figs. 16.20–16.32). Make a detail drawing of an assigned machine part. Draw all necessary views. Give a detail title with suitable notes concerning material, number required, etc.

Fig. 16.20   Guide bracket.

Fig. 16.21   Link.

Fig. 16.22   Handle block.

Fig. 16.23   Angle bracket.

FILLETS & ROUNDS R3

Fig. 16.24   Caster frame.

Fig. 16.25   Slide bracket.

Fig. 16.26   Stabilizer bracket.

Fig. 16.27   Gear-shifter link.

Fig. 16.28   Gear shifting fork.

Fig. 16.29 Shifter guide.

Fig. 16.30 Slide block.

Fig. 16.31 Shifter arm. (*Dimension values
shown in [ ] are in millimeters.*)

Fig. 16.32 Shaft hanger.

SECTION NUMBER	A	B	C	D	E
1	1.00	2.00	3.00	1.88	1.62
2	.94	1.88	2.82	1.75	1.38
3	.62	1.82	2.44	1.62	1.12
4	0.00	1.88	1.88	1.50	.88
5	.68	2.12	1.44	1.38	.62
6	1.00	—	—	1.25	—

14. (Fig. 16.33). Make a two-view freehand detail sketch of the fan spindle. Determine dimensions by transferring them from the drawing to one of the given scales, by means of the dividers. The material is SAE 1045, CRS. The limits for the spindle may be found in Table 22, in the Appendix. The last figure in the bearing number is the bearing bore number given in the tables. Use a free running fit between spindle and felt retainer, spindle and felt retaining washer, and spindle and cone-clamp washer.

15. (Fig. 16.33). Make a complete two-view detail drawing of the fan pulley. It is suggested that a half-circular view and a full-sectional view be shown. Determine the dimensions as suggested in Problem 14. Housing limits for the given bearings will be found in Table 22 in the Appendix.

16. (Fig. 16.34). Make a detail drawing of an assigned part of the pipe stand.

17. (Fig. 16.35). Make a fully dimensioned drawing of an assigned part of the tool holder.

18. (Fig. 16.36). Make a detail working drawing of an assigned part of the flexible joint. Compose a suitable title, giving the name of the part, the material, etc.

19. (Fig. 16.37). Make a detail working drawing of an assigned part of the stud bearing. Compose a suitable title, giving the name of the part, the material, etc.

20. (Fig. 16.38). Make a detail drawing of an assigned part of the conveyor take-up unit.

21. (Fig. 16.39). Make a detail drawing of an assigned part of the Simplex ball-bearing screw jack.

22. (Fig. 16.40). Make a detail drawing of an assigned part of the bench arbor press.

23. (Fig. 16.41). Make a detail drawing of an assigned part of the gear pump.

24. (Fig. 16.42). Make a detail drawing of an assigned part of the air cylinder.

25. (Fig. 16.43). Make a detail drawing of an assigned part of the bench grinder.

26. (Fig. 16.44). Make detail drawings of the carrier and the holder. Make an assembly drawing showing all of the parts in their relative positions.

27. (Fig. 16.45). Make a complete set of working drawings of the adjustable attachment. The complete set should consist of detail drawings of the individual parts and an assembly drawing complete with a bill of material.

28. (Fig. 16.46). Make a two-view assembly drawing of the cup center, using the given details. Use the schematic symbol for screw threads. Study the pictorial drawing carefully before starting the views.

29. (Fig. 16.47). Make an assembly drawing of the radial engine unit, using the given details. It is suggested that one piston be shown in full section so that the relative positions of the parts will be revealed.

30. (Figs. 16.48–16.50). Make an assembly drawing of the hand-clamp vise, using the given details. Use the schematic symbol for screw threads.

31. (Figs. 16.51–16.53). Make an assembly drawing of the blowgun.

32. (Figs. 16.54–16.56). Make an assembly drawing of the hand grinder.

33. (Figs. 16.57–16.59). Make an assembly drawing of the right-angle head, using the given details.

Fig. 16.33 Fan assembly.

Fig. 16.34 Adjustable pipe stand.

Fig. 16.35 Tool holder.

Fig. 16.36 Flexible joint.

Fig. 16.37 **Stud bearing.** (*Courtesy Ross Gear Division of TRW, Inc.*)

Fig. 16.38 **Conveyor take-up unit.**

PC.NO.	NAME	QUAN.	MATERIAL	PC.NO.	NAME	QUAN.	MATERIAL
1	STANDARD	1	MALL. IRON	5	GROOVE PIN	3	$\frac{7}{32}$ D.$\times\frac{5}{8}$ STEEL ROD
2	SCREW	1	S.A.E. 1120 FORGING	6	LEVER BAR	1	REROLLED RAIL STK.
3	CAP	1	S.A.E. 1045 FORGING	7	$\frac{7}{8}$ DIA. BALL BEARING	1	STD.
4	THRUST WASHER	1	S.A.E. 2315				

Fig. 16.39 Simplex ball-bearing screw jack. (*Courtesy Templeton, Kenly & Co.*)

Fig. 16.40 Bench arbor press.

PART NO.	NAME	MATERIAL	NO. REQ.
1	BASE	C.I.	1
2	TABLE	C.I.	1
3	TURNTABLE	C.I.	1
4	RAM	S.A.E. 1045	1
5	RAM HEAD	S.A.E. 1040	1
6	SPINDLE	S.A.E. 1045	1
7	HANDLE	C.R.S.	1

Fig. 16.41 Gear pump.

GEAR DATA

NUMBER OF TEETH (N)——13
DIAMETRAL PITCH (D.P.)——10
OUTSIDE DIAMETER (OD)——38
PRESSURE ANGLE——14½°

M5×0.8 ×10 BRASS
MACHINE SCREWS
6 REQD

R30
R24
33
23
HEX 22
ACROSS FLATS
⌀ 38.3–38.4
2 PLACES
R5
17
28
33
④
HALF OF COVER REMOVED IN THIS VIEW

BRASS PIN ⌀2.5 × 25
STAKE ON TWO SIDES
ONE GEAR ONLY

¼ PIPE THD

PACKING
M20×1.5 THD
9 DEEP
30°
⌀13
3
14
6
25
⑧
⑦

GASKET
THICK

RUNNING
FIT
②
⌀13
R8
18
30 AS CAST
5
0.8
2
18
R3
8
R10
27
37
⑤
①
④
19
18
9
R13
2
11
11
10
22
2
LIGHT FORCE
FIT
⌀12
⌀20
#TAPER PIN
19
6
65
⑥
③
26
30°
10
R0.8
MAX.
⌀22
13
⌀32
⌀64
6 RIBS EQUALLY SPACED
THICK

PC	NAME	NO.	MAT'L
1	HOUSING	1	C.BRO.
2	HOUSING COVER	1	C.BRO.
3	PULLEY	1	ALCOA 195-T6
4	GEAR	2	BRO
5	GEAR STUD	1	BRASS
6	SHAFT	1	BRASS
7	NUT	2	BRO
8	PACKING NUT	2	BRO

Fig. 16.42 Air cylinder.

Fig. 16.43 Bench grinder.

Fig. 16.44 Tool holder.

Fig. 16.45 Adjustable attachment.

Fig. 16.46 Cup center details.

Fig. 16.47 Radial engine details.

11	NO. 6 TAPER PIN		1
10	SPECIAL MACHINE SC.	M.S.	4
9	SPECIAL FLAT HD. SC.	M.S.	2
8	LINK	C.R.S.	1
7	COUPLING	C.R.S.	1
6	WEDGE	C.R.S.	1
5	JAW PLATE	C.R.S.	2
4	SHOE	C.R.S.	1
3	HAND SCREW	C.R.S.	1
2	JAW	C.I.	1
1	BASE	C.I.	1
PART NO.	NAME OF PART	MATERIAL	NO. REQ.

Fig. 16.48 Hand clamp vise.

FILLETS & ROUNDS R3

PC #1 BASE
C.I. 1 REQ'D.

Fig. 16.49 Hand-clamp-vise details.

Fig. 16.50 Hand-clamp-vise details.

Fig. 16.51 Blowgun. (*Courtesy A. Schrader's Son Mfg. Co.*)

Fig. 16.52 Blowgun details.

Fig. 16.53 Blowgun.

Fig. 16.54 Hand grinder details. (See Fig. 16.56.)

KEYWAY $\frac{1}{8}$ WIDE × $\frac{1}{16}$ DEEP

.6245
.6255 REAM

GEAR
C.I. I REQ'D.

48 TEETH
12 DIA. PITCH
14$\frac{1}{2}$ PRES. ANGLE

#405 WOODRUFF KEYSEAT
$\frac{1}{2}$-13 UNC-2A (FOR HEX. JAM NUT)

.499
.498

$\frac{1}{16}$ × 45° CHAMFER

2.320
2.325

.6240
.6235

GEAR SHAFT
C.R.S. I REQ'D.

$\frac{1}{32}$ CHAMFER

#405 WOODRUFF KEYSEAT

12 TEETH
12 DIA. PITCH
I PITCH DIA.
14$\frac{1}{2}$ PRES. ANGLE

GRIND

$\frac{1}{16}$ × 45° CHAMFER
$\frac{1}{2}$-13 UNC-2A (FOR HEX. NUT)
BREAK ALL CORNERS

.6240
.6235

.738
.742

.780
.784

PINION SHAFT
C.R.S. I REQ'D.

LINE REAM IN ASSEMBLY

.6245
.6255

.8770
.8760

SHAFT BUSHING
BRO. 4 REQ'D.

KEYWAY $\frac{1}{8}$ WIDE × $\frac{1}{16}$ DEEP
ONE FLANGE ONLY

.4995
.5005 REAM

WHEEL FLANGE
C.R.S. 2 REQ'D.

KEYWAY $\frac{1}{8}$ WIDE × $\frac{1}{16}$ DEEP

.4995
.5005 REAM

SECTION C-C

4 SLOTS $\frac{1}{16}$ WIDE

CRANK
C.I. I REQ'D.

$\frac{3}{16}$ DRILL-90° C'SK TO $\frac{11}{32}$ D
FILLETS & ROUNDS $\frac{1}{8}$ R

BREAK ALL CORNERS

CLAMP CAP
C.R.S. I REQ'D.

#24 DRILL & REAM FOR #1×1
TAPER PIN WITH CLAMP
SCREW IN POSITION

KNURL

.3745
.3755 REAM

BREAK ALL CORNERS

CLAMP KNOB
C.R.S. I REQ'D.

#24 DRILL & REAM FOR #1×1 TAPER PIN
WITH CLAMP KNOB IN POSITION

.3735
.3725

$\frac{1}{2}$-13 UNC-2A

CLAMP SCREW
C.R.S I REQ'D.

FILLETS & ROUNDS $\frac{1}{8}$ R

TOOL REST
C.I. I REQ'D.

$\frac{17}{64}$ DRILL-2 HOLES

$\frac{1}{4}$ × $\frac{1}{4}$ CARRIAGE BOLT

$\frac{1}{4}$ WING NUT

TOOL REST ASSEMBLY

FILLETS & ROUNDS $\frac{1}{8}$ R

METAL BAND

RIVET IN ASSEMBLY

HANDLE PIN
C.R.S. I REQ'D.

$\frac{7}{32}$ DRILL-82° C'SK TO $\frac{13}{32}$ D
2 HOLES (FOR #10 × $\frac{3}{8}$ FT. HD. MACH. SC.)

HANDLE
WOOD I REQ'D.

TOOL REST ARM
C.I. I REQ'D.

Fig. 16.55   Hand grinder details.

Fig. 16.56   Hand grinder.

Fig. 16.57 Right-angle-head details. (*Courtesy R. C. Haskins Co.*)

Fig. 16.58 **Right-angle-head details.**

Fig. 16.59 Right-angle-head details.

PART
5
THE COMPUTER

FRONT SUSPENSION-LOWER ARM

PHOTO OF ACTUAL PARTS

**TOP VIEW
DEFORMED AND UNDEFORMED**

**PLOTTED
PERSPECTIVE VIEW**

**SIDE VIEW
DEFORMED AND UNDEFORMED**

NASA computer program aids American designers "This photograph of the lower arm in the front suspension of a Ford Motor Company vehicle shows the actual part (upper left) and a grid system in various views of the same part as it appears on a computer plot produced by the NASA-developed NASTRAN program. The picture shows plots of the deformed (stresses) and undeformed (not stressed) equipment. The NASTRAN program has given Ford engineers capability to analyze structures, particularly under dynamic loading, which has not been previously available to them. More than 70 industrial firms, universities, laboratories and government agencies are now using NASTRAN, a NASA-developed computer program, to solve their structural engineering problems. The program is presently being used in more than 185 different applications, ranging from suspension units and steering linkages on automobiles to the design of power plants and skyscrapers. At least 55 more uses of NASTRAN are currently planned." (*Photo and quoted statement courtesy National Aeronautics and Space Administration*)

Language of the Computer (Input). Punched cards, punched tapes, and magnetic tapes (Fig. 17.4) with information recorded as magnetized spots (called *bits*) are the conventional means for making contact with an electronic computer in the only language that the computer recognizes, the language of electrical impulses. The information supplied by these patterns of holes or magnetic spots are translated by a *sensing element* or *feeler* into the on–off binary language of the computer. Up-to-date punched-tape and punched-card sensing elements use photoelectric cells that transmit current when light reaches them. Paper tapes are punched by an input tape-punching device.

On magnetic tape, data are recorded in continuous parallel tracks. Usually, there are either six or eight data tracks and one checking track, depending on the format. Clocking mechanisms and changes in magnetic polarity indicate separate bits in the tracks. (A bit is a binary digit, either 0 or 1.) In the tape drive, a magnetic read–write head simultaneously reads a bit at a time from each of the tracks. The six or eight bits read represent a complete character. (Eight bits, called a *byte*, can represent two decimal digits or one character.) Using binary-coded decimal coding, the letters of the alphabet as well as decimal numbers may be recorded on magnetic tape. The BCD format also provides for punctuation marks and permits the use of special characters.

Punched cards, such as the one shown in Fig. 17.4, are stacked into a *card reader*, which senses the cards and translates the information furnished by the perforations into the pulse language of the computer.

To permit more efficient use of a computer, the pulses punched into the paper may be transferred first to a magnetic tape and then from it to the computer. Magnetic tapes can deliver magnetized pulses at the rate of 180,000 characters (or 360,000 decimal digits) per second, a stretch of time in which a human can say only "one and two" at the usual word rate of a speaking voice.

17.4

Computer Storages. Every computer must have a memory system; that is, it must have repositories for the storage of information. Input and output storages receive data needed for calculation and

(a) Punched Card

(b) Paper Tape—eight channel code

(c) Magnetic Tape—seven-bit alphanumeric code

Fig. 17.4 Data recording media—punched card, paper tape, and seven-track magnetic tape. (*Courtesy International Business Machines Corporation*)

note the results before these results are typed out by the high-speed printer. There are also intermediate (short) storages that receive results that the computer will need for continued calculation. A special component for intermediate storage, the accumulator, receives totals to be returned immediately to the matrix.

The three categories of computer storage are (1) the main memory; (2) auxiliary storage, which includes direct-access devices such as magnetic disc units and magnetic drums; and (3) bulk storage, involving the use of punched cards and magnetic tape. (See Fig. 17.1)

17.5

Computer Output. Up to this point, we have focused our attention on how information is fed into electronic computers by means of an input unit and to some extent on how the computer digests data and makes the necessary calculations. All of this action would quite naturally be useless, however, if the computer could not express the results of the calculations (Fig. 17.3). The apparatus needed for this purpose is called an *output unit*. Several types of units are available to satisfy different requirements.

One such unit is the tape punch having perforating pins that are actuated by the electrical impulses of the computer. A tape-punch unit is capable of punching 50 or more characters per second in the paper tape. Punched tapes are deciphered by a printer that converts every character into either a letter or a number. Output card punches are also available. Punched-card information may be converted to printed copy through the computer or by using special auxiliary equipment.

When the results obtained represent an intermediate step and there is no need to read the results at the time, the pulses issued by the computer are picked up on magnetic tape to be held in reserve, say, for use a month later. When it becomes necessary to learn what is on a magnetic tape, it may be run through a complicated electronic device that sends the magnetic impulses to a high-speed output printer. This high-speed printing device is almost a miracle in itself because it can print a full line at a time (up to 132 characters) at a rate of 1100 lines per minute.

Information may be transferred, without changing the code, from magnetic tapes to punched tapes or cards and, as also might be expected, from punched cards to magnetic tapes. This is accomplished by devices known as *converters*.

17.6

Communicating with a Computer. Improvement in programming is gradually eliminating the formidable barrier that has existed between the computer and its prospective users. In the years ahead computers will become commonplace and they will have capabilities that will make it possible for them to be used by almost anyone. The day when each of us can consult with his own computer lies just ahead. With this thought in mind, it will be shown how man and computer can work together as partners, with man in control of the creative aspect of the design process. To instruct the computer, the user needs little more than a working knowledge of basic graphics, since the programming barrier is fading away to a great extent in this area.

17.7

Computer Graphics. In addition to making extensive use of conventional computer techniques for engineering and scientific analysis, the engineering team can, with the aid of one of several graphic data processing systems, bring a large-scale digital computer to bear on those problems that can best be analyzed and finally formalized by using and modifying graphical representation. This great step forward in automated graphics makes it possible for an engineer or designer to do at least a part of his creative work on a graphic input–output cathode-ray console that is connected directly to the computer (see Fig. 17.5). The means for man-machine communication is not in the future; it is available now and several practical systems are in daily use for design and production work in the aerospace and automotive fields.

The direct man–computer interaction that this system affords leads to quick answers for even small problems and permits concepts to be evaluated and tested and then accepted or rejected. This relationship in which the computer has become man's immediate partner in creative design became possible only when a way was found to make the computer correctly interpret drawings and, if desired, make a drawing after the intentions of its human partner had been made known. Properly used, these systems can relieve both the engineer and the man at the drawing board of endless calculations and of much of the tedious work connected with making layout and detail drawings. Such systems bring together the talents and creativity of the user and the power and speed of the digital computer in conjunction with digitized plotters or cathode-ray-tube consoles with appropriate supporting equipment.

In general, the term computer-aided design implies that the computer assists the designer in analyzing and modifying a previously created design within new design parameters that he, the designer, has established. Almost any designer or draftsman can use graphic data processing effectively and can analyze, modify, and distribute design information even though he may have only a limited knowledge of computer programming. This is true because the designer at the console and the computer are in two-way communication while using the designer's own graphic language of lines, symbols, pictures, and words.

The computer system described thus does not replace the man, nor does it eliminate the requirement that a designer have a working knowledge of engineering graphics. The graphic language, so fundamental to the design process, has now become a computer language. As such, it has proved to be a reliable means for a conversational type of man–machine communication.

The GM DAC–1 (Fig. 17.5) system that

will be discussed in Secs. 17.8–17.10 is no longer in general use. However, full descriptions of its operations and capabilities have been given here in this text in order to point up the potential of a complete system of this type. Ordinarily, it is much more practical for an industrial organization to assemble a needed graphic system using units now available on the market than to develop a special "one-of-kind" system that may not justify the cost and may soon be out-dated. In a sense, it can be said that DAC–1 was developed as somewhat of a laboratory experiment to determine the potential of a total system of computer graphics in the area of design. As such, it represented a great leap forward at that time.

17.8
Main Elements of the DAC–1 Digital Graphic Input–Output System. The main elements of the GM graphic data processing system shown in Fig. 17.6 are as follows:

1. A large-scale digital computer system (with supporting software) that is capable of storing vast amounts of information and retrieving specifically desired information rapidly.

2. One or more graphic consoles with viewing screens.

3. A film recorder (35 mm) for "hard copy" of graphic information that has been computer generated.

4. A 35 mm film scanner for the input of data in graphical form. This unit of the system converts images recorded on

Fig. 17.5 Man–machine communication.
"At the General Motors Research Laboratories, using the new GM DAC-I system (Design Augmented by Computers), a research engineer checks out a computer program that allows him to modify a design 'drawing.' A touch of the electric 'pencil' to the tube face signals the computer to begin an assigned task, in this case, 'Line Deletion,' where indicated. The man may also instruct the computer using the keyboard at right, the card reader below the keyboard, or the program control buttons below the screen. Hundreds of special computer programs, written by GM computer research programmers, are needed to carry out these studies in man–machine communications."

microfilm to digital form for further processing.

17.9
DAC–1 Display Console. It is at the display console that the man–machine conversation takes place. The principal feature of the console is the cathode-ray-tube

Fig. 17.6 GM DAC–1 graphic processing components. (*Courtesy General Motors Corporation*)

(CRT) screen, on which the computer-programmed graphic and alphanumeric information is displayed (Fig. 17.5). Close at hand on the console are the keyboards and the light pencil, which provide convenient means of entering and modifying rapidly displayed computer-programmed graphic and alphanumeric information. At the console unit the users have direct access to data (previously stored in the system's main or auxiliary storage units in digital form) that they may study and then modify and redisplay.

Fig. 17.7 DAC–1 image processing unit. (*Courtesy General Motors Corporation*)

17.10

GM DAC–1 Image Processing Unit. The DAC–1 computerized design and drafting system provides for both the input and output of data in a graphical form. A direct output for engineering and manufacturing use can be in the form of hard copy, a drawing prepared by an on-line plotter, or a programmed tape that can be used to actuate an automated production machine. At this point in our discussion our attention will be directed to the special image processor that enables the computer to read and generate drawings (Fig. 17.7). The use of tapes with plotters adaptable for off-line operation will be discussed later in this chapter.

The image processing unit provides means for the input and output of data in graphical form. Primarily, this component contains a CRT photo-recorder, projectors, a CRT photo-scanner, an input camera, and the rapid film processing equipment needed to develop 35mm film. When the film is developed within the recorder, the image can be viewed on a rear projection screen that is located for easy viewing by the user (Fig. 17.6).

After an input copy has been inserted into the image processor on a tray and then photographed and developed, the 35mm negative is positioned so that the CRT photo-scanner can "read it." The film scanner converts the image furnished by the 35mm negative (light lines on dark background) directly to digital data that is sent either to the computer or to memory storage. Basically, the near miracle is accomplished via an electronic beam that, when it determines whether each of many addressable points is above or below a set level of light intensity, notes the result in computer (digital) language. The digital information thus developed may now be either modified or stored permanently.

From the high-speed recorder of the image processor, on 35mm film, come the permanent records of drawings, charts, and parts lists that are needed by designers and draftsmen. The film, exposed to a high-resolution CRT and developed within the recorder, can be viewed at development on a rear projection screen, as stated previously. Finally, conventional working copies (on paper) may be readily obtained using modern viewer–copier equipment and recorder-produced exposures mounted in aperture cards. It was the development of image processing units, such as the DAC–1 described, that extended the capability of computers to the point where they can accept, interpret, analyze, relate, and finally produce needed output in graphic form.

17.11

IBM 2250 Display Unit (Fig. 17.8). Images are generated on the CRT at the console by computer-programmed positioning and deflection of the CRT's electronic beam to the more than one million possible points (grid format of 1024 by 1024 addressable points) of the 300 mm × 300 mm effective display area. The beam is deflected to each point on the screen that is addressed by the program, either intensifying it or not intensifying it, as directed. The image as we see it on the face of the CRT (screen) is produced by the beam hitting a phosphor coating, causing the coating to glow for just an instant. Operating under this condition, the information normally displayed fades rapidly (within a fraction of a second); therefore, the display must be regenerated approximately 30–40 times

every second for a flicker-free representation of high resolution. A buffer is used to regenerate the display independently of the computer.

Alphanumeric symbols may be formed by synthesizing the characters from a series of individually programmed dots or line segments, either horizontal or vertical. To improve computer efficiency, an optional character generator can be used for computer-independent formation of all alphanumeric symbols.

The position-indicating light pencil, when used in conjunction with the keyboards and programs in storage, provides the means whereby the image displayed can be altered (Fig. 17.5). It may be used at will to create, remove (erase), enlarge, or rearrange any part of a displayed image that, as a whole, may consist of points, line segments of any length and orientation, simple and complex geometric shapes, and appropriate graphical symbols and statements. The light pencil (receiving device), having a photocell ele-

Fig. 17.8 Example of a CRT display.
The IBM 2250, an advanced graphics display unit, permits users to exchange visual information with IBM's lowest-cost computing system, the desk-sized IBM 1130 (right). This particular unit was developed especially for engineers and designers who need fingertip access to their computers. The display enables them to work with charts, diagrams, drawings, or printed letters and numbers directly on the face of the television-like screen. With the electronic "light pen" the designer can revise images, add or delete lines, or change dimensions. He also can change information and images on the screen through the 1130 and 2250 keyboards (note programmed function and alphanumeric keyboards). A civil engineer, for example, could use the system to help design a building or a bridge, or an electrical engineer could use it to diagram new circuits. As the designer works he can call on the 1130 to perform calculations and update information previously stored in the system. If the user wants a more powerful computer, he can link his combined 1130/2250 system to a larger IBM System/360 through standard communication lines. (*Courtesy International Business Machines Corporation*)

ment that senses a spot of light directly under its point, signals the system. In turn, based on where the cathode-ray beam was scanning at that instant, the system sets a register, which indicates the position of the light pencil on the face of the display. When the control system feeds this positional information to the computer, the computer in turn relates the identified portion of the image to the equivalent digital representation stored in memory.

After a picture has once been generated, either the man with his light pencil or the computer may direct the other to reconsider a portion of the image. The computer may react, as programmed, to any portion of the display by placing an X at the point in the display that the man should consider. Thus, the machine suggests while the man directs. Each reacts to the other.

The programmed function keyboard (Fig. 17.8, bottom left) consists of program control keys, status indicator lights, and sensing switches for use with interchangeable (plastic) descriptive overlays. An overlay identifies a series of program functions contained in the computer memory system. When a key is depressed, the computer receives a signal to act on the image as directed by the program subroutine associated with the key selected. For example, the signal given might direct the computer to either enlarge, reduce, or delete a portion of the programmed image displayed. Any of these actions can be performed by the computer in a fraction of a second. To complete the same task manually might take minutes or even hours.

With the alphanumeric typewriter-type keyboard shown in Fig. 17.8, the user can compose statements consisting of letters, numbers, or symbols and can perform editing functions. The message, as composed, is displayed on the screen for verification. It may then be transferred on demand to the computer's main memory. The statement or message composed could be a parts list, a dimension value, an electrical value for a circuit component, a descriptive note, or a drawing title.

Through the use of graphic data processing as described, the designer can create a complete digital description of a design within the system.

17.12

Design Programs. It has been stated in Section 17.7 that the designer can use a computer-aided design system even though he may know little about computer programming. This statement can be true only if it is possible for him to select from a library of previously created design programs those programs suitable for the design functions required for the specific job that faces him. Since it is not to be expected that he will be able to do this on all occasions, the designer or draftsman is fortunate if he has at least a basic understanding of how programs are prepared.

Programs and related design data are stored on tapes or dics files as digital descriptions and operational routines. Sets of words and numbers may represent standard graphic items, such as dimensions, part descriptions, and even formulas. In time, when graphic data processing systems have wider use and have been improved, more standard computer programs will become available and the tedious task of writing and developing programs may, to a great extent, be relegated to the past.

17.13

How a Graphic Data Processing System Is Used. At present, how do we use an existing graphic data processing system?

First, the designer writes out statements that give a short description of his problem. In the case of auto-body design, the problem is very often a descriptive geometry problem requiring drawings and sketches as guides. These statements are then key-punched on cards and entered into the computer memory. With this much accomplished, the designer, at the display console, is ready to command the computer to perform the functions requested. In executing the program, the computer checks each statement and displays an error comment if any error is detected. When and if this happens, the designer inspects the statement, makes the needed corrections, and instructs the computer to proceed.

Once the program has been executed, the computer should respond by displaying a simple line drawing, as shown in Fig. 17.5. In inspecting the solution to determine whether or not it is acceptable, the engineer using DAC–1 may enlarge any portion and view it from any angle. If not satisfied, he may then modify the design by entering other statements, by changing statement parameters, or by simply adding or deleting lines. When he is satisfied, he can, as has been explained in Sec. 17.10, direct the computer to produce either permanent copy or control tapes for automatic drafting machines or machine tools. Since the program is general, it may now be placed in the library of design programs for use at a later time for a specific job.

17.14
Design of Electronic Circuits. Using a graphic data processing system, a designer has almost unlimited flexibility in designing an electronic circuit because so little time is needed to rearrange and analyze components. For example, if a designer should wish to see what would happen if a component were changed (or eliminated) in an electronic circuit that has been displayed on the screen along with the waveform that circuit would produce, he can get an immediate answer. By entering the new value and working with the light pencil and the program in storage, he can see on the screen almost immediately the new waveform that would result from the change. At the same time, the old circuit has been stored in the computer's memory and may be recalled if needed.

17.15
Achievements of Graphic Input–Output Systems. Computers with graphic input–output systems are capable of performing marvelous feats. For example, either a single feature or a related group of features in a design display may be rotated, repositioned, enlarged, or reduced in size at will. When special attention must be given to an individual feature requiring development, that feature may be shown enlarged. This may be done with the rest of the graphic image out of view, that is, not showing on the screen. If need be, a standard element, such as a rivet or a cap screw that has been drawn and stored away in computer memory, may be added by pushing an appropriate button. Once the task of adding necessary details has been completed, the feature can again be brought back to scale and the complete image shown as before.

The first program to provide a graphical link with the computer was developed at Massachusetts Institute of Technology's Lincoln Laboratory some years ago. The first program, known as SKETCHPAD, proved beyond any doubt that a computer system could be developed that would interact with a creative designer.

It was found that a program could be written for an animated movie showing a small automobile moving along a winding road. Other programs were developed to show parts in motion, which, for example, made it possible to show linkages in action. SKETCHPAD proved to be an excellent draftsman, since, as straight lines, circles, and curves were all drawn freehand, irregularities could be removed at a push of a button.

Programs exist for the creation of exotic shapes (airfoils, hydrofoils, etc.). A program, prepared at MIT under the supervision of Professor Coons, shows that by varying chords a shape can be changed accordingly and then reevaluated. In a motion picture showing actual use of this program, the shape displayed was rotated in alternate directions and tilted at different angles to permit the operator to view it from several directions, even from the rear.

In computer-aided design, new developments occur with increasing frequency and the state-of-the-art advances steadily despite the fact that new achievements come only at great expense. As of today, we are moving away to some extent from gibberish geometric languages and English-like statements in using sketches to talk to the computer. In the near future, voice control may be added to the SKETCHPAD concept, and drawings and rough sketches coupled with the voice will be accepted methods for input. When this point has been reached, automated drafting and design will be an actuality and a computer graphic system will not be as inconvenient to use as it is now.

B □ Automated drafting

17.16
Digital Drafting Systems. Present models of automated drafting machines now on the market are digitally controlled. That is, the input (instructions) to the control system of the drafting machine is entered in digital form on punch cards, punch tape, or magnetic tape. The information supplied may be in either straight binary or binary-coded decimal form, the latter format being the most widely accepted.

A few of the many fine automated drafting machines in use today are the Gerber Plotting System (Fig. 17.22), the Orthomat (Universal Drafting Machine Corp.), the Coradomat, the CalComp System (California Computer Products, Fig. 17.9), and the IBM with the 1627 Plotter (Figs. 17.14–17.16).

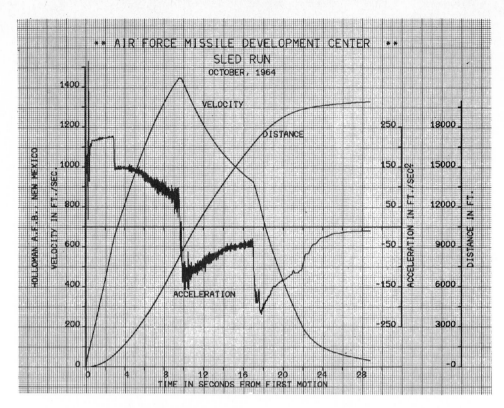

Fig. 17.9 Example of computer–plotter output.
Plotters make rough sledding easier at *Holloman*. Shown above is the plot of dynamic time history for a rocket sled run at the Holloman Air Force Base high-speed track. To cope with the growing volume of computed data, the Air Force Missile Development Center at Holloman turns out a large percentage of its scientific and test data in this form. A general plot program and two CalComp Model 565 plotters are used to produce fully annotated linear or logarithmic charts, ready for direct reproduction. The pair of 565's is operated on-line with a CDC 3600 160A computer system. A time-share program is used for simultaneous full-speed operation of the plotters, three line printers, a card reader, a card punch, and a data link. (*Courtesy California Computer Products, Inc.*)

The all-digital design approach of high-speed XY plotting and drafting machines provides for flexible internal programming so that the plotter will execute on command a circle, a dashed line, a center line, and so forth. The Gerber Plotting System produces high-quality displays of computer-generated digital information. The Orthomat can automatically translate complex formulas into reliable line drawings.

When one is thinking of drafting machines, automated drafting may be defined as being the drawing of lines (ink or pencil) or the formation of scribe (groove) marks on a working surface with little or no human intervention (see Fig. 17.10). The medium used for the drawing may be paper, plastic, film, metal, or any material with a scribe coat.

Of most interest to the user of an all-digital electronic plotter are the control system and the drafting equipment itself.

Basically, the control system consists of a punch card or tape reader (punch or magnetic) and a translator. The translator converts input information received from the reader into a series of command signals to the servomotors of the drawing head. The drawing head responds as directed, checks its position, and then indicates that the command, as given, has been performed. Informing the control system when a command has been accomplished is made possible by a *feedback device*, a mechanical–electrical system designed to provide mechanical error correction.

Depending on the model, a drafting machine can have either a horizontal, vertical, or tilting table. Although the horizontal table requires more floor space than a vertical table and cannot be viewed as readily, there are other advantages to its use that cause it to be favored for very accurate work. A tilting table can be posi-

Structural deviations from load stress are depicted from before-load and after-load configuration plots.

Fig. 17.10 Analysis load configuration of the prime maneuverable reentry vehicle.
The data processing center of this company is presently utilizing a plotting system to study load configurations on the prime maneuverable reentry vehicle. A mathematical structure is shown being plotted in what is considered an unloaded configuration (see plot above). Then, theoretical load configurations are applied through the computer to different parts of the structure. New plotting tapes are generated and replots are made of the new load structure. By visually comparing before- and after-load plots, company engineers are able to determine deviation and stress of the structure due to loading. (*Courtesy Martin Marietta Corporation*)

tioned horizontally or it can be adjusted to be almost vertical.

A drawing head has a turret (revolving) mechanism that may have four, six, or eight stations, depending again on the model. The stations accommodate the pens, pencils, or scribes that can be selected and used under program control. Ordinarily Rapidograph-type pens are used; however, ballpoint-type pens serve better for drawing at high speed. For wide lines, felt marker pens have been used and found to be satisfactory.

A rotary scriber is primarily used for printed circuitry. The turret has the same basic station arrangement as the turret for the normal drafting machine. How-

ever, instead of pens at each of the positions there are small end-mill-type tools that cover a range of sizes.

The two types of control systems are the incremental system and the absolute system. At the present time, the incremental with its delta motion is the most widely used. This system requires the ΔX and ΔY values for the distance from the "where-now" location to the next position. The somewhat different approach provided by the absolute system requires that the X and Y coordinates of the next point be related to the very first, that is, the starting point (set point or zero point).

In addition to the XY-coordinate information that is needed to meet position requirements, there are control commands associated with the drawing of a line segment, such as: move to origin, move and plot, pen up and move, pen down and move, rotate turret, draw dotted line, pause and change pen, and so forth.

Circles and curves must be approximated with straight lines. For these to be drawn accurately, acceleration and deceleration blocks and feed numbers must be calculated. These data can best be calculated by a computer.

The drafting machine (plotter) shown in Fig. 17.10 is suitable for a wide range of applications from contour map generation to automatic drafting and graphical verification of machine-tool tapes. A four-station drawing head is provided, with each station accommodating a standard holder for either an inking or a scribing device. The carriages are positioned by means of digital servo systems. The effective working area is 60 in. (1525 mm) on both the X and Y axes for all styli (marking devices).

All control functions, operator functions and input–output devices are contained in the separate control unit, which includes a general-purpose computer. The computer includes a 4K core memory and high-speed arithmetic unit.

The computer continuously monitors the position of the stylus carriage (marking device carrier) and controls the drive system as required by operator control and input–output devices. The standard input-output device is an ASR–33 Teletypewriter with paper tape reader and punch (magnetic tape input optional). Additional controls have been provided for manual entry. The digital servo control signals from the computer are processed within the control unit to form suitable outputs to the servomotors.

Performance capabilities

Straight-line interpolation is performed digitally. The format may be coordinate, delta, or incremental.

Stored in prime form in core memory are 64 alphanumeric characters and symbols. These characters, having a prime height of .064 in. (1.62 mm), may be drawn by any of the styli, scaled up to 1.024 in. (26 mm), and rotated to eight angular orientations. With the X and Y coordinates of the first character of a line specified, subsequent characters are spaced to the right to form lines of up to 120 characters per line.

Accuracy relative to reference point is .002 in. (0.05 mm) over the entire working area for both static and dynamic operation; repeatability is .001 in. (0.025 mm). The encoder resolution is .00025 in. (0.0064 mm) for the X-axis and .000375 in. (0.0095 mm) for the Y-axis.

A few of the optional features available to users are (1) additional characters and symbols, up to 255 in number; (2) interpolation of circular, elliptical, or other conic sections; (3) routing head in lieu of a stylus carriage; (4) photographic head suitable for generation of printed-circuit work in lieu of stylus carriage; (5) digitizing head for operation with CRT monitor in lieu of stylus carriage (see Sec. 17.22); and (6) a drive system for paper advance.

At the present time, there are several recognized computer programming systems in general use. Although all of them no doubt meet the needs for which they were created, only six will be mentioned here. They are APT (Automatically Programmed Tools), which is particularly suited for use in engineering education; AUTOPROMPT, developed by IBM; ALLADIN, developed by the Allison Division of General Motors; AUTODRAFT, which represents the joint efforts of IBM and North American Aviation, Inc.; UDRAFT–8, developed by the Universal Drafting Machine Corporation; and TRIDM, developed and tested in the classroom at North Carolina State University. Both APT and AUTOPROMPT were developed primarily for NC (numerically controlled) machine tools. ALLADIN, AUTODRAFT, and TRIDM are drafting language systems.

In preparing a program, it is necessary for the draftsman to describe the shapes that are to be drawn. A description may be either in digital form or an equation given in the vocabulary of the system being used. The program that is prepared using these descriptions becomes, as one might expect, a complete sequence of coded instructions that must be followed by the drafting machine. This encoded information is then keypunched and fed to the computer. The computer, in turn, performs the calculations necessary to digitize it for either punched-tape or magnetic tape output.

A plot of acceleration, velocity, and distance curves is shown in Fig. 17.11.

17.17

TRIDM. TRIDM is an advanced version of a computer-graphics language originally

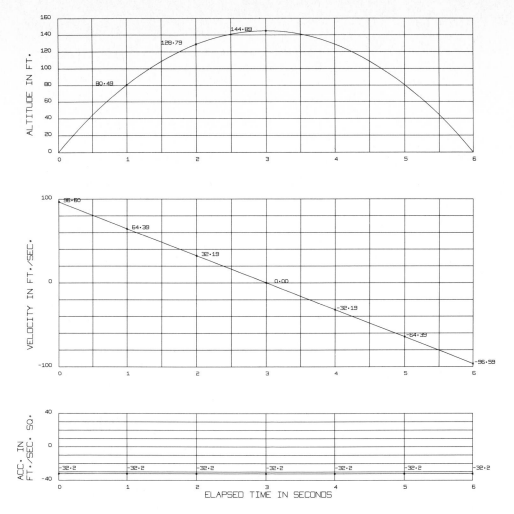

Fig. 17.11 Acceleration, velocity, and distance curves.
These plots were done on a digital plotter connected on-line with an IBM 1130 computer.

developed for small computers such as the IBM 1130 by Bertram Herzog of the University of Michigan. The improvements effected at North Carolina State University in his DRAWL 1130 largely in terms of data structure, subroutine simplification, and input–output options, resulted in the version known as TRIDM. At this time both Herzog and North Carolina State University have waived rights to TRIDM. Herzog has retained rights to DRAWL 70 and DRAWL 71 for larger computers, while North Carolina State retains all rights to TRIDMN, a later, more efficient, and more versatile version of TRIDM.

For a computer-graphics language to meet the requirements of engineering education, it must be natural and must be applicable to a wide range of design and information display problems. TRIDM, utilizing command keywords that have been developed through classroom testing

and are natural for the designer, offers an unusual set of options to information users. TRIDM has been used in the design of a geodetic dome structure and a space shuttle vehicle as well as for information displays relating to studies such as highway routing alternatives, the mapping of drug reactions within the human body, and the evaluation of educational programs. A person closely identified with this work is Byard Houck in the field of engineering graphics at North Carolina State University. Houck prepared the program given in Fig. 17.12 and supplied the plotter-prepared drawings in Fig. 17.13.

The student will find it worthwhile to read through the program and relate the given comment statements to the plotter-prepared drawings. Note that the top view was drawn first, and then the front and right side views follow in order. Next, well down in the program there is a state-

```
**** NCSU'S TRIDM FOR OS/360/370          VER 3  MOD 03      REV  1 JUNE 72    BY LLR ****

$ SIZE0          12.00    10.00    0.00    SIZE CARD FOR PRINTER-PLOT
$ OBJSOSB         0.00     0.00    0.00    THIS IS A SUPPORT BLOCK-LOCATED AT ORIGI.
OBJS   1 HAS    44 POINTS
$ AXISO           2.00     3.00    0.00    SET AXES FOR ORTHOGRAPHIC VIEWS
$ DRAW2SB         0.00     0.00    0.00    TOP VIEW (VIEW FROM +Y)
DRAW   1
X-AXIS = 2.00    Y-AXIS = 3.00    N= 0
THE DRAWING HAS BEEN COMPLETED.
$ DRAW3SB         0.00     0.00    0.00    FRONT VIEW(VIEW FROM -Z)
DRAW   1
X-AXIS = 2.00    Y-AXIS = 3.00    N= 0
THE DRAWING HAS BEEN COMPLETED.
$ DRAW1SB         0.00     0.00    0.00    RIGHT SIDE (VIEW FROM +X)
DRAW   1
X-AXIS = 2.00    Y-AXIS = 3.00    N= 0
THE DRAWING HAS BEEN COMPLETED.
$ AXISO           3.50     4.00    0.00    SET AXIS FOR 3-VIEW ORTHOGRAPHIC
$ SHIFOF1        -3.00    -3.50    0.00    FRONT VIEW LOCATION
MATRIX   -1
$ TRANOB1         0.00     0.00    0.00    MOVE OBJECT FOR FRONT VIEWING
TRANS   -1 = MATRIX   -1(   1).
$ COMBOC1         0.00     0.00    0.00    TOP VIEW SETUP
$ ROTA1         -90.00     0.00    0.00    ROTATE OBJECT--TOP VIEWING
$ SHIFO          -3.00     0.50    0.00    SHIFT OBJECT -- TOP VIEWING
$ ECOMO           0.00     0.00    0.00
MATRIX  -17
$ TRANOT1         0.00     0.00    0.00    TOP VIEW LOCATION
TRANS   -4 = MATRIX  -17(   1).
$ COMBOC2         0.00     0.00    0.00    RIGHT SIDE VIEW SETUP
$ ROTA2          90.00     0.00    0.00    ROTATE OBJECT--RIGHT SIDE VIEW
$ SHIFO           0.50    -3.50    0.00    SHIFT OBJECT--RIGHT SIDE VIEW
$ ECOMO           0.00     0.00    0.00
MATRIX   -33
$ TRANOR1         0.00     0.00    0.00    RIGHT SIDE VIEW LOCATION
TRANS   -7 = MATRIX  -33(   1).
$ ASMBOA1         0.00     0.00    0.00    LOCK ALL VIEWS INTO COMMON PICTURE
ASMB  -10 HAS    3 PARTS
$ DRAWOA1         0.00     0.00    0.00    TAKE 3-VIEW ORTHOGRAPHIC VIEW
DRAW -10
X-AXIS = 3.50    Y-AXIS = 4.00    N= 0
THE DRAWING HAS BEEN COMPLETED.
$ COMBOC3         0.00     0.00    0.00    BEGIN ORIENTATION FOR PERSPECTIVE
$ SHIFO          -2.50    -1.50    0.00    MOVE TO LEFT OF Y-AXIS AND CENTER ON X
$ ROTA2          50.00     0.00    0.00    ROTATE +50 ABOUT Y
$ ROTA1         -22.00     0.00    0.00    ROTATE -22 ABOUT X
$ PERS3           5.00     0.00    0.00    PERSPECTIVE FROM -5
$ECOMO            0.00     0.00    0.00    PERSPECTIVE SETUP
$TRANOP1          0.00     0.00    0.00
TRANS  -14 = MATRIX  -49(   1).
$ DRAWOP1         0.00     0.00    0.00    PERSPECTIVE VIEW
DRAW -14
X-AXIS = 3.50    Y-AXIS = 4.00    N= 0
THE DRAWING HAS BEEN COMPLETED.
$ SHIFOS1         2.50     2.00    0.00    PLACE PERSPECTIVE IN UPPER RIGHT QUADRAN
$ TRANOP2         0.00     0.00    0.00    PLACE PERSPECTIVE IN UPPER RIGHT QUADRAN
TRANS  -17 = MATRIX  -65(  -14).
$ ASMBOA2         0.00     0.00    0.00    LOCK ALL VIEWS INTO COMMON PICTURE
ASMB  -20 HAS    4 PARTS
$ DRAWOA2         0.00     0.00    0.00    DRAW ORTHOGRAPHIC & PERSPECTIVE COMBINED
DRAW -20
X-AXIS = 3.50    Y-AXIS = 4.00    N= 0
THE DRAWING HAS BEEN COMPLETED.
$ SCALOS2         0.50     0.00    0.00    REDUCE SCALE
MATRIX   -81
$ TRANOA3         0.00     0.00    0.00    REDUCE SCALE
TRANS  -25 = MATRIX  -81(  -20).
$ DRAWOA3         0.00     0.00    0.00    REDUCED SCALE-ORTHOGRAPHIC/PERSPECTIVE
DRAW -25
X-AXIS = 3.50    Y-AXIS = 4.00    N= 0
THE DRAWING HAS BEEN COMPLETED.
```

Fig. 17.12 Print-out of the program statements needed for preparing the orthographic and perspective drawings shown in Fig. 17.13. (Note that the top line reads NCSU's TRIDM.)

ment BEGIN ORIENTATION FOR PERSPECTIVE. This is the point at which the drawing of the perspective started. Finally, after the drawing shown had been completed another drawing was made in response to the program statement REDUCE SCALE. A half-size representation was drawn with the views and the perspective.*

17.18

IBM 1130 Data Presentation System. The IBM 1130 Data Presentation System is a computer program. The basic 1130 system for DPS consists of a 1627 plotter and an IBM 1442 card read/punch. The system can be used by anyone, even by persons who may have no previous computer knowledge. If the user desires a graph, chart, or whatever, he or she merely feeds the computer a limited number of command phrases that describe the representation wanted and the format of the files where the data is stored. The computer provides the chart, graph, etc., that is needed. The graphic representation requested is drawn by the computer's 1627 plotter (Fig. 17.14). The 1627 plotter receives numerical data (plotting coordinates) from the computer and converts the information into graphic form. DPS processes any data that can be presented graphically and which have been recorded on punched cards or in the 1130 disc file. The Data Presentation System is designed in three modules. The most advanced level is called the "graphic report generator." This is a fully developed user-oriented system that requires no additional user program to construct a graph illustrating data in an 1130 file. A few of the characteristics that illustrate the versatility of the graphic report generator are: (1) as many as eight scales can be drawn in any location; (2) information may be selected from several data sets for the same plotted display; (3) scales may be linear, logarithmic, or polar; (4) scales are automatically set to keep all data on the plot; and (5) data may be plotted as point plots, line plots, and logarithmic plots. Circles and spirals may also be generated. As can be observed in Fig. 17.15, the system is capable of constructing straight-line figures or perspective views of geometric shapes.

* TRIDM was given to all persons who attended the Computer-Graphics Summer School that was held at the time of the Annual Meeting of the American Society for Engineering Education in Lubbock, Texas, in June 1972. Another approach to computer-graphics was presented at this same meeting by Clarence E. Hall of Louisiana State University. Those who may be interested in Hall's method of presenting computer-graphics to engineering students will find his method presented in detail in Vol. 36, No. 2 (Spring 1972) issue of *Engineering Design Graphics*, published by the Division of Engineering Design Graphics of ASEE.

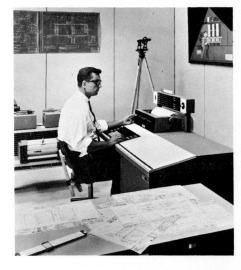

Fig. 17.14 IBM 1130 Data Presentation System.
The 1627 plotter may be observed to the left of
the man, at about the level of the chair-seat.
(*Courtesy International Business Machines
Corporation*)

**Fig. 17.13 Drawings prepared by a plotter following the instructions given by
the program shown in Fig. 17.12.**

From tabular results calculated by the
computer, the plotter can be made to
trace the shape of a cam being designed.
The laborious work involved in structural
detailing can be handled by the 1130
computer and the 1627 plotter. For example, in one design organization the system
is used to calculate the dimensions for a
steel member and then the 1627 produces
the needed drawing with the required dimensions appearing along the view. The
time needed to produce a detail sheet is
reduced as much as 40% by this method,
thus making additional man-hours available for technical design. Figure 17.16
shows a modified standard drawing prepared by the 1627 plotter.

**17.19
UDRAFT–8.** The Universal Drafting Machine Corporation now has available a
software program for automatic drafting
that can be used by persons having no

**Fig. 17.15 Examples of graphic output—1130
Data Presentation System.** (*Courtesy
International Business Machines Corporation*)

knowledge of computer programming. The easily understood language, known as UDRAFT–8, can be used to produce tapes for drawings to be made on Universal Drafting Machine Corporation's Orthomat Systems. A feature of UDRAFT–8 is that it can be implemented on the drafting system computer, thus permitting the system to prepare and edit input tapes. By using simple statements, combined alphanumeric capability, macro library, and other needed features, the designer or draftsman can prepare complex drawings with minimal effort and at a great saving in time.

Fig. 17.16 Modified standard. (*Courtesy International Business Machines Corporation*)

Fig. 17.17 Detail drawing produced by a plotter (metric). (*Courtesy Ford Motor Company*)

17.20
CRT Plotting Systems (Fig. 17.6). The CRT microfilm printer–plotter has undergone gradual improvement since the first system was placed in use in 1960. During the first years, the cost of CRT plotters was so high that they could be found only at aerospace and government research centers. However, extensive research and development work has led to the recent announcement of several economical systems with the flexibility needed for high-speed computer output, data analysis, automatic drafting, and circuit board design. Use of CRT printer–plotters is expected to rise rapidly.

Input data, decoded and displayed on a high-resolution cathode-ray tube, are photographed using a high-speed 35 mm magazine camera. This film is fed from a continuous roll that may be up to 400 ft. (122 m) in length.

The average plotting time for a typical plot is less than 1 sec. Should a "quick-look" hard copy be needed, the auxiliary hard-copy camera can be used to obtain copy directly on 9-in. (228 mm) paper within 4 sec. This is accomplished by an integrally mounted processor that is relatively inexpensive.

17.21
Electrostatic Printer/Plotters. Gould Plotmaster Systems employing high-speed electrostatic printer/plotters are versatile on-line or off-line printing and plotting systems for IBM System/360 and System/370 computer systems. Plotmaster Systems plot data many times faster than the conventional pen plotters discussed in Secs. 17.16–17.19. Designed expressly for minicomputers, the series 5000 Systems (Fig. 17.18) offers models that plot up to 400 times faster than pen plotters and print almost twice as fast as line printers for large computers. Solid-state electronics makes the entire series very reliable and easy to maintain. In addition to a basic PLOT graphics package, specialized software is available for a wide range of applications. ENGINEERING SOFTWARE/FAST-DRAW consists of 80 FORTRAN-CALLABLE routines readily usable by a design engineer or draftsman familiar with FORTRAN coding techniques. The user needs only to specify key parameters defining a geometric figure and the program will automatically perform the geometric and trigometric calculations to produce the desired construct. FAST-DRAW is a general-purpose drafting package for producing computer-generated graphic output for a wide variety of engineering graphics applications from simple illustrations to highly sophisticated engineering drawings, two-dimensional or perspective projections of three-dimensional configurations.

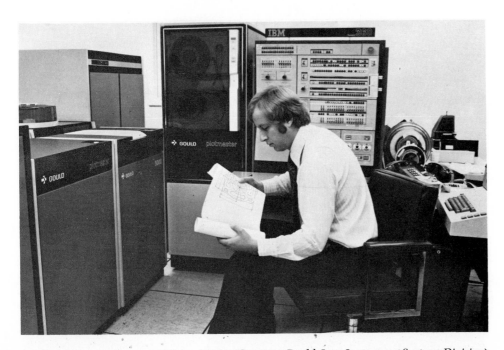

Fig. 17.18 Gould Electrostatic Printer/Plotter (*Courtesy Gould, Inc., Instrument Systems Division*)

Computer-aided
design and
automated drafting

3 1/4

1 5/8

3/8 D

PUNCH
4 HOLES

1 5/8

1/8 R   1/8 R

2 Ø

1/2 Ø

3 1/2

3/4

1 1/2

2 3/4

1/2

1/8 R

1/4

3/8 R

GASKET
PUNCH FROM 1/16 CORK

```
C
C
CC   INITIALIZE PLOT
     CALL INIT ('SIZE',14.,10., 'ORIG',7.,5., 'FACT',2.0)
C
CCCCCCCCCCCCCCCCCCCCCCCCC   DRAW THE GASKET   CCCCCCCCCCCCCCCCCCCCCCCCCCCC
C
CC   RESERVE SPACE FOR 50 POINTS, 50 LINES, AND 50 CIRCLES
     CALL RESERV (50, 50, 50)
C
CC   DRAW THE 1-1/2" DIA CENTER CIRCLE
     CALL CIRC ('CEN',0.,0., 'RAD',0.75)
C
CC   DRAW THE GASKET OUTLINE.THE CENTER OF THE 4 CORNER CIRCULAR ARCS IS
C    1-5/8 - 3/8 OR  1-1/4 HORIZONTALLY AND VERTICALLY FROM THE CENTER.
C    DEFINE THIS ARC IN THE 1ST QUADRANT.
     CALL DEFINE (CIR(1), 'RAD',0.375, 'CEN', PNT(1.25,1.25),
    2                             'SANG',0., 'FANG',90.)
     ANG=90.
     CALL MOVEA (CIR(1))
     DO 10 N=1,4
     CALL DRAW (CIR(1))
     CALL ROTAT(1, 'AB', ANG)
     CALL LINEA(STRPNT(CIR(1)))
10   ANG=ANG+90.
     CALL CLEAR(1)
C
CC   DRAW THE 4 HOLES (3/8" DIA) AT THE FOUR CORNERS.
     CALL DEFINE(CIR(1), 'CEN',PNT(1.75, 0.0), 'RAD',0.1875)
     ANG=45.
     DO 20 N=1,4
     CALL ROTAT(1, 'AB', ANG)
     CALL MOVEA(1.75,0.0)
     CALL DRAW (CIR(1))
20   ANG=ANG+90.
     CALL CLEAR(1)
C
CC   DEFINE THE RIGHTMOST IRREGULARLY SHAPED HOLE
C
C    FIRST DEFINE THE 2" DIAMETER CIRCLE:
     CALL DEFINE(CIR(1), 'CEN',PNT(0.,0.), 'RAD',1.0)
C
C    THEN DEFINE A 45 DEG LINE PASSING THRU THE POINT (X,0.),
C    WHERE X=0.25 * SQRT(2.)
     X=0.25*SQRT(2.)
     CALL DEFINE(LIN(1), PNT(X,0.0), PNT(X+1.0,1.0))
C
C    DEFINE LINE #2 AS A HORIZONTAL LINE, 3/4" ABOVE THE ORIGIN
     CALL DEFINE(LIN(2), PNT(0.,.75), PNT(2.,.75))
C
C    DEFINE LINE #3 AS A VERTICAL LINE 1-3/8" TO THE RIGHT OF THE ORIGIN
     CALL DEFINE(LIN(3), PNT(1.375,+2.), PNT(1.375,-2.))
C
C    DEFINE LINE #4 AS A HORIZONTAL LINE, 3/4" BELOW THE ORIGIN
     CALL DEFINE (LIN(4), PNT(2.,-.75), PNT(0.,-.75))
C
C    DEFINE A (-)45 DEG LINE PASSING THRU THE POINT (X,0.),
C    WHERE X=0.25 * SQRT(2.)
     CALL DEFINE(LIN(5), PNT(X+1.0,-1.0), PNT(X,0.0))
C
C    DEFINE THE 6 ROUNDING ARCS (1/8" RAD) WITHIN THE IRREGULAR HOLE
     CALL CIRTCL(CIR(2), 0.125, 'OUT','HMAX',CIR(1), 'HMAX',LIN(1))
     CALL CIRTLL(CIR(3), 0.125, 'HMAX',LIN(1), 'VMIN',LIN(2))
     CALL CIRTLL(CIR(4), 0.125, 'VMIN',LIN(2), 'HMIN',LIN(3))
     CALL CIRTLL(CIR(5), 0.125, 'HMIN',LIN(3), 'VMAX',LIN(4))
     CALL CIRTLL(CIR(6), 0.125, 'VMAX',LIN(4), 'HMAX',LIN(5))
     CALL CIRTLC(CIR(7), 0.125, 'HMAX',LIN(5), 'OUT','HMAX',CIR(1))
C
C    REDEFINE CIRCLE ARC #1 AS BEING TANGENT TO CIR(7) AND CIR(2)
     CALL CIRTCC(CIR(1), 1.0, 'OUT',CIR(7), 'OUT','HMIN',CIR(2))
C
CC   NOW DRAW THE IRREGULAR HOLE IN 4 PLACES
     ANG=0.0
     DO 30 N=1,4
     CALL MOVEA(CIR(1))
     CALL DRAW (CIR(1),CIR(2))
     CALL LINEA(CIR(3))
     CALL DRAW (CIR(3))
     CALL LINEA(CIR(4))
     CALL DRAW (CIR(4))
     CALL LINEA(CIR(5))
     CALL DRAW (CIR(5))
     CALL LINEA(CIR(6))
     CALL DRAW (CIR(6))
     CALL LINEA(CIR(7))
     CALL DRAW (CIR(7))
     ANG=ANG+90.
30   CALL ROTAT(1, 'AB', ANG)
     CALL CLEAR(1)
C
C
C    THE DRAWING OF THE GASKET IS COMPLETED. THE REMAINDER OF THIS PROGRAM
C    DOES CENTERLINES, DIMENSIONING, AND ANNOTATION.
C

                                  |
                                  |
                               (ETC)
                                  |
                                  |

C
CC   ANNOTATE 'PUNCH 4 HOLES', 'GASKET', AND 'PUNCH FROM 1/16 CORK'
     CALL TXSIZ(0.1)
     CALL TXPLT(2.250,2.000,0.,0, 'PUNCH@','4 HOLES@')
     CALL TXPLT(0.,-2.40,0.,0, 'PUNCH FROM 1/16 CORK@')
     CALL TXSIZ(.125)
     CALL TXPLT(0.,-2.2, 0., 0, 'GASKET@')
C
CC   TERMINATE PLOT
     CALL FINI
     STOP
     END
```

Fig. 17.19 Gasket drawing prepared using FAST-DRAW. (*Courtesy Gould, Inc., Instrument Systems Division*)

The drawing in Fig. 17.19 was prepared using FAST-DRAW. The entire geometry of the gasket (exclusive of dimension lines and annotations) was described using only 51 FAST-DRAW statements.

17.22

Digitizing. The computer can be made to "see" drawings, sketches, and photographs but it can be made to do so only through reading either a deck of cards or a roll of paper or magnetic tape produced by a digitizer (Fig. 17.20). What it "sees" as input must be in digital form and a drawing of any complexity must be received by the computer as a series of digits, that is, pairs of coordinate points. In the early days digitizing was a difficult and time-consuming task because it was necessary to "break-up" a drawing into straight-line segments, the "break-up" theory being based on the fact that two points define a straight line. The readings for pairs of digits for straight-line segments were then manually established or directly read off of graph paper. All readings were carefully recorded to be subsequently punched into a deck of cards that

Fig. 17.20 CalComp RP1 Positioner listing the digitized pad positions on a teletypewriter.
(*Courtesy California Computer Products Inc.*)

would be read into the computer via a card reader.

Manual digitizers soon followed. These in their simplest form resemble a drafting board with a movable rail that is equipped with a cursor, mounted so as to be free to move in the X and Y directions. With the drawing mounted on the board, the operator first positions the cursor with the cross-hairs centered over the end point of a line segment and then pushes a button to punch the coordinate values of the point onto a card or paper tape. Digitizers of this type utilize cable driven, rotary, optical encoders. These encoders send out electrical pulses that are translated into coordinate measurements by a control unit.

At the present time there are a wide variety of digitizers on the market with optional features. Of importance to designers in the electronic field is the grid recognition capability of some of these models since electronic (printed circuit) artwork generation is usually done so that features are dimensioned to integer multiples of some grid spacing.

Even though manual digitizing is satisfactory in every way for drawings composed mainly of straight lines and discrete points, manual digitizers require considerable painstaking effort when circular

arcs and other mathematically defined curves are being prepared for the computer through digitizing. For this reason, computer programs have been developed that permit curves and repetitive patterns to be described in easily prepared coded statements that can be punched into cards to be read and interpreted by the computer in accordance with programmed instructions.

While manual digitizing coupled with part programming by means of statements works well for most types of graphics, limitations are encountered when the computer must be made to "see" some exotic curve that cannot be mathematically defined.

New forms of automatic digitizers are now available. For example, one digitizer has a special head that locks onto a line being digitized and follows along it automatically while sending out digital information as directed by the program of the mini-computer control unit. Another type of automatic digitizer uses a *scanner*. In this case, the graphic representation is mounted around a drum that spins while a photosensitive receiver travels across the spinning drum in an axial direction while converting the light level differences being received into data for computer processing. The graf/pen shown in Fig. 17.21 uses

Fig. 17.21 Graf/pen Sonic Digitizer—gp3. (*Courtesy SAC—Science Accessories Corporation*)

hypersonic impulses generated at the point of the stylus as its means of measurement. The times required for the sound wave to reach two linear microphone sensors are converted into X and Y coordinates. The sensors detect each sonic impulse and convert the impulse to electrical signals for the control unit. This widely used digitizer may be used wherever data in graphical form must be entered into a digital system. In using the graf/pen the operator merely traces the line. Graf/pen units are available that provide digital readings in either the English or the metric system.

The Gerber 2075/OLF-DMS, shown in Fig. 17.22, combines automatic digitizing and high-speed, large-area drafting in one complete system. Its capabilities include expansive data and program file management, display, editing, input–output, and numerical control tape production. These capabilities make this system ideal for processing high quantities of large-scale drawings and NC machine tool tapes. Hardware components include: (1) stored program drafting/digitizing control, (2) punched tape reader, (3) tape punch, (4) teletypewriter, (5) operator control console, (6) drafting/digitizing table, and (7) automatic digitizing with optical line follower (Fig. 17.23). Together with Gerber's proprietary drafting and digitizing programs, these components form a system that generates drawings from a data base, converts graphical data to digital form, and creates and verifies NC tapes.

The optical line follower, shown mounted on the table Y-axis (Fig. 17.22), contains the light source for illumination of the digitizing surface, a line-sensing device, a television camera, and a three-station tool holder (Fig. 17.23). While the line-sensing device transmits signals to the digitizing and line following logic housed within the control, the television camera transmits a magnified view of the area being digitized to the monitor at the operator's console.

Drafting is facilitated, without hardware change, by the integral three-station tool holder. A wide range of ink pens, ballpoint pens, fiber-tip markers, and scribes may be used on a variety of media.

Fig. 17.22 Automatic Drafting/Digitizing System. (*Courtesy The Gerber Scientific Instrument Company*)

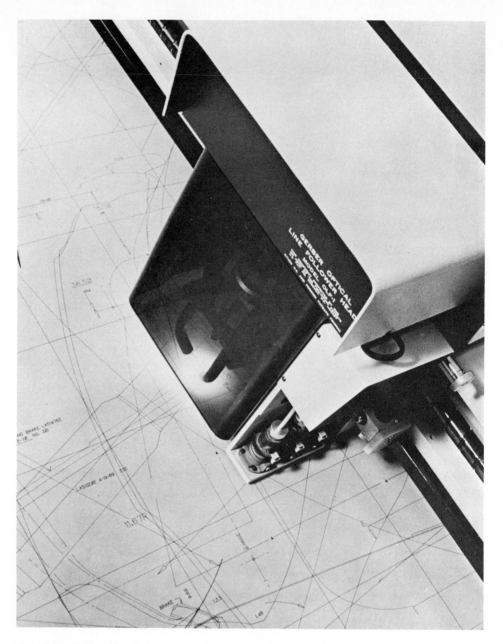

Fig. 17.23 Optical Line Follower (see Fig. 17.22). (*Courtesy The Gerber Scientific Instrument Company*)

17.23

Use of the Computer in Automobile Body Design. The Ford Motor Company has turned to interactive graphics to deal effectively and quickly with the complex arbitrarily curved surfaces that are encountered in body design. In taking this step, a working combination of man, computer, CRT, and light pen was assembled to facilitate the tedious handling, correcting, and manipulating of digital data from the clay model (Fig. 17.25) in the process of moving from the model to the machining of a production stamping

die. In Fig. 17.24, the designer at the CRT console is using knowledge that he has acquired in the field of "human engineering." This is another application of interactive graphics in body design.

The needed digital definition of the clay model is obtained by using the Surface Scanner (Alpha 3d) shown in Fig. 17.25. These data are cut in X-, Y-, and Z-coordinate planes and along three-dimensional lines and panel boundaries. Data can be obtained in either the English inch or the metric system.

The surface scanner is servo-driven.

Fig. 17.24 Designer at the CRT console.
(*Courtesy Ford Motor Company*)

Fig. 17.25 Surface Scanner. (*Courtesy Ford
Motor Company*)

While following a specified coordinate section on the model surface, it can be moved either manually or automatically at a speed of 100 inches (2540 mm) per minute. Accuracy at this maximum velocity is specified as being less than $\pm.010$ in. (0.25 mm). The specially designed electromechanical probe used has a contact pressure of only 5 grams, so it will not penetrate the clay surface.

The digital output of the surface scanner is transmitted by Dataphone to a central computer to be processed for verification drafting of sections, character lines, and boundary lines on a numerically controlled drafting machine (Fig. 17.26). Also, the data are processed for transmittal on a removable disc pack to body engineering for surface development using interactive computer graphics as previously mentioned at the beginning of this discussion. Here the designer with his light-pen input adjusts and manipulates these three-dimensional data for a better and more attractive design. To achieve desired results, he may create new data by altering character lines and boundary lines. This, of course, requires some alteration of the sections between them in order to obtain good proportioning.

When designs have been completed, the digital data are then formatted for draft output and extracted for NC machining.

17.24
Artwork Generators. Artwork generators, which are now widely used in the electronics field, can be said to be high-precision automatic drafting machines. These machines are equipped with a photo-exposure head. They produce finished graphic drawings utilizing a beam of light focussed on a photosensitive emulsion surface. The servo commands that direct the movement of the light beam are fed to the generator control unit by means of punched paper tape or magnetic tape. If desired, the artwork generator (drafting machine) may be operated directly on-line with the computer. However, most artwork generators are operated off-line.

17.25
Applications of Plotting and Digitizing Systems. Computer-operated drafting machines are widely used for the preparation of perspective and axonometric drawings and integrated-circuit-board layouts. They do mapping, perform lofting, and plot impeller, airscrew, and wing sections. By

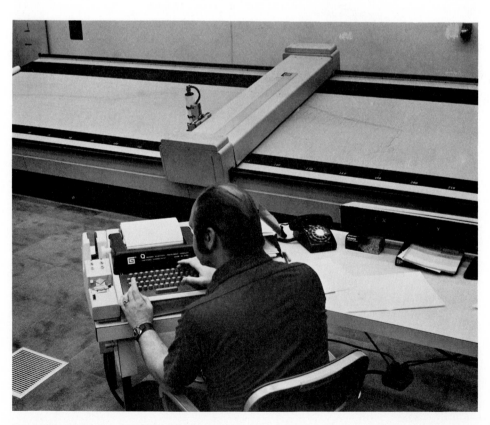

Fig. 17.26 Verification drafting. (*Courtesy Ford Motor Company*)

utilizing their now recognized capabilities, drafting systems can prepare a perspective of an aerodynamic configuration composed of station lines or plot a series of compound surfaces and almost immediately present a perspective of these same surfaces for inspection and evaluation. This has been demonstrated using AUTOPROMPT as a drafting language.

To recreate the maneuvers of aircraft, animated motion pictures have been made with the cockpit and terrain features both included to provide orientation and to relate most of what a pilot would see from his position in the plane. Of course, a higher degree of realism could have been obtained by making a motion picture from the plane doing the maneuver, but only at a much higher cost. The automated drawing was done by a Mosely plotter and a little final rendering was added to bring about some realism. The creation of these highly complex three-dimensional scenes for a rough movie-type sequence has been largely the work of several outstanding engineering "graphicians" at The Boeing Company. The results of their work appear in a book entitled *Computer Graphics in Communication.** The author, William A. Fetter, was a member of a small group of persons assigned to the project. Some of the work of these men included maneuver simulation of a bank and turn and then a dive and pullout, pilot views of landing on a CVA–19 under simulated heavy seas, and pilot views of landing at an airport.

Three-dimensional graphs with scales

*William A. Fetter, *Computer Graphics in Communication* (New York: McGraw-Hill, Inc., 1966).

and rotated perspective views have been prepared using an Orthomat plotter. Properly programmed, other plotters can perform the same tasks.

Now coming to the attention of people in industry is the application of plotting systems to the graphic interpretation of machining routines that have been programmed by a computer. The graphic interpretation provides a quick check and eliminates the need for running tests of routines on a numerically controlled machine tool. As stated elsewhere in this chapter, drafting machines can also produce drawings to verify cutter-tool paths specified by machine control tapes. For quality control, a drafting machine can be used as an aid in establishing checkpoints on inspection tapes and in making scribed optical comparator charts and boards.

Typical plotter prepared design drawings have been shown in Figs. 17.27–17.30.

17.26

The Future. At some point in the future it may be economically feasible to develop a system that will computer-generate detail drawings by software techniques from a design layout. Plans are now being made for systems that will permit a designer to prepare his design layout in collaboration with a suitably programmed computer. With the layout completed and checked, the next logical step will be to have the computer generate the needed detail and assembly drawings, parts list, and machine tool tapes with little, if any, human intervention. Only when we have reached the point where this can be accomplished can we say that we have real "automatic drafting."

Fig. 17.27 Structural drawing prepared on a plotter.
Made using the CONSTRUCTS System of automated drafting. (*Courtesy Meiscon Division, Control Data Corporation*)

Fig. 17.28 Highway route plan and profile.
Program systems have been developed to produce civil engineering plans for highway,
sewer, and drainage construction. The route plan and profile sheet shown was produced
on a CalComp 702 in less than 25 minutes. All of the lines and lettered information
were generated, except the title block at the bottom. The keypunch operator interprets
the field notes and punches the required information into cards. This information is
then fed into the computer to produce a magnetic tape which contains the needed
drafting instructions. Johnson & Anderson, Inc., started to develop their system in late
1967. Their initial success encouraged further work toward the complete system that
they now use to make their drawing at much less cost than by manual methods. To
improve the total system further, they modified their method of taking field notes and
changed the graphic read-out procedure. Increased efficiency is expected in the near
future. (*Courtesy Johnson & Anderson, Inc., Consulting Engineers, Pontiac,
Michigan*)

Fig. 17.29 Plot of a subdivision.

The computer program designated SAMPS (Subdivision and Mapping System) plots a complete subdivision map with bearings, distances, and other information ordinarily given on a map of this type. As developed by PMT Associates of Sacramento, California, the program can also be used for plotting control networks, for surveying jobs, and for primary control of aerial photographs. The program provides for choice of plotting scale, rotation of plotting axes, plotting of lines and points, plotting of a north arrow, and annotation of distances and bearings. The subdivision plot shown was drawn on a CalComp Model 702 flatbed plotter operating on-line with an IBM 1130 computer. (*Courtesy of PMT Associates-Engineers, Land Surveyors, Planners*)

RIDGECREST – UNIT NO. 3

SACRAMENTO COUNTY, CALIFORNIA

WORK ORDER 6801-11

SCALE 1'' = 50'

1.250 DIA. CAMFOLLOWER

10-5/8

9-3/8 R

1-5/8

.625 RISE IN 20°

0267AJ-20-0625

5R

STAMP DIRECTION OF
ROTATION AND PART NO.

60° DWELL

.625 FALL IN 20°

0267AJ-20-0625

4-3/8 R REF

260° DWELL

1.500 - 1.501 DIA.

USE .005 SHIM
WHEN BORING

1-5/32

2-5/16

11/32 DRILL 25/32 DP.
1/2 C BORE 5/16 DP.
2 HOLES

F (.257) DRILL THRU
5/16 - 18 TAP THRU
2 HOLES ITEM = 1

**Fig. 17.30 Westinghouse Electric Corporation uses a CalComp plotter on-line with an IBM 360
computer to aid in the design of special-purpose cams.**
The layout drawing shown was prepared using a computer program developed at the Lamp
Division of this company. It is reported that the typical cam design now requires approximately 5
minutes card punching time, 2 minutes computer time, and 5 minutes or less plotting time, for
a total of 12 minutes. Manual methods for the same task usually required 25–30 hours.
(*Courtesy Westinghouse Electric Corporation*)

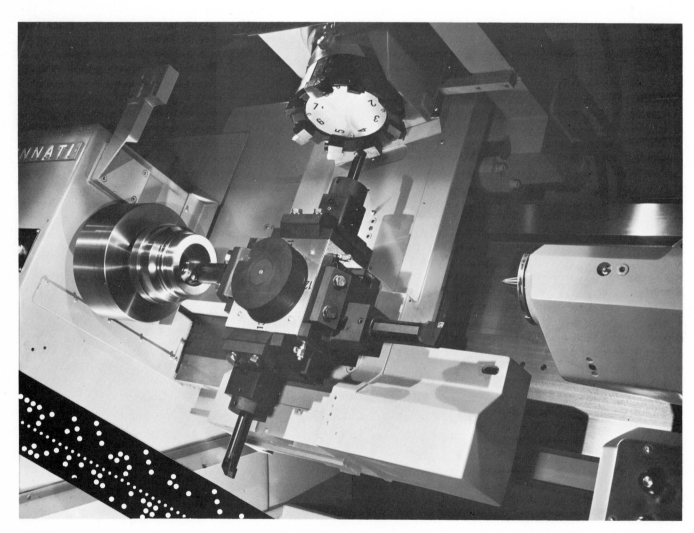

With the coming of numerical control (NC) American industry stands at the threshold of a new era. In some industries, numerical control now starts at the product design level, where engineers and technologists utilize the computer for creating mathematical models that may ultimately become line drawings prepared on a plotter (see Chapter 17). Under a total system concept, now a practical reality, numerical control will be utilized from the very early stages of the design through drafting, manufacturing, and quality control. We are entering the electronic age—an age that may be identified in history as the "second industrial revolution." Graphic methods have now assumed new importance and the graphic language has become a means for communication with a computer. (*Courtesy Cincinnati Milacron, Inc.*)

numerically controlled machine tools 18

18.1
Introduction. In use at the present time are production machines that follow directions given on punched tapes and gauges that measure electrically to 0.025 or 0.050 micrometer (1 or 2 microinches). The use of tapes and computers by industry does not mean that the draftsmen and engineers will have less work to do and that there will be fewer of them. It does mean, however, that the men assigned to both areas must keep themselves informed and up-to-date on late developments. The draftsman should learn how to dimension the drawing of a part to meet the requirements for programming the machine or machines that will perform the shop operations (Fig. 18.8). He must have a more thorough understanding of basic fundamentals than in the past and at the same time be willing to accept change.

The fact must be recognized that automation is here and the only question the technically trained man of today can ask himself is whether or not he, as an individual, can adjust to new knowledge and different requirements.

18.2
Numerically Controlled Equipment. The term *numerical control* (NC), as applied to automated production machines, denotes a method of electronically controlling the operation and motions of a machine tool. This method may be considered relatively new since the first of

these machines was placed in operation only a little more than a decade ago. Early machine tools with built-in intelligence were numerically controlled milling and profiling machines that provided faster and less expensive means for producing aircraft parts. Now, numerical control has been applied to other machine tools, such as lathes, drilling and boring machines, welding and flame cutting machines, punching machines, and inspection devices. The day may be near when the designer, with sketches as a guide, will reduce his ideas to APT (Automatically Programmed Tools) language and then have a computer complete his design and prepare the instruction tape needed for machining (read Chapter 17).

There are two basic types of numerical control: (1) contouring or continuous path and (2) positioning or point-to-point. In the case of the contouring system, generally applied to milling machines, the path is continuously controlled. Normally, the path consists of very short straight-line segments, which approximate specified circular or curved lines. This system is used for machining items such as dies, gears, cams, and pieces having contoured surfaces. For most types of contour machining, the computer is used for instruction calculations.

The cutting tool is controlled mainly with respect to its final position for the point-to-point system as it is primarily applied to drilling and boring operations (Fig. 18.8). The path that a tool may take when directed from one point to another when drilling is relatively unimportant since the tool is not in contact with the workpiece during the movement. When the operations are not complex, the instruction tape may be prepared without the aid of a computer.

Numerically controlled equipment has either a two-axis or a "three-and-more-axis" system. As one might expect, the two-axis system is for two-dimensional machining, while the three-and-more-axis system makes three-dimensional machining possible.

A numerically controlled machine tool must have complete instructions to perform an operation at a specified location. This is accomplished by the machine control unit that contains the data processing equipment needed to translate programmed information into operating commands (electrical pulses) for the machine-tool system with its servomechanism drives and related feedback and measuring devices. Basically, the control unit consists of an input section, an interpolating or command-generating section, and a servoloop (drive) section. The input section, with tape reader and temporary storage capabilities, decodes and processes the programmed information for eventual transfer to the interpolating section (Fig. 18.1). The instructions, presented as a series of holes punched in tape (either paper or aluminum–Mylar), are read by a tape reader that may be either photoelectric or mechanical. Provided with both temporary and active storage areas, the complete control unit can receive new information from tape for temporary storage while using information in active storage. The information transferred into the interpolating section (active storage) is processed by logic circuitry that periodically sends signals (pulses) to the servoloop section. The digital or pulse information received by the servoloop section is, in turn, converted to an analog form and directed to the servomechanisms of the machine tool. Each command pulse (converted to an analog voltage) will create a definite movement of a slide, the actual movement itself being accomplished by some form of auxiliary equipment, such as a hydraulic motor, a hydraulic cylinder, or an electric motor (ac or dc) attached to the machine tool itself.

As stated before, the information on the control tape must provide complete instructions in a logical sequence. With a print or a sketch at hand, the parts programmer prepares a process (program) sheet (Fig. 18.4). On this sheet he lists the movements and operations necessary to machine the piece to the dimensions specified. This information, in order, is then punched on the control tape by a machine (tape punch) that resembles an electric typewriter. As the operator types the digits, symbols, or letters, the device translates the information into a binary-coded-decimal form and at the same time punches the tape (see Fig. 17.4). Manually programming a machine tool to machine a part is not always an easy task. It may be both frustrating and time consuming, for records show that it may require as many as 250–400 hr to prepare a program for a particularly complicated piece.

Numerical control may be the most important metalworking development of the twentieth century. It will grow rapidly in use, and as metalworking assumes a new aspect, new job patterns will result that

Fig. 18.1 Cintimatic Vertical Machining Center automatically drills, taps, mills, and bores in one setting. Work table positioning accuracy is ±0.025 mm (±.001 in.) in 610 mm (24 in.) at 10 m/min. (400 ipm.) and its repeatability is ±0.013 mm (±.0005 in.). (*Courtesy Cincinnati Milacron, Inc.*)

will require higher skills and considerably more technical training for the individual. It has been predicted that by 1980, NC machines may do up to 80% of the work now done by general-purpose machines (Fig. 18.2). Numerical control may well revolutionize our national economy and ultimately bring about great changes in our social structure. These changes must be faced knowing that there is no turning back from our plunge into the electronic age.

18.3
Use of Cartesian Coordinates (Fig. 18.3).
The standard Cartesian coordinate system is used to designate machine-tool axes and is the basis for all programming for numerical control. Most numerical-control systems use only the first quadrant for dimensioning purposes in accordance with the recommendation given in the EIA (Electronics Industry Association) standards. This practice of considering the workpiece to be positioned in a right-handed Cartesian system provides an accepted frame of reference that enables a programmer to relate a workpiece to almost any numerically controlled machine tool in general use. The geometry of the workpiece is described within the framework of X, Y, and Z axes by relating all

critical points of the part to the origin. The position of a point is established by giving its distances from the origin, measured parallel to the mutually perpendicular axes.

Although the angular orientation and the location of the origin is not the same for all machines (see Figs. 18.1 and 18.2), the X and Y axes usually lie in a horizontal plane. In the plane of the X and Y axes, measurements taken along the X axis to the right are plus (+), while those taken to the left in the opposite direction are considered to be minus (−). The Z axis, perpendicular to both the X and Y axes, is always parallel to the axis of the cutting tool. Finally, the longest motion of which a machine is capable is usually in the direction of the X axis. See Fig. 18.1.

In preparing programs one must always remember that the machine axis designation is always based on a coordinate system associated with that particular machine.

18.4
Programming Process and Manual Tape Preparation (Fig. 18.4). It is essential for those who are concerned with numerical control, as it applies to design, drafting and production, to have a thorough understanding of the programming of

Fig. 18.2 Tape (NC)-controlled turning machine that automatically performs all turning operations, including facing, grooving, contouring, threading, boring, drilling, and reaming. The Acramatic control for the lathe shown above features: (1) complete integrated circuit electronics; (2) four-digit, feed-rate coding in either millimeters or inches per minute; (3) automatic acceleration and deceleration without special coding; (4) incremental programming; (5) 300-row-per-second tape reader; (6) buffer storage; (7) manual data input; (8) linear interpolation; (9) circular interpolation; and (10) threading. The last two features are optional. (*Courtesy Cincinnati Milacron, Inc.*)

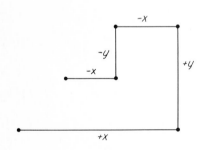

Fig. 18.3 Cartesian coordinate system.

numerically controlled equipment. The programming process is known as *part programming*. Part programming, however, may involve the use of million-dollar computers and computer programs in the process of producing needed numerical-control tapes.

The usual steps in the programming process are as follows:

1. Make an analysis of the part and consider its relation to the machine tool.

2. Consider the tooling required to produce the workpiece. That is, determine the type of holding device, the style and size of cutters, and so forth.

3. Determine the operations required and the sequence in which they must be performed.

Some parts may be programmed exceedingly fast because of their simplicity. This may be done with the aid of a desk calculator and trigonometry tables. In the case of the simplest NC machine tools, which have only the X and Y distances controlled by perforated tape, parts of the total operation may be performed by the machinist manually in accordance with instructions that are included in the program manuscript. After deciding on the machining sequence, the programmer must list on the program sheet (see Fig. 18.4) all tool movements, operations, tool selections, speeds, feeds, coolant control, and tool changes. He must remember as he prepares the program sheet that he alone controls every phase of the manufacturing process.

A portion of the programming for the workpiece shown in Fig. 18.4 is given for sequences numbered H046–N056 (see column 1). The sequence numbers show the order in which the machining operations are to be performed. Column 2 (from the left), preparatory function, is for programming the operating mode, such as drill, tap, mill, or bore. The function code G81 calls for the machine to drill.

The X and Y columns are for listing the hole locations. These are programmed in positive numbers in the first quadrant. The 0 position for the X and Y axes corresponds to the left-front corner of the table. It should be noted that, aside from

SEQ. NO.	PREP. FUNCT.	X POSITION	Y POSITION	MISC. FUNCT.	POS. NO.	FEED MM/REV.	SPEED RPM	DEPTH OF CUT	TOOL REMARKS
HO46	G81	X42000	Y15000	M57	30	.13	1060	35	17 MM DRILL
NO47			Y35500		31				
NO48		X28000			32				
NO49		X22500			33				
NO50		X9000			34				
NO51			Y15000		35				
NO52		X22500			36				
NO53		X28000			37				
NO54		X16000	Y25000	M06	38				
NO55	G84	X16000	Y25000	M58	20	1.19	490	20	M6 X 1 TAP
NO56		X28000	Y15000		21				

Fig. 18.4 Program sheet showing a portion of the programming for the workpiece shown (metric). *(Courtesy Cincinnati Milacron, Inc.)*

calling for a drilling cycle, the sequence number H046 also calls for the table to move to the hole position 420.00 mm along the X axis and 150.00 mm along the Y axis.

Entries in the column headed "Miscellaneous Function" are for such items as specifying what depth of collar to use. The code M57 calls for the use of depth stop No. 7.

The sequence numbers that follow through N054 call for the table to move to the hole positions called for in the drilling cycle. At sequence number N054, the code M06 (see column 5) stops the automatic drilling cycle. At this point the tool, feed, or speeds are changed as called for in the manuscript instructions. In looking ahead to sequence number N055, we see that preparatory function G84 has been specified. This command cycle code calls for a tapping sequence that instructs the machine to move rapidly to gage height, feed to depth, reverse spindle, feed out, and then move to upper limit. No other instructions are needed to perform this operation.

Sequence number N056 calls for a move to a new hole position in the same tapping sequence.

By following the solid black arrows in Fig. 18.5, it can be noted that the next step is punching the information given on the program form (process sheet) into the tape. Before this is done, however, it is wise to have the completed draft checked by a second programmer. If no errors are found, the manuscript is turned over to a typist, who enters the programmed numerical commands on a Friden Flexowriter or similar machine having a keyboard much like that of an electric typewriter. As the operator types, two outputs result. One is in the form of a master document that is called a print-out. This is an exact copy of the manuscript that was typed. The other output, made simultaneously, is a control tape with a series of perforations. See Fig. 18.9. A second tape may then be made using the data tape verification unit and the first tape. The automatic retyping of the process sheet produces a second tape. Mechanical proofreading is thus provided since the second tape will

MACHINE CONTROL UNIT
DATA TAPE
PREPARATION

PROCESS
SHEET

BLUEPRINT

MACHINE CONTROL UNIT
DATA TAPE VERIFICATION

MANUAL
TAPE
PREPARATION

PARTS PLANNER

MACHINE

PART
PROGRAM
MANUSCRIPT

COMPUTER
PROGRAM

PUNCH CARD
OR
TAPE DATA
PREPARATION

COMPUTER
PROCESSED
TAPE
PREPARATION

COMPUTER

NC PROGRAMMING
SYSTEM AND POST
POSTPROCESSOR
PROGRAM.

COMPLEX PROGRAMMING

Fig. 18.5 Simple and complex programming. (*Courtesy Bendix Corporation*)

not punchout if it differs in any way from the first tape.

The tape used is standard 1-in. (25.4-mm)-wide paper or Mylar-coated tape. Using a word address format, this eight-track tape is perforated in accordance with EIA Standard RS–273 (Fig. 18.6). Tab and delete codes are permitted. The code perforations are punched across the 1-in. (25.4-mm) width of the tape. The eight positions that are available for code holes are called *channels* (Fig. 18.6). Tape feed holes, located between the third and fourth channels, ensure the proper positioning of the tape in the tape punch as well as in the tape reader. The upper portion of the tape shown in Fig. 18.7 shows a block containing X and Y commands.

18.5

Computer-assisted Programming. Computer programs, requiring only minimal input using a problem-oriented language, offer the most reasonable approach to contour machining (Fig. 18.11). The work of the part programmer in this case is considerably simplified since the computer assumes the tedious task of making the necessary computations as well as the logical decisions. It is not even necessary for him to have intimate knowledge of the specific codes required for a particular control unit. With accurate postprocessing, all control units can be programmed by using the same problem-oriented language (APT–ADAPT–REMAPT) for the part program input to a computer. However, since both APT and ADAPT are basically computer programs, the part programmer and computer programmer will find it necessary to work closely together. This is true because the part programmer, using the standard Cartesian system, defines his part geometrically as being made up of planes, circles, conicals, and so forth.

18.6

Time-shared Computer Services. Even in very small industrial plants a parts programmer can now take advantage of a million-dollar computer to prepare control tapes for NC machine tools in a few minutes. This has been made possible through a time-sharing concept now in operation all over the country. A teletypewriter unit that can be installed on the programmer's desk gives him access to his own personal large computer. The computer is linked to the teletypewriter by public telephone lines. In some cities, it is possible for users to dial local telephone numbers to gain access to remote computers. This can be done without incurring long-distance charges. The programmer, at the keyboard of his teletypewriter, can use library programs stored in the computer to describe and analyze part geometry, calculate X and Y coordinates, generate data for machine-tool controllers, and receive as computer-output punched EIA coded tape at the terminal. It is possible at the present time to prepare NC tape for two- and three-axis machine tools. The programs available can also be used to prepare control tapes for contouring operations.

18.7

Preparation of NC Tapes. The part programs needed to produce the workpieces shown in Figs. 18.8 and 18.10 were pre-

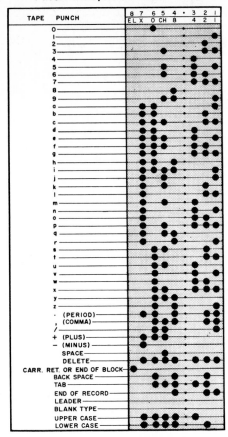

I INCH WIDE, EIGHT TRACK TAPE

Fig. 18.6 EIA standard coding.

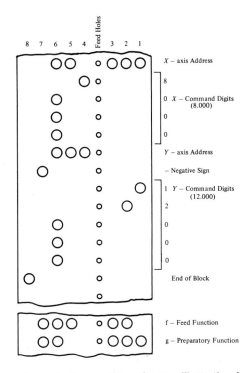

Fig. 18.7 Two portions of a tape illustrating format. Upper portion shows block containing X and Y commands.

UPPER PLATE
H.R.S 1 REQ'D

Engineering Drawing

Fig. 18.8 Positional dimensioning using X and Y coordinates. This plate was drilled and tapped on the NC machine shown in Fig. 18.1.

pared utilizing a computer and APT post-processing.

The first twenty lines in Fig. 18.9 are computer input statements for drilling and tapping the upper plate shown in Fig. 18.8. The remainder of the print-out (lines 01–N38) gives X and Y positions, tool sizes, specific instructions, and so forth. Omitted from the complete print-out are the POSTPROCESSOR SUMMARY (slide motion recap and machining recap), the CINDIV GENERATOR ASSEMBLY NUMBER LISTINGS, and the LIST OF REQUIRED GAUGES. Each statement of the print-out in Fig. 18.9 contains a unit of information that is complete in itself as an instruction related to a specific machining operation.

Contouring control, used for milling, turning, and similar operations, is much more complex than point-to-point control. This is so because the workpiece must be completely described and the path of the cutting tool must be rigidly controlled along a tight course.

The adjustment cone shown in Fig. 18.10 was produced from a solid 76-mm bar on the turning machine shown in Fig. 18.2. The input statements that ultimately led to the preparation of the needed control tape are given in Fig. 18.11. Not shown are the 300 or more lines of computer output that followed the no errors detected-processing continuing statement. The APT POSTPROCESSOR portion alone consisted of 72 lines.

```
CARDS PART=JR  000028         83
PARTNO FIG 19.8
PPRINT MACHN= 10VC-1000 PROG'D BY J.M. RITENOUR
PPRINT LOAD PART IN VISE - CENTER OF PART TO BE ON X 0 AND Y 0.
CALL/STP,X0=0,Y0=0,MACHN=38,CLS=2.54,XS=254,YS=254,MET=1,TLIST=1
CALL/MTLF
DATA=DXX(1)/-125,-95,-95,-125,-125,-95,-35,35,-50,-35,35,0,35,35,50,$
35,35,0,95,125,125,95,95,125,25,-25/
DATA=DYY(1)/38,38,0,0,-38,-38,-35,35,0,35,35,50,35,35,0,-35,35,-50,$
-38,-38,0,0,38,38,0,0/
CALL/RDPAT,NP=20,K=1,N=1              $$(20)12.7 HOLES
CALL/RDPAT,NP=2,K=2,N=21             $$(2) M10 X 1.5 THRDS
PPRINT ALL TOOLS ARE METRIC.
DATA=D(1)/12.5,8.8,10/
DATA=TAN(1)/38020125,38020088,38080010/
CALL/TLLD2,TN=1,TP=2,DTH=12.7,FP=1
CALL/TLLD2,TP=3,DTH=12.7,FP=2
CALL/TLLD8,TP=3,P=1.5,DTH=12.7,FP=2,ORD=1
CALL/FIN
CALL/TOOLS,GLIST=1,ETNO=3
```

```
CCCC  MMMM      MMMM          C I N C I N N A T I   A P T   P O S T P R O C E S S O R
CCCCC MMMMM    MMMMM
CCC       MMM MMM                              C I N A C 6   52
CCC       MMM MMM     C I N C I N N A T I
CCCCC MMMMM    MMMMM           CINCINNATI ACRAMATIC CNC MACHINING CENTER
CCCC  MMMM      MMMM  M I L A C R O N
```

	ABSX	ABSY	ABSZ	MMPM	RPM	TIME	CLRC

```
     DATE 08/14/79    TIME 1255    SNUMB 7370T
FIG 19.8
MACHN= 10VC-1000 PROG'D BY J.M. RITENOUR
LOAD PART IN VISE - CENTER OF PART TO BE ON X 0 AND Y 0.
ALL TOOLS ARE METRIC.
* * * * * * * * * * * * * * * * * * * * * * * * * *
TOOL   +01.=    TWIST DRILL, DIAM=        +012.5000
* * * * * * * * * * * * * * * * * * * * * * * * * *

   LEADER/   1.00
```

	ABSX	ABSY	ABSZ	MMPM	RPM	TIME	CLRC
!$ O1 G00 F13 T1 M06$						.359	287
O2 G81 X-125000 Y38000 Z-15388 R0 F208 S698 T2 M13$	-125.000	38.000	-15.388	207	698	.093	358
N3 X-95000$	-95.000	38.000	-15.388	207	698	.102	359
N4 Y0$	-95.000	.000	-15.388	207	698	.103	360
N5 X-125000$	-125.000	.000	-15.388	207	698	.102	361
N6 Y-38000$	-125.000	-38.000	-15.388	207	698	.103	362
N7 X-95000$	-95.000	-38.000	-15.388	207	698	.102	363
N8 X-35350 Y-35350$	-35.350	-35.350	-15.388	207	698	.107	364
N9 X-50000 Y0$	-50.000	.000	-15.388	207	698	.103	365
N10 X-35350 Y35350$	-35.350	35.350	-15.388	207	698	.103	366
N11 X0 Y50000$	.000	50.000	-15.388	207	698	.103	367
N12 X35350 Y35350$	35.350	35.350	-15.388	207	698	.103	368
N13 X50000 Y0$	50.000	.000	-15.388	207	698	.103	369
N14 X35350 Y-35350$	35.350	-35.350	-15.388	207	698	.103	370
N15 X0 Y-50000$	.000	-50.000	-15.388	207	698	.103	371
N16 X95000 Y-38000$	95.000	-38.000	-15.388	207	698	.112	372
N17 X125000$	125.000	-38.000	-15.388	207	698	.102	373
N18 Y0$	125.000	.000	-15.388	207	698	.103	374
N19 X95000$	95.000	.000	-15.388	207	698	.102	375
N20 Y38000$	95.000	38.000	-15.388	207	698	.103	376
N21 X125000$	125.000	38.000	-15.388	207	698	.102	377
N22 G00 Z3000$	125.000	38.000	3.000	207	698	.000	446
O23 G00 X125000 Y38000 Z2540 F208 S698 M13$	125.000	38.000	2.540	207	698	.000	450

```
* * * * * * * * * * * * * * * * * * * * * * * * * * *
TOOL   +02.=    TWIST DRILL, DIAM=        +008.8000
* * * * * * * * * * * * * * * * * * * * * * * * * * *
```

	ABSX	ABSY	ABSZ	MMPM	RPM	TIME	CLRC
O24 G00 F208 T2 M06$						.239	465
O25 G81 X25000 Y0 Z-14622 R0 F201 S992 T3 M13$	25.000	.000	-14.622	200	992	.113	528
N26 X-25000$	-25.000	.000	-14.622	200	992	.104	529
N27 G80 R2540$	-25.000	.000	-14.622	200	992	.001	602
N28 G00 Z5540$	-25.000	.000	5.540	200	992	.000	604
O29 G00 X254000 Y254000 Z5080 F201 S992 M13$	254.000	254.000	5.080	200	992	.069	608
N30 G01 M00$							611

```
     TAPE LENGTH =   1.20 METERS         MACHINING TIME =   2.952

OPERATOR - PLEASE CLEAN AND PREPARE ALL HOLES TO BE TAPPED
```

	ABSX	ABSY	ABSZ	MMPM	RPM	TIME	CLRC
O31 G00 X254000 Y254000 Z2540 F201 S992 M13$	254.000	254.000	2.540	200	992	.001	624

```
* * * * * * * * * * * * * * * * * * * * * * * * * * *
TOOL   +03.=    STD TAP, SIZE=    +010.0000-       +001.50
* * * * * * * * * * * * * * * * * * * * * * * * * * *
```

	ABSX	ABSY	ABSZ	MMPM	RPM	TIME	CLRC
O32 G00 F201 T3 M06$						.239	643
O33 G84 X-25000 Y0 Z-20200 R0 F380 S253 M13$	-25.000	.000	-20.200	379	253	.193	693
N34 X25000$	25.000	.000	-20.200	379	253	.133	694
N35 G00 Z3000$	25.000	.000	3.000	379	253	.000	716
O36 G00 X25000 Y0 Z2540 F380 S253 M13$	25.000	.000	2.540	379	253	.000	718
N37 X254000 Y254000 Z5080$	254.000	254.000	5.080	379	253	.067	722
N38 G01 M02$							767

```
   LEADER/   1.00

     TAPE LENGTH =   1.65 METERS         MACHINING TIME =   3.587
```

Fig. 18.9 Computer print-out of input data used to prepare the tape needed to drill and tap the upper plate shown in Fig. 18.8. (*Courtesy Cincinnati Milacron, Inc.*)

3×45°CHAMFER

ADJUSTMENT CONE
SAE 1035   2 REQ'D.

Engineering Drawing

Finished Part

Fig. 18.10 Engineering drawing of adjustment cone.

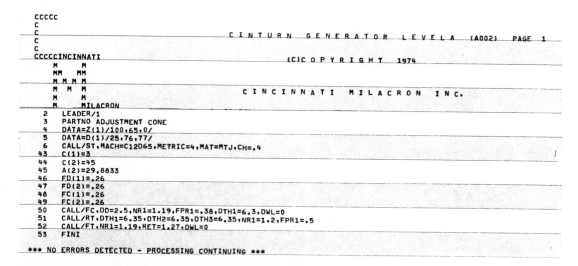

```
CCCCC
C
C                                          CINTURN GENERATOR LEVELA  (A002)  PAGE 1
C
C
CCCCCCINCINNATI                              (C)C O P Y R I G H T  1974
   M      M
   MM    MM
   M M  M M
   M  M  M                     C I N C I N N A T I   M I L A C R O N   I N C.
   M     M
   M     MILACRON
    2   LEADER/1
    3   PARTNO ADJUSTMENT CONE
    4   DATA=Z(1)/100,65,0/
    5   DATA=D(1)/25,76,77/
    6   CALL/ST,MACH=C12D65,METRIC=4,MAT=MTJ,CH=.4
   43   C(1)=3
   44   C(2)=45
   45   A(2)=29.8833
   46   FD(1)=.26
   47   FD(2)=.26
   48   FC(1)=.26
   49   FC(2)=.26
   50   CALL/FC,DD=2.5,NR1=1.19,FPR1=.38,DTH1=6.3,DWL=0
   51   CALL/RT,DTH1=6.35,DTH2=6.35,DTH3=6.35,NR1=1.2,FPR1=.5
   52   CALL/FT,NR1=1.19,RET=1.27,DWL=0
   53   FINI

*** NO ERRORS DETECTED - PROCESSING CONTINUING ***
```

**Fig. 18.11 Print-out of statements used to prepare the control
tape for production of the adjustment cone.** Contour control
has been utilized here. (*Courtesy Cincinnati Milacron, Inc.*)

PART 6

GRAPHIC
METHODS
FOR
ENGINEERING
COMMUNICATION,
DESIGN,
AND
COMPUTATION

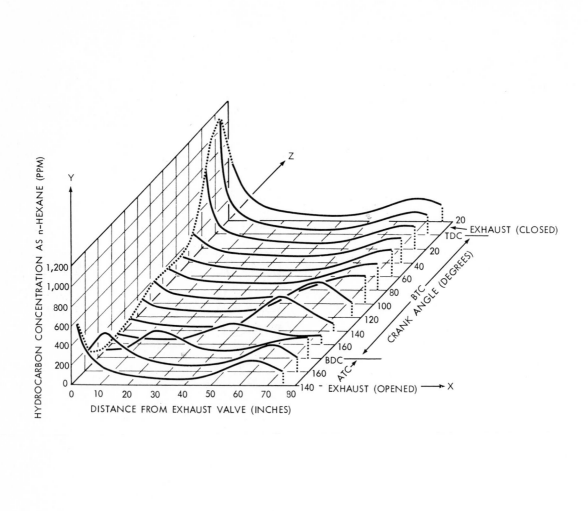

Three-dimensional graph. The curves show the hydrocarbon concentration in the exhaust pipe as functions of crank angle and distance from the exhaust valve. Distance from the exhaust valve is plotted along the X-axis; engine crank angle, beginning with the time the exhaust valve opens, along the Z-axis; and hydrocarbon concentration is plotted along the Y-axis. Three-dimensional charts are used to show the relationship among three variables. Their construction is based upon pictorial projection. (*Courtesy General Motors Corporation; reprinted from* General Motors Engineering Journal *with permission*)

graphic methods for engineering communication and computation 19

A □ Graphs and charts

19.1

Introduction. A properly designed graphical representation will convey correlated data and facts to an average individual more rapidly and effectively than a verbal, written, or tabulated description, because a visual impression is easily comprehended and requires less mental effort than would be necessary to ascertain the facts from complex tables and reports (Figs. 19.1 and 19.22). It is because of this that diverse kinds of graphs and charts have been developed to present scientific, statistical, and technical information. Note how quickly the relationship presented by the line graph in Fig. 19.2 can be interpreted.

Engineers, even though they are concerned mainly with technical graphs, should be familiar also with the popular forms, for every industrial concern frequently must prepare popular types of graphs to strengthen their relationship with the public.

It is impossible to treat exhaustively the subject of graphical representation in a single chapter. Only a few of the most common forms used to analyze economic, scientific, and technical data can be discussed in detail. Many of the principles followed in the construction of engineering graphs, however, apply to the other types.

Those who are interested in further study of graphical methods for problem solving in the area of graphical mathematics (graphical algebra and graphical calculus) will find full coverage in Chapter 20.

Fig. 19.1 Engineering graph. (*Courtesy General Motors Corporation*) Reprinted from the *General Motors Engineering Journal* with permission.

As much drafting skill is required in the execution of a graph as in making any other type of technical drawing. Good appearance is important and can be achieved only with the help of good lettering and smooth, uniform, and properly contrasted lines.

19.2
Classification of Charts, Graphs, and Diagrams. Graphs, charts, and diagrams

DESIGN STRESS VS STRESS RELIEF ANNEAL
CDA-510 PHOS BRONZE-SPRING TEMPER

Fig. 19.2 Engineering graph prepared for publication. (*Courtesy Bell Laboratories*)

may be divided into two classes in accordance with their use and then further subdivided according to type. When classified according to use, the two divisions are, first, those used for strictly scientific and technical purposes and, second, those used for the purpose of popular appeal. The classification according to type is as follows:

1. Rectilinear charts
2. Semilogarithmic charts
3. Logarithmic charts
4. Barographs and area and volume charts
5. Percentage charts
6. Polar charts
7. Trilinear charts
8. Alignment charts (nomographs)
9. Pictorial charts

19.3
Quantitative and Qualitative Charts and Graphs. In general, charts and diagrams are used for one of two purposes, either to

read values or to present a comparative picture relationship between variables. If a chart or graph is prepared for reading values, it is called a *quantitative* graph; if prepared for presenting a comparative relationship, it is called *qualitative*. Obviously, some charts serve both purposes and cannot be classified strictly as either type. One of these purposes, however, must be predominant. Since a number of features in the preparation depend on the predominant purpose, such purpose must be determined before attempting to construct a graph.

19.4
Ordinary Rectangular Coordinate Graphs. Most engineering graphs prepared for laboratory and office use are drawn on ruled, rectangular graph paper and are plotted in the first quadrant (upper-right-hand), with the intersection of the X (horizontal)-axis and Y (vertical)-axis at the lower left used as the zero point or origin of coordinates. The paper is ruled with equispaced horizontal and vertical lines, forming small rectangles. The type most commonly used for chart work in experimental engineering is $8\frac{1}{2} \times 11$ in. and is ruled to form $\frac{1}{20}$-in. squares [Fig. 19.3(a)], every fifth line being heavy. Another type of paper frequently used, which is suitable for most laboratory reports in technical schools, has rulings that form 1-mm and 1-cm squares [Fig. 19.3(b)]. Other rulings run $\frac{1}{10}$, $\frac{1}{8}$, or $\frac{1}{4}$ in. apart. Ordinarily the ruled lines are spaced well apart on charts prepared for reproduction in popular and technical literature (Figs. 19.6 and 19.8). The principal advantage of having greater spacing between the lines is that large squares or rectangles tend to make the graph easier to read. Ready printed graph papers are available with various rulings in several colors.

Ordinary coordinate line graphs are used extensively because they are easily constructed and easily read. The known relationship between the variables is expressed by one or more continuous lines, which may be straight, broken, or curved.

The graph in Fig. 19.4 shows the approximate barometric pressure at different heights above sea level.

A graphical representation may be drawn easily and correctly if, after the required data have been assembled, careful consideration is given to the principles of curve drawing discussed in the following sections.

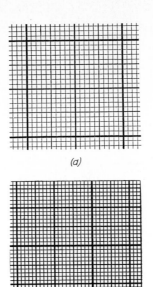

Fig. 19.3 Types of graph paper.

Fig. 19.4 Rectangular graph.

19.5
Determination of the Variables for Ordinate and Abscissa. The independent variable, the quantity arbitrarily varied during the experiment, usually is chosen for the abscissa (Fig. 19.5). Certain kinds of experimental data, however, such as a stress–strain diagram (Fig. 19.6), are plotted with the independent variable along the ordinate.

19.6
Selection of Suitable Scales. Since the slope of the curve, in its entirety as well as at intermediate points, provides a visual

Fig. 19.5 Independent and dependent variables.

Fig. 19.6 Stress–strain diagram.

impression of the degree of change in the dependent variable for a given increment of the independent variable, care must be exercised in selecting the scales. The choice of scales is the controlling factor in creating the correct impression of the relationship to be shown by the graph.

The range of the scales should be such as to ensure full and effective use of the coordinate area in attaining the objective of the graph. Should a visual comparison of plotted magnitudes be desired, the zero line should be included. That is, should the chart or graph be qualitative, both the ordinate and abscissa generally should have zero value at the intersection of the axes, as in Figs. 19.2 and 19.6. However, if the chart is quantitative, the intersection of the axes need not be at the origin.

In the case of arithmetic scales, the scale numbers on the graph and the space between coordinate rulings should gener-

ally correspond to 1, 2, or 5 units of measurement that may be multiplied (or divided) by 1, 10, 100, 1000, and so forth. Other units could be used except for the fact that they create situations wherein it becomes difficult to interpolate values. For example, one square should equal one of the following:

0.01	0.1	1	10	100	etc.
0.02	0.2	2	20	200	etc.
0.04	0.4	4	40	400	etc.
0.05	0.5	5	50	500	etc.
etc.	etc.	etc.	etc.	etc.	etc.

Along the horizontal (independent variable) scale, the values increase from the left to the right, while those along the vertical (dependent variable) scale should increase from bottom to top.

19.7

Locating the Axes and Marking the Values of the Variables. On graphs prepared for laboratory reports and not for publication, the axes should be located 30 mm (1 in.) or more inside the border of the coordinate ruling (Fig. 19.7). When selecting the scale units and locating the axes, it should be remembered that the abscissa may be taken either the long way or short way of the coordinate paper, depending on the range of the scales.

The use of many digits in the scale numbers is to be avoided. This can be easily accomplished by using a suitable designation in the scale caption such as: —IN THOUSANDS.

Examples

FORCE-kN
VELOCITY-km/h
PRESSURE, MM OF HG $\times 10^{-5}$;
RESISTANCE, THOUSANDS
OF OHMS.

The numbers should read from the bottom when possible (Fig. 19.7). For the sake of good appearance, they never should be crowded. Always place a cipher to the left of the decimal point when the quantity is less than unity.

Usually, only the heavy coordinate lines are marked to indicate their values or distance from the origin, and, even then, the values may be shown only at a regular selected interval (Fig. 19.7). The numbers should be placed to the left of the Y-axis and just below the X-axis.

When several curves representing different variables are to appear on the same

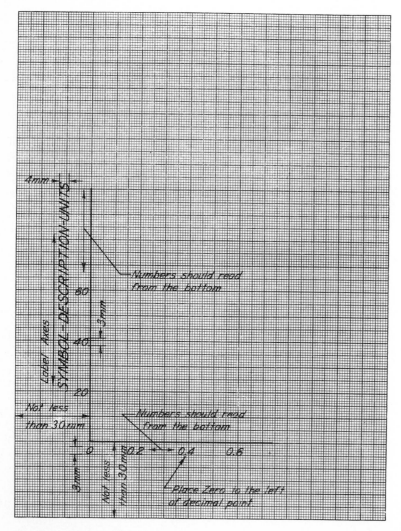

Fig. 19.7 Graph construction.

graph, a separate axis generally is required for each variable (Fig. 19.8). In this case, a corresponding description should be given along each axis. The axes should be grouped at the left or at the bottom of the graph, unless it is desirable to place some at the right or along the top.

19.8
Indicating Plotted Points Representing the Data. If the data represent a set of experimental observations, the plotted points of a single-curve graph should be marked by small circles approximately 3 mm (.1 in.) in diameter (Fig. 19.9). The following practice is recommended: open circles, filled-in circles, and partially filled-in circles (○ ● ◒) rather than crosses, squares, and triangles should be used to differentiate observed points of several curves on a graph. Filled-in symbols may be made smaller than those not filled in.

Mathematical curves are frequently drawn without distinguishing marks at computed positions.

19.9
Drawing a Curve. Since most physical phenomena are continuous, curves on engineering graphs usually represent an average of plotted points (Fig. 19.10). Discontinuous data should be plotted with a broken line, as shown in Fig. 19.11.

It is preferable to represent curves by solid lines. If more than one curve appears on a graph, differentiation may be secured by varied types of lines; but the most important curve should be represented by a solid one. A very fine line should be used for a quantitative curve if values are to be read accurately. A heavy line 0.7 mm (.03 in.) in width is recommended for a qualitative curve. It should be observed in Figs. 19.10 and 19.11 that the curve line does not pass through open circles.

When several curves are to appear on a graph, it is recommended that a solid line be used for the most important curve. Differentiation between the curves may be secured by using different types of lines for the other curves (dashed, etc.). In addition, the curves should bear suitable designations and, if practicable, be identified by brief labels placed horizontally or along the curves rather than by letters and numbers that require the reader to refer to a key (Fig. 19.8).

19.10
Labeling the Scales. Each scale caption

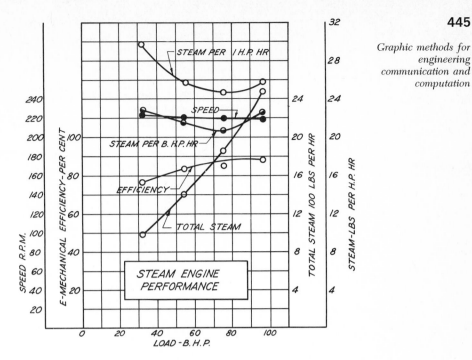

Fig. 19.8 Representation of several curves on a graph.

should give a description of the variable represented and the unit of measurement. The captions on engineering graphs frequently contain an added identifying symbol, such as

N-EFFICIENCY-PERCENT
P-OUTPUT-HP

All lettering should be readable from the bottom and right side of the graph (not the left side). When space is limited, standard abbreviations should be used, particularly for designating the unit of

Fig. 19.9 Identification symbols.

Fig. 19.10 Continuous curve.

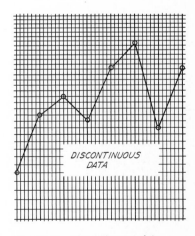

Fig. 19.11 Discontinuous data.

STRESS - STRAIN DIAGRAM
FOR
COMPRESSION
IN
CAST IRON

Fig. 19.12   Title.

Fig. 19.13   Logarithmic graph.

Fig. 19.14   Coordinate graph.

measurement. To avoid confusing the reader, the draftsman should use only recognized word contractions.

19.11

Titles, Legends, Notes, and So On. The title of a graph should be clear, concise, complete, and symmetrical. It should give the name of the curve, the source of the data, the date, and other important information (Fig. 19.12). It should be so placed that it gives a balanced effect to the completed drawing (Fig. 19.8). In addition to the title, a wiring diagram, pictorial diagram, formula, or explanatory note is often necessary to give a clear picture of the nature of the experiment. For example, if there is any great irregularity in the plotted points or a condition that may have affected the values as shown by the data, a note of explanation should be given. A legend or key is sometimes included to explain a set of curves in greater detail.

In commercial practice, alcohol is often used to clear a rectangular area of coordinate lines in order that the title may be printed in an open space.

19.12

Procedure for Making a Graphical Representation in Ink.

1. Select the type of coordinate paper.

2. Determine the variables for ordinate and abscissa.

3. Determine the scale units.

4. Locate the axes and mark the scale values in pencil.

5. Plot the points representing the data. [Many draftsmen ink the symbol (○ ◑), indicating the points at this stage.]

6. Draw the curve. If the curve is to strike, an average among the plotted points, a trial curve should be drawn in pencil. If the curve consists of a broken line, as is the case with discontinuous data, the curve need not be drawn until the graph is traced in ink.

7. Label the axes directly in ink.

8. Letter the title, notes, and so on. The title should be lettered on a trial sheet that can be used as a guide for lettering directly in ink on the graph.

9. Check the work and complete the diagram by tracing the curve in ink.

19.13

Logarithmic Graphs. Logarithmic coordinate graphs are constructed on prepared paper on which the parallel horizontal and parallel vertical rulings are spaced proportional to the logarithms of numbers (Fig. 19.13). This type of graph has two principal advantages over the ordinary coordinate type. First, the error in plotting or reading values is a constant percentage, and, second, an algebraic equation of the form $y = ax^b$ appears as a straight line if x has a value other than zero. The exponent b may be either plus or minus.

The equation for a falling body, $D = \frac{1}{2}gt^2$, is represented in Figs. 19.13 and 19.14. A practical application of interest to engineers is in the design of drop hammers. In this equation, based on uniform accelerated motion, t represents time in seconds and D the distance traveled in t seconds by a freely falling body with no initial velocity. Observe that the plotted points form a parabolic curve on ordinary coordinate graph paper and a straight line on logarithmic paper. To draw the line on the graph in Fig. 19.13, it is necessary to calculate and locate only two points, while in Fig. 19.14, several points must be plotted to establish the location of the corresponding curved-line representation. The line on Fig. 19.13 has a slope of 2:1, because the exponent of t is 2. Therefore, the line could be drawn by utilizing one point and the slope, instead of plotting two points and joining them with a straight line.

Log paper is available with rulings in one or more cycles for any range of values to be plotted. Part-cycle and split-cycle papers may also be purchased.

19.14

Semilogarithmic Graphs. Semilogarithmic paper has ruled lines that are spaced to a uniform scale in one direction and to a logarithmic scale in the other direction (Fig. 19.15). Charts drawn on this form of paper are used extensively in scientific studies, because functions having values in the form of geometric progressions are represented by straight lines. In any case, the main reason for the use of semilogarithmic paper is that the slope of the resulting curve indicates rate of change rather than amount of change, the opposite being true in the case of curves on ordinary coordinate graph paper. Persons who are interested may determine the rate of increase or decrease at any point by measuring the slope. A straight line indicates a constant rate of change. In commercial work this form of paper is generally called *ratio paper*, and the charts are known as *rate-of-change charts*.

As previously stated, the choice of a type of graph paper depends on the information to be revealed. Curves drawn on uniform coordinate graph paper to illustrate the percentage of expansion or contraction of sales, and so on, present a misleading picture. The same data plotted on semilogarithmic paper would reveal the true rate of change to the business management. For this reason, semilogarithmic paper should be used whenever percentage of change rather than quantity of change is to be shown. In scientific work, when the value of one variable increases in a geometric progression and the other in an arithmetic progression, this form is valuable.

19.15

Bar Charts. Bar charts or barographs are used principally in popular literature covering economic and industrial surveys. They are a simple diagrammatic form giving a pictorial summary of statistical data and can be easily understood by the average person. Logarithmic and uniform coordinate graphs are less suited for this purpose, because few people know the procedure for reading curves or understand their picture qualities.

Whenever values or quantities are illustrated, as in Fig. 19.16, by consecutive heavy bars whose lengths are proportional to the amounts they represent, the resulting representation is called a *bar chart*.

The bars on this type of diagram may be drawn either horizontally or vertically, but all should start at the same zero line.

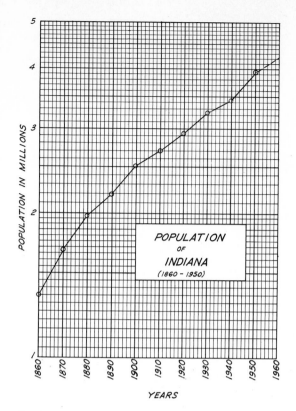

Fig. 19.15 Semilogarithmic chart.

Their lengths should be to some fixed scale, the division values of which may be given in the margin along the bottom or left side of the graph. When it is necessary to give the exact values represented, the figures should be placed along each bar in a direction parallel to it. To place the values at the end gives the illusion of increasing the length of the bars. Usually, the names of the items are lettered to the left of the vertical starting line on a horizontal chart and below the starting line on a vertical chart.

19.16

Area (Percentage) Charts. An area diagram can be used profitably when it is desirable to present pictorially a comparison of related quantities in percentage. This form of representation illustrates the relative magnitudes of the component divisions of a total of the distribution of income, the composition of the population, and so on. Two common types of the various forms of area diagrams used in informative literature are illustrated in Figs. 19.17 and 19.18. Percentages, when represented by sectors of a circle or subdivisions of a bar, are easy to interpolate.

The pie chart (Fig. 19.17) is the most popular form of area diagram, as well as

Fig. 19.16 Bar chart.

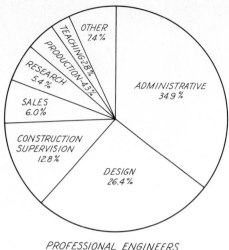

PROFESSIONAL ENGINEERS
THE KIND OF WORK THEY DO

Fig. 19.17 Pie chart.

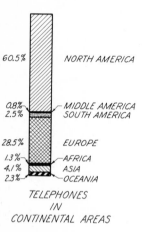

TELEPHONES
IN
CONTINENTAL AREAS

Fig. 19.18 Percentage bar chart.

the easiest to construct. The area of the circle represents 100% and the sectors represent percentages of the total. To make the chart effective, a description of each quantity and its corresponding percentage should be lettered in its individual sector. All lettering should be completed before the areas are crosshatched or colored if this is to be done. The percentage bar chart shown in Fig. 19.18 fulfills the same purpose as the pie chart. The overall area of the bar represents 100%. Note that each percentage division is crosshatched in a different direction. The descriptions may be placed on either side of

the bar; the percentages should be on the bar or at the side.

19.17
Polar Charts. Certain types of technical data can be more easily plotted and better represented on polar coordinate paper. Polar charts are drawn by self-recording instruments, polar diagrams, and plotted polar curves representing various kinds of scientific data are very common. Polar curves are used to represent the intensity of diffused light, intensity of heat, and so on. The polar chart in Fig. 19.19 gives, in terms of candlepower, the intensity of light in two planes.

19.18
Trilinear Charts. Trilinear charts are used principally in the study of the properties of chemical compounds, mixtures, solutions, and alloys (Fig. 19.20). Basically this is a 100% chart the use of which, owing to its geometric form, is limited to the investigation of that which is composed of three constituents or variables. Its use depends on the geometric principle that the sum of the three perpendiculars from any point is equal to the altitude. If the altitude represents 100%, the perpendiculars will represent the percentages of the three variables composing the whole.

The ruling can be accomplished conveniently by dividing any two sides of the triangle into the number of equal-percentage divisions desired and drawing through these points lines parallel to the sides of the triangle.

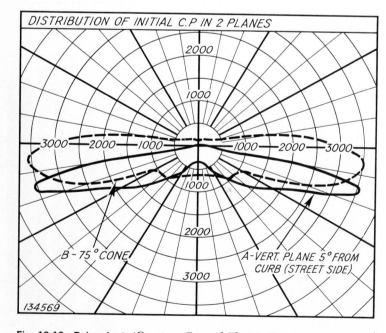

Fig. 19.19 Polar chart. (*Courtesy General Electric Company*)

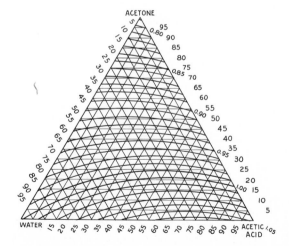

Fig. 19.20 Trilinear chart. (*Courtesy American Chemical Society*)

19.19
Chemical Engineering Charts. Figure 19.21 shows a type of flow chart that must be prepared frequently by chemical engineers in industrial practice.

19.20
Pictorial Charts. Pictorial charts are quite generally used to present data in reports prepared for nontechnical readers. Usually, such charts present comparisons of populations (Fig. 19.22), expenditures, costs, and so forth. Stacks of silver dollars may represent expenditures; sizes of animals can represent livestock production; and human figures can present employment data.

B □ Empirical equations

19.21
Empirical Equations. In all phases of engineering work considerable experimentation is done with physical quantities, and the engineer (or engineering technologist) in his work is usually the person most concerned with the behavior of quantities in relation to one another. Often it is known that the subject of the experiment obeys some physical law which can be expressed by a mathematical equation, but the exact equation is unknown. Then the person performing the experiment is faced with the task of finding an equation to fit the data that has been obtained. The three articles that follow discuss means of arriving at an equation from a graphical study of the data. An equation determined in this manner is an *empirical equation*. Since the unknown law may be quite complex, a single empirical equation to fit the whole range of data may not exist. However, in the majority of such cases a series of the various empirical equations with limited coverage (parameters) can be found.

19.22
Equations of the Form $y = a + bx$. If the plotted points, representing the data, lie in what appears to be a straight line when plotted on rectangular coordinate paper, the equation of the data is a linear or first-degree equation of the form $y = a + bx$, where b is the slope of the line and a is the y-intercept ($x = 0$).

After it has been decided that the relationship between quantities is linear, the next step is to draw the best average

Fig. 19.21 Flow chart of ammonia-soda operations. (*Courtesy* Chemical Industries Magazine)

straight line that will be representative of the data (see Fig. 19.23). This line, extended if necessary to the y-axis, establishes the y-intercept ($x = 0$) which will be the value of a in the equation. The slope can be determined from any two points along the line. However, for accuracy the points should be selected as far apart as possible. The value of b is expressed as

$$b = \frac{y_2 - y_1}{x_2 - x_1}$$

Thus, for the line in Fig. 19.23, having selected two points as illustrated,

$$b = -\frac{3 - 1}{6 - 3} = -\frac{2}{3} \quad \text{(slope)}$$

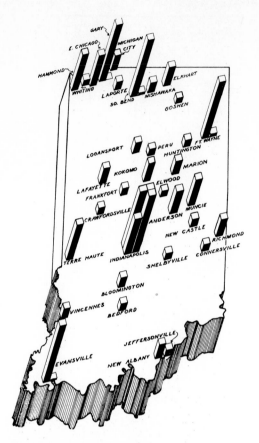

Fig. 19.22 **Population chart.** (*Courtesy Indiana State Planning Board*)

With the slope now known and the value of *a* having been read directly on the *y*-axis as 5, the equation of the line can be written as

$$y = 5 - \frac{2}{3} x$$

If it is not reasonable to include $x = 0$ in the plot of the data, then a pair of simultaneous equations are set up using two points on the line:

$$y_1 = a + bx_1$$
$$y_2 = a + bx_2$$

This system can then be solved for *a* and *b*.

19.23

Equations of the Form $y = ax^b$. If the data plots very nearly in a straight line on logarithmic graph paper, the equation of the data is a power equation of the form: $y = ax^b$. Power equations of this type, where one quantity varies directly as some power of another, appear as either parabolic or hyperbolic curves when plotted on rectangular coordinate paper, depending upon whether the exponent is positive or negative. For positive values of *b* (except for unity) the curves are parabolic; negative values produce hyperbolic curves.

When the equation is placed in logarithmic form and rewritten as log y = log $a + b$ log x, it is now in the same form as the equation of a straight line. In this case, when $x = 1$, then $y = a$ because log $1 = 0$.

Hence, after the data have been plotted (y vs. x) on logarithmic paper as in Fig. 19.24, a representative (average) straight line should be drawn. The line extended to the y-axis establishes *a* at the y intercept ($x = 1$). The value of *a* can be read directly, since the spacing is logarithmic. As before, two points along the line are used

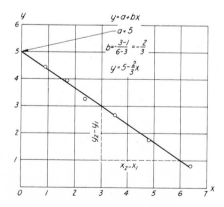

Fig. 19.23 Determination of empirical equations of the form $y = a + bx$ (first degree).

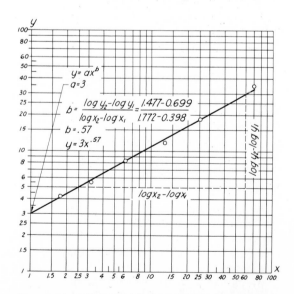

Fig. 19.24 Determination of empirical equations of the form $y = ax^b$ (power).

in the equation that follows to determine b:

$$b = \frac{\log y_2 - \log y_1}{\log x_2 - \log x_1}$$

For the two points selected along the line in Fig. 19.24,

$$b = \frac{1.477 - 0.699}{1.772 - 0.398} = .57$$

Thus, the equation is $y = 3x^{.57}$.

If $x = 1$ is not included in the plot of the data, then one may resort to a pair of simultaneous equations,

$$\log y_1 = \log a + b \log x_1$$
$$\log y_2 = \log a + b \log x_2$$

In solving for a (also b) the solution gives the value for log a. The real value of a (and b) must be found in log tables.

19.24

Equations of the Form $y = ab^x$. If the data plots as a nearly straight line on semilogarithmic graph paper (Fig. 19.25), the equation of the data is an exponential equation of the form $y = ab^x$. This equation may be rewritten as log y = log a + x log b, which once again is in the same form as the equation of a straight line. In this case, when $x = 0$, then $y = a$.

With the data for the exponential equation plotted directly on semilogarithmic graph paper, one must again draw the most representative line along the path of the plotted points. This line extended to the y-axis ($x = 0$) determines the value of a. Again, two points must be selected along the line and their coordinates substituted in the equation

$$\log b = \frac{\log y_2 - \log y_1}{x_2 - x_1}$$

Then, with the logarithm of the value of b known, the real value of b as needed for the equation can be determined from log tables.

For the two points selected for the line in Fig. 19.25,

$$\log b = \frac{1.778 - 1.176}{5 - 2}$$
$$= \frac{0.602}{3} = .201$$

Hence,

$$b = 1.59$$

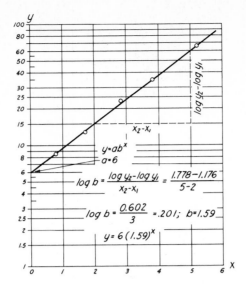

Fig. 19.25 Equations of the form $y = ab^x$.

With the value of a read on the y-axis as 6, the formula for the data may be written as $y = 6 (1.59)^x$.

If $x = 0$ is not included in the plot of the data, then one must solve the simultaneous equations

$$\log y_1 = \log a + x_1 \log b$$
$$\log y_2 = \log a + x_2 \log b$$

Then, since the solution gives both a and b in terms of logarithms, log tables must be used to determine the real values of a and b.

When natural logarithms are available, the equation $y = ab^x$ may be changed to $y = Ae^{mx}$. This latter equation, in turn, may be written as ln y = ln A + mx, which is still the equation of a straight line. The value of A can be determined as before at $x = 0$, and

$$m = \frac{\ln y_2 - \ln y_1}{x_2 - x_1}$$

where the value of m is found directly.

When $x = 0$ is not included, the equations to be used are

$$\ln y_1 = \ln A + mx_1$$
$$\ln y_2 = \ln A + mx_2$$

C □ Alignment charts

19.25

Alignment Charts (Nomographs). The purpose of alignment charts is to eliminate many of the laborious calculations necessary to solve formulas containing three or more variables. Such a chart is often complicated and difficult to construct, but if it can be used repeatedly, the labor involved in making it will be justified. In the commercial field, these charts appear in varied forms, which may be very simple or very complicated (Fig. 19.26).

Briefly stated, the simplest form of alignment chart consists of a set of three or more inclined or vertical scales so spaced and graduated as to represent graphically the variables in a formula. The scales may be divided into logarithmic units or some other types of functions, depending upon the form of equation. As illustrated in Fig. 19.27, the unknown value may be found by aligning a straightedge to the points representing known values on two of the scales. With a scale or triangle so placed, the numerical value representing the solution of the equation can be read on the third scale at the point of intersection.

Since alignment charts in varied forms are frequently prepared by engineers and technologists, it is desirable that students studying in the fields dealing with the sciences have some knowledge of the fundamental principles underlying their construction. However, in any brief treatment, directed toward a beginner, it is impossible to explain fully the mathematics involved in the construction of the many and varied types (Fig. 19.26). Therefore, our attention here must be directed toward an understanding of a few of the less complicated straight-line forms with the hope that the student will gather sufficient knowledge to construct simple charts for familiar equations (Fig. 19.34).

19.26

Forms of Alignment Charts. Examples of some of the forms that alignment charts may have are shown in outline in Fig. 19.26. Examples of forms of proportional charts are illustrated in (*a*), (*b*), and (*c*). Miscellaneous forms are shown in (*d*), (*e*), and (*f*). One may obtain information needed for the construction of proportional-type charts, concurrent scale charts, four-variable N-charts, and charts having a curved scale from any of the several books on nomography that are listed in the Bibliography of this text.

19.27

Construction of Alignment Charts. In this limited study, the explanation for the constructions will be based on the principles of plane geometry. The two forms to be considered for formulas are the parallel-scale chart and the Z-chart, also called an N-chart (Fig. 19.27).

Without giving thought at this time to the geometry underlying the construction of alignment charts and to the selection of scales, the methods that might be used to construct simple charts for graphical addition and subtraction and graphical multiplication and division might well be considered [see Fig. 19.27(*a*) and (*b*)].

A parallel-scale alignment chart of the type shown in (*a*), prepared for the purpose of making additions and subtractions, could be constructed as follows:

Step 1. Draw three vertical straight lines spaced an equal distance apart.

Equation: $\dfrac{f(t)}{f(u)} = \dfrac{f(v)}{f(w)}$

Proportional Chart
(a)

Equation: $f(t) + f(u) = f(v)/f(w)$

Proportional Chart
(b)

Equation: $f(t) + f(u) = \dfrac{f(v)}{f(w)}$

Proportional Chart
(c)

Equation: $f(u) + f(v) f_1(w) = f_2(w)$

Curved-scale Chart
(d)

Equation: $\dfrac{1}{f(t)} + \dfrac{1}{f(u)} = \dfrac{1}{f(v)}$

Concurrent-scale Chart
(e)

Equation: $f(t) f(u) + f(v) f_1(w) = f_2(w)$

Combination Chart
(f)

Fig. 19.26 Some common chart forms.

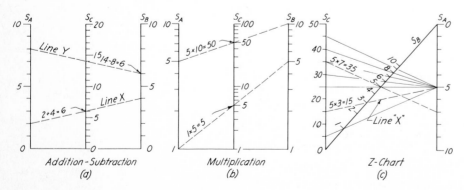

Addition - Subtraction
(a)

Multiplication
(b)

Z-Chart
(c)

Fig. 19.27 Parallel-scale chart and Z-chart.

Step 2. Draw a horizontal base line. This line will align and establish the origins (0) of the three scales.

Step 3. Using an engineer's decimal or metric scale, mark off a series of equal lengths on scales S_A and S_B. Start at the base line in each case. Mark the values of the graduations upward on both scales starting with 0 at the base line.

Step 4. Mark off on the S_C scale a series of lengths that are half as long as those on scales S_A and S_B. Number the graduation marks starting at the base line.

In using this chart to add two numbers, say 2 and 4, one may align the ruling edge of a triangle through 2 on the S_A scale and 4 on the S_B scale, then, read their sum at the point where the edge of the triangle crosses the S_C scale (see line X). To subtract one number from another, say 8 from 14, the edge of the triangle should be placed so as to pass through 8 on scale S_A and 14 on S_C. The difference, read on the S_B scale, will be 6, as shown by line Y.

If logarithmic scales are used for this form of chart as in (*b*) instead of natural scales as in (*a*), a chart for multiplication and division results, for the log of the product of two numbers is equal to the sum of the logs of the factors. Thus, by a method of addition a product can be obtained.

Necessary information for the construction of logarithmic scales is given in Sec. 19.29.

A Z-chart, (also called an N-chart), which will give the product of two numbers, is shown in (*c*). For example, if line X is assumed to represent the edge of a triangle so placed as to pass through 5 on the S_A scale and 3 on the S_B scale, it can be seen that $3 \times 5 = 15$.

A simple Z-chart that has been prepared solely for straight multiplication will have outside vertical scales of uniform spacing. The length of the scales and the spacing of the graduation marks can be arbitrarily determined as long as the chart will fit on the paper that is available and provided that the graduations cover the desired range in each case. The two vertical scales begin with 0 (zero) value at opposite ends of the diagonal scale so that the values of the graduations read in increasing magnitude upward on one scale and downward on the other.

19.28

Definitions. Before starting a discussion on the construction of scales, it is necessary that the student have an understanding of the meaning of the following terms and expressions that are commonly used when constructing alignment charts for solving equations:

Constant. A quantity whose value remains unchanged in an equation.

Variable. A quantity capable of taking values in an equation. A variable is designated by some letter, usually one of the latter letters of the alphabet.

Function of a variable. A mathematical expression for a combination of terms containing a variable, usually expressed in abbreviated form as $f(x)$ which is understood to mean "function of *x*." An equation usually contains several functions of different variables, such as $f(r) + f(s) = f(t)$; or $f(u) \cdot f(v) = f(w)$.

Functional modulus. A proportionality multiplier that is used to bring a range of values of a particular function within a selected length for a scale. For instance, with the upper and lower limits of a function known and a definite length L chosen for the scale, the value of the functional modulus (*m*) can be found by dividing L by the amount of the difference between the upper and lower limits of the function. The scale equation for determining m may be written

$$ m = \frac{L}{f(u_2) - f(u_1)} $$

where $f(u_2)$ and $f(u_1)$ are the upper and lower limits.

Scale. A graduated line that may be either straight or curved. When the graduation marks are equally spaced, that is, when the distance between marks is the same for equal increments of the variable as the variable increases in magnitude, the scale is known as a uniform scale. When the lengths to the graduation marks are laid off to correspond to scale values of the function of a variable, the scale is called a *functional scale*.

19.29

Construction of a Functional Scale. Let it be supposed that it is necessary to construct a functional scale, 125 mm in length, for $f(u) = u^2/2$ with u to range from 0 to 10. It will be found desirable to make the necessary computations by steps and to record the scale data in tabular form (Fig. 19.28).

Step 1. Record the values of u in the table.

u	0	1	2	3	4	5	6	7	8	9	10
u^2	0	1	4	9	16	25	36	49	64	81	100
$u^2/2$	0	0.5	2.0	4.5	8.0	12.5	18.0	24.5	32.0	40.5	50.0
$m(u^2/2)$	0	1.25	5	11.25	20	31.25	45	61.25	80	101.25	125

Fig. 19.28 Table.

Step 2. Compute the values of the function.

Step 3. Determine the functional modulus m.

Step 4. Multiply the recorded values of the function by the functional modulus.

In this case and in many other cases, the functional modulus may be chosen by inspection. For this problem, the overall length (L) of the scale will be 125 mm when $m = 2.5$. If the scale equation is used to determine the functional modulus, then

$$m = \frac{L}{f(u_2) - f(u_1)} = \frac{125}{(10^2/2) - 0}$$
$$= \frac{125}{50} = 2.5$$

All that now remains to be done to construct the scale is to lay off the computed distances along the line for the scale and mark the values at the corresponding interval points (Fig. 19.29).

Although a logarithmic scale might be constructed in this same manner, much time can be saved by using the graphic method shown in Fig. 19.30 for subdividing the scale between its end points. To apply this method to a scale that has already been laid off to a predetermined length with the end points of the range (say 1 to 10) marked, the steps of the construction are as follows:

Step 1. Draw a light construction line through point 1 making any convenient angle with the scale line.

Step 2. Using a printed log scale, mark off points on the auxiliary line.

Step 3. Draw a line through the 10 point on the construction line and the 10 point on the scale; then, through the remaining points draw lines parallel to this line through the 10s. These will divide the scale in proportion to the logarithms of numbers from 1 to 10.

19.30
Parallel-scale Charts for Equations of the Form $f(t) + f(u) = f(v)$. An alignment chart that is designed for solving an equation that can be set up to take this form will have three parallel functional scales that may be either uniform or logarithmic depending upon the equation. More information will be presented later concerning parallel-scale charts with logarithmic scales. In this section, attention will be directed to charts having scales with uniform spacing.

Before one can start constructing the chart, he must determine by calculation certain necessary information. First, he must determine how the scales are to be graduated, and second, he must calculate the ratio for the scale spacing.

To be competent to design parallel-scale charts, one must have a full understanding of the geometric basis for their construction. The explanation to follow is associated with the line layout in Fig. 19.31. Three parallel scales S_A, S_B, and S_C are shown with the origins t_0, u_0, and v_0 on line AB. Line (1) and line (2) are drawn parallel to AB through points v and u, respectively, on the isopleth. By similar triangles (shown shaded),

$$\frac{L_t - L_v}{a} = \frac{L_v - L_u}{b}$$

Now if

$$m_t = \frac{L_t}{f(t) - f(t_0)}$$

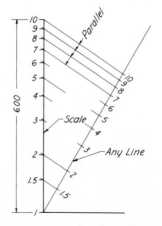

Fig. 19.30 Graduating a log scale.

Fig. 19.29 Functional scale for $u^2/2$ (distances in millimeters).

then

$$L_t = m_t[f(t) - f(t_0)]$$

When the function of t_0 is zero, the equation becomes

$$L_t = m_t f(t)$$

Similarly, when v_0 is zero,

$$L_v = m_v f(v)$$

and, when u_0 is zero,

$$L_u = m_u f(u)$$

Substituting these values,

$$\frac{m_t f(t) - m_v f(v)}{a} = \frac{m_v f(v) - m_u f(u)}{b}$$

Collecting terms,

$$m_t f(t) + \left(\frac{a}{b}\right) m_u f(u)$$

$$= \left(\frac{a}{b}\right) m_v f(v) + m_v f(v)$$

$$= m_v \left(1 + \frac{a}{b}\right) f(v)$$

But $f(t) + f(u) = f(v)$ only if the coefficient of the three terms are equal; therefore,

$$m_t = \left(\frac{a}{b}\right) m_u = \left(1 + \frac{a}{b}\right) m_v$$

and

$$\frac{m_t}{m_u} = \frac{a}{b} \qquad (1)$$

Now, since

$$m_t = \left(1 + \frac{a}{b}\right) m_v$$

and

$$\frac{a}{b} = \frac{m_t}{m_u}$$

then

$$m_t = \left(1 + \frac{m_t}{m_u}\right) m_v$$

$$m_v = \frac{m_t}{\left(1 + \dfrac{m_t}{m_u}\right)}$$

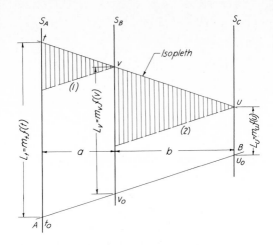

Fig. 19.31 Geometric basis for construction of parallel-scale alignment charts.

and, finally,

$$m_v = \frac{m_t m_u}{m_t + m_u} \qquad (2)$$

Now suppose that it is desired to construct a chart having the form of $t + u = v$ and that t is to have a range from 0 to 10 and u from 0 to 20 (Fig. 19.32). It has been determined that the

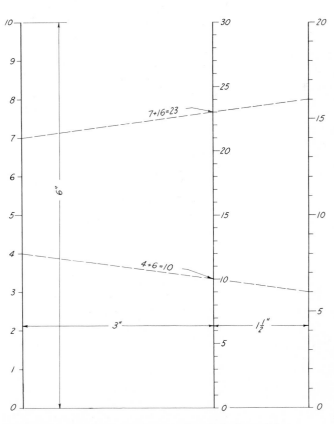

Fig. 19.32 Alignment chart for equation of the form $f(t) + f(u) = f(v)$.

scale lengths should be 6 in. Then

$$m_t = \frac{6}{10.0} = \frac{6}{10} = .6$$

$$m_u = \frac{6}{20.0} = \frac{6}{20} = .3$$

and

$$m_v = \frac{m_t m_u}{m_t + m_u} = \frac{.6 \times .3}{.6 + .3}$$

$$= \frac{.18}{.90} = .20 \qquad (2)$$

To determine the ratio of the scale spacing,

$$\frac{m_t}{m_u} = \frac{a}{b} = \frac{.6}{.3} = \frac{2}{1} \qquad (1)$$

For convenience, distance a between scales can be made 3 in. Distance b must

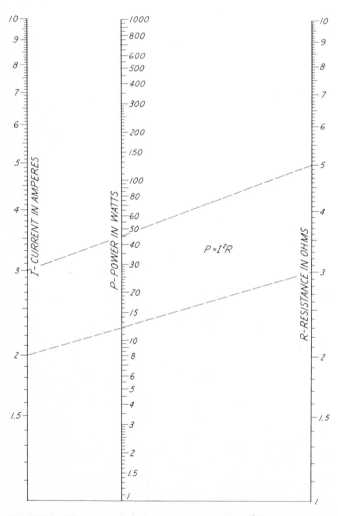

Fig. 19.33 Alignment chart for the equation $P = I^2R$.

then be $1\frac{1}{2}$ in. to satisfy the proportion of 2 to 1. The total width of the chart will be $4\frac{1}{2}$ in.

Since in this particular case the modulus for each of the scales is in full tenths and the scales are to be uniformly divided, an engineer's scale may be used to mark off the scales of the chart. With other conditions, it would be necessary to prepare either a table such as the one shown in Fig. 19.28 or to divide the line between its end-points using a geometric method.

19.31

Parallel-scale Charts for Equations of the Form $f(t) \cdot f(u) = f(v)$. Equations of this form may be rewritten so as to take the form $f(t) + f(u) = f(v)$ by using logarithms for both sides of the equation. For example, suppose that it is desirable to prepare a chart for $M = Wl$, a formula commonly used by designers and engineers. In this equation, M is the maximum bending moment at the point of support of a cantilever beam, W is the concentrated load, and l is the distance from the point of support to the load.

Given:

$$M = Wl$$

Rewritten:

$$\log M = \log W + \log l$$

Another example would be the formula for determining the discharge of trapezoidal weirs.

Given:

$$Q = 3.367\, Lh^{3/2}$$

Rewritten:

$$\log Q = \log 3.367 + \log L$$
$$+ 1.5 \log h$$

In this last equation, Q is discharge in cubic feet per second, L is the length of the crest in feet (width of weir), and h is the observed head (depth of water).

For the purpose of our discussion, the formula $P = I^2 R$ will be used where P is power in watts, I is current in amperes, and R is resistance in ohms (Fig. 19.33). It has been determined that I must vary from 1 to 10 amp and R from 1 to 10 ohms. A chart of this type might be used to determine the power loss in inductive windings. The length of the scales is to be 5 in.

The steps for the construction are as follows:

Step 1. Write the equation in standard form.

$$\log P = 2 \log I + \log R$$

Step 2. Determine the moduli m_I and m_R for the outside scales.

$$m_I = \frac{5}{2 \log 10 - 2 \log 1} = \frac{5}{2} = 2.5$$

$$L_I = 2.5(2 \log I) = 5 \log I$$

$$m_R = \frac{5}{\log 10 - \log 1} = \frac{5}{1} = 5$$

$$L_R = 5 \log R$$

Step 3. Determine m_P and L_P for the P-scale.

$$m_P = \frac{2.5 \times 5}{2.5 + 5} = \frac{12.5}{7.5} = \frac{5}{3}$$

$$L_P = \frac{5}{3} \log P$$

Step 4. Determine the ratio for the spacing of the scales.

$$\frac{m_I}{M_R} = \frac{2.5}{5} = \frac{1}{2} \qquad \text{(ratio)}$$

Step 5. Draw three vertical lines 1 in. and 2 in. apart and add a horizontal base line. By using these selected values, the ratio for spacing will be maintained and the chart will have good proportion.

Step 6. Graduate the scales for I and R using the method shown in Fig. 19.30. In this particular case, both logarithmic scales will be alike and will range from 1 on the base line upward to 10.

Step 7. Graduate the P-scale. By substituting values in the equation, it will be found that P will range from 1 W to 1000 W; therefore, the scale will be a three-cycle logarithmic scale, and the graphic method may be used for locating the graduation marks.

Figure 19.34 shows another parallel-scale alignment chart. This particular one could be used to determine the volume of a cylindrical tank when the diameter and height are known.

The formula for the volume of a cylinder, as given in the illustration, may be rewritten as

$$\log V = \log (\pi/4) + 2 \log D + \log H$$

Suppose that it has been decided, as in this case, that both the diameter (D) and the height (H) are to vary from 1 meter to 10 meters, and that the length of the two outside scales is to be 15 cm. For the moment, the constant term $\log (\pi/4)$ can be ignored, for it can be accounted for at a later time by shifting the V-scale for the volume upward until it is in the position for furnishing a correct reading for a particular value as computed using the formula.

The moduli can be determined as previously explained.

$$m_D = \frac{15}{2 \log 10 - 2 \log 1} = 7.5$$

and

$$L_D = 7.5(2 \log D) = 15 \log D$$

$$m_H = \frac{15}{\log 10 - \log 1} = 15$$

and

$$L_H = 15 \log H$$

$$m_V = \frac{m_D \times m_H}{m_D + m_H} = \frac{7.5 \times 15}{7.5 + 15}$$

$$= \frac{112.5}{22.5} = 5$$

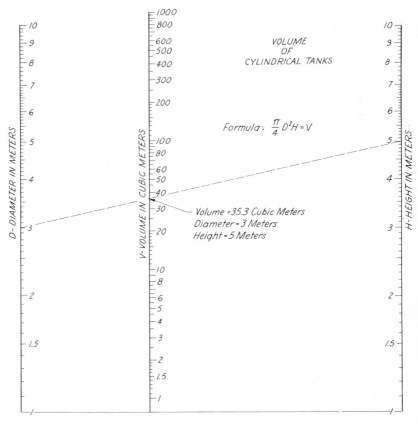

Fig. 19.34 Alignment chart for $V = (\pi/4)D^2H$ (volume of a cylinder).

and

$$L_V = 5 \ (\log V)$$

Scale spacing ratio:

$$\frac{m_D}{m_H} = \frac{7.5}{15} = \frac{1}{2}$$

For convenience, the scales can be spaced at 5 cm and 10 cm, giving a chart that is a square in form.

The scales for D and H may be graduated by using the graphical method illustrated in Fig. 19.30. The scale for V will be a three-cycle scale with a 15 cm range of length between the 1 and 1000 values. Since the value of V at the base line must result from the substitution in the equation of the values of 1 and 1 for the other two scales, the resulting value of .7854 cu m is the volume at the base line. The most convenient procedure to follow in graduating the V-scale is to start with the .7854 value at the base line. By so doing, the constant term, $\log (\pi/4)$, is taken into account. Should one desire to determine the distance to be laid off along the V-scale from the base line to the 1, it will be found to be equal to 5 times the difference between the logs of the numbers in this particular case, for the value of m_v is 5.

19.32

Three-scale Alignment Chart—Simplified (Graphical) Construction. After the student has acquired a thorough understanding of the theory of alignment charts, which includes a full knowledge of the geometric basis underlying their construction, he will be in a position to simplify his construction work and make it purely graphical through the use of tie-lines. Of course, this can only be done after the chart form has been identified, the equation converted, and the ranges for the variables (as needed) have been decided upon. The use of tie-lines to locate check points on the third scale of a three-scale alignment chart is shown in Fig. 19.35.

Given

$$M = \frac{wl^2}{12}$$

the equation for the bending moment (maximum) of a beam, having a uniformly distributed load and fixed at both ends. In this equation, w is the load in pounds per foot of length and l is the span in feet.

It was decided that the alignment chart would fulfill most needs if w had limits from 10 to 200 lb per ft and l had limits from 5 to 25 ft.

Equation rewritten:

$$\log w + 2 \log l = \log M + \log 12$$

The steps for the construction of the chart, with only minor calculations that may be performed mentally, are as follows:

1. Draw two vertical lines (say 5 in. long) at any convenient distance apart.

2. Graduate the w-scale (left-hand line) from the lower point 10 to the uppermost point 200. (In this case, use was made of the method shown in Fig. 19.30 to locate the scale values between 10 and 200.)

3. Graduate the l-scale, with the lower point as 5 and the point at the top of the scale as 25. (Scale values located as in step 2.)

4. Using a value of 10 for l and 500 for M yields $w = 60$ for the tie-line from the 60 value on the w-scale to the 10 value on the l-scale. Again, with $l = 20$ and $M = 500$, w is found to be 15 for the second tie-line through $M = 500$. The intersection of these two tie-lines is a check point on the third vertical scale, a point that establishes not only the position of the scale line but the 500 value as well.

5. Locate a second check point on the M-scale by drawing a third tie line. In this case it was found to be convenient to use $w = 200$ and $l = 12$ to yield $M = 2400$.

6. Graduate the M-scale as shown (read Sec. 19.29).

19.33

Graphical Construction of a Z-chart. A simplified graphical construction may be employed to construct a Z-chart (N-chart) such as the one shown in Fig. 19.36, for the equation $V = 2.467 \ Dd^2$, which gives the volume of a torus, a solid taking the form illustrated in Fig. 3.61. The given equation was considered to be so arranged that the vertical V- and D-scales would be uniformly graduated. The d-scale on the diagonal representing a variable to a power, is nonuniformly graduated. For

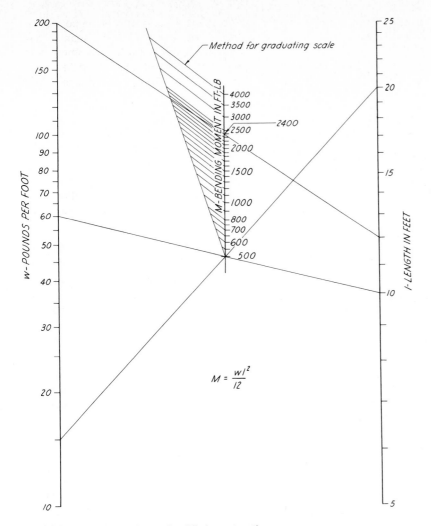

Method for graduating scale

M-BENDING MOMENT IN FT-LB

4000
3500
3000
2500 — 2400
2000
1500
1000
800
700
600
500

w-POUNDS PER FOOT

200
150
100
90
80
70
60
50
40
30
20
10

l-LENGTH IN FEET

25
20
15
10
5

$$M = \frac{wl^2}{12}$$

Fig. 19.35 Alignment chart—simplified construction.

this form of chart, one should recall that the vertical scales begin at zero and run in opposite directions (see Sec. 19.27). The diagonal connecting these two zero points completes the usual N-shape.

The steps in making the construction are as follows:

> **Step 1.** Draw the scale lines for the variables V and D at any convenient distance apart, in this case 9 cm. Then, lay off the D-scale to some selected length, say 15 cm, and locate the zero point of V scale opposite the 10 of the D-scale.

> **Step 2.** Graduate the V-scale uniformly, being certain that the range is sufficient to make possible a full reading when using the maximum values of the other scales.

> **Step 3.** Graduate the D-scale for the selected range of values, in this case from 0 to 10 cm.

> **Step 4.** Graduate the d-scale by selecting one convenient value of D and substituting it repeatedly in the original equation,

each time with a different value of d which is to be marked on the diagonal scale. In so doing, values of V are found that establish the ends of tie-lines from the pole point (10) selected on the D-scale. For example, when the values 10 and 1 are substituted for D and d, respectively, in the given equation, V is found to be 24.67 cu cm. With this known, a tie-line drawn from the pole to 24.67 on the V-scale, will locate $d = 1$ on the diagonal scale. The volumes for a range of d from 1 to 5 cm are shown in the table in Fig. 19.36. These values were used to draw the tie-lines shown.

19.34

Four-variable Relationship—Parallel-scale Alignment Chart. A four-variable parallel-scale alignment chart may be constructed for an equation of the form $f(t) + f(u) + f(v) = f(w)$ when the equation has been rewritten as follows:

$$f(t) + f(u) = f(k)$$
$$f(k) + f(v) = f(w)$$

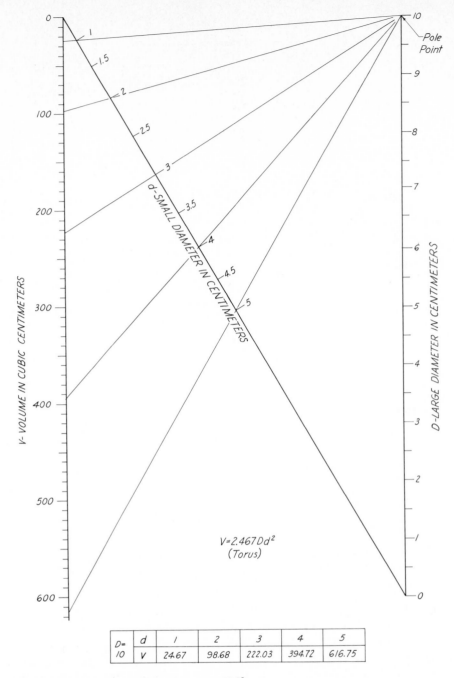

Fig. 19.36 Chart for equation $V = 2.467\,Dd^2$.

$D=$	d	1	2	3	4	5
10	V	24.67	98.68	222.03	394.72	616.75

When rewritten in this form, the four-variable chart can then be constructed as two three-scale nomographs with the k-scale common to both. The k-scale, which serves as a pivot line in using the chart, need not be graduated, but it must be remembered that it does have a modulus and does represent a function in both equations.

In Fig. 19.37, a four-variable alignment chart is shown for the addition of numbers. For this particular construction, it was arbitrarily decided that t should have

a range from 0 to 5, u from 0 to 10, and v from 0 to 20. The procedure for the construction of this chart is as follows:

Step 1. Draw the vertical lines (stems) for the t- and u-scales at any convenient distance apart and, by means of the engineer's scale or graphically, graduate t from 0 to 5 and u from 0 to 10.

Step 2. Draw the two intersecting tie-lines between the t- and u-scales that determine the position of the pivot line k at their point of intersection (7). Intersecting tie-lines could have been drawn

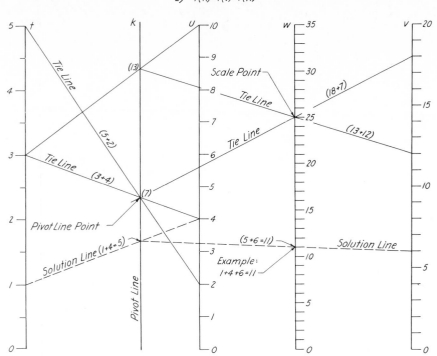

Fig. 19.37 Alignment chart for the equation $f(t) + f(u) + f(v) = f(w)$.

from other graduation marks along the t- and u-scales as long as the sum of the values at the ends of one of the selected tie-lines equals the sum of the values at the ends of the other tie-line intersecting it.

Step 3. Draw a third tie-line to locate another check point along the pivot line k. In this case, the tie-line was drawn from 3 on the t-scale to 10 on the u-scale to locate the mark for 13 on the k-scale.

Step 4. Decide upon the best location for the v-scale, then draw the stem line and graduate it uniformly from 0 to 20.

Step. 5. Using the 7 and 13 value marks that have now been established on the k-scale, draw the two tie-lines ($18 + 7 = 25$ and $13 + 12 = 25$) that will determine the location of the w-scale at their point of intersection as well as the position of the 25 graduation mark.

Step 6. Draw the w-scale and graduate it from 0 to 35, the 35 value representing the largest sum obtainable from the use of the t-, u-, and v-scales as they have been graduated for this particular chart.

The procedure for finding the sum of the numbers 1, 4, and 6 is illustrated by the broken lines extending between the scales. First, a straightedge can be laid across the t- and u-scales through the 1 and 4 values, as shown, to determine the position on the pivot line (k) that the graduation mark would have that represents their sum, in this case 5. Then the straightedge should be shifted so that its edge will pass through the 5 position on the pivot line and the graduation mark for the 6 value on the v-scale. The sum of the three given numbers can then be read as 11 on the w-scale.

The four-scale alignment chart, shown in Fig. 19.38, for the computation of simple interest, may be easily constructed once the equation $I = PRT$ has first been written in the logarithmic form of

$$\log P + \log R + \log T = \log I,$$

and then rewritten as

$$\log P + \log R = \log k \qquad (1)$$

and

$$\log k + \log T = \log I \qquad (2)$$

The equation as first rewritten is in the form $f(t) + f(u) + f(v) = f(w)$ and as

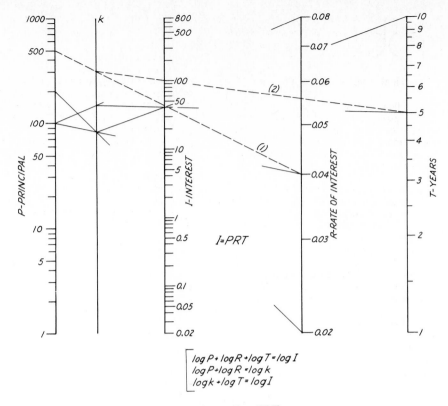

$$\begin{bmatrix} \log P + \log R + \log T = \log I \\ \log P + \log R = \log k \\ \log k + \log T = \log I \end{bmatrix}$$

Fig. 19.38 Alignment chart for the equation $I = PRT$.

arranged in (1) and (2) in the form of

$$f(t) + f(u) = f(k)$$
$$f(k) + f(v) = f(w)$$

Now, since equations (1) and (2) have the same form as those for the previous problem, the alignment chart for the equation $I = PRT$ is constructed similarly except that logarithmic scales must be used in this case in place of natural (uniformly divided) scales. As before, tie-lines were used to position the k- and I-scales and to

locate needed check points. The procedure for reading the chart is illustrated by the broken lines, numbered (1) and (2).

The reader is urged to follow through the construction of the alignment chart for the equation $I = PRT$ mentally, step by step. At the start, however, he must recognize that the k-scale belongs first to the PRk chart and then to the kTI three-scale chart that together make the complete chart for the four variables $P, R, T,$ and I.

Problems

Technical graphs and charts

The following problems have been designed to emphasize the fundamental principles underlying the preparation and use of technical graphs and charts. If desired, the tables given in the Appendix may be used to prepare a graph or chart in metric.

1. Determine the values for the following equations, as assigned, and plot the curve in each case for quantitative purposes.

Parabola	$Y = 4x^2$, x from 0 to 5
Ellipse	$Y^2 = 100 - 2x^2$
Sines	$Y = \sin x$, x from 0° to 360°
Cosines	$Y = \cos x$, x from 0° to 360°
Logarithms	$Y = \log x$, x from 1 to 10
Reciprocals	$Y = 1/x$, x from 1 to 10

2. The freezing temperatures for two common antifreeze solutions for various compositions are given below.

Prestone			Denatured Alcohol		
% by vol.	Temp. (°F)	Temp. (°C)	% by vol.	Temp. (°F)	Temp. (°C)
10	25	−4	10	25	−4
20	16.5	−8.5	20	17.5	−8
25	11	−11.5	25	14	−10
30	5	−15	30	5	−15
35	−3	−19.5	35	−1	−18
40	−12	−24.5	40	−11	−24
45	−25	−31.5	45	−18	−28
50	−38	−39	50	−25	−31.5
55	−47	−44	55	−32	−35.5
			65	−45	−43

Prepare a chart for these data, mainly quantitative in character, from which the required per cent volume can be read for any desired freezing temperature. Use the type of paper shown in Fig. 19.3(b).

3. Approximate barometric pressures at different heights above sea level are given below. Prepare a qualitative chart for the given data on rectangular coordinate paper.

Note that the curve would be straight on semilogarithmic paper.

H, Altitude (miles)	B, Barometric pressure (in. Hg)
0	29.92
1	24.5
2	20.0
3	16.2
4	13.45
5	11.0
6	8.9
7	7.28
8	5.95
9	4.87
10	4.0

4. Data on rate of growth are frequently plotted on semilogarithmic paper, because the slope of the curve then represents the rate of growth. On semilogarithmic paper, plot the data for the enrollment in a university.

Place the vertical axis 1 unit from the left edge and the horizontal axis at the extreme bottom of the page ruling. Letter in an appropriate title.

Also draw the curve of the general trend and enter on the chart, just below the title, what the enrollment would be in 1975 if the same rate of growth is maintained.

Year	Enrollment
1941	5,745
1942	5,831
1943	5,906
1944	6,238
1945	6,984
1946	7,313
1947	6,011
1948	5,112
1949	4,838
1950	4,621
1951	10,221
1952	11,320
1953	10,210
1954	10,101
1955	9,871
1956	9,211
1957	9,071
1958	9,647
1959	10,101
1960	10,456
1961	11,132
1962	11,656
1963	12,231
1964	12,910
1965	13,621
1966	14,222
1967	15,011
1968	15,976
1969	16,211
1970	18,013

5. In a hydraulics laboratory, the construction of a quantitative curve that would give the weight of water contained in tubes of various diameters and lengths was desired. This was accomplished by filling tubes of known diameters with water to a depth of 1 ft and observing the weight of water thus added. The water was kept at a temperature for maximum density and the following data were obtained:

D = diameter of tube (inches)	W = weight of 1-ft column of water
2	1.362
2½	2.128
3	3.064
3½	4.171
4	5.448

D = diameter of tube (inches)	W = weight of 1-ft column of water
$4\frac{1}{2}$	6.895
5	8.512
$5\frac{1}{2}$	10.299
6	12.257
$6\frac{1}{2}$	14.385
7	16.683
$7\frac{1}{2}$	19.152
8	21.790

On a sheet of graph paper [Fig. 19.3(*b*)], plot the above data. Place the axes 3 cm in from the edges. Letter the title in any convenient open space.

6. Owing to uncontrollable factors, such as lack of absolute uniformity of material or test procedure, repeated tests of samples of material do not give identical results. Also, it has been observed in many practical situations that
a. Large departures from the average seldom occur.
b. Small variations from average occur quite often.
c. The variations are equally likely to be above average and below average.

The foregoing statements are borne out by the accompanying data showing the results of 4000 measurements of tensile strength of malleable iron.

On a sheet of coordinate graph paper [Fig. 19.3(*a*) or (*b*)] prepare a graph showing frequency of occurrence of various strength values as ordinates and tensile strength as abscissa. Draw a smooth symmetrical curve approximating the given data.

Range of tensile strength values (pounds per square inch)	Number of observations
Under 45,000	0
45,000–45,999	1
46,000–46,999	2
47,000–47,999	3
48,000–48,999	6
49,000–49,999	20
50,000–50,999	232
51,000–51,999	376
52,000–52,999	590
53,000–53,999	740
54,000–54,999	771
55,000–55,999	604
56,000–56,999	383
57,000–57,999	184
58,000–58,999	60
59,000–59,999	20
Over 60,000	8
	4000

7. On a sheet of paper of the type shown in Fig. 19.3(*b*), using India ink, plot a curve to represent the data given above. (*Note:* For stress–strain diagrams, although the load is the independent variable, it is plotted as ordinate, contrary to the general rule as given in Sec. 19.5. Figure 19.6 shows a similar chart. In per-

forming tests of this nature, some load is imposed before any readings of elongation are taken.

It is suggested that the label along the abscissa be marked "Strain, 0.00001 in. per in." then fewer figures will be required along the axis.

Stress (pounds per square inch)	Strain (inches per inch)
3,000	.0001
5,000	.0002
10,000	.00035
15,000	.00054
20,000	.00070
25,000	.00090
30,000	.00106
32,000	.00112
33,000	.00130
34,000	.00140

8. Make a vertical multiple bar chart showing the enrollment at _____ University from 1961 to 1970. Obtain data from Problem 4.

9. Make a semilogarithmic graph showing the enrollment of your school for the last 20 years.

Alignments charts

The following problems have been designed to emphasize the fundamental principles underlying the preparation and use of alignment charts.

10–21. Prepare an alignment chart for the given equation. Chart scales should have a sufficient number of division marks to enable the user to obtain some reasonably accurate results (readings). In each problem the range for two of the variables has been given. The range of the third variable must make possible the use of the full range of each of the other two variables.

10. Construct an alignment chart for the multiplication of numbers from 1 to 100 (read Sec. 19.27 and study Fig. 19.27).

11. Construct an alignment chart of the form $t + u = v$. Let t vary from 0 to 10 and u from 0 to 15 (read Sec. 19.30 and study Fig. 19.32).

12. Construct an alignment chart for determining the area of a triangle. The student is to determine for himself the range for each of the scales.

13. Construct an alignment chart for determining the volume of a cylinder. The diameter is to range from 1 to 5 m and the height from 1 to 10 m. The volume is to be in cubic meters (read Sec. 19.31).

14. Make an alignment chart for determining the volume of a paraboloid, $V = 1/8\pi ab^2$, where a is the length (measured along the axis) and b is the diameter of the base. Let a vary from 1 to 20 in. and b from 1 to 10 in.

15. Construct an alignment chart for the maximum bending moment at the point of support of a cantilever beam, formula $M = Wl$. W is the concentrated load at the end of the beam and l is the distance from the point of support to the load (read Sec. 19.31). Let W vary from 100 to 1000 lb and l from 10 to 20 ft. M will be in ft-lb.

16. Make an alignment chart for the discharge of trapezoid weirs, formula $Q = 3.367 \, Lh^{3/2}$. Q is discharge in cubic feet per second; L is length of crest in feet (width of weir); and h is the head (depth of water). Let L vary from 1 to 10 ft and h from .5 to 2 ft (read Secs. 19.31 and 19.32).

17. Make an alignment chart for the formula $P = I^2R$ as explained in Sec. 19.31. Let I vary from 1 to 20 amp and R from 1 to 10 ohms.

18. Make an alignment chart for the formula $R = E/I$, where E is the electromotive force in volts; I is current in amperes; and R is resistance in ohms. Let R vary from 1 to 20 ohms and I from 1 to 100 amp (read Secs. 19.31 and 19.32).

19. Make an alignment chart for the formula $I = bd^3/36$, where I is the moment of inertia of a triangular section; b is the length of the base in inches; and d is the depth of the section (altitude of triangle). Let b vary from 1 to 10 in. and d from 1 to 10 in. (read Secs. 19.31 and 19.32).

20. Make an alignment chart for the formula $M = wl^2/8$, where M is the bending moment in foot-pounds, w is the load in pounds per foot, and l the length of span in feet. Let w vary from 10 to 200 lb per ft and l from 5 to 25 ft (see Fig. 19.35).

21. Make an alignment chart (Z-chart) for the equation for the volume of a right circular cylinder, $V = \pi r^2 h/144$, where V is the volume, r is the radius of the base in millimeters, and h is the height in millimeters. Let r vary from 0 to 50 mm and h from 0 to 500 mm.

$$y = Ae^{mx}$$

$$y = ab^x$$

$$y = ax^b$$

$$y = a + bx$$

$$y = 3x^{.57}$$

This manned reusable space vehicle will be used for a wide variety of missions. The Shuttle will consist of a reusable orbiter, mounted "piggyback" at launch on a large expendable propellant tank and two recoverable and reusable solid-propellant rockets. The various stages of launch have been illustrated. Graphics played a very large role in the development of the Shuttle. (*Courtesy National Aeronautics and Space Administration*)

graphical mathematics 20

A □ Graphical algebra

20.1

Introduction. Persons who have a thorough understanding of the many graphical methods at hand frequently find it more desirable to use a graphical procedure for making a calculation than to resort to an algebraic approach, particularly when repeated calculations of the same nature are required. These people, no matter in what phase of engineering endeavor they may be found (research, design, development or production), have discovered that a graphical method can be used to present a needed visual relationship of the variables, save valuable time, minimize the possibility of error, and provide a sure way to explain the given calculation.

The student must realize that graphical methods, while theoretically precise, give only approximate solutions. The accuracy is limited by the size of the paper, the tools available, and the skill of the "graphician." However, in many problems the initial data is also approximate in that it is dependent upon the visual reading of instruments provided with graphical scales. With careful construction, a graphical solution can often be more accurate than the initial data, particularly if one has made a proper choice of the scale and a wise selection of tools. On certain occasions, where extreme accuracy is needed, a magnifying glass and pricker may be used for laying out measurements with a scale. A magnifying glass is also helpful in establishing the exact intersection of lines.

As stated before, when the solution has been accomplished by certain graphical methods, that solution may often be used for subsequent problems at a great saving in time. Furthermore, in most cases the solution by graphical methods can be arrived at in a shorter time than through the use of pure mathematical methods.

It is desirable for an engineer to be well versed in both the graphical and mathematical sciences and he should be aware of the limitations of both so that he can exercise sound judgment as to which method to use for any given situation.

20.2

Simultaneous Equations. Since simultaneous equations are frequently encountered in engineering work and there are times when the algebraic solution of a pair of simultaneous equations is complicated

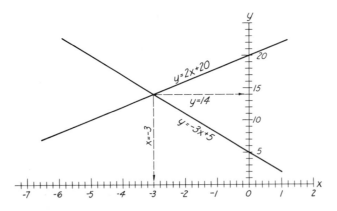

Fig. 20.1 Solution of linear simultaneous equations.

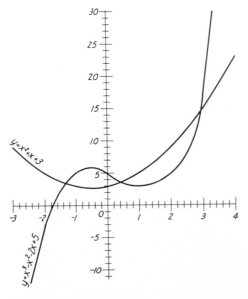

Fig. 20.2 Solution of higher-order equations.

and difficult, if not almost impossible, it is necessary for an engineer to know how to find the solution by graphical means. Situations that make a graphical approach desirable arise when the data have been obtained as the result of an experiment and when the equations, although known, are so complicated that the algebraic solution would be too time consuming. In some cases, the graphical solution may, by giving the roots to the desired accuracy or by showing where the roots are located, permit a saving of time in the mathematical solution. Figure. 20.1 shows the solution of two linear simultaneous equations. Figure 20.2 shows the solution of two higher-order equations. The equations are

$$y = x^2 + x + 3$$
$$y = x^3 - x^2 - 2x + 5$$

The mathematical solution of this pair of equations would be very time consuming. The graphical solution is largely a matter of plotting. It should be noted that there are three roots that satisfy both equations.

20.3

Determination of the Roots of Quadratic Equations. There are three methods of solution that can be used for the determination of the roots (points where $y = 0$) of quadratic equations, each method having its own advantages. First, the equation can be plotted directly as illustrated in Fig. 20.3. The roots are the points where the curve crosses the X-axis. If the curve crosses the X-axis, there are two real and different roots; if the curve is tangent to the X-axis, there are two real but equal roots; and, finally, if the curve does not cross the X-axis, there are two imaginary roots. Imaginary roots are always in pairs, equal but opposite in sign. The advantage of this method is that an immediate picture of the conditions of the equation is obtained. This method is applicable for any order equation. However, a new curve must be drawn for each equation.

Second, the equation can be rewritten as two simultaneous equations, one being the equation of a straight line. For example, the equation

$$y = x^2 + Ax - C$$

can be rewritten as

$$y = x^2$$
$$y = -Ax + C$$

This pair of equations can be solved as explained in Sec. 20.2 and the x-values of their solution will be the roots of the original quadratic equation. Figure 20.4 shows this method applied to the same quadratic equation as represented in Fig. 20.3. If the straight line cuts the curve, there are two real and different roots; if the straight line is tangent to the curve, there are two real but equal roots; and finally, if the straight line does not cut the curve, there are two imaginary roots. The advantage of this method is that the curve representing $y = x^2$ need be drawn only once. Each subsequent quadratic is solved by drawing a different straight line.

For the third method, based on geometric construction, the quadratic must be written in the form

$$y = x^2 + Ax + C$$

The steps to be followed in making the construction shown in Fig. 20.5 are: (1) Draw a set of X- and Y-axes and lay off A along the X-axis, with opposite sign, to locate point a. (2) At a erect a perpendicular and lay off C with the same sign to locate c. Finally, (3) Connect c to a point at plus one unit on the Y-axis with a straight line and draw the circle that has this line as a diameter. The points where this circle cuts the X-axis are the roots that are sought. Again there is the possibility of real and different; real but equal; and imaginary roots. The existing condition would be revealed by the position of the circle. The advantage of this method is that all construction is done with a straightedge, scale, and compass. No irregular curves need be drawn with the likelihood of errors in construction. However, a completely new construction is required for each equation.

B □ Graphical calculus

20.4

Graphical Calculus. In solving engineering problems, it is frequently desirable and often necessary to present a graphical analysis of empirical data. Even though it is often possible to make an evaluation through the use of analytical calculus, a graphical representation is more meaningful because it is pictorial in character. Graphical integration and differentiation are particularly desirable for problems for which only a set of values are

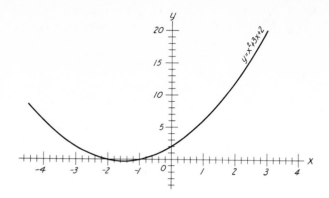

Fig. 20.3 Graphical determination of roots of the equation $y = x^2 + 3x + 2$.

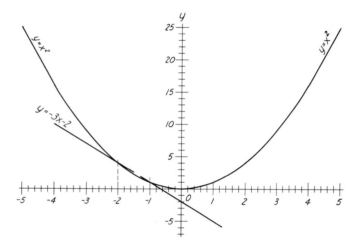

Fig. 20.4 Graphical determination of roots of the equation $y = x^2 + 3x + 2$.

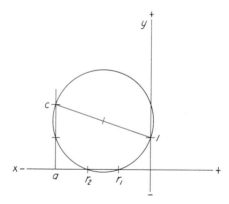

Fig. 20.5 Graphical determination of roots of the equation $y = x^2 + 3x + 2$.

known, or for curves that have been produced mechanically, as in the case of steam engine indicator diagrams, or if the results cannot be determined by the analytical methods of calculus.

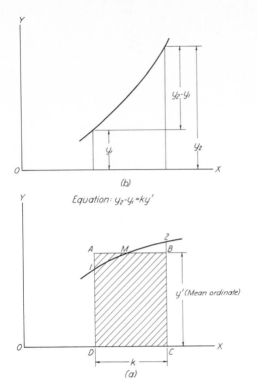

Fig. 20.6 Illustration of the principle of integration.

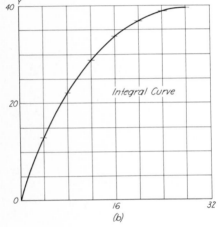

Fig. 20.7 Integration of a curve.

The following sections are devoted to the graphical rules and methods for determining derived curves. Discussion of the interpretation of results has been intentionally omitted since interpretation is not usually graphical and, therefore, not within the scope of this text.

20.5

Graphical Integration In deriving curves of a higher order, the principle is applied that the area bounded by two successive ordinates, the curve, and the axis is equal to the difference in magnitude of the corresponding ordinates of the integral curve. Figure 20.6 illustrates this principle of graphical integration. In (a) an increment of a curve is shown enlarged. The area under the curve will be approximately equal to the area of the shaded rectangle $ABCD$ when the line AB is drawn so that the area $AM1$ above the curve is approximately equal to the area $MB2$ below the curve. With a little practice one will find it easy to establish a line such as AB quite accurately by eye if a strip of celluloid or a triangle is used through which the curve can be seen.

By applying the principle of graphical integration to a series of increments, an integral curve may be drawn as shown in Fig. 20.7. At this point, it should be recognized that since the difference between successive ordinates represents increase in area, the difference between the final ordinate and the initial ordinate represents the total area between these ordinates that is bounded by the curve and the X-axis.

The scale selected for the Y-axis of the integral curve need not be the same as the scale for the given curve.

Portions of a lower-order curve that are above the X-axis are considered to be positive whereas areas below with negative ordinates are recognized as negative [see Fig. 20.8(a)]. Since the negative area between any two ordinates on the lower-order curve represents only the difference in the length of the corresponding ordinates on the integral curve, the length of y_7 is less than the length of y_6 by an amount equal to the negative area. Also, because areas represent only differences in length of successive ordinates of the integral curve, the initial point on the integral curve might have any value and still fulfill its purpose. For example, either integral curve shown in Fig. 20.8(b) is a satisfactory solution for the curve in (a).

Figure 20.9 shows the derived curves for a falling drop hammer. It is common practice, when drawing related curves, to place them in descending order as shown; that is, the lower-order curve is placed below.

In (a) the straight line represents a uniform acceleration of 9.81 m per sec per sec, which is the acceleration for a freely falling body. The initial velocity is 0. The units along the X-axis represent time in seconds, and the units along the Y-axis represent acceleration in meters per second per second.

Since the acceleration is uniform, the velocity will be a straight line of constant slope, (b). The length of the last ordinate is equal to the total area under the acceleration curve, 7.85 m per sec.

The distance curve, which is obtained by integrating the velocity time curve, is shown in (c). The length of the ordinate at any interval point is equal to the total area below the velocity curve between the origin and the point $[D = A = \frac{1}{2} (9.81T)T]$. See Figs. 20.6, 20.7, and 20.8.

20.6

To Integrate a Curve by the Ray Polygon Method. An integral curve may be drawn by a purely graphical process known as

Fig. 20.8 Integration of a curve.

Fig. 20.9 Derived curves for a falling drop hammer.

the ray polygon method. This method of integrating the area under a curve is illustrated in Fig. 20.10. Divide the X-axis into intervals and draw ordinates at the division points. Then, select the pole point P at some convenient location that will make the distance d equal to any number of the full units assigned to the X-axis. The selection of the number of units for the distance d determines the length of the scale along a Y-axis for the integral curve. To establish relationship between y_1 and y_0 the following equation based on similar right triangles can be written:

$$y_1:k = y_0:d$$
$$y_1 \cdot d = k \cdot y_0$$
$$y_1 = \frac{k \cdot y_0}{d} = \frac{k}{d} \cdot y_0$$

Determine the mean ordinate for each strip and transfer its height to the Y-axis as length OA, OB, OC, and so forth. Draw rays from P to points A, B, C, D, E, and F.

To construct the integral curve, start at

O and draw a line parallel to PA cutting the first vertical through R at R'. Through R' draw a line parallel to PB until it cuts the second vertical through S at S'. Repeat this procedure to obtain points T', U', and so on. Points R', S', T', U', V', and W' are points on the required integral curve. In Fig. 20.10 the integral curve is constructed on the same coordinate axes as the lower-order curve.

20.7
Pole Method Applied to the Construction of an Integral Curve.
The first of the several examples that have been selected to illustrate the use of the *pole method* for

Fig. 20.10 Use of the ray polygon.

the construction of an integral graph, has a sloping straight line, plotted on rectangular coordinates, as the given curve of lower order (see Fig. 20.11). It should be noted in this case that the integral graph has been placed directly above the lower graph and in projection with it so that the same intervals may be used for both. As shown, ordinates were first drawn to divide the chart area into vertical strips. With this done the horizontal lines were drawn next, one for each of the strips at mean-ordinate height, to form the five rectangular areas that are assumed to approximate the area under the curve (sloping straight line). Then, these mean-ordinate heights must be projected horizontally to the Y-axis. With the pole distance k determined and the pole point P

located along the X-axis extended, ray lines are drawn from P to the points on the Y-axis representing the heights of the mean ordinates. The manner in which the distance k affects the steepness of the integral curve, as illustrated in Fig. 20.13, will be discussed later. The first point on the integral curve was found by drawing a line through the origin parallel to the ray for the corresponding interval below, in this case the lowest ray. The second segment is then drawn parallel to the ray for the second interval through the end of the first segment. Other intervals are treated similarly and then the smooth curve, that is, the integral curve, is drawn through the points obtained. Since the finished curve has in reality been drawn through the end points of chords by this method, most persons identify this construction procedure by calling it the *chordal method.*

In Fig. 20.12, the chordal method using a pole has been applied to the integration of an irregular curve. The intervals between the ordinates need not be equal as shown. In fact, the number and spacing of the ordinates will usually be determined by the shape of the curve and one soon discovers that it is wise to space ordinates closer together where there is a sharp change or a reversal of curvature and farther apart where the curve is flat. For this illustration, an example was purposely selected that would have a portion of the curve below the X-axis. This area below represents negative area that must be subtracted from the positive area above the X-axis. When a pole is used, this sub-

Fig. 20.11 Pole method.

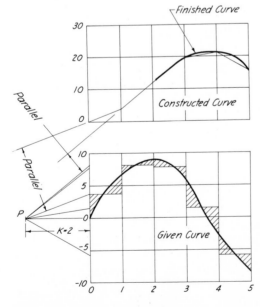

Fig. 20.12 Use of the pole point.

traction is taken care of graphically by the negative slope of the rays for the intervals at the location where the curve lies below the X-axis.

Ordinarily, the modulus of the ordinate (Y-axis) scale of the integral curve must be smaller than the modulus of the given (lower-order) curve if the graph for the integral curve is to be of reasonable size. When the height of the derived curve is somewhat near the same as the height of the given curve, the general overall appearance of the graphs will be satisfactory and the scales will ordinarily be large enough to be read with reasonable accuracy. At the very beginning, when it becomes necessary to determine the maximum value to be read on the ordinate scale for the integral curve, one must first remember that the value will be equal numerically to the value for the total area under the given curve. This value may be found either by adding the areas of the individual rectangles or by calculating the area of a single large rectangle that is estimated to be equal to the total area under the curve. Once this maximum value is known it is relatively easy to select a value for k that will give good proportion to the integral curve and at the same time permit one to make accurate readings.

When the pole distance used is equal to the length of one abscissa unit as shown in Fig. 20.13(a), the same distance represents the same number of units on both scales. In (b), where a pole distance of 2 has been used, ten units are represented on the scale of the constructed curve by the same distance that represents 5 units on the scale of the curve below. As can be easily seen, a three-unit pole distance reduces the height of the constructed curve still more (c). When the pole distance is one-half unit, the constructed graph becomes very high, too high to be appropriate in the case shown in (d). However, there may be conditions encountered now and then when the use of a pole distance of one-half unit is desirable. An area graph is shown in Fig. 20.14, where a five-unit pole distance was used to construct the work (integral) curve.

An interesting application of graphical integration is shown in Fig. 20.15. Let it be supposed that an irregularly shaped plot of land, lying along a lake shore and a road, is to be divided into three equal lots. The area to be divided was first plotted (meters against meters) on the lower graph. Its boundary lines in addition to the shore line at the north, are the X and

Y axes, representing property lines, and the right-of-way line of S.R. 26. It was decided that the division lines were to run due north and south, perpendicular to the south property line. With the lower curve drawn, the intergral curve was constructed by the pole-and-ray method as shown and the total area was found to be 11,850 square meters, an area that, in this particular case, can be divided evenly into three lots, each having 3,950 square meters. Horizontal lines, drawn through points representing 3,950 and 7,900 values on the scale of the area graph to the area curve and thence downward to the shore line, located the end points of

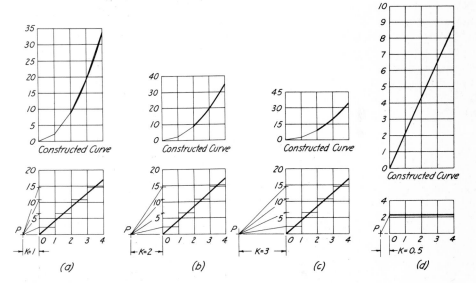

Fig. 20.13 Pole distance and steepness of the curve.

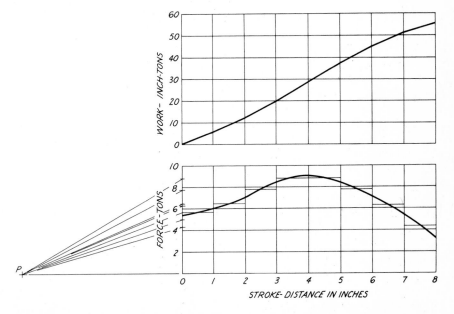

Fig. 20.14 Example of graphical integration.

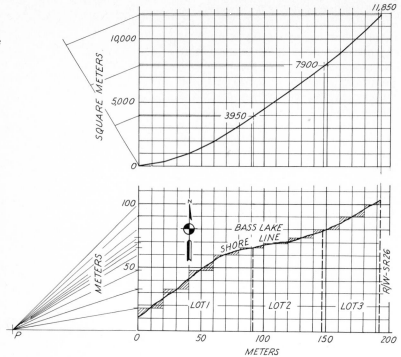

Fig. 20.15 Example of graphical integration.

the division lines for the lots. A graphical method was used to divide the area as shown to the left of the area scale.

20.8
Graphical Differentiation. Curves of a lower order are derived through the application of the principle that the ordinate at any point on the derived curve is equal to the slope of a tangent line at the corresponding point on the given curve. The slope of a curve at a point is the tangent of the angle with the X-axis formed by the tangent to the curve at the point. For all practical purposes, when constructing a derivative curve, the slope may be taken as the rise of the tangent line parallel to the Y-axis in one unit of distance along the X-axis, or the slope of the tangent equals y_1/k as shown in Fig. 20.16.

Figure 20.16 illustrates the application of this principle of graphical differentiation. The length of the ordinate y_1' at point A' on the derived curve is equal to the slope y_1/k at point A on the given curve as shown in (a). When the slope is zero as at point C, the length of the ordinate is zero and point C' lies on the X-axis for the derived curve. When the slope is negative, as shown at D, the ordinate is negative and lies below the X-axis.

The graph shown in Fig. 20.17 is composed of segments of straight lines. Since

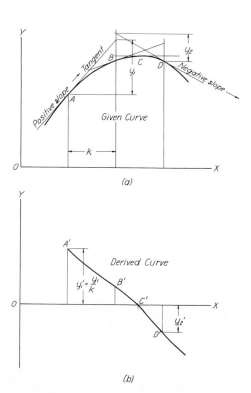

Fig. 20.16 Illustration of the principle of differentiation.

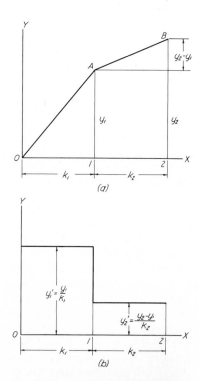

Fig. 20.17 Graphical differentiation.

the slope is constant for the interval 0–1, the derivative curve in the interval is a horizontal line. In the interval 1–2, the slope is also constant but of a lesser magnitude. Thus, the derivative curve is composed of straight line segments as shown in (b).

At this point in the discussion of graphical calculus it becomes possible to determine the relationship between the principles of integration and differentiation and to show that one is derived from the other.

From inspection of the graphs shown in Fig. 20.17, equations may be formulated as follows:

By the principle of differentiation,

$$y_2' = \frac{y_2 - y_1}{k_2}$$

in the interval 1–2, where

$$\frac{y_2 - y_1}{k_2}$$

represents the slope of AB. The area under the curve in (b) in the interval 1–2 is equal to

$$y_2' \cdot k_2 = \frac{y_2 - y_1}{k_2} \cdot k_2$$
$$= y_2 - y_1 \quad \text{(integral curve)}$$

and

$$y_2' = \frac{y_2 - y_1}{k_2} \quad \text{(differential curve)}$$

In constructing a derivative curve, the determination of the tangent lines is often difficult because the direction of a tangent at a particular point is usually not well defined by the curvature of the graph. Two related schemes that may be used for constructing tangents are shown in Fig. 20.18(a) and (b). In (a) the tangent is drawn parallel to a chord of the curve, the arc of which is assumed to approximate the arc of a parabola. A sufficiently accurate location for the point of tangency T_1 may be determined by drawing a line from the midpoint of the chord to the arc parallel to an assumed direction for the axis of the parabola. When working with small segments of the curve, one may assume the axis to be either horizontal or vertical.

A more accurate construction is shown in (b), where the tangent is drawn parallel to two parallel chords. The point of tan-

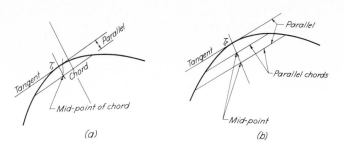

Fig. 20.18 Construction of a tangent line.

gency T_2 is determined by connecting the midpoints of the chords and extending this line to the curve. This line determines the direction of the axis.

The construction in (b) using two chords to establish a tangent is applicable to any curve that may be approximated by a portion of a circle, ellipse, parabola, or hyperbola.

Since a tangent is assumed to be parallel to a chord, it is common practice to use chords instead of tangents for constructing a derivative curve, as shown in Fig. 20.19(a). The slope is plotted on an ordinate located midway in the corresponding interval of the derived curve.

A derivative curve can also be drawn using the ray polygon method as explained in Sec. 20.6 in reverse (see Fig. 20.20). The lines PA, PB, and PC of the ray polygon are drawn parallel to the tangents at points T_1, T_2, and T_3. Point Q is found by drawing a line horizontally from point A to the ordinate through the point of contact of the tangent parallel to PA. Points R and S are found similarly.

20.9
Drawing a Derivative Curve by the Pole Method.
Before attempting to construct a derivative curve by the pole method, the

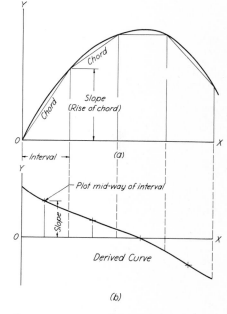

Fig. 20.19 Use of chords in place of tangents.

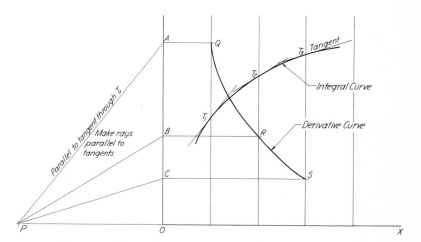

Fig. 20.20 Ray polygon method.

beginner must have a clear understanding of the "slope law" as it applies to the differentiation, and he must be able to think of the derived curve being sought as the curve of slopes (slope locus) for the given curve. If deemed necessary, the first part of the discussion given in the previous section should be reviewed before continuing. In making a start, the given curve should first be analyzed and certain facts noted concerning the slope at specific points along the curve. For example, in Fig. 20.21, it should be recognized that the slope for the first and second intervals is constant and positive, a fact indicated by the single straight line sloping upward across both strips. In the third interval (2 to 3), the curve has a constant zero slope; while again, for the fourth and fifth intervals, the slope is constant and positive. Finally, the curve for the last three intervals has a constant negative slope, because the straight line, this time downward, indicates decreasing values.

A pole distance must now be selected, remembering that the distance in abscissa units determines the ordinate scale ratio for both curves. Then, with the location of the pole point P established, rays are drawn, parallel to the straight lines of the given curve until they intersect the Y-axis for the derived curve. These points of intersection, projected horizontally, determine the derived curve that is composed of straight lines in this particular case.

(a)

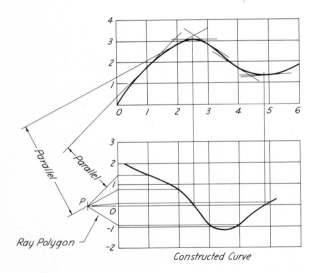

Fig. 20.21 Derivative curve constructed by the pole method.

Fig. 20.22 Use of tangents.

(b)

Fig. 20.23 Chordal method.

In Fig. 20.22, the derived curve (slope locus) was obtained from the given (upper) curve by using tangents. Through the pole point P, a ray was drawn parallel to each of the several tangent lines. Then, the intersection point of each ray with the Y-axis was projected horizontally across to the corresponding ordinate as shown. The drawing of a smooth curve through these points completed the graph.

Chords were used instead of tangent lines in Fig. 20.23. The procedure that is followed here may be thought of as being the reverse of the chordal method illustrated in Fig. 20.11. As the initial step for the construction, chords of the given curve were drawn as shown. In selecting these chords, an effort was made in every case to see that the curve would be nearly symmetrical about a perpendicular bisector of the subtended chord. Also, as is appropriate and desirable, shorter chords were used where the curvature is sharper. After the pole point P had been located and the axes established for the derivative curve, rays were drawn through P, parallel to the chords, to an intersection with the Y-axis. Next, the tangent point (such as T_1) was located for each of the subtended arcs of the given curve, using the method shown below in (b). In doing this it was assumed that the chords would be parallel to tangents at points, such as T_1, where the perpendicular bisector of the chord would cut the subtended arc. These tangent points, when projected downward to corresponding horizontal lines drawn from the points along the Y-axis, at the rays, determine the derivative curve. For example, in (b) point U_1 in interval A is found by projecting T_1 downward to a horizontal line drawn through the point of intersection of the Y-axis and the related ray through P. The related ray in this case is the one that is parallel to the chord of the arc of the given curve in interval A. The desired derivative curve was completed by drawing a smooth curve through U_1 and the several other points shown, all of which were located in the same manner.

When using the chordal method to determine a derivative curve, some people frequently prefer to use equally spaced ordinates and then confine each chord to a single interval. Although this procedure proves to be quite satisfactory in many cases, there are times when desired results cannot be obtained for there are either no points or too few points given to draw a smooth curve at critical locations.

In such cases, some extra construction will be needed. However, the easiest and best solution would have resulted from a careful choice of chords of the given curve at the start.

Figure 20.24 shows the differentiation of a curve using chords instead of tangents. The given distance–time curve was plotted from data obtained for a passenger train leaving a small station near a large city. The velocity and acceleration curves reveal that the train moves with a constant acceleration for approximately 100 sec until it reaches a velocity of 55 miles per hr. From this point it travels with a constant velocity toward its destination.

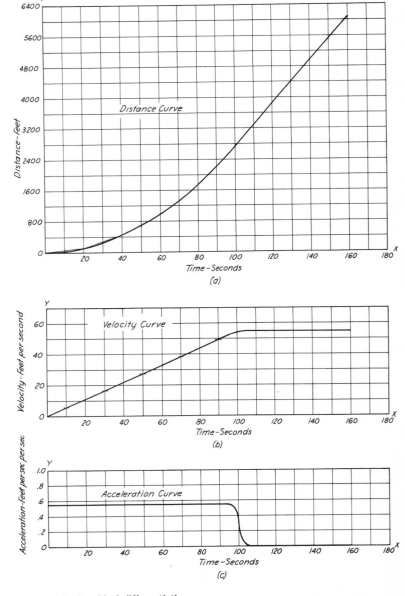

Fig. 20.24 Graphical differentiation.

Problems

Graphical algebra

1. Find the roots of the simultaneous equations

$$y = 2x^2 - x + 4$$
$$y = x^2 + 3x + 9$$

2. Find the roots of the equation by the method shown in Fig. 20.4.

$$y = x^2 + 2x - 3$$

3. Find the roots of the equation by the method shown in Fig. 20.5.

$$y = 3x^2 + 6x + 3$$

4. Find the roots of the equation by the method shown in Fig. 20.4.

$$y = 3x^2 + 12x + 12$$

Graphical calculus

The following problems have been designed to emphasize the fundamental principles underlying graphical integration and differentiation and to offer the student an opportunity to acquire a working knowledge of the methods that are commonly used by professional engineers.

5. A beam 3 m long is uniformly loaded at 30 kg per meter as shown in Fig. 20.25. Plot distance along the X-axis. Draw (1) the integral curve to show the shearing force, and (2) the second integral curve to show the bending moment. Note: Under the SI system a force is given in Newtons. Read Sec. 9.23.

Fig. 20.25

6. A beam 15 ft long and supported at both ends is loaded uniformly at 18 lb per foot as shown in Fig. 20.26. Plot distance in feet along the X-axis and load along the Y-axis. Draw (1) the integral curve to show the shearing force, and (2) the second integral curve to show the bending moment.

Fig. 20.26

7. Plot the points given in the table and draw a smooth curve. Construct the derivative curve. Write the equation of the derivative curve. Equation of given curve $y = \frac{1}{2}x^2 + 10$.

x	0	1	2	3	4	5	6	7	8	9	10
y	10	10.5	12	14.5	18	22.5	28	34.5	42	50.5	60

8. Construct the distance–time, velocity–time, and acceleration–time curves for an automobile moving as follows: time = 10 sec, acceleration = 1.5 m per sec per sec throughout the interval, initial velocity = 0.

9. The passenger train for which derived curves are shown in Fig. 20.24 is brought to a stop with a constant negative acceleration of 1.0 ft per sec per sec. Before applying the brakes the train was traveling with a constant velocity of 55 ft per sec. Construct the curves showing acceleration–time, velocity–time, and distance–time relationships.

PART 7

DESIGN
AND
COMMUNICATION
DRAWING IN
SPECIALIZED
FIELDS

IGNITOR

FUEL NOZZLE

EXHAUST

COMBUSTOR

TRUCK ACCESSORY
DRIVE

POWER
TURBINE

REDUCTION GEAR

COMPRESSOR
TURBINE

AIR
INLET

COMPRESSOR

REGENERATOR

STARTER
MOUNTING

OIL
SUMP

FUEL CONTROL
DRIVE

LUBE PUMP AND
REGENERATOR DRIVE

New research prototype gas turbine engine, a low-pressure, regenerative power plant.
The compressor in the 375-horsepower engine turns at 37,500 revolutions per minute,
the output shaft at only 3000 revolutions per minute. It weighs 1700 pounds installed,
with a length of 40 inches and a height of 39 inches. The reader should note the machine elements that can be readily seen. A few of these elements will be discussed in
this chapter. (*Courtesy Ford Motor Company*)

design of machine elements—gears, cams, and linkages

21

21.1

Gears. It is important for the designer of mechanical systems to know the general proportions and nomenclature pertaining to gearing. In Fig. 21.1 the nomenclature for bevel gears is shown. It will be noted that, in general, the definitions pertaining to gear-tooth parts can be represented in a right section of the gear.

The theory of gears is a part of the study of mechanisms. In working drawings of gears and toothed wheels it is necessary to draw at least one tooth of each gear. Some of the terms used in defining gear teeth are shown in Fig. 21.2.

Two systems of generating tooth curves are in general use, the *involute system* and the *cycloidal system*. The curve most commonly used for gear-tooth profiles is the involute of a circle.

An *involute* is the curve generated by a point on a straightedge as the straightedge is rolled on a cylinder. It also may be defined as the curve generated by a point on a taut string as the string is unwrapped from a cylinder. The circle from which the involute is developed is called the *base circle*.

A method of constructing an involute curve is shown in Fig. 21.3. Starting with point 0 on the base circle, divide the base circle into a convenient number of equal arcs of length 0–1, 1–2, 2–3, and so forth. (Where the lengths of the divisions on the base circle are not too great, the chord can be taken as the length of the arc.) Draw a tangent to the base circle at point 0, and divide this line to the left of 0 into equal parts of the same lengths as the arcs. Next, draw tangents to the circle from points 1, 2, 3, and so on. With the center of the base circle *O* as a pivot, draw

Fig. 21.1 Bevel-gear nomenclature.

Fig. 21.2 Spur-gear nomenclature.

Fig. 21.3 Involute tooth.

concentric arcs from 1', 2', 3', and so forth, until they intersect the tangent lines drawn from 1, 2, 3, and so forth. The intersection of the arcs and the tangents are points on the required involute curve, such as 1'', 2'', 3'', and so forth. The illustration in Fig. 21.3 shows the portion XY of the tooth outline as part of the involute curve.

The *cycloidal system*, as the name implies, has tooth curves of cycloidal form. A *cycloid* is the curve generated by a point on the circumference of a circle as the circle rolls on a straight line. If the circle rolls on the outside of another circle, the curve generated is called an *epicycloid;* if it rolls on the inside of another circle, the curve generated is called a *hypocycloid.* In Fig. 21.4, let R be the radius of the

fixed circle and r be the radius of the rolling circle. Draw through a a circle arc, AB, concentric with the fixed circle. Lay off on the rolling circle a convenient number of divisions, such as 0–1, 1–2, 2–3, and so forth; then divide the fixed-circle circumference into divisions, of the same length, such as 0–1', 1'–2', 2'–3', and so on. Through these points on the fixed circle, draw radii and extend them to intersect the arc AB, thus producing points a_1, a_2, a_3, and so on. These points will be the centers of the successive positions of the rolling circle. Draw the positions of the rolling circle, using the centers a_1, a_2, a_3, and so forth. Next draw, on the rolling circle with the center O of the fixed circle as the pivot point, concentric arcs through points 1, 2, 3, and so forth. The intersection of these arcs with the rolling circles about a_1, a_2, a_3, and so forth, determine points, such as 1'', 2'', 3'', and so forth, on the epicyclic curve. The illustration in Fig. 21.4 shows XY of the tooth outline as part of the epicyclic curve.

The hypocyclic-curve construction is the same as that for the epicyclic curve. In the construction of the hypocyclic curve, if the rolling circle has a diameter equal to one-half of the diameter of the fixed circle, the hypocyclic curve thus generated will be a radial line of the fixed circle.

Gear terms

1. The addendum circle is drawn with its center at the center of the gear and bounds the ends of the teeth. See Fig. 21.2.

2. The dedendum circle, or root circle, is drawn with its center at the center of the gear and bounds the bottoms of the teeth. See Fig. 21.2.

3. The pitch circle is a right section of the equivalent cylinder the toothed gear may be considered to replace.

4. Pitch diameter is the diameter of the pitch circle.

5. The addendum is the radial distance from the pitch circle to the outer end of the tooth.

6. The dedendum is the radial distance from the pitch circle to the bottom of the tooth.

7. The clearance is the difference between the dedendum of one gear and the addendum of the mating gear.

8. The face of a tooth is that portion of the tooth surface lying outside the pitch circle.

9. The flank of a tooth is that portion of the tooth surface lying inside the pitch circle.

10. The thickness of a tooth is measured on the arc of the pitch circle. It is the length of an arc and not the length of a straight line.

11. The tooth space is the space between the teeth measured on the pitch circle.

12. Backlash is the difference between the tooth thickness of one gear and the tooth space on the mating gear, measured on the pitch circles.

13. The circular pitch of a gear is the distance between a point on one tooth and the corresponding point on the adjacent tooth, measured along the arc of the pitch circle. The circular pitches of two gears in mesh are equal.

14. The diametral pitch is the number of teeth per inch of pitch diameter. It is obtained by dividing the number of teeth by the pitch diameter.

15. The face of a gear is the width of its rim measured parallel to the axis. It should not be confused with the face of a tooth, for the two are entirely different.

16. The pitch point is on the line joining the centers of the two gears where the pitch circles touch.

17. The common tangent is the line tangent to the pitch circles at the pitch point.

18. The pressure angle is the angle between the line of action and the common tangent.

19. The line of action is a line drawn through the pitch point at an angle (equal to the pressure angle) to the common tangent.

20. The base circle is used in involute gearing to generate the involutes that form the tooth outlines. It is drawn from the center of each pair of mating gears tangent to the line of action.

21. When two gears mesh with each other, the larger is called the *gear* and the smaller the *pinion*.

It should be noted that *circular pitch* is a linear dimension expressed in inches, whereas *diametral pitch* is a ratio. There must be a whole number of teeth on the circumference of a gear. Thus it is necessary that the circumference of the pitch

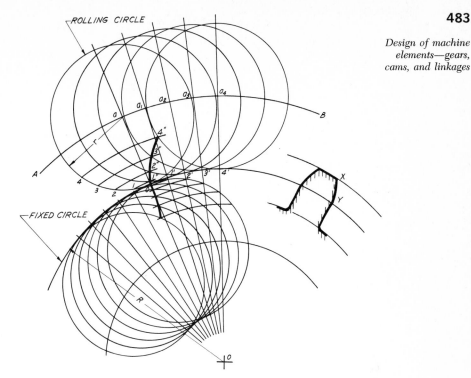

Fig. 21.4 Tooth curves—cycloidal form.

circle, divided by the circular pitch, be a whole number.

For circular pitch, let P' be the circular pitch in inches, D the pitch diameter, and T the number of teeth. Then

$$TP' = \pi D, \qquad T = \frac{\pi D}{P'},$$

$$P' = \frac{\pi D}{T}, \quad \text{and} \quad D = \frac{TP'}{\pi}$$

For diametral pitch, let P be the diametral pitch, D the pitch diameter, and T the number of teeth. Then

$$T = PD, \qquad D = \frac{T}{P},$$

$$\text{and} \qquad P = \frac{T}{D}$$

The Brown and Sharpe $14\frac{1}{2}°$ involute system has been adopted as one of the American standards and is commonly known as the $14\frac{1}{2}°$ *composite system.* The tooth proportions of this system are given in terms of the diametral pitch P and circular pitch P'.

Pressure angle $= 14\frac{1}{2}°$.

Addendum (inches) $= 1/\text{diametral pitch} = 1/P$.

484

*Design of machine
elements—gears,
cams, and linkages*

Dedendum (inches) = addendum plus clearance = $(1/P) + .05P'$.

Clearance = .05 × circular pitch = $.05P'$.

Whole depth of tooth = 2 × addendum + clearance = $2 \times (1/P) + .05P'$.

Working depth of tooth = 2 × addendum = $2 \times (1/P)$.

Thickness of tooth = circular pitch/2 = $P'/2$.

Width of tooth space = circular pitch/2 = $P'/2$.

Minimum radius of fillet = clearance = $.05P'$.

In the above calculations the backlash is zero. Actually, however, it is common practice to provide backlash, and this is accomplished by using standard cutters and cutting the teeth slightly deeper than for standard teeth.

21.2

To Lay Out a Pair of Standard Involute Spur Gears. The following facts are known regarding the laying out of a pair of standard spur gears: (1) number of teeth on each gear—large gear 24, small gear 16; (2) diametral pitch = 2; (3) pressure angle = $14\frac{1}{2}°$.

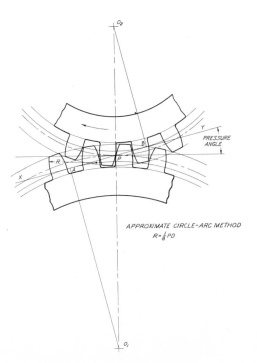

APPROXIMATE CIRCLE-ARC METHOD
$R = \frac{1}{8} PD$

Fig. 21.5 To draw a pair of spur gears.

To draw a pair of spur gears, determine the pitch diameters as follows:

$$D = \frac{T}{P} = \frac{24}{2} = 12 \text{ in.} \qquad \text{(for large gear)}$$

$$D = \frac{T}{P} = \frac{16}{2} = 8 \text{ in.} \qquad \text{(for small gear)}$$

In Fig. 21.5, with radii O_1P and O_2P equal to 6 and 4 in., respectively, draw the pitch circles and, through P, draw the common tangent. Draw the line of action XY at an angle of $14\frac{1}{2}°$ to the common tangent. Drop perpendiculars from the centers O_1 and O_2, cutting the line of action at A and B, respectively. O_1A and O_2B are the radii of the base circles that can now be drawn.

From Sec. 21.1, determine the addendum and dedendum of the teeth, and draw in the respective addendum and dedendum circles.

Divide the pitch circle of the smaller gear into 16 equal parts and the pitch circle of the larger gear into 24 equal parts, which will give the circular pitch. Assuming that no allowance is made for backlash, bisect the circular pitch on each of the gears, which will give 32 equal divisions on the small gear and 48 equal divisions on the large gear.

At any point on the base circle of each gear, develop an involute (see Fig. 21.3) and draw in the curves between the base and addendum circles through alternate points on the pitch circles. This produces one side of all the teeth in each gear. The curve for the other side of the tooth is the reverse of the side just drawn. The part of the tooth between the base and dedendum circles is part of a radial line drawn from the base circles to the centers of the gears. The tooth is finished by putting in a small fillet between the working depth and dedendum circles.

21.3

Dimensioning Gears. On a detail drawing, a gear may be represented by a one-view section, except for the larger sizes, where it is necessary to show spokes, and for the small sizes, when a full description of some feature would not be given. A second view was drawn in Fig. 21.6 to reveal the shape of the keyway. In dimensioning, it is recommended that the dimensions be given on the view or views, as shown in Figs. 21.6 and 21.7, and that the cutting data be incorporated in an accompanying table.

21.4

Working Drawings of Gears. Figure 21.6 shows a working drawing of a helical gear. In practice it is customary to show one view in section and just enough of the circular view to supply needed information for the shop. Individual teeth are never shown on the circular view unless the drawing is to be used for display purposes. In some drafting rooms where practice requires a full circular view, the addendum and root circles are drawn using a phantom line and the pitch circle is given as a center line.

It should be noted that the dimensions for the gear blank are given on the sectioned view, while the necessary data for cutting the teeth are given in a table. This is in accordance with the practice recommended in Sec. 21.3.

A working drawing of a spiral bevel gear is shown in Fig. 21.7. The same practices apply as for spur gears. Before starting to prepare a working drawing of a bevel gear one should study Fig. 21.1, which illustrates bevel-gear nomenclature.

Figure 21.8 shows a working drawing of a worm.

21.5

Cams. A *cam* is a plate, cylinder, or any solid having a curved outline or curved groove that, by its oscillating or rotating motion, gives a predetermined motion to another piece, called the *follower*, in contact with it. The cam plays a very impor-

tant part in the operation of many classes of machines. Cam mechanisms are commonly used to operate valves in automobiles and stationary and marine internal combustion engines. They also are used in automatic screw machines, clocks, locks, printing machinery, and in nearly all kinds of machinery that we generally regard as "automatic machines." The applications of cams are practically unlimited, and

CUTTING DATA			
NO. OF TEETH	32	PITCH DIAM.	4.250
CHOR. ADD. — ADD.	.125	DIAM. PITCH	8.0 NORM.
DEDENDUM	.145	PRESS ANGLE	20°
FULL DEPTH	.270	HELIX ANGLE	19°45'RH.
CENTER DIST.	4.251	BACKLASH	.004-.008

MAT.- N.E. 8645
HARDNESS ROCKWELL "C" 38-42

Fig. 21.6 Detail drawing of a helical gear. (*Courtesy Fairfield Mfg. Co.*)

CUTTING DATA			
NO. OF TEETH	28	CIRC. PITCH	.5048
CHOR. ADD. — ADD.	.148	FACE ANGLE	45°7'
DEDENDUM	.155	PITCH ANGLE	42°5'
FULL DEPTH	.303	ROOT ANGLE	39°26'
CONE DISTANCE	3.358	PRESS ANGLE	14½°
PITCH DIAM.	4.502	SPIRAL ANGLE	RH.33°52'
DIAM. PITCH	6.22	BACKLASH	.004-.008

FINISH ALL OVER
MAT- S.A.E. 4620
HARDNESS-ROCKWELL "C"
58 MIN.

Fig. 21.7 Detail drawing of a spiral bevel gear. (*Courtesy Fairfield Mfg. Co.*)

Fig. 21.8 **Detail drawing of a worm.** (*Courtesy Warner & Swasey Co.*)

Fig. 21.9 Disc cam and follower.

The most common type of cam is the disc or plate cam. Here the cam takes the form of a revolving disc or plate, the circumference of the disc or plate forming the profile with which the follower makes contact. In Figs. 21.9 and 21.10, two simple examples of a disc cam and follower are shown. In Fig. 21.9, the cam is given a motion of rotation, thus causing the follower to rise and then return again to its initial position. In cams of this type it is necessary to use some external force, such as the spring, to keep the follower in contact with the cam at all times. Contact between the follower and the cam is made through a roller, which serves to reduce friction. It is sometimes necessary to use a flat-faced follower, instead of the roller type, an example of which is shown in Fig. 21.10. The follower face that comes in contact with the cam is usually provided with a hardened surface, to prevent excessive wear.

Another type of cam is one in which the follower is constrained to move in a definite path without the application of external forces. See Fig. 21.11. In this type, two contact surfaces of the follower bear on the cam at the same time, thus controlling the motion of the follower in two directions.

their shapes or outlines are found in wide variety.

All cam mechanisms consist of at least three parts: (1) the cam, which has a contact surface either curved or straight; (2) the follower, whose motion is produced by contact with the cam surface; and (3) the frame, which supports the cam and guides the follower.

21.6

Design of a Cam. The design of a cam outline is governed by the requirements with respect to the motion of the follower.

Fig. 21.10 Disc cam and follower.

Fig. 21.11 Cam and follower.

487

*Design of machine
elements—gears,
cams, and linkages*

In the layout of a cam, the initial position, displacement, and character of the motion of the follower are generally known. It is convenient to make first a graphical representation of the follower movement, a procedure that is called *making a displacement diagram*. This is a linear curve in which the length of the diagram represents the time for one revolution of the cam. The height of the diagram represents the total displacement of the follower; the length is made to any convenient length and is divided into equal time intervals, the total representing one rotation of the cam.

In Fig. 21.12 is shown a displacement diagram in which the follower rises 2 in. during 180° of rotation of the cam, then rests for 30° and returns to its initial position for the remainder of the cam revolution. Cam outlines should be designed to avoid sudden changes of motion at the beginning and end of the follower stroke. This can be accomplished by having a uniformly accelerated and decelerated motion at the beginning and end of the constant-velocity curve. The construction for uniformly accelerated motion is shown in Fig. 21.12. On a line, OX, making any convenient angle with OA, mark off any unit of length in this figure equal to Oa. The next point, b, is found by marking off, from 0, 4 units of length. Point c is found by marking off 9 units of length. Next, project the intersection (point s) of time unit 3 and the constant-velocity line over to line OA, thus locating point t. Connect points c and t with a straight line and draw parallel lines from a and b intersecting line OA. From these intersections draw lines parallel to ts, intersecting the time-unit lines 1 and 2, respectively. These intersections are points on the displacement curve. With uniformly decelerated motion, the series of points are laid off in the reverse order, such as

9-4-1. It will be noted that the units are laid off according to the square of the time unit. Thus, if there were 4 time units, the acceleration curve would be laid off according to the ratio of 1, 4, 9, 16, and the deceleration, 16, 9, 4, 1.

The construction for the displacement diagram for simple harmonic motion is shown in the same figure. A semicircle is drawn as shown, the follower displacement being used as a diameter, and is then divided into a convenient number of parts equal to the number of cam displacement units. Horizontal projection lines are drawn from the semicircle, and the intersections of these lines with the cam displacement lines are points on the displacement curve. Thus, the projection of point 15 on the semicircle to time-unit line 15 locates one point on the displacement curve for simple harmonic motion.

The next step is that of finding the cam profile necessary to produce these movements. The construction is shown in Fig. 21.13. Select a base circle of convenient size, and on it lay off radial lines according to the number of time units of cam displacement. Draw line OB extended to W, and on it lay off the distances y_1, y_2, y_3, and so forth, obtained from the displacement diagram, from the center of the roller shown in the starting position, thus locating points B_1, B_2, B_3, and so forth. With O as a center, draw arcs B_1-B_1', B_2-B_2', B_3-B_3', and so forth, and at B_1', B_2', B_3', and so forth, draw in the circles representing the diameter of the roller. To complete the cam outline, draw a smooth curve tangent to the positions of the roller.

21.7
Linkages.* The link is the most common machine element. A link is a rigid bar that

---

*Secs. 21.7–21.11 prepared by Wesley L. Baldwin, Mechanical Engineering Technology, Purdue University.

Fig. 21.12 Displacement diagram.

Fig. 21.13 Construction for cam profile.

transmits force and velocity. An assemblage of links that produces a prescribed motion is called a *linkage*. Since linkages appear in nearly all phases of mechanical design, a knowledge of linkage design and analysis is necessary for all those designers who may be involved in the creation of useful mechanisms. See the page facing the beginning of this chapter.

21.8
Position and Clearance Analysis. In the preliminary design stage, a design layout man is frequently called upon to find the position of each link as the machine operates. When the layout has been completed, he must then check to be sure that the linkage does not interfere with any other moving machine elements. Sufficient clearance must be left between rapidly moving machine parts to ensure that any deflection in the parts will still allow adequate clearance.

Mathematics, primarily geometry and trigonometry, can be used to calculate accurately both the position and clearances. On the other hand, graphical solutions provide sufficient accuracy with a minimum amount of effort and time. A graphical solution also allows the designer to visualize the problem easily and rapidly. Figure 21.14 shows how the layout man can analyze clearances between moving machine parts.

21.9
Four-bar Linkage. The most common type of linkage found in machines is called 4-bar linkage. This linkage is made up of four pin-connected links. Figure 21.15 shows a typical 4-bar linkage. Notice that link 1 is the distance between the two pin connections attached to the ground. The 4-bar linkage has one degree of freedom. This means that once the position of a link is defined with respect to another link, the remaining links are also in a fixed position. For example, if the angle θ is defined in Fig. 21.15, then the angles β and ψ can have only one value.

The 4-bar linkage is also important to the designer because many other linkages are analyzed in a similar manner. The slider crank and the scotch yoke, shown in Fig. 21.16, are examples of this type of mechanism.

489

*Design of machine
elements—gears,
cams, and linkages*

Fig. 21.14 Graphical solution for the clearances between moving machine elements. (*Courtesy General Motors Corporation*)

21.10

Velocity Analysis of a Link.* A link does not have to be straight. It may be curved so that it does not interfere with another moving element of a mechanism. For analysis, however, we can replace a curved link by a straight link connecting the two ends. Figure 21.17 shows a curved link represented by a straight link. When the link is in motion, the velocity of points A and B will be different. The velocities of these two ends will be designated V_A and V_B, respectively. Because velocity is a vector quantity, we may resolve each vector, V_A and V_B, into two components. Consider the components of V_A and V_B parallel to and perpendicular to the straight line connecting points A and B. In Fig. 21.17 the components parallel to the straight line connecting A and B are designated V_A^t and V_B^t. The perpendicular components are designated V_A^n and V_B^n.

Because the link is solid, V_A^t and V_B^t must be equal. If V_A^t and V_B^t were not equal, then the line joining A and B would have a changing length. The difference in velocity between points A and B must then be due entirely to the difference be-

Fig. 21.15 Typical 4-bar linkage.

Fig. 21.16 Examples of mechanisms that are similar to 4-bar linkages—(a) scotch yoke and (b) slider crank. (*Extracted from* Kinematics and Linkage Design *with the permission of Allen S. Hall, Professor of Mechanical Engineering, Purdue University*)

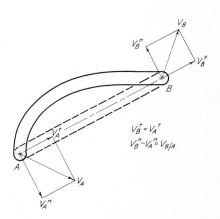

$$V_B^t = V_A^t$$
$$V_B^n - V_A^n = V_{B/A}$$

Fig. 21.17 Link in motion.

tween $V_A{}^n$ and $V_B{}^n$. This difference between $V_A{}^n$ and $V_B{}^n$ is called the relative velocity between points A and B. The relative velocity is designated $V_{B/A}$, when $V_A{}^n$ is subtracted from $V_B{}^n$.

Since $V_A{}^t$ and $V_B{}^t$ are equal, $V_{B/A}$ is also equal to the difference between V_A and V_B. The difference between V_A and V_B, called the relative velocity $V_{B/A}$, lies in a direction perpendicular to the straight line joining A and B. The magnitude of $V_{B/A}$ is the same as the difference between $V_A{}^n$ and $V_B{}^n$. Figure 21.17 shows these relationships.

21.11
Velocity Analysis of a 4-bar Linkage. A typical 4-bar linkage has link 2 driven at a constant angular velocity ω_2. To calculate the angular velocity, ω_4, of link 4:

1. Draw the linkage to scale in the desired position.

2. Calculate the velocity of point A.
$$V_A = \omega_2 \,(O_2 A)$$
V_A is the velocity of point A and is perpendicular to $(O_2 A)$, ω^2 is the angular velocity of link 2 measured in radians/second, and $(O_2 A)$ is the length of link 2.

3. Construct vector V_A to scale.

4. From the tip of V_A construct a line perpendicular to link 3. This line lies in the direction of $V_{B/A}$.

5. From the tail of vector V_A construct a line perpendicular to $(O_4 B)$ to the line constructed in step 3. This line lies in the direction of V_B.

6. The line drawn in step 5 is the scaled velocity of point B.

7. Calculate ω_4.
$$\omega_4 = \frac{V_B}{O_4 B}$$

Problems

491

*Design of machine
elements—gears,
cams, and linkages*

Note: With the approval of his instructor, the student may convert the inch measurements in the problems that follow to millimeters by using Tables 3–9 in the Appendix. Since it is common practice under the metric system to use full millimeters for all dimension values that are not critical, the student should do likewise. For example, 3 in. converts to 76.2 millimeters. The student should use 76 millimeters.

Prob. no.	Gear				Pinion				Pressure angle
	Circular pitch (in.)	Diametral pitch	Pitch diameter	No. of teeth	Circular pitch (in.)	Diametral pitch	Pitch diameter	No. of teeth	
1	1.31			24	1.31			16	$14\frac{1}{2}°$
2		2		20		2		14	$14\frac{1}{2}°$
3		2.5	10			2.5	8		$14\frac{1}{2}°$
4	1.0			30	1.0			20	$14\frac{1}{2}°$
5	2.0			18	2.0			12	$14\frac{1}{2}°$
6		3	8			3	6		$14\frac{1}{2}°$

Gears and cams

Gear problems
1–6. Following the method shown in Sec. 21.2, lay out a pair of standard involute spur gears as assigned from the given table. The pinion is the driver.

Cam problems

Cam Data

Diameter of cam shaft	$1\frac{1}{4}$ in.
Diameter of cam hub	$2\frac{1}{4}$ in.
Diameter of roller	1 in.
Keyway	$\frac{1}{4} \times \frac{1}{8}$ in.
Diameter of base circle	$2\frac{3}{4}$ in.
Follower displacement	2 in.
Scale: full-size	
Cam rotation: as noted	

Determine points on the cam profiles at intervals of 15°.

7. Using the data above, design a plate cam to satisfy the following conditions: (a) a rise of 2 in. in 180°, with constant velocity, except for uniform acceleration for the first 30° and uniform deceleration for the last 45°; (b) rest 30°; and (c) return with simple harmonic motion. Use clockwise cam rotation.

8. Same as Problem 7, except that the follower is of the flat-face type and is $2\frac{1}{2}$ in. wide.

9. Using the data for Problem 7, design a plate cam to satisfy the following conditions: (a) rise of 2 in. during 180°, the first 45° of which is uniformly accelerated motion, the next 60° being constant velocity, and the last 75° of rise being uniformly decelerated motion; (b) rest 15°; and (c) return to starting position with simple harmonic motion. Use counterclockwise cam rotation.

10. Same as Problem 9, except that the follower is of the flat-face type and is $2\frac{1}{2}$ in. wide.

11. Using the data for Problem 7, design a plate cam to satisfy the following conditions: (a) rise to 2 in. during 150°, by simple harmonic motion; (b) rest 30°; and (c) return to starting position during remainder of the revolution, with uniformly accelerated and decelerated motion, the value of the deceleration being twice that of the acceleration. Use clockwise cam rotation.

12. Using the data for Problem 7, except that the follower is to be of the flat-face type, $2\frac{1}{2}$ in. wide, design a plate cam to satisfy the following conditions: (a) rise of 2 in. with simple harmonic motion, in 120°; (b) rest 30°; (c) return in 150°, with constant velocity, except for uniform acceleration for the first 45° and uniform deceleration for the last 30° of fall; and (d) rest the balance of the revolution. Use counterclockwise cam rotation.

Linkage problems*

13. A 4-bar linkage has link 1 = 6 in., link 2 = 2 in., link 3 = 8 in., link 4 = 5 in. Draw the linkage in all positions for $\theta = 0°$ to 360° in 30° increments. Plot a graph of θ vs. ψ.

14. If link 2 in Problem 13 has a constant angular velocity of 10 radians per sec, find the angular velocity of link 4. Use the 12 positions indicated in Problem 13. Plot θ vs. ω_4.

*Prepared by Wesley L. Baldwin, Mechanical Engineering Technology, Purdue University.

The Mole was used for rock tunneling in constructing the 157.7-km Metro System in our nation's capital. This massive machine has a rotating cutter head 5.8 m in diameter with 45 disc cutters. The cutting head, which may turn from 2 to 4 revolutions per minute, is driven by six electric motors. Hydraulic rams force the cutting head against the rock face. The Mole weighs 258 metric tons and is 91.4 m long. Many well-dimensioned drawings were needed for the construction of this giant machine. (*Courtesy The Robbins Co. and Washington Metropolitan Transit Authority*)

electronic drawings* 22

22.1

Introduction. Different kinds of drawings are needed in the design, manufacture, and use of electronic equipment (Fig. 22.1). The preparation of these drawings, particularly schematic and connection diagrams, requires a working knowledge of specialized symbolic forms and conventions. These conventions vary somewhat according to the product and the field of application, which, broadly speaking, may be classified as communications and commercial equipment, military, scientific research, or industrial control.

Engineering sketches are prepared first to show the layout of a proposed system or circuit or the features of existing equipment. Finished drawings are then made by a draftsman who follows the instructions given in the sketch. Usually, many details are left to the draftsman's discretion; however, it is the engineer's responsibility to make sure that the final work is correct and complete. In all instances, electronic drawings should be prepared in accordance with the relevant technical standards and specifications so that they will be acceptable for the particular applications.

22.2

Electronic Drawings and Diagrams. The most specialized work in electronic drafting concerns the preparation of symbolic diagrams. In contrast to mechanical drawings, which represent objects, symbolic diagrams impart technical information in abstract form. Since these diagrams are intended to show the function

* Prepared by George Shiers, Santa Barbara City College.

Fig. 22.1 Pictorial diagram. (*Courtesy Allied Radio Corp.*)

of a system or a circuit, they lack intrinsic dimensions and, in general, do not show physical details of the parts.

Important examples of these specialized drawings are as follows: block diagrams, which show the overall organization of a system; schematic diagrams, which show the component parts and electrical details of a circuit; and connection diagrams, which show the wiring and connections between the component parts of an assembly. Other drawings and artwork are needed in the design and processing of printed circuits.

All these drawings are interrelated with the physical assembly and with each other. In many instances further material, such as lists, tables, and charts, may be needed to supplement assembly drawings and electronic diagrams, particularly for production breakdowns, test procedures, and service manuals.

22.3

Graphical Standards. Technical publications dealing with standardized engineering practices are available to guide the engineer and draftsman in the preparation of electronic diagrams. These standards have been established by technical committees representing professional societies, trade associations, government agencies, and various manufacturers and users. The most useful standards for electronic purposes are those that concern graphical symbols, letter symbols, reference designations, abbreviations, color codes, and electrical diagrams, as given in the following list.

ANSI—American National Standards Institute, Inc.
- Y14.15—Electrical Diagrams
- Y32.2 —Graphical Symbols for Electrical Diagrams
- Z10.1 —Abbreviations for Scientific and Engineering Terms
- Z10.5 —Letter Symbols for Electrical Quantities
- Z32.13—Abbreviations for Use on Drawings

EIA—Electronic Industries Association
- GEN–101—Color Coding for Numerical Values
- GEN–102—Preferred Values
- REC–108—Color Coding

IEEE—The Institute of Electrical and Electronics Engineers, Inc.
(These publications were formerly issued by The Institute of Radio Engineers, Inc.)
- 51 IRE 21.S1—Standards on Abbreviations of Radio-Electronic Terms
- 54 IRE 21.S1—Standards on Graphical Symbols for Electrical Diagrams
- 57 IRE 21.S1—Standards on Letter Symbols and Mathematical Signs
- 57 IRE 21.S2—Standards on Reference Designations for Electrical and Electronic Equipment
- 57 IRE 21.S3—Standards on Graphical Symbols for Semiconductor Devices

MIL—U.S. Government Printing Office
- MIL–STD–12 —Abbreviations for Use on Drawings
- –15 —Electrical and Electronic Symbols
- –16 —Electrical and Electronic Reference Designations
- –122—Color Code for Chassis Wiring for Electronic Equipment
- –283—Letter Symbols for Electrical and Electronic Quantities

NEMA—National Electrical Manufacturers Association
- IC–1959—Industrial Control

NMTBA—*National Machine Tool Builders' Association*

Electrical Standards for Industrial Equipment

Machine Tool Electrical Standards

These and related standards have been developed in the interest of clear documentation and accurate terminology. Their use promotes efficient engineering methods leading to economies in time, materials, and labor. An acquaintance with these standards, therefore, forms an essential part of the background of engineers, draftsmen, technicians, and supervisors engaged in the electronics industry.

22.4

Graphical Symbols. A graphical symbol is a geometrical design that represents an electronic device or component part in a circuit. Most symbols are composed of two or more basic elements, each of which represents a functional part of the device. Some commonly used electronic symbols are shown approximately full-size, with their letter designations, in Fig. 22.2.

Graphical symbols should be drawn in proportion to one another and with clear details so that the significance will be unmistakable. Symbol templates are widely used for drawing schematic diagrams. Printed designs on adhesive sheets (appliqués) are also employed, particularly for ink work and drawings prepared for publication. These drafting aids save time, avoid tedious construction of individual symbols, and promote uniformity.

22.5

Reference Designations. Component designations, such as R1, C3, etc., are added to each symbol to indicate the class of component and its position in a circuit. These designations are assigned during the development stages of a product and usually appear as component markings on the actual equipment.

Other designations or symbol references may be required in a drawing to show the part value, type, electrical rating, or specific characteristics of a component. Examples are 15K, for a 15,000-Ω (ohms) resistor, where K = 1000; 6.8MH (or mH), for a coil with an inductance of 6.8 millihenrys; 2/400, as applied to a 2-μF (microfarad—MF) capacitor with a maximum working rating of 400 V (volts) dc; and 5–25, as applied to the range in micromicrofarads ($\mu\mu$F, or MMF) of an adjustable capacitor.

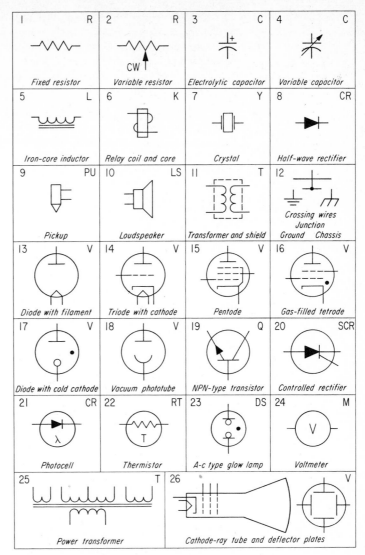

Fig. 22.2 Graphical symbols with reference designations.

The following examples are with reference to Fig. 22.2.

Item 2. The letters CW indicate the direction of the slider when the shaft is rotated clockwise. An arrowhead may be used instead.

Item 3. The plus sign indicates a polarized capacitor, such as an electrolytic type.

Item 4. The arrowhead signifies variability.

Item 5. Parallel lines indicate a magnetic core.

Item 11. Broken lines signify a shield or screen, either electrical or magnetic, or a mechanical enclosure. A broken line that connects two or more symbols also denotes a mechanical linkage.

Item 12. Crossing lines that are not connected are drawn straight; a loop or saddle at the crossover point is not necessary. The junction dot is optional and may be eliminated provided the junction is obvious. Connections to a ground point or to a chassis or frame are distinguished by different symbols.

Item 15. Grids in electron tubes are numbered starting with the one nearest the cathode. A suppressor grid (No. 3) may be connected to the cathode internally, as shown, or to a separate pin in the tube base.

Item 16. Gaseous tubes are distinguished from high-vacuum devices by a solid dot conveniently located in the symbol, as shown.

Item 19. For *PNP*-type transistors the arrowhead on the emitter is reversed.

Item 21. Conductive type of photocell employing a semiconductor. "Lambda" signifies light sensitivity.

Item 22. The letter T denotes temperature sensitivity.

Item 24. Meter function is shown by an appropriate letter designation.

22.6
Data Sheets. Manufacturers' data sheets, technical bulletins, and product manuals contain full particulars of each class of device. These publications, such as tube, diode, transistor, and rectifier handbooks, should be consulted for basing diagrams, lead arrangements, socket and terminal

Fig. 22.3 Typical block diagrams: (A) computer system; (B) control system with special symbols. [*From G. Shiers,* Electronic Drafting (*Englewood Cliffs, N.J.: Prentice-Hall, Inc.*)]

connections, polarities, part codings, and the like.

22.7
Block Diagrams. Rectangular and other symbols in block diagrams may represent integral parts of a system, self-contained units, complete circuits, or single functional stages, according to the purpose of the drawing. The blocks are customarily located and joined with flow lines so that the overall progression is from left to right. Auxiliary items, such as power sources, are placed below the major blocks. In general, all blocks should be drawn the same size, with other symbols in proportion, and arranged in an orderly manner with equal spacings between adjacent blocks and flow lines.

The lines to each block also follow a similar sequence: input on the left, output on the right, with auxiliary lines usually at the bottom. In more complex drawings, where the flow paths may be somewhat circulatory, the symbols should be located to preserve the most logical sequence in the clearest possible manner with a minimum of crossovers.

Two representative block diagrams are shown in Fig. 22.3. In the computer system diagram (A) the blocks are drawn large enough to accommodate the lettering. The paths for control signals and data are distinguished by different line symbols, and a legend is furnished accordingly. The flow directions are indicated by arrowheads, which may be placed either on or beside the lines.

The control system diagram (B) includes some special block-type symbols, also some lettering adjacent to the symbols. Graphical symbols may be used in block diagrams, particularly for auxiliary items, such as antennas, grounds, switches, and various input and output devices.

22.8
Schematic Diagrams. A schematic diagram shows the electrical functions of a circuit without regard to the physical layout of the actual parts. Circuit components, wiring devices, and the interconnections are represented in a schematic by graphical symbols. Suitable part designations and other lettering must be furnished for reference. These diagrams are used in designing and developing circuits, in manufacturing and testing a product, and for installing and servicing complete equipment.

Fig. 22.4 Part of a complex drawing showing a typical schematic pattern. [*From G. Shiers, Electronic Drafting (Englewood Cliffs, N.J.: Prentice-Hall, Inc.)*]

Part of a complex schematic, shown in Fig. 22.4, illustrates how a large amount of circuit data can be presented in an orderly manner. In laying out this kind of diagram, care is required during the planning stages to ensure proper distribution of the parts and to avoid undue crowding, wasted space, or excessive crossovers. The finished diagram should show the circuit in a logical sequence proceeding from left to right, as for block diagrams.

22.9

Schematic Projection. Most electronic circuits consist of several functional blocks, or stages. Each stage normally contains an active device, such as a tube or transistor, and the associated circuit components. Usually, stages are connected in cascade, whereby the output from one stage is fed to the input of the succeeding stage. In general, the symbols representing a single stage are grouped together. Each stage requires the four following basic connections: signal input and output, power supply line, and common line.

Using *horizontal* projection, stages are located between the power bus line and the common line, as shown in Figs. 22.4 and 22.6. When the *vertical* method of projection is used, the power bus is omitted and the circuit layout is arranged

somewhat differently with the power supply lines located vertically between the stages, as shown in Fig. 22.5. In both methods the diagram may be *complete*, showing all bus lines and common circuit paths (Fig. 22.6); or these lines may be isolated, with terminating arrowheads and designations to show the respective connections (Fig. 22.5). As a rule, horizontal

Fig. 22.5 Typical engineering sketch, marked up to show symbol grouping. [*From G. Shiers, Electronic Drafting (Englewood Cliffs, N.J.: Prentice-Hall, Inc.)*]

Fig. 22.6 Trial layout: (A) preliminary layout for a two-stage amplifier showing typical spacings; (B) through (J), layout errors to be avoided. [*From G. Shiers,* Electronic Drafting (*Englewood Cliffs, N.J.: Prentice-Hall, Inc.*)]

Symbols may be divided where they refer to a multifunction device, such as a dualtriode, multicircuit switch, or a multiple-contact relay. Symbols for each part of a device, with suitable references, are placed in the respective circuit paths. Clarity is improved, and drafting time is saved by the elimination of long lines and the reduction of crossovers.

Figure 22.4 shows the use of split symbols where each half of tubes V405, 406, and 407 are drawn apart to allow space for the interstage circuits. These sections may be located anywhere in a diagram, as for V206 in the bottom circuit. This diagram also shows the omission of tube heaters, again in the interest of simplicity and clarity. Conventionally, tube heaters and associated items, such as low-voltage pilot lamps, are drawn as subcircuits below the main part of a diagram.

22.11
Modular Layouts. Because of the abstract nature of the data and the "dimensionless" form, schematic diagrams are best laid out according to a modular plan. Cross-section or quadrille paper with eight rulings per inch is suitable for developing most diagrams. Symbol sizes, symbol locations, and the needed drawing areas can be estimated or specified easily and accurately in terms of grid units with this method. If necessary, a large diagram may be broken down into sections containing several stages. Each section may then be developed separately in trial layouts, revised, and fitted together ready for the final drawing.

22.12
Trial Layouts. Figure 22.6(A) is an example of a trial layout, with typical spacings in grid units marked on each axis. The shaded areas represent spaces for required lettering. The solid dots denote tube pin numbers. Symbols should be aligned on common grid lines but staggered if required to permit adequate lettering. Common layout faults to be avoided are also shown. These consist of lines too close to symbols (B), (C), (D), and (G); symbols too close (E), (F), and (J); inconsistent symbols (E); insufficient lettering space (H); and symbols off-center (J).

The grid units shown in Fig. 22.6 represent typical dimensions suitable for most schematic layouts. Thus, lengthwise: four–six units for a capacitor, eight units for a resistor or inductor, and nine or ten

projection is commonly used for military and some industrial applications, whereas the vertical projection method is favored for most commercial and radio communication equipment.

22.10
Circuit Analysis. In the early stages of electronic design suitable components are selected and connected to achieve a certain result. An engineering sketch, such as Fig. 22.5, is drawn to show the proposed circuit. Component types and values may be given in this sketch or in a separate parts list. With the parts mounted on a temporary base, or "breadboard," the circuit is then tested and modified as may be required. As soon as the circuit features have been established a finished schematic is drawn from the engineering data.

The first step in laying out a schematic diagram is to analyze the circuit sequence and electrical details so that the technical requirements and data are clearly understood. Major elements common to all schematic diagrams that should be recognized are as follows: (1) graphical symbols; (2) symbol grouping; (3) flow paths for signals and power; (4) operating functions; and (5) reference designations, electrical values, and similar identities.

It may be helpful to mark up a sketch temporarily in order to show the preferred symbol grouping, as indicated by the broken lines in Fig. 22.5. In this diagram the functional blocks are marked ①–⑦. Flow paths for the signal ⑧ and the operating power ⑨ are shown, also mechanical linkage ⑩ between the variable capacitors, which actually consist of two units in a common assembly.

units between the plate and cathode connections of a tube, or between the emitter and collector connections of a transistor. Horizontal spacings will depend on the specific circuits, symbol arrangements, amount of lettering, and the method of projection.

Sufficient space should be allotted for auxiliary circuits, such as filters, and for switches, terminal blocks, connectors, and the related lettering. If the lettering size corresponds with the grid scale it will be in proportion to the symbols. Complex diagrams containing several hundred symbols, or other diagrams designed for a compact space, as in equipment manuals, can be drawn to a smaller *scale*, such as $\frac{1}{10}$ in., with lettering to correspond. Using a modular plan and trial layout as described, a finished diagram may be drawn to an enlarged or reduced scale by selecting a suitable grid background.

22.13

Placement of Designations. The parts in a schematic diagram have to be identified for descriptive purposes, cross references, circuit testing, and parts replacement. Numbers should be assigned in sequence from left to right and top to bottom. Lettering should be placed close to and, where possible, centered on the respective symbol.

Typical forms of reference designations and their locations are shown in Fig. 22.7. Several conventions are given, as follows: (A) part designation, R41; ohms value, 1800; wattage rating, 10; type, wire-wound (WW); (B) terminal numbers; (C) lead colors, T3; (E) terminal markings, C29; test point voltage, TP3; relay K12 coil resistance, 500 Ω.

22.14

Drawing Notes. Repetitious lettering in schematic diagrams may be avoided if items with common characteristics are explained in the general notes. Typical examples are as follows: (1) All resistors are half-watt except where specified. (2) Capacitor values given decimally are in microfarads; all others are in micromicrofarads. (3) All tolerances are ±10%, except where specified. (4) Components marked with an asterisk (*) apply only to Model No. *XYZ*.

Other significant data, such as control settings, meter types, test conditions, and calibration procedures, are generally included in these notes, along with explanation of special symbols or critical values.

Test points, test voltages, and waveform pictures may be given as supplementary references in the body of the diagram. A reference table [see Fig. 22.7(F)], is generally furnished for the more complex diagrams. Such a table shows the extent of part numbering and any numbers in a sequence that are not used or that have been deleted through revisions. Since lettering is an important part of symbolic diagrams, all notations should be carefully compiled and checked to ensure clarity and completeness and to avoid ambiguity.

22.15

Industrial Schematics. Because most industrial control circuits contain electrical devices that frequently serve specialized functions, these diagrams may have to be laid out somewhat differently from the more usual radio, television, or instrument schematics. Typical examples of the different symbols required for industrial diagrams are those that denote circuit breakers, solenoids and actuators, timers, rotary machines, power tubes, relays, contactors, and special-purpose switches. The approved graphical symbols for these devices (some of which are alternative forms) are given in the respective standards. A few of these symbols are shown full-size in the numbered blocks in Fig. 22.8.

Several variables, particularly time, motion, and sequence, are involved in most industrial processes. Other factors, such as heat, light, color, weight, size, thickness, and alignment, may enter into the control or operation of equipment and the treatment of materials. Consequently, industrial schematics, or elementary dia-

Fig. 22.7 Reference designations. [*From G. Shiers,* Electronic Drafting (*Englewood Cliffs, N.J.: Prentice-Hall, Inc.*)]

Fig. 22.8 Industrial-type graphical symbols and a typical elementary diagram of an industrial control circuit.

grams, as they are also called, should be laid out in a logical sequence according to the functions of the individual circuits.

22.16
Control Circuit Layouts. The type of circuit format commonly used for industrial diagrams is shown in Fig. 22.8. Because of the layout, with individual circuits drawn between vertical supply lines, this is sometimes referred to as a *ladder diagram*. The power source is placed at the top with the circuits that follow located more or less in the order in which the operating sequence progresses.

As will be seen, contactor and relay coils and the associated contacts are separated, and the symbols for them are drawn in the circuits where they functionally belong. Circuit lines are evenly spaced in the interest of neatness and clarity. Similarly, symbols are evenly located and staggered to permit lettering.

Usually, all control devices are connected to the left-hand bus, while operating coils, lamps, solenoids, etc., are connected to the right-hand common line. In

some layouts, the control circuits and the power circuits may be separated and drawn as two diagrams on the same sheet. Industrial diagram layouts may be prepared according to the modular principles previously described in Sec. 22.11.

22.17
Industrial Designations. Although basically serving the same purpose, the method of designating components in industrial diagrams differs somewhat from communications practice. Device designations are comprised of a standard letter or letter sequence that denotes the part and its particular function. Thus, with reference to Fig. 22.8: TR—timing relay; CRM—control relay master; FR—feed relay; CR—control relay; and SOL—solenoid. Other variations to be found include ET—electron tube; P—potentiometer; RH—rheostat; and SS—selector switch. Two or more similar devices are differentiated by prefix numbers, for example, 1V, 2V, 1R, and 2R. Component types and values may be given within a symbol, for example, 1200 for a resistor value in ohms; or G for a lamp with a green jewel.

The positions of component parts in a diagram are given with reference to the circuit "level," which is identified by a number on the left-hand side. Thus, the solenoid on level 16 in Fig. 22.8 has associated contacts, which are shown on levels 12 and 13. The UP contact is normally closed, as denoted by the slant line in the symbol, and by the underlined reference number. The other numbers adjacent to the wire lines identify individual wires. Corresponding numbers are marked on the wires in the equipment, as discussed in Sec. 22.20.

22.18
Connection Diagrams. A drawing that shows the wiring between components is known as a connection, or wiring, diagram. A similar drawing showing the external connections between self-contained units is referred to as an interconnection diagram. These drawings are used for original wiring and assembly, circuit tracing, equipment installation, testing, and maintenance.

The best method of presenting wiring and connection data depends on the wiring method, type of equipment, and the purpose of the drawing. The most common types of connection diagrams are presented in Fig. 22.9. Separate line diagrams are not needed for all equipment;

sometimes a wiring list provides the necessary information. Again, the wiring may be superimposed on a plan view of the assembly, or it may be shown in a pictorial view (Fig. 22.1)

22.19

Wires and Cables. Particulars concerning wire types and sizes are frequently included with connection data. Typical examples of wires and cables with the respective graphical symbols are shown in Fig. 22.10.

Stranded conductors are used for most chassis wiring. The wire size may be specified by the number of strands and the strand size or by the equivalent American Wire Gauge. For example, 7/30 (22) signifies a conductor with seven strands of No. 30 gauge equivalent in cross-sectional area to a solid wire, size AWG 22. These conductors are usually covered with a thermoplastic insulating material, such as polyvinyl chloride (PVC), in single colors or with a colored stripe, as shown in Fig. 22.10(A).

A solid conductor of bare tinned copper (BTC) wire is generally used for rigid bus bars or for short connections (jumpers) between adjacent terminals. The other types shown in Fig. 22.10 are a twisted pair (B), shielded cable (C), multiconductor cable (D), and a coaxial cable (E). The outer metallic braids serve as shields to minimize radiation or interference.

22.20

Wire Coding. Wire markings are necessary for original wiring inspection, testing, and troubleshooting. These markings may consist of insulation colors or number–letter sequences stamped on or attached to the wires. Such identities are given by code in line diagrams. Wire grades may be designated by a code letter that denotes type of insulation, voltage rating, temperature limits, etc.

The standard color code universally employed for electronic applications consists of color markings and their numerical equivalents. This code, with the two-letter abbreviations, is as follows: 0, black (BK); 1, brown (BN); 2, red (RD); 3, orange (OR); 4, yellow (YL); 5, green (GN); 6, blue (BL); 7, violet (VI); 8, gray (GY); and 9, white (WH). A single-letter code is sometimes used, as shown in Fig. 22.13(B).

Wire-coding methods are based on the need to identify specific wires; circuit function; wire size, type, and number; wire destination; and terminal connection. In

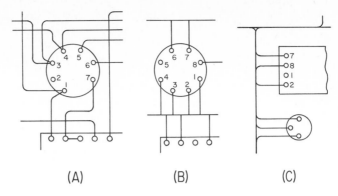

Fig. 22.9   Basic line diagrams: (A) point to point; (B) base line; (C) trunk-line or highway. [*From G. Shiers,* Electronic Drafting *(Englewood Cliffs, N.J.: Prentice-Hall, Inc.)*]

the case of multicolored wires, for example, a code such as 14–22–A would signify a brown wire (1) with a yellow tracer (4), size 22, segment A. A two-section code showing the wire destination and the wire details, such as R7/3–C20/6, would signify a blue wire (6), size 20, type "C," connected to terminal 3 on resistor R7.

22.21

Wiring Methods. Basic methods for interconnecting components within a self-contained assembly are as follows: direct or point-to-point wiring; grouped or cable-form wiring; and printed circuits. These methods may be combined in a single assembly.

An example of point-to-point wiring will be seen in Fig. 22.1, where small parts are suspended by their leads and other wires take the shortest route between terminals. This type of wiring, which can be visually traced quite easily, is suitable for simple assemblies and special circuits that operate at high frequencies.

In other applications, particularly military, industrial control, scientific research, and high-grade commercial equipment, all parts must be individually mounted with the wiring properly secured. Usually, small parts are grouped together on component boards, with the wires tied in bundles or cableforms. These wiring assemblies, also called cable harnesses, are usually prefabricated on a wiring jig. Wire markings are essential in order to distinguish the wires and trace the connections.

22.22

Connection Diagram Layouts. The component parts in a connection diagram are represented in simplified outline by blocks or pictorial symbols. Sufficient detail should be included to enable the reader to

Fig. 22.10   Typical wires and cables with symbols. [*From G. Shiers,* Electronic Drafting *(Englewood Cliffs, N.J.: Prentice-Hall, Inc.)*]

identify the parts and the terminations. Graphical symbols may also be used. Although these drawings need not be done to scale, the layout should be in approximate agreement with the assembly, with similar placement of terminals, connections, and identifying features. These diagrams should be developed using a modular layout, as described in Sec. 22.11.

To reveal every component, perpendicular surfaces, such as the sides of a box-type chassis, are turned 90° and developed in the major plane. Connection diagrams therefore present a plan view of the wiring in which the turned surfaces are

Fig. 22.11 Point-to-point wiring diagram of a panel layout. [*From G. Shiers*, Electronic Drafting (*Englewood Cliffs, N.J.: Prentice-Hall, Inc.*)]

Fig. 22.12. Typical base-line diagrams: (A) partial chassis wiring; (B) terminal-strip interconnections. [*From G. Shiers*, Electronic Drafting (*Englewood Cliffs, N.J.: Prentice-Hall, Inc.*)]

normally placed adjacent to the respective edges. Connections to certain components, such as multigang switches, may require a detailed view or a supplementary view of the part. If necessary, a symbol may be divided for clarity. In such instances, the sections should be suitably aligned and properly labeled, with specific references given to clarify the views.

22.23
Point-to-point Diagrams. In a point-to-point diagram, each wire is shown by a separate line. These diagrams show the wiring side of an assembly, such as for the back view of a control panel presented in Fig. 22.11. The wire lines are spaced apart for clarity and to permit lettering and therefore do not necessarily represent the actual wire routing. Wires, connections, and component parts must be suitably identified. Other references should correspond with panel or chassis markings, terminal numbers or letters, and schematic designations.

22.24
Base-line Diagrams. A connection diagram may be simplified if the wire lines from the components are terminated into common base lines. These lines, drawn horizontally and vertically between component symbols, may or may not represent the routing of the actual wires.

Figure 22.12(A) is a base-line diagram, which partially shows the wiring between a potentiometer (R5), a tube socket (V3), a component board, a test point, and related items. The wire identities showing wire numbers and segments, for example, 5A, 5B, and so on, are inserted in the wire lines. The actual connections would be given in a wiring list, which would also specify other wire details, such as wire type, size, length, and type of terminations.

The diagram shown in Fig. 22.12(B) refers to the terminal wiring on a chassis designated "C." In this example, the wire coding includes wire size, color, and destination or address, for example, a red (R) wire, size 16, on terminal 20, is connected to terminal 9 on chassis "B." The other end of this wire would be designated C20 (R16) in the diagram and marked accordingly.

22.25
Highway Diagrams. In another type of connection diagram the feeder lines from the component terminals are merged into

a single trunk line, or highway. The routing for each lead should be indicated by a slant line, as shown in Fig. 22.13(A), or by radial turnoffs, as shown in (B).

Since the wire identities are lost by grouping, a code number is required to show the destination. With reference to Fig. 22.13(A), for example, the lead connected to terminal 2 on item 2 is designated 6/E(7), which signifies a violet wire (7), which is connected at the other end to terminal E on item 6. Such items, for example, component boards, tube sockets, switches, connectors, and so on, may be arbitrarily numbered, as shown, or they may be designated according to the class of item using standard letter symbols (such as V5, TB2, etc.) which agree with the schematic references.

Figure 22.13(B) shows separate highways for associated wires, a method that may be preferred because it facilitates circuit tracing. Thus, highway A carries wires from left to right to terminals 30, 24, 12, 8, and 6; also wires to the right from terminals 28, 16, and 6. These (and perhaps other wires) are carried upward to other parts of the drawing, a fact that is not specifically shown in (A).

22.26

Printed Circuits. Savings in space and weight and other advantages may be realized if conventional wiring is replaced by a "printed circuit." Such a circuit may be produced by depositing a conductive pattern on a dielectric base, by stamping or etching copper foil, or by special processes, which may include the formation of circuit components with the conductive pattern.

A suitable grade of insulating material, such as phenolic stock with copper foil bonded to the sheet, is used for general applications. Glass, ceramic, and synthetic dielectrics are used for special purposes. Typically, the circuit pattern is reproduced on the foil by photographic means, followed by an etching process. A large-scale black-and-white drawing showing the pattern in silhouette is prepared as a master layout. A photographic negative is made from this artwork.

In the usual construction, components are mounted on the plain side of the board, opposite the etched pattern. Component leads are inserted through holes or eyelets in the conductor pattern, the connections being made either by hand soldering or by dipping the wiring side of the assembled board in a bath of molten solder.

22.27

Conductor Patterns. Conductor patterns must be compatible with various mechanical and electrical requirements, notably, size, shape, and layout of the board; circuit connections; and electrical values. Adequate contact areas at the terminal

Fig. 22.13 Typical trunk-line or highway diagrams showing a component board and associated wiring: (A) single trunk; (B) separate highways. [*From G. Shiers,* Electronic Drafting *(Englewood Cliffs, N.J.: Prentice-Hall, Inc.)*]

points and interconnections must be incorporated in the design. Several basic conductor shapes are shown in Fig. 22.14. Conductor patterns should be drawn solid black with clean edges, free-flowing lines, and smooth contours. Fillets should be provided at junctions and land areas.

A conductor width of .031 in. is commonly used for general-purpose boards where the current is not more than a few hundred milliamperes. The spacing between conductors depends on the working voltage; .062 in. is usual where the peak voltage difference does not exceed 300 V. Conductor locations and spacings are also governed by electrical factors, such as capacitance, intercircuit coupling, shielding, and other requirements, which may be critical in certain designs.

Fig. 22.14 Basic conductor shapes for printed circuits. [*From G. Shiers*, Electronic Drafting (*Englewood Cliffs, N.J.: Prentice-Hall, Inc.*)]

22.28

Printed-circuit Layouts. Quadrille or cross-section paper is ideal for sketching printed-circuit layouts. The pattern can be drawn freehand in pencil, using the grid lines, with the divisions representing a suitable scale. Terminal pads should preferably be located on standard .1-in. intersections, with increments of .025 in., according to the space requirements and the component dimensions.

At this stage the objective is to develop connecting paths as short and direct as possible without any crossovers. Component parts may have to be reoriented or interchanged to provide access for every conductor. When an obstruction is unavoidable, however, a jumper or insulated wire is used as a bridge at the crossover point. Several sketches are generally needed before a feasible pattern can be established.

22.29

Printed-circuit Artwork. Master layouts for printed-circuit work are made to an enlarged scale, which may range from 2:1 to 20:1, depending on the size of the part and the intricacy of detail. This enlargement facilitates drawing and improves the accuracy of the final full-size copy. Accurate locations of the terminals and other center points are essential because land areas, pads, edge-connector strips, and so on, must match the holes and cutouts in the board. One major dimension, placed outside the board area, should be given for reference.

Some important details of master layouts are shown in Fig. 22.15. If compo-

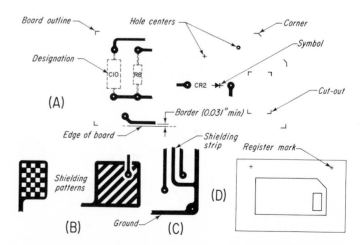

Fig. 22.15. Printed circuit artwork: (A) master layout details; (B) shielding patterns; (C) shielding strips and ground; (D) overlay and register marks. [*From G. Shiers*, Electronic Drafting (*Englewood Cliffs, N.J.: Prentice-Hall, Inc.*)]

nent designations, part symbols, and other lettering are given on the master drawing they will appear in copper on the board. Such markings, sometimes alternatively printed or silkscreened on the board, are valuable service aids. Other markings, for example, hole centers and cutouts, may be needed for experimental models, hand assemblies, and small-quantity productions.

Plastic drafting film, glass cloth, and similar materials with good dimensional stability are preferred for master layouts. The drawing may be done in black India ink or with adhesive drafting aids. These aids, or appliqués, especially developed for printed-circuit work, consist of black tapes and precut matching shapes, such as el-

bows, tees, junctions, and pads, as shown in Fig. 22.14. In another method, the surface of a specially prepared film may be scribed and stripped to produce either a positive or negative master.

Separate drawings are required for double-sided boards, silkscreen lettering stencils, and for making press tools and jigs used in quantity productions. In such cases, each drawing is aligned by means of register marks, as shown in Fig. 22.15(D), to ensure satisfactory matching of the important features. Printed-circuit artwork must be accurately and skillfully prepared because, unlike other technical drawings, the qualities of the drawing will be fully reproduced in the finished product.

Problems

Suitable layouts for the following problems should be developed freehand on opaque cross-section paper. Finished drawings should be made on plain vellum with a cross-section paper beneath it to serve as a grid background. Templates with appropriate cutouts should be used for drawing symbols. Some of the problems should be finished with adhesive drafting aids and in India ink in order to provide acquaintance with these techniques. The publications listed in Sec. 22.3 should be consulted when selecting symbols and designations.

1. Draw a block diagram representing the circuit of the chassis shown in Fig. 22.1. Letter the blocks as follows: (1) pentagrid converter; (2) IF (intermediate-frequency) amplifier; (3) detector, AVC, audio amplifier; (4) power amplifier; and (5) half-wave rectifier. Include graphical symbols for the antenna (to block 1) and the loudspeaker (from block 4). Show block 5 as an auxiliary circuit, and include the power line.

2. Draw a schematic diagram of Fig. 22.5. Use horizontal projection and a $\frac{1}{8}$-in. grid scale.

3. Make a schematic diagram of the control panel wiring shown in Fig. 22.11.

4. Draw a schematic diagram to correspond with the parts and circuit of the chassis shown in Fig. 22.1. Use a symbol similar to item 11 of Fig. 22.2 with separate variable tuning cores for the IF transformers. The connections are as follows: blue-plate, red-B plus, green-grid, white-return, or common-line. The antenna and oscillator sections of the tuning capacitor (not shown) are ganged. The line switch and the volume control also are ganged. Use vertical projection, as shown in Fig. 22.5, and a $\frac{1}{8}$-in. grid scale.

5. Make a schematic diagram of the "400" series circuit shown in the middle level of Fig. 22.4. Show the circuit from V404B up to the grid of V407A. Renumber all symbols starting

from 1 for each class. Include all designations and appropriate notes. Use horizontal projection and a $\frac{1}{10}$-in. grid scale.

6. Redraw the timing circuit (1V, 2V, from the input to level 8) shown in Fig. 22.8, using horizontal projection and a suitable scale. Show a double-wound transformer, similar to item 25 in Fig. 22.2 with two separate secondary windings, instead of T. Show a linecord terminated by a 3-pin plug, with the third pin grounded. Use zigzag resistor symbols. Capacitor designated 2C is an electrolytic type. Relay TR has two sets of contacts: single-pole, normally open; single-pole, double-throw. Include a terminal block with five terminals suitably numbered and identified for these connections.

7. The underside view of a small chassis ($7 \times 5\frac{1}{4} \times 2$ in.) is shown in Fig. 22.16. This unit includes a power supply consisting of power transformer T; full-wave vacuum rectifier $V1$, type 5U4–GB; filter choke L; triple-unit electrolytic filter capacitor $C1$; 5-W resistor $R1$; and part of the circuit shown in Fig. 22.4. This circuit contains tube V406 ($V2$) and the associated components from C412 to pin 2, V407. With reference to Fig. 22.4, the respective component designations are C2–C412 through C5–C415 and R2–R433 through R8–R439. Chassis connections are made via a 10-wire terminal strip (TB).

Make a connection diagram to correspond with the circuit described, and the following connections. Use either the point-to-point or the base-line method. [*Note:* Component board connections are denoted thus: (T)–top, (B)–bottom.]

TB1–T9–V2/4–V2/5. TB2–T10–V2/9. TB3–C2(B). TB4–T7–C4(B)–chassis lug (ground). TB5–R1B–LB–C1B. TB6–R7(T)–R1A–C1C. TB7–R8(T). TB8 (spare). TB9–T1. TB10–T2. T3–V1/8. T4–LA–C1A. T5–V1/2. T6–V1/4. T8–V1/6.

8. Make a working sketch of a schematic to correspond with the circuit given in Problem 7. Use quadrille paper with eight divisions per inch. Include all designations and part values (in addition to those shown in Fig. 22.4) as follows: Power transformer primary, 1, 2; 5-V heater winding, 3, 4 (center tap), 5; 325–0–325-V secondary, 6, 7 (center tap), 8; 6.3-V heater winding, 9,10. L, 20H; C1A, C1B, 20 mF; C1C, 10 mF; R1, 1800 Ω, 5-W, wire-wound. Show a divided terminal block with the power line input on the left and the other connections on the right.

9. A printed-circuit board with the required land areas and strip connections is shown in Fig. 22.17. Make a master layout drawing of this board twice full size to suit a conductor width of .031 in. A minimum clearance of .050 in. should be maintained between conductors and between a conductor and the edge of the board. Land dimensions in inches are ID, .040; OD, .110. The strip connector widths of .10 in. should be maintained for .20 in. All tolerances

Fig. 22.16 Chassis assembly.

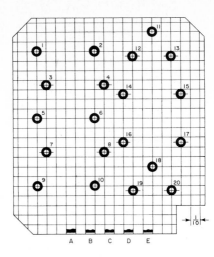

Fig. 22.17 Printed circuit board.

on the conductor pattern are ±0.015 in. The connections are as follows: A-1-5-9, B-2-4-12-14, C-8-10-19, D-17, E-13-20, 3-15-16-18, and 6-7-11. The grid pattern and land numbers are for reference only.

10. Make a working sketch of a schematic to correspond with the printed circuit given in Problem 9. The components are as follows (terminal connections are with reference to Fig. 22.17):

R1, 9-10; R2, 5-6; R3, 16-17; R4, 14-15; R5, 1-2. C1, 7-8; C2, 3-4. Transistors are *PNP*-type: *Q*1, base 18, emitter 20, collector 19; *Q*2, base 11, emitter 13, collector 12. Label the strip connections as follows: *A*, negative 15 V; *B*, output; *C*, trigger; *D*, positive; *E*, ground.

Research prototype of a quadruped machine, fabricated by General Electric Company engineers. It was developed for material handling under the most severe conditions. By means of an advanced control system, the machine mimics and amplifies the linear movements of the operator. The right front leg of the unit is controlled by the operator's right arm, its left front leg by his left arm, its right rear leg by his right leg, and its left rear leg by his left leg. Units of this type require considerable welding. (*Courtesy General Electric Company*)

welding drawings 23

23.1

Welding Processes. For convenience, the various welding processes used in commercial production may be classified into three types: pressure processes, nonpressure processes, and casting processes. The nonpressure processes are arc welding and gas welding. Metallic arc welding is the joining of two pieces of metal through the use of a sustained arc formed between the work and a metal rod held in a holder (Fig. 23.1). The intense heat melts the metal of the work and at the same time heats the end of the electrode, causing small globules to form and cross the arc to the weld. In gas welding, the heat is produced by a burning mixture of two gases, which ordinarily are oxygen and acetylene. The weld is formed by melting a filler rod with the torch flame, along the line of contact, after the metal of the work has been preheated to a molten state. This method is essentially a puddling process, in that the weld is produced by a small moving molten pool that is maintained by the flame constantly directed on it. Resistance welding is a pressure process, the fusion being made through heat and mechanical pressure. The work is heated by a strong electrical current that passes through it until fusion temperature is reached; then pressure is applied to create the weld.

The forms of resistance welding are projection welding, seam welding, spot welding, and flash welding. In spot welding, the parts are overlapped and welds are made at successive single spots. A seam weld is similar to a spot weld, except that a continuous weld is produced. In projection welding, one part is embossed and welds are made at the successive

Fig. 23.1 **Arc welding.** (*Courtesy Lincoln Electric Company*)

Fig. 23.2 Types of welds.

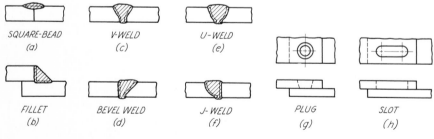

SQUARE-BEAD (a) V-WELD (c) U-WELD (e) PLUG (g) SLOT (h)

FILLET (b) BEVEL WELD (d) J-WELD (f)

Fig. 23.3 Types of welds.

LAP JOINT (SINGLE FILLET) (a) BUTT JOINT (SINGLE VEE) (b) SQUARE TEE JOINT (c) CORNER JOINT FULL OPEN (d) EDGE JOINT (SQUARE) (e)

Fig. 23.4 Types of welded joints.

projections. In making a flash weld, the two pieces to be joined are held end to end in jaws and act as electrodes. At the right instant, after the facing metal has been heated by the arc across the gap, the power is shut off and the two ends are forced together to cool in a fused position.

Thermit welding can be considered a casting process, in that molten iron is run into a mold built around the parts at the point at which they are to be connected. The liquid metal is obtained from a mixture of finely divided iron oxide and aluminum, which is ignited in a crucible. In the chemical reaction that takes place, the oxygen passes from the iron oxide to the aluminum, leaving free molten iron that flows into the mold around the preheated parts forming the joint. The metal of the members being welded fuses with the liquid metal and forms a weld when the joint is cool.

23.2

Types of Welds. Figure 23.2 illustrates in pictorial some of the various types of welds. Cross-sectional views of the fundamental welds that are commonly encountered are shown in Fig. 23.3.

23.3

Classification of Weld Joints. Welded joints are classified in accordance with the method of assembly of the parts at a joint. See Figure 23.4.

23.4

Working Drawings of Welded Parts. Figure 23.5 shows parts that have been constructed by welding rolled plates and bar stock. It should be noted that each joint is completely specified through the use of a welding symbol. A careful study will show that the drawings, except for the absence of fillets and rounds and the fact that properly composed welding symbols are directed to the necessary joints, are very much like casting drawings.

A satisfactory welding design may be produced by a competent designer who possesses a fair amount of ingenuity, and the necessary drawing can be made by any draftsman who has a thorough understanding of the use of the symbols recommended in ANSI Y32.3–1969.

23.5

Arrow-side and Other-side Welds (Fig. 23.6). In the present system the joint is the basis of reference. Any joint the welding of which is indicated by a symbol will

(a)

(b) BELL CRANK STEEL

Fig. 23.5 Welding drawing.

always have an *arrow-side* and an *other-side*.

On a drawing, when a joint is represented by a single line (see Fig. 23.6) and the arrow is directed to the line, the arrow-side is considered to be the near-side for the reader of the drawing. The side opposite becomes the other-side. In applying fillet and groove welding symbols, where the arrow portion connects the welding symbol reference line (shaft) to one side of the joint, this side is accepted as the arrow-side of the joint. For plug, slot, and projection welding symbols, where the arrow from the reference line must necessarily be directed to the outer surface of one of the two members at the center line of the required weld, the member to which the arrow points is recognized as being the arrow-side member. This, of course, would be the near-side to the reader. In general, it can be said that spot, seam, and flash or upset symbols have no arrow-side or other-side significance. However, supplementary symbols, sometimes used with them, can have such significance.

23.6

Welding Symbols. An enlarged drawing of the approved welding symbol is shown in Fig. 23.7, along with explanatory notes that indicate the proper locations of the marks and size dimensions necessary for a complete description of a weld.

The arrow is the basic portion of the symbol, as shown in Fig. 23.8(a). It points toward the joint where the required weld is to be made, as in Fig. 23.8(b).

If the weld is on the arrow-side, the symbol indicating the type of weld is placed below or to the right of the base line, depending on whether that line is

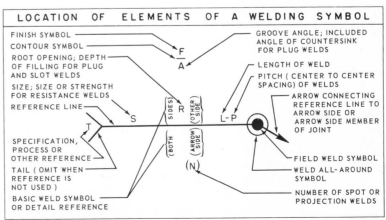

Fig. 23.6 Arrow-side and other-side welds. The arrow-side is considered to be the near side.

Fig. 23.7 Basic welding symbol.

(a)

(b)

Fig. 23.8 Welding arrow.

OTHER SIDE — Other Side Symbol — Arrow-Side Symbol — ARROW-SIDE

(a)

Perpendicular leg of symbol always drawn to left hand

Other Side Symbol — OTHER SIDE — ARROW-SIDE

(b)

Fig. 23.9 Location of welding symbols.

— Weld all around symbol

Fig. 23.10 Weld-all-around symbol.

— Size at side of symbol

5

Fig. 23.11 Method of specifying the size of weld (metric).

½

Fig. 23.12 Dimensioning a weld.

12 / 8

Fig. 23.13 Dimensioning a weld (metric).

— ACTUAL THROAT — FACE — SIZE (LEG)

(a)

INCLUDED ANGLE — FACE — ROOT FACE — ROOT OPENING

(b)

GROOVE ANGLE — ROOT RADIUS — ROOT OPENING

(c)

Fig. 23.14 Welding terms.

S = SIZE   O = ROOT OPENING

S/O 45°
SINGLE BEVEL
(a)

S/O 60°
SINGLE-V
(b)

S/O 30°
SINGLE-U
(c)

S/O 30°
SINGLE-J
(d)

Fig. 23.15 Single-groove welds and symbols.

S/O 60°
DOUBLE-V
(a)

S/O 40°
DOUBLE BEVEL
(b)

S/O 30°
DOUBLE-U
(c)

S/O 20°
DOUBLE-J
(d)

Fig. 23.16 Double-groove welds and symbols.

horizontal or vertical (Fig. 23.9). If the weld is located on the other-side, the symbol should be above or to the left.

To indicate that a weld is to be made all around a connection, as is necessary when a piece of tubing must be welded to a plate, a weld-all-around symbol, a circle, is placed as shown in Fig. 23.10.

The size of a weld is given along the base of the arrow, at the side of the symbol, as shown in Fig. 23.11. If the welds on the arrow-side and the other-side of a lap joint are the same size, only one dimension should be given (Fig. 23.12). If they are not the same size, each dimension should be placed beside its associated symbol (Fig. 23.13).

The welding terms associated with the specification of the size of welds are illustrated in Fig. 23.14.

Figure 23.15 shows the common types of single-groove welds and the related symbol for each. The symbols for double-groove welds are illustrated in Fig. 23.16.

The same symbol is used for both plug and slot welds. As illustrated in Fig. 23.17, a hole is formed in one member to receive the weld. When the hole or slot is in the member on the arrow-side, the symbol must be placed below the reference line of the symbol arrow; if in the other-side member, the symbol must be placed above the reference line.

The needed size specifications for plug and slot welds are placed as shown in Fig. 23.18. In the case of plug welds, the smallest diameter of the hole, if countersunk, is placed at the left of the symbol. The angle of countersink is added below the symbol, as shown in (a). However, the angle may be omitted if it is user's standard. When the depth of filling of a plug weld is to be less than complete, the depth of filling is entered inside the symbol. If no value is given, it is understood that the depth of filling is to be complete. To complete the size specifications, the pitch (distance between centers) of plug welds is added to the right of the symbol as shown in (a). When plug welds are to be approximately flush without finishing, a flush-contour symbol is added to the weld symbol. If the welds are to be made flush by some mechanical means, a standard user's finish symbol should be given with the flush-contour symbol.

The depth of filling of slot welds is indicated within the symbol in the same manner as for plug welds. Then, as noted in Fig. 23.18(b), all size and location dimensions must be shown on the drawing.

23.7

Gas and Arc Welding Symbols. To satisfy the need for a standard group of symbols that could be understood in all manufacturing plants, the American Welding Society recommended in 1940 a set of conventional symbols so designed that each symbol resembled in a general way the type of weld it represented. Figure 23.19 shows a condensed table of symbols. With one exception the symbols shown are the same as those first proposed by the American Welding Society.

23.8

Resistance Welding. Figure 23.20 shows the symbols for the four principal types of resistance welding. The method of specifying resistance welds differs somewhat from the methods used for arc and gas welds.

In the case of the spot-weld symbol shown in Fig. 23.20, the size of the weld is given to the left of the weld symbol as a diameter. If desired, the minimum acceptable shear strength in Newtons or in pounds per spot may be entered in the same location. The pitch of welds is specified at the right of the symbol. Should it be desirable to indicate a specific number of welds for a joint, the number is given in parentheses either above or below the weld symbol.

Projection welds are specified in the same manner as spot welds. The size may

Fig. 23.17 Plug and slot welds.

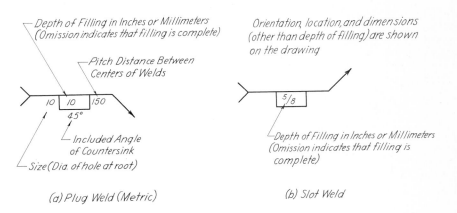

Fig. 23.18 Size specification—plug and slot welds.

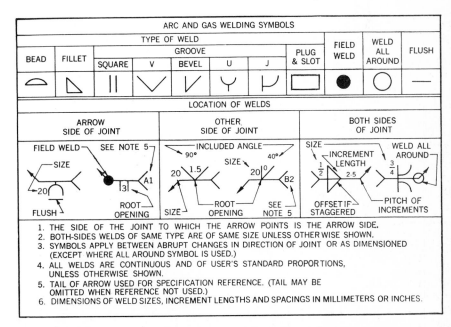

Fig. 23.19 American National Standard arc and gas welding symbols. Examples for arrow-side and other-side of joint show dimensions in millimeters.*

*Based on ANSI Y32.3–1969.

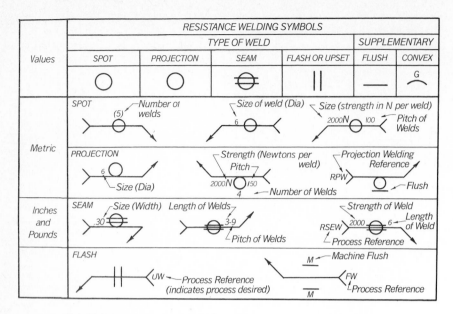

Values	RESISTANCE WELDING SYMBOLS					
	TYPE OF WELD				SUPPLEMENTARY	
	SPOT	PROJECTION	SEAM	FLASH OR UPSET	FLUSH	CONVEX
	○	○	⊕	‖	—	⌒G

Fig. 23.20 American National Standard resistance welding symbols.*

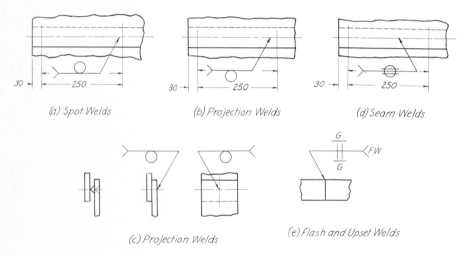

(a) Spot Welds (b) Projection Welds (d) Seam Welds

(c) Projection Welds   (e) Flash and Upset Welds

Fig. 23.21 Resistance welding—spot, projection, seam, and flash and upset welds.

Fig. 23.22 Bracket-welded design.   Fig. 23.23 Link-welded design.

* Based on ANSI Y32.3–1969.

also be given as a diameter or the minimum acceptable shear strength in Newtons or pounds per weld entered in its place. If it is desired that the exposed surface of one member be flush, the flush symbol may be added.

The size of a seam weld is its width that is given decimally to the left of the weld symbol in hundredths of an inch, as shown in Fig. 23.20. If it is thought to be desirable to specify the minimum acceptable shear strength in pounds or Newtons per linear unit, this may be substituted for the size of the weld at the left of the weld symbol. The weld length is placed at the right of the weld symbol. For intermittent seam welding, both the length of the weld and the pitch is placed in this position as illustrated.

Flash and upset weld symbols have been shown in Fig. 23.20. No dimensions are shown. However, process reference (FW or UW) must be given in the tail of the symbol to indicate the process to be used. When flash and upset welds are to be made flush by some mechanical means, the complete symbol would consist of the weld symbol along with contour and machining symbols as shown. The use of symbols for resistance welding in conjunction with welding dimensions has been illustrated in Fig. 23.21.

23.9

Welded Machine Parts. Many machine parts often can be constructed of welded rolled shapes at a much lower cost than if they were cast. This is due to the fact that the cost of the preparation of patterns is completely eliminated, less material is required, and labor costs are lower. A welded part is sometimes more desirable for a particular mechanism, because steel is stiffer, stronger in tension, and more resistant to fatigue stresses and sudden impact. Also, aside from the production of new parts, welding can be used to make a machine part to replace a broken cast part when it is necessary to place a machine back in operation in the shortest possible time. Figure 23.22 shows a part that is constructed of plates. Figure 23.23 shows the construction of a link using plates and round bar stock.

As previously stated, a designer is limited only by his own ingenuity. Parts of all shapes and sizes may be produced of readily available rolled forms. Simple bearings, levers, cranks, clevises, gear arms, and even cams can be quickly and easily made.

Problems

Two problems offering experience in the preparation of welding drawings are given in Figs. 23.24 and 23.25. Others may be had by redesigning many of the cast parts (given at the end of Chapter 5, and in some of the other chapters) in such a way that they may be made of welded steel shapes. The student will find in these problems an opportunity to exercise some of his own ingenuity.

Note: With the approval of his Instructor, the student may convert the inch measurements in Prob. 2 to millimeters by using Appendix Tables 3–9. Since it is common practice under the metric system to use full millimeters for all dimension values that are not critical, the student should do likewise. For example, in Fig. 23.25 the 1 in. value converts to 25.4 millimeters. The student should use 25 millimeters. Metric drill sizes are given in Tables 23 and 49 in the Appendix.

1. Prepare a welding drawing of the bracket shown in Fig. 23.24.

2. Prepare a welding drawing of the caster bracket shown in Fig. 23.25. The length of the tubing is $2^{11}/_{16}$ in.

3. Make a two-view orthographic drawing of the object shown in Fig. 23.2. The dimensions are to be assumed. The plates are $3/_8$ in. thick. Show the correct specification for each type of weld.

4. Make a three-view detail drawing of the bracket shown in Fig. 23.22. The dimensions are to be assumed. Show the correct specifications for the welds.

5. Make a two-view detail drawing of the link shown in Fig. 23.23. The dimensions are to be assumed. Show correct specifications for the welds.

Fig. 23.24 Bracket.

Fig. 23.25 Caster bracket.

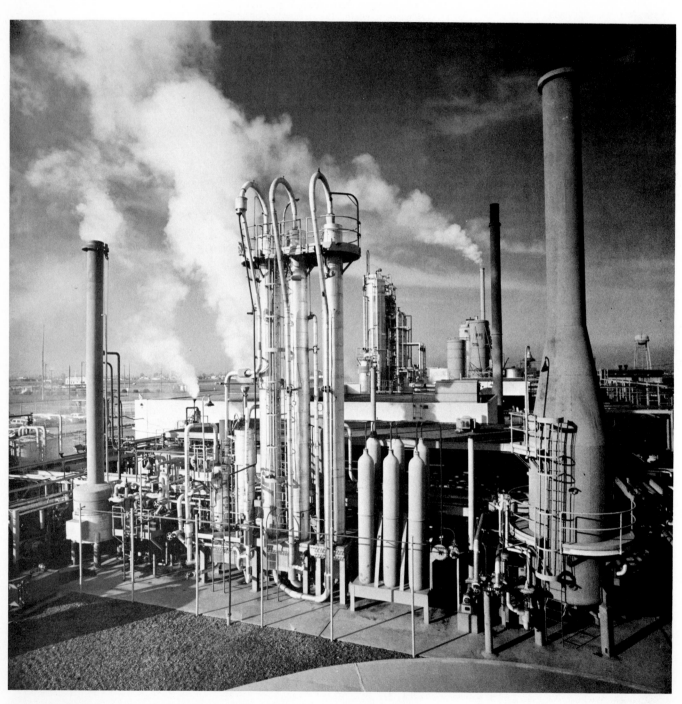

Pipes and tubes are used for the transmission of liquids and gases through the various steps in a chemical process system. (*Photo courtesy of Proctor and Gamble.*)

piping drawings and models 24

A □ Pipe fittings— pipe drawings

24.1

Since piping is used in all types of construction for conveying fluids and gases such as oil, water, steam, and chemicals, some knowledge of it is essential not only for the draftsman making drawings but for the engineer who must select and use pipe in the design of machines, power plants, water systems, and so on. There are so many types of fittings and materials used for various purposes that only the most common can be discussed briefly in this chapter. Additional information may be obtained from publications of research associations and from the catalogs of manufacturers.

24.2

Pipe Materials. Cast-iron pipe is suitable for underground gas and water mains, plumbing lines, and low-pressure steam systems.

Steel pipe is used chiefly where high temperatures and high pressures are encountered. The addition of such alloys as nickel, chromium, and the like makes the steel pipe more resistant to corrosion at high temperatures.

Seamless brass pipe is the most satisfactory type for hot-water lines, condenser tubes, and so on, but it is expensive and, therefore, is used only when conditions justify the extra cost.

For lines having turns and bends in inaccessible locations, copper tubing is frequently used. But copper pipe, even though it is flexible and can resist the corrosive action of chemicals, is not always practical. It cannot be used in any system subject to high temperatures and repeated stress.

Lead and lead-lined pipe is widely used for chemical work, particularly where piping is subject to the action of acids.

Galvanized pipe (ordinary iron pipe that has been dipped in molten zinc to prevent rust) is suitable for lines conveying drinking water.

24.3

Special Pipe and Tubing. Special-purpose pipe and tubing, made from a variety of materials such as stainless steel, brass, bronze, aluminum, and plastics, are now being manufactured. Plastic pipe and plastic-lined pipe are being used extensively in the chemical industry in place of metal pipe because plastic does not corrode and has high resistance to a wide range of chemicals. Plastic pipe is easily bent and resists weathering. In the heavy weights it may be threaded. The principal disadvantages are its higher cost and temperature–pressure limitations. However, the higher price will be partially offset by lower installation costs. Where strength is needed, a metal pipe lined with plastic may be used.

Polyvinyl chloride (PVC) is the basic plastic material most widely used for pipe. PVC pipe is very light, has low flow resistance, and can be readily bent and easily assembled by solvent cementing.

24.4

American National Standard Pipe Thread (NPT) for General Use. The American National Standard pipe taper thread, illustrated in Fig. 24.1, is similar to the ordinary American National Standard thread and has the same thread angle; but it is tapered $\frac{1}{16}$ in. per in., to ensure a tight joint at a fitting. The crest is flattened and the root is filled in so that the depth of the thread is .80P. The number of threads per inch for any given nominal diameter can be obtained from Table 43 in the Appendix.

The distance a pipe enters a fitting is fixed for any nominal diameter. Numerical values for this distance may be determined from Table 43.

An American National Standard straight pipe thread, having the same number of threads per inch as the taper thread, is in use for pressure-tight joints for couplings, for pressure-tight joints for grease and oil fittings, and for hose couplings and nipples. This thread may also be used for free-fitting mechanical joints. Usually a taper external thread is used with a straight internal thread, for pipe material is sufficiently ductile for an adjustment of the threads.

The NPT threads are for general use. When taper pipe threads are used and the joints are made up wrench-tight with a lubricant or sealer, they will be pressure tight.

24.5

Specification of Threads. In specifying pipe threads, the ANSI recommends that the note be formulated using symbolic letters as illustrated in Fig. 24.2. For example, the specification for a 1-in. standard taper pipe thread should read:

1–NPT

The letters NPT, following the nominal diameter, indicate that the thread is American National Standard (N), pipe (P), taper (T) thread.

Continuing with the same scheme of using letters, the specification for a 1-in. straight pipe thread would read

1–NPS[American National Standard (N)–pipe (P)–Straight (S)]

The form of note given in Fig. 24.2(*b*), reading 1 AM STD PIPE THD, is quite commonly used in practice.

The letter symbols that have been adopted by the ANSI have the following significance:

N—American National Standard
P—pipe
T—taper

$A = D - (0.05D + 1.1)P$
$B = A + 0.0625F$
$E = (0.80D + 6.8)P$
Depth of Thread = 0.80P

An Internal Thread is known as a Female Thread
An External Thread is known as a Male Thread

Fig. 24.1 American National taper pipe thread.

Fig. 24.2 Representation and specification of pipe threads.

S—straight
C—coupling
H—hose coupling
L—lock nut
M—mechanical joints
R—railing
F—fuel and oil
I—intermediate
SPL—special

See Sec. 14.9 for specification of metric taper threads.

24.6

Threads for Joints and Couplings. Pipe threads are used for several types of joints. The significance of the letters designating the threads for these joints is given in Sec. 24.5. Where the letter S appears in the designation, a straight thread is indicated.

NPTR—taper pipe thread for rail fittings. These threads are the same as the NPT threads except the external thread is shortened to permit use of the larger end of the pipe thread.

NPSC—internal straight pipe threads in pipe fittings. These threads are straight threads of the same thread form as the NPT thread shown in Fig. 24.1. NPSC threads form a pressure-tight joint when assembled with an external taper (NPT) pipe thread using a lubricant or sealer.

NPSH—straight pipe thread for loose-fitting mechanical joints for hose couplings. Ordinarily, straight internal and external threads are used for hose-coupling joints. One of several standards for hose threads is based on the American National Standard pipe thread.

NPSL—straight pipe threads for loose-fitting mechanical joints with lock nuts. These threads are straight threads of the same form as the NPT thread. The external NPSL thread has the largest diameter that can be cut on standard pipe. Usually, straight internal threads are used with these straight external threads.

NPSM—straight pipe threads for free-fitting mechanical joints for fixtures.

24.7

Dryseal (American National Standard) Pipe Threads (NPTF, NPSF, and NPSI). With these threads, truncation at the crest and root is controlled to ensure metal-to-metal contact coincident with, or prior to, flank contact. This contact at root and crest makes a pressure-tight joint to prevent spiral leakage. Although a lubricant or sealer is not required, the use of a lubricant to minimize galling is not objectionable.

NPTF—taper pipe thread. This series of threads covers both external and internal threads. These threads are suitable for pipe joints in practically every type of service.

PTF–SAE SHORT—this series conforms to the NPTF series except that (1) the full thread length of the external thread is shortened by one thread at the small end, and (2) the full thread length of the internal thread is shortened by eliminating one thread at the large end.

NPSF—straight pipe thread (internal only) for fuel and oil. This series of threads is generally used for soft and ductile materials, which will adjust to the taper of external threads in assembly.

NPSI—intermediate straight pipe thread (internal only). This series of threads is suited for hard or brittle materials where the section is heavy and where there will be little expansion with the external taper threads in assembly.

Dryseal threads are specified as follows:

$\frac{1}{4}$—18 DRYSEAL NPTF
$\frac{1}{8}$—27 DRYSEAL NPSI

24.8
Drawing Pipe Threads. The taper on a pipe thread is so slight that it will not attract attention on a drawing unless it is exaggerated. If it is shown at all, it is usually magnified to $\frac{1}{8}$ in. per in.

Pipe threads are generally represented by the same conventional symbols used for ordinary American National Standard thread. See Fig. 24.2.

24.9
Specification of Wrought-iron and Steel Pipe. The standardized weights commonly used are the standard, extra-strong, and double-extra-strong. All are specified by the nominal inside diameter.

The nominal inside diameter of standard pipe is less than the actual diameter, because early manufacturers made the wall thickness greater than necessary and, in correcting, took all of the excess from the inside to avoid altering the size

of openings in fittings. Metal was added to the inside to increase wall thicknesses for the extra-strong and double-extra-strong. As a result, all three weights of pipe for any given nominal diameter have the same outside diameter and can be used with the same fittings.

Wrought-iron or steel pipe greater than 12 in. in diameter is specified by giving the outside diameter and the thickness of the wall.

Fig. 24.3 illustrates the relative wall thickness of 1-in. standard, extra-strong, double-extra-strong pipe.

24.10
Sizes of Wrought-iron, Steel, and Cast-iron Pipe. The standard-weight pipe is used for normal pressures. It may be purchased in sizes ranging from $\frac{1}{8}$ to 12 in. (nominal diameter). Pipe is received threaded on both ends with a plain coupling attached.

Extra-strong pipe, designed for steam and hydraulic pressures over 125 lb per sq in., is also manufactured in sizes $\frac{1}{8}$–12 in.

Double-extra-strong pipe, designed for extremely high pressures, is furnished in nominal diameters from $\frac{1}{2}$ to 8 in. in the same lengths as the extra-strong.

Cast-iron pipe, in sizes ranging up to 48 in., can be used for pressures up to 350 lb per sq in.

24.11
Pipe Fittings. Fittings are parts, such as elbows, tees, crosses, couplings, nipples, flanges, and so on, that are used to make turns and connections. They fall into three general classes: screwed, welded, and flanged. See Figs. 24.4–24.6.

Fig. 24.3 Comparison of different weights of 1 in. wrought-iron pipe.

O.D. 1.315	O.D. 1.315	O.D. 1.315
I.D. 1.043	I.D. .949	I.D. .577
STANDARD	EXTRA STRONG	DOUBLE EXTRA
(a)	(b)	(c)

0.136   0.183   0.369

TEE   SERVICE TEE   90°ELBOW   45°ELBOW   REDUCING ELBOW   90°STREET ELBOW

45°Y-BRANCH   CROSS   CAP   R&L COUPLING   REDUCER   HEX. BUSHING

BUSHING   CLOSE NIPPLE   SHORT NIPPLE   LONG NIPPLE   PLUG   BEND

Fig. 24.4 Screwed fittings.

Fig. 24.5 Flanged fittings.

TEE   90°ELBOW   45°ELBOW

Fig. 24.6 Welded fittings.

In small piping systems and for house plumbing, screwed fittings are generally used.

Welded fittings are used where connections are to be permanent. They are manufactured of forged seamless steel having the same thickness as the pipe. In this type of construction, the weld is depended on to seal the joint and to carry the pipeline stresses. Many miles of line having welded fittings are giving satisfactory service to pipeline corporations.

Flanged fittings are used in large piping systems where pressures are high and the connection must be strong enough to carry the weight of large pipes. Table 47 in the Appendix gives the dimensions for American National Standard cast-iron flanged fittings. Several types of flanges and flanged joints are shown in Figs. 24.7 and 24.8.

Fig. 24.7 Flanges.

Fig. 24.8 Flanges.

24.12

Screwed Fittings (Fig. 24.4). Straight sections of pipe are connected by a short cylindrical fitting (threaded on the inside), which is known as a *coupling*. A *right-and-left coupling*, which can be recognized by the ribs on the outside, is often used to close a system. A *union* is preferable, however, where pipe must be frequently disconnected.

A *cap* is screwed on the end of a pipe to close it.

A *plug* is used to close an opening in a fitting.

A *nipple* is a short piece of pipe that has been threaded on both ends. If it is threaded the entire length, it is called a *close nipple;* if not, it is called a *short* or *long nipple.* Extra-long nipples may be purchased.

A *bushing* is used to reduce the size of an opening in a fitting when It would be inconvenient to use a reducing fitting.

Tees, crosses, and *laterals* form the connections for lines and branches in a piping system.

By standardizing the screwed fittings, the American National Standards Institute has eliminated many difficulties that would arise if each manufacturer produced the varied sizes of elbows, tees, laterals, and so on, according to his own specifications. The adopted dimensions, now recognized by all manufacturers, will be found in Tables 44–46 in the Appendix. See Fig. 24.13.

24.13

Specification of Fittings. A fitting is specified by giving the nominal inside diameter

of the pipe for which the openings are threaded, the type of fitting, and the material. If it connects more than one size of pipe, it is called a reducing fitting, and the largest opening of the through run is given first, followed in order by the opposite end and the outlet. Figure 24.9 illustrates the order of specifying reducing fittings. If all of the openings are for the same size of pipe, the fitting is known as a straight tee, cross, and so on. A straight fitting is specified by the size of the openings followed by the name of the fitting (2-in. tee, 4-in. cross, etc.).

24.14

Unions. Screwed or flanged unions connect pipes that must be frequently disconnected for the purpose of making repairs. In many cases, screwed unions are used for making the final closing connection in a line. The union illustrated in Fig.

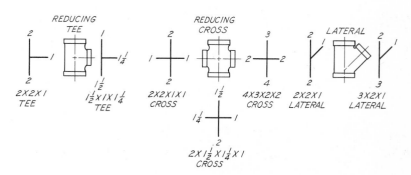

Fig. 24.9 Specification of fittings.

521

24.10(*a*) is made up of three separate pieces. The mating parts, *A* and *B*, are screwed on the ends of the two pipes. The third part, the nut, draws them together so that *A* and *B* will be against the gasket, *D*, to ensure a tight joint. In systems having pipes more than 2 in. in diameter, screwed unions are not generally used, because the stronger and more substantial flange unions, such as the one shown at (*c*), become desirable. A screwed union with a ground metal seat is shown at (*b*).

24.15

Valves. Valves are used in piping systems to stop or control the movement of fluids and gases. A few of the many forms are illustrated in Fig. 24.11. Of these, the globe valve and gate valve are the two types most frequently used.

Globe valves are used for throttling steam, in both high- and low-pressure steam lines, and to regulate the passage of other fluids. Their design, however, creates a slight retardation to the flow, because the fluid is forced to make a double turn and pass through the opening at 90° to the axis of the pipe. The valve disc is raised or lowered to stop or regulate the flow through a circular opening.

A gate valve allows a straight-line movement of a fluid and offers only slight resistance to the flow. Since the disc moves completely out of the passage and leaves a full opening, this type of valve is particularly suitable for water lines, oil lines, and the like.

A swing-type check valve permits movement in one direction only and prevents any back flow. It will be noted from a study of this valve that the design makes the action automatic. Such valves are used in feedwater lines to boilers. The ball-check valve is preferred for heavy liquids.

The dimensions of the valves given in Fig. 24.11, as well as those for many special types, may be found in the catalogs of manufacturers.

SCREWED UNION
(a)

GROUND JOINT UNION
(b)

FLANGED UNION
(c)

Fig. 24.10 Unions.

24.16

Piping Drawings. Since standard pipe and fittings can be purchased for almost any purpose, a piping drawing usually shows only the arrangement of a system in some conventional form and gives the size and location of fittings. The drawing may be a freehand sketch, single-line diagram, double-line diagram, or pictorial diagram. Occasionally, when conditions necessitate the design of special valves or the redesign of an existing type, complete working drawings are made.

Single-line drawings or sketches are made in orthographic projection or are drawn as though the entire system were swung into one plane (Fig. 24.12). On

GLOBE VALVE

SWING CHECK VALVE

GATE VALVE

HORIZONTAL LIFT CHECK VALVE

FOUR-WAY STEAM COCK
BALL CHECK VALVE

Fig. 24.11 Valves. (*Courtesy Crane Co.*)

STEAM MAIN

GATE VALVE

TEE

ELBOW

BOILER

Fig. 24.12 Single-line drawing.

these drawings, single lines represent the runs of pipe, regardless of variations in diameters; conventional symbols are used for the fittings. A developed single-line sketch is frequently used for repair work, small jobs, and for making studies and calculations. For more complicated small-scale layouts, a single-line diagram drawn in orthographic projection is more suitable.

Double-line diagrams are drawn when many similar installations are to be made at the plants of various purchasers of pumps, manufacturing equipment, heating equipment, and so on (Fig. 24.13).

A diagrammatic isometric layout (Fig. 24.14) showing a piping system in space reveals the changes in direction and the difference in levels more clearly than does any other type of line diagram. Pictorial diagrams are often used for preliminary layouts.

24.17

Dimensions on Piping Drawings. The rules for dimensioning working drawings apply to piping drawings. Fittings and pipes are always located by giving center-to-center distances, because the determi-

Fig. 24.13 Double-line drawing.

Fig. 24.14 Diagrammatic isometric layout. (*Courtesy Grinnell Company*)

nation of pipe lengths is generally left to the pipe-fitter. Notes should be used to specify the nominal size and type of each fitting and the nominal size of the pipe in each run. In addition, it is good practice to indicate, on a flanged valve, the diameter of the handwheel and its distance above the center of the fitting when wide open. It may be necessary to give overall dimensions for other apparatus, if the maximum space to be allowed is important.

The date and a descriptive title should be given on all piping sketches and drawings.

24.18
Conventional Symbols. Figure 24.15 shows a few of the conventional symbols for fittings that have been approved by the American National Standards Institute.

B □ Process models

24.19
Piping Models (Fig. 24.16). Plant models with extensive piping installations usually portray complete process systems. The use of piping models varies widely from company to company. At times, some companies build models only after complete sets of design drawings, including piping assembly drawings, have been prepared. These conventional drawings show the piping along with related mechanical equipment, structural steel components, electrical conduit, and so forth. This usage makes the model an instrument that provides the means for review and complete checking of the entire system before final confirmation of the total design. However, owing to somewhat different design and construction habits, other companies build models without first preparing assembly piping drawings. In this case, the piping system is designed directly on the model with the aid of flow diagrams. When this practice is followed every pipeline and all piping accessories, such as metering instruments, valves, traps, strainers, and so forth, must be placed in proper location. When almost all design work has been done directly on the model and only a few drawings, flow charts, and sketches have been prepared, the model with all related systems shown ultimately becomes the means for communicating design information to the construction crews in the field (Fig. 24.17).

Since in most cases it is difficult to coordinate an entire processing system that involves complicated piping, all phases of the total system should be shown on the model along with the necessary piping. If all of the related mechanical and electrical equipment, along with minor structural components, platforms,

GRAPHICAL SYMBOLS FOR PIPE FITTINGS	FLANGED	SCREWED	BELL & SPIGOT	WELDED	SOLDERED
1) BUSHING					
2) CAP					
3) CROSS—STRAIGHT SIZE					
4) ELBOW — 45 DEGREE					
— 90 DEGREE					
— TURNED DOWN					
— TURNED UP					
5) LATERAL					
6) PLUG					
7) TEE — STRAIGHT SIZE					
— OUTLET UP					
— OUTLET DOWN					
8) UNION					
9) REDUCER — CONCENTRIC					
— ECCENTRIC					
10) CHECK VALVE — STRAIGHT WAY					
11) GATE VALVE					
12) GLOBE VALVE					
13) ANGLE VALVE GLOBE — ELEVATION					
GLOBE — PLAN					

Fig. 24.15 Conventional symbols for fittings.*

*Based on ANSI Z32.23–1949 (R1953).

ladders, and supports, are properly positioned with related items such as ducts, chutes, instruments, electrical conduits, and lights, and each pipe is traced out and checked to see that requirements are met, it is most likely that all phases of the total system will fit together at the construction site.

When the model is nearing completion, the design should be examined critically for inaccuracies and faulty representation. In reviewing the model at this stage those responsible for the project should determine whether the proposal fulfills the requirements of the process and the most economical design has been adopted.

Piping models are built to scales that vary from $\frac{1}{4}$ in. to the foot to 1 in. to the foot. When conventional design and working drawings have been prepared and accuracy of the model is not important, a small-size model prepared to a scale of either $\frac{1}{4}$ or $\frac{3}{8}$ in. to the foot proves to be satisfactory. However, if the design is to be worked out on the model, the scale should be at least $\frac{3}{4}$ in. to the foot.

When the representation of each piece of equipment has been mounted in its proper location on the basic model, consisting essentially of a baseboard and a replica of the structural design, the piping can be installed through and around the formed background (Fig. 24.16). The two techniques that are commonly employed for the piping are known as (1) the *fine-wire* method and (2) the *true-scale* technique. When the fine-wire method is used, a fine wire with fibre discs or rubber sleeves mounted along its length is used to represent the pipe. The wire, normally $\frac{1}{16}$ in. in diameter, represents the center line of the pipe, while the outside diameter of the discs or sleeves indicates the outside diameter of the pipe.

Because fine-wire models sometimes prove unsatisfactory due to built-in inaccuracies and poor representation, many companies are now using the true-scale technique for building true-scale models (Fig. 24.17). In building these models, plastic rod or plastic tubing, having outside diameters equal to the scaled diameters of the various sizes of standard pipe, is used for the piping. A full line of true-scale piping and piping components of plastic may be purchased ready for use. Plastic, as a model material, is easy to work, light in weight, and may be readily joined using solvent cements. It is for these reasons that it is so widely used for models.

Fig. 24.16 Typical equipment layout model showing processing. (*Courtesy E. I. du Pont de Nemours Company*)

After a model has been reviewed and approved it is color-coded, tagged, and made ready for shipment to the plant construction site. On models that have been prepared primarily for confirmation and approval, pipelines are coded to indi-

Fig. 24.17 Complete true-scale production engineering model ready for use at the construction site. (*Courtesy Procter and Gamble*)

cate particular systems to which the lines belong. That is, water lines might be shown in blue, while lines carrying a specified chemical might be presented in green. In the case of models for which there are few, if any, conventional drawings, pipelines are coded to indicate their construction specification.

Three-dimensional models enable designers to discover errors, detect interferences, and determine the most economical design more readily than would be possible from drawings alone. In addition, models prove invaluable to those directing construction work in the field. However, drawings will always be needed since not all details can be incorporated and indicated on a model.

Problems

1. Make a freehand sketch (on $\frac{1}{8}$-in. grid paper, if it is available) of a 1-in. nipple connecting, a 1-in. cast-iron elbow, and a $2 \times 2 \times 1$-in. malleable-iron tee. The distance between centers of fittings is to be 6 in. Enter neatly, in draftsman's style, the length of the nipple to the nearest $\frac{1}{8}$ in.

2. Make a freehand sketch (on $\frac{1}{8}$-in. grid paper) of a $1 \times 1 \times 1$-in. malleable-iron tee and a 1-in. cast-iron elbow joined by a length of pipe. The distance between centers of fittings is 4 in. Enter the length of the connecting pipe to the nearest $\frac{1}{8}$ in.

3. Make a single-line multiview sketch of the portion of a piping system shown in Fig. 24.18.

4. Make a double-line developed drawing of the portion of a piping system shown in Fig. 24.18. Use 2-in. pipe and screwed fittings. Select a suitable scale. Determine the measurements by transferring distances from the drawing to the open-divided scale in the figure.

5. Dimension the drawing of Problem 4.

Fig. 24.18 Pictorial line diagram.

6. Make a detail (working) drawing of an assigned part of the air cleaner shown in Fig. 24.19. Determine the dimensions by transferring them from the drawing to the accompanying scale by means of the dividers (study the pictorial representation).

Fig. 24.19 Air cleaner. (*Courtesy A. Schrader's Son Mfg. Co.*)

Bridge over Houston ship canal. (*Courtesy Bethlehem Steel Corporation.*)

structural drawings* 25

25.1

Although structural drawings are prepared in accordance with the general principles of projection, they differ somewhat from machine drawings in certain practices. These differences, which have gradually developed due to the type of raw material used and methods of fabrication, have become established drawing-room customs that are recognized universally throughout the industry and must be understood and adhered to by every prospective structural engineer.

Steel structures vary widely and include almost everything fabricated from rolled shapes and plates.

Fabrication consists in shearing, flame cutting, punching, bending, forging, and machining, then fitting and aligning the parts, and finally permanently fastening the assembly by bolting, riveting, or welding. Although small roof trusses and girder bridges may be assembled as complete units in the shop, the size of most structures makes necessary the fabrication of subassemblies and shipment in knock-down form.

Sections of the principal shapes (angles, I-beams, channels, wide-flange sections, and plates) are shown in Fig. 25.1.

The dimensions of the various standard shapes and other available information required by a structural detailer are given in structural steel handbooks published by different manufacturers and by the American Institute of Steel Construction.

*Prepared in collaboration with R. S. Green, Ohio State University, and M. T. Ward, American Bridge Company.

Fig. 25.1 Structural shapes.

25.2

Equipment of a Structural Draftsman. The equipment needed by a structural draftsman is the same as for any other line of industrial drafting, with a few additions. Smaley's tables of trigonometric functions and logarithms and Smaley's tables of slopes and rises are a necessity. A copy of the structural handbook published by the American Institute of Steel Construction must be readily available for constant use.

Some companies furnish each draftsman with a book that gives drawing-room standards. Included are typical drawings illustrating the arrangement of views, approved dimensioning practices, and notes for various types of structures. Some information also may be given about plant equipment.

It is suggested that the structural draftsman in training obtain a copy of the latest edition of the AISC text, *Structural Steel Detailing*. Mastery of the instructions in this excellent text should qualify one who is starting in this field to do acceptable work in structural detailing with any structural steel fabricator.*

A □ Classes of structural drawings

25.3

Classes of Structural Drawings. Most of the large steel fabricators maintain a design office and a detailing office. The former prepares design drawings and estimates costs in the preparation of bids and frequently serves in a consulting capacity on designs furnished by a customer. The detailing office, which is usually located at the fabricating plant, orders material and prepares shop and erection plans from the design sheets.

* *Structural Steel Detailing*, New York: American Institute of Steel Construction, Inc., 1966.

A set of specifications covering special conditions, unit stresses, materials to be used, and so forth, is considered as part of the design information.

Shop detail drawings show all of the information necessary for shop fabrication.

Erection plans, which are prepared primarily for use in the field, consist of line diagrams giving dimensions, shipping marks, and notes in sufficient detail to guide the erector in assembling the parts to complete the finished structure.

25.4

Design Drawings. Design drawings usually are line diagrams showing the shape of a structure, the principal dimensions, structural sections, and in some cases the stresses to be used in detailing the connections. Figure 25.2 is an example of a design drawing.

For the use of the layout man a set of design drawings may contain elaborate design details, showing the type of connections, thickness of gusset plates, and the number of rivets. Figure 25.3 is the design detail for joint L_0 of the truss shown in Fig. 25.2. Design drawings of this type, however, are furnished only for important or complicated structures. Quite often the design of connections becomes a detailer's task.

A design drawing of a light industrial building is shown in Fig. 25.4. On this drawing the designer has given the detailer the information necessary to prepare the shop drawings for the structural frame. The composite plan view shows the top and bottom chord bracing and the braced bays that require sway frames. Although the size of the eave struts has been indicated on the plan, their location is given on the typical wall section. The size and location of the purlins and girts have been shown. It should be noted that sag rods have been used to align the purlins and girts.

The cross section, shown at the far right, has been taken in the direction of

Fig. 25.2 Stress sheet.

Fig. 25.3 Design detail of joint L_0 (Fig. 25.2).

the 60-ft width of the structure. This view shows the sizes of the columns and truss components. It may be noted that the designer has given the axial tension and compression stresses in the truss and knee braces. This is information needed by the detailer for developing adequate connections. Two sets of stresses have been indicated: those produced by gravity (vertical) loads and those that might result from wind. Wind stresses have been preceded by a $\pm$ to indicate either tension or compression, since the wind may blow from either direction against the building. The listing of stresses, as was done in Fig. 25.4, is important because the detailer must know if any of the double-angle truss members will be subject to both tension and compression.

Design drawings of trusses should show those dimensions that are needed to establish the necessary working points. However, the exact position of working points for diagonal bracing is sometimes left for the detailer to determine.

25.5

Layouts. The first step in the development of structural steel detail drawings is the drawing of the layout sheets. These are intermediate drawings that are used only in the drafting room of the fabricating shop. Layouts are used for ordering material, obtaining early approval of details, and coordinating the work of the several draftsmen who may be employed on the project. As a general rule, layouts are made only when the complexity of the work demands a carefully scaled picture. For example, layouts would be necessary for all types of skewed work and for truss joints. Layouts may be drawn to any appropriate scale, but usually the detail drawing and layout scales are the same. They are usually made on bond paper that is thin enough to permit blueprinting. Layouts are not completely dimensioned, but the layout person may indicate any dimensions he wishes to be used.

Figure 25.5 is a layout of the cross frame of a deck-girder bridge. The detailer

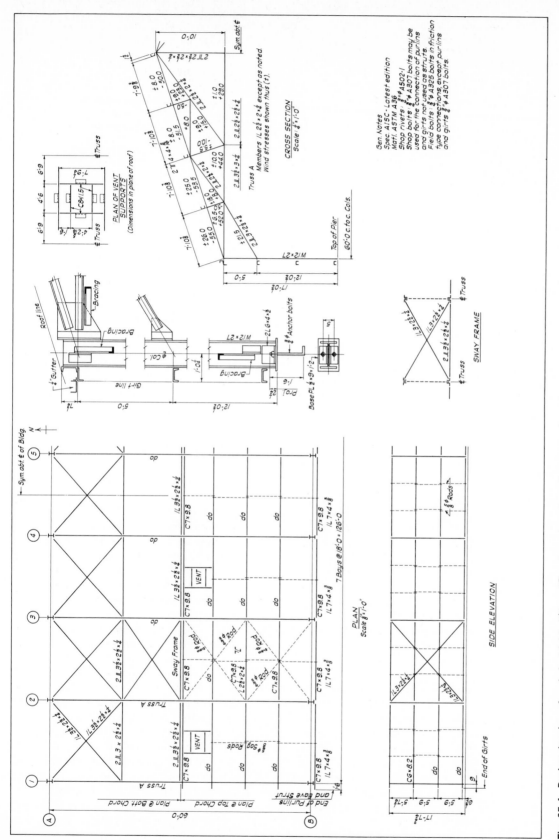

Fig. 25.4 Design drawing of an industrial building. (*Courtesy American Institute of Steel Construction, Inc.*)

Fig. 25.5 Layout for internal cross frame.

individually, as is the practice for machine work. Figure 25.7 is a detail drawing of a cross frame. It describes each plate and main member and shows the relations of the various elements of the structure to one another. When the structure is too large to be completely assembled in the shop and shipped in one piece, an assembly or erection diagram becomes necessary.

The scales in general use are not large enough to permit direct scaling of dimensions. One of the scales most commonly used in structural work is $\frac{3}{4}$ in. = 1 ft. Often structural members are too long to be drawn to scale and yet be contained on the sheet of drawing paper. In this event, the transverse dimensions and the details are drawn to one scale and the longitudinal dimensions are shortened or drawn to a smaller scale.

Structural steel drawings incorporate a few practices of projection that differ from other types of work. In all structural work the view of the structure that corresponds to the front view in a machine drawing is termed the *elevation* of the structure. The view that corresponds to the top view in a machine drawing is called the *plan view* of the structure. The view below the cross frame in Fig. 25.7 is a plan view of the lower-chord member and is not a bottom view. Likewise, the two views at the bottom of the sheet in Fig. 25.3 are not bottom views of the joint but are sectional views looking downward. The left view shows the location of the anchor bolts and the right view shows both the top and the bottom of the member L_0–L_1 where it frames into the joint. Shopmen prefer the use of bottom sections because the elevation of a piece, while being fitted, corresponds to the elevation on the drawings. Any fitting or inspection of the bottom is more easily accomplished by looking down than by crawling under and looking upward.

Angular dimensions on structural drawings are shown as slopes that are expressed in inches per foot. A slope triangle is a right triangle constructed with its hypotenuse on the gage or working line whose slope is to be shown. The longer of the two legs is always given as 12 in., and the length of the shorter leg determines the slope. These slope triangles are not drawn to scale but are constructed in any convenient size (Fig. 25.7).

Structural drawings are currently made in pencil on translucent bond paper. Pencil work requires particular care on the

takes layouts of this kind, in addition to the design drawings for the project, and completely details the structure.

25.6
Detail Drawings. The making of the detail drawing is the final step in the process of creating structural steel working drawings (Fig. 25.6). These drawings must be clear and concise, to enable persons in the shop to do their portion of the work efficiently. To ensure accuracy, a thorough check of all arrangements and dimensions is made in the drafting room by a checker.

Parts to be riveted together in the shop are detailed in their assembled positions in the structure, instead of being detailed

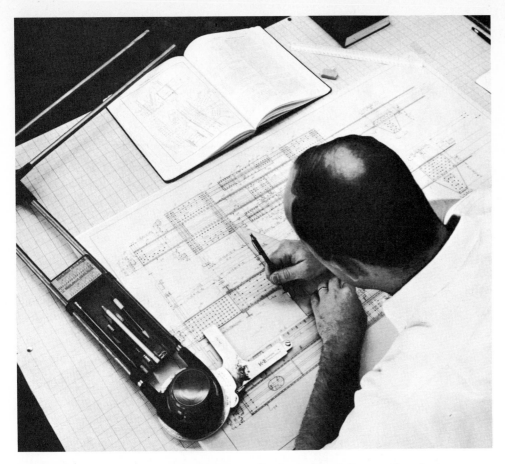

Fig. 25.6 Detailer at work. (*Courtesy American Institute of Steel Construction, Inc.*)

part of the draftsman to make the lines black enough to blueprint clearly. The outlines of the structural members should be wider than the gage lines and dimension lines because the contrast makes the blueprints easier to read.

Pencil tracings on special cloth or ink tracings on cloth are required by certain customers.

Even when roof trusses are to be transported to the site in more than one piece, the several component sections are usually detailed in their assembled position. Although the truss shown in Fig. 25.8 might not be too large to be shipped as a single piece, the details, as prepared, show the usual separation that would be used for the fabrication and shipment of a large truss.

25.7
Procedure for Making a Layout of a Gusset Plate. The general procedure for making the layout of a gusset plate that connects two members of a roof truss to the bottom chord member is given in Fig. 25.9. Each member is composed of two angles.

Step 1. Calculate the slopes of the diagonal members and draw in the working lines (Fig. 25.9). Use a scale of $1\frac{1}{2}$ in. = 1 ft or 3 in. = 1 ft. (Ordinarily the working lines will be gage lines.)

Step 2. Determine the correct dimensions and gages (Fig. 25.10) and draw in the lines representing the outstanding and perpendicular legs of the angles on the proper sides of the working lines. Draw in the clearance line at a preferred distance above the bottom chord angle and cut the diagonal angles perpendicular to their axes so that the corners fall on the clearance line.

Step 3. Locate an initial rivet in each diagonal at the desired edge distance from the sheared edge of the member. Since it is customary in structural drawing to give the distance from the working point to the first rivet to the nearest $\frac{1}{4}$ in., it usually will be found necessary to change either the edge distance or the clearance to meet this requirement. Locate the remaining rivets in each diagonal, using minimum spacing (3 × diameter of rivet), so that the plate will not be larger than is absolutely necessary.

Fig. 25.7 Detail drawing of a cross frame.

Step 4. Draw in the edges of the gusset plate, after giving some consideration to the factors involved in an economical treatment of the design. The points to be observed are as follows: (1) Allow not less than minimum edge distance *e* from the center of each rivet to the nearest edge of the plate (Fig. 25.11). (2) Allow no corners of the plate to project beyond the angle. (3) Design the plate so that there will be a minimum number of cuts, for each cut increases the labor cost. (4) Make at least two edges parallel at a distance apart equal to a standard plate width, so that unnecessary cuts and material waste may be avoided.

The shorter dimension usually is considered to be the width and the longer dimension, the length. If the longer dimension is across the plate between par-

allel sides, however, it should be given as the width, because the plate may be cut from a long plate of that width. The length dimension usually is given to the nearest $\frac{1}{4}$ in. The dimensions of a plate are always those of a rectangle from which the plate may be cut.

B □ Notations and dimensioning practices

25.8
Structural Definitions and Notations

$'$ = foot or feet
$''$ = inch or inches
ϕ = diameter

Fig. 25.8 Riveted truss. (*Courtesy American Institute of Steel Construction, Inc.*)

Fig. 25.9 Steps for making layout of gusset.

Fig. 25.10 Gage distances.

LEG	8	7	6	5	4	3½	3	2½	2	1¾	1½	1⅜	1¼	¾
G₁	4½	4	3½	3	2½	2	1¾	1⅜	1⅛	1	⅞	⅞	¾	⅝
GAGE G₂	3	2½	2½	2	—	—	—	—	—	—	—	—	—	—
GAGE G₃	3	3	2¼	1¾	—	—	—	—	—	—	—	—	—	—
MAX. RIVET	1⅛	1	1	1	1	⅞	⅞	¾	⅝	½	⅜	⅜	¼	¼

Fig. 25.11 Edge distance *e*.

Dia. of Rivet	Extreme Minimum "e"	Usual Minimum "e"	Preferred
3/8	9/16	3/4	3/4
1/2	3/4	1	1
5/8	15/16	1⅛	1¼
3/4	1⅛	1¼	1½
7/8	1 5/16	1½	1¾
1	1½	1¾	2

\# = pound or pounds
∠ = angle
I = I-beam
⊔ = channel
WF = wide-flange section

Anchor bolt. A bolt used to fasten steel girders, columns, trusses, and so on, to masonry.

Beam. A horizontal structural member (usually an I-beam or WF beam).

Bearing plate. A plate used under the end of a truss, beam or girder, to increase the bearing area.

Chip. To chip off projecting parts with a chisel (pneumatic chisel).

Clearance. A general term applying to an opening or space between two adjacent pieces, without which interference would result (Fig. 25.9).

Column. A general term for a vertical member supporting beams or trusses.

Countersink. The operation of chamfering the edges of a hole to receive the conical head of a bolt, rivet, or screw.

Detail. To prepare a working structural drawing. (Working drawings are called details.) (Fig. 25.7).

Driving clearance. The distance from the center of a rivet to the nearest obstruction that would interfere with the driving of the rivet.

Edge distance. The distance from the center of a rivet or hole to the edge of the member (Fig. 25.9).

Erection diagram. An assembly diagram drawn to aid the erector in placing members of the structure in their proper positions.

Fabrication. The shop work of converting rolled shapes to complete structural members.

Field clearance. Minimum distance between unfinished edges that abut when erected at the site. Usually $\frac{1}{2}$–$\frac{3}{4}$ in.

Field rivet. A rivet driven into the structure at the site of construction.

Filler. A plate or washer used to fill up space between two surfaces.

Flame cut. To cut by hand or by a machine-guided oxyacetylene torch.

Flange. A general term for the outstanding part of a member (Fig. 25.1).

Gage. The distance from the back of an angle or channel to the center line of a row of rivets, or the distance between the center lines of two rows of rivets (Fig. 25.10).

Gage line. The center line of a row of rivets (Fig. 25.10).

Girder. A horizontal member built up of plates and angles.

Grip. The combined thickness of members connected by a rivet.

Gusset plate. A connection plate used to connect several members of a truss (Fig. 25.9).

Layout. A preliminary scale drawing made in the detailing department prior to detailing, for the purpose of ordering material and coordinating the work of the several detailers (Fig. 25.5).

Leg. The name for either of the two flanges of an angle (Fig. 25.1).

Line diagram. A drawing in which each member is represented by a single line (Fig. 25.2).

Pitch. The center-to-center longitudinal distance between adjacent rivets (Fig. 25.10).

Plate. A flat piece of structural steel having a rectangular cross section (Fig. 25.1).

Punch. To make a hole by forcing a nonrotating tool through the material.

Ream. To enlarge and finish a punched hole, using a rotating fluted cutter.

Rivet. A cylindrical rod of steel that is used to fasten together members of a steel structure. It has one head formed when manufactured; the other is formed after the rivet is in position.

Shape. The structural term for rolled steel having any cross section (except a steel plate) (Fig. 25.1).

Shop clearance. Minimum distance between unfinished abutting edges of members assembled in the shop. Usually $\frac{1}{4}$ in.

Shop drawing. A working drawing made for the shop. Commonly called a *detail drawing* (Fig. 25.7).

Slope. The inclination of a line designated by a slope triangle expressed in inches of rise to a base of 12 in. (Fig. 25.7).

Span. The center-to-center distance between the supports of a beam, girder, or truss.

Staggered rivets. Rivets spaced alternately in parallel rows (Fig. 25.10).

Stitch rivets. Rivets spaced at intervals along a built-up member to cause the component parts to act as a unit.

Stress sheet. A drawing having a line diagram on which are recorded the stresses in the main members of a structure (Fig. 25.2).

Truss. A rigid framed structure, in the form of a series of triangles, which acts as a beam (Fig. 25.8).

Web. The thin portion between the flanges of a member (Fig. 25.1).

Weldment. Any welded assembly or subassembly (Fig. 25.18).

Working point. The point of intersection of working lines (usually gage lines) (Fig. 25.8).

25.9

Sizes of Standard Members. The following structural specifications and abbreviations are those adopted by the American Institute of Steel Construction.

Plates. Width (in inches) × thickness × length. (Pl 15 × $\frac{3}{8}$ × 1′-10.) If it is a connection plate on a truss, cross frame, and so on, which is fabricated in the shop, the specification will be followed by the letters *pa, pb, pc,* or *pd,* and so on, which indicate the location of the plate. (Pl 15 × $\frac{3}{8}$ × 1′-10 *pa*.)

Angles—equal legs. Size of leg × size of leg × thickness × length. (∠ 3$\frac{1}{2}$ × 3$\frac{1}{2}$ × $\frac{3}{8}$ × 7′-6.)

Angles—unequal legs. Size of leg shown × size of outstanding leg × thickness × length. (∠ 4 × 3 × $\frac{5}{16}$ × 24′-7.)

I-Beams. Depth of I-weight per foot × length. (12 I 31.8 × 15′-4.)

Channels. Depth of -weight per foot × length. (10 ⊔ 15.3 × 15′-10.)

Wide-flange sections. Depth of WF-weight per foot × length. (24 WF 74 × 12′-6.)

	SHOP RIVETS												FIELD RIVETS			
BUTTON HEADS	COUNTERSUNK AND CHIPPED			COUNTERSUNK NOT CHIPPED			FLATTENED TO $\frac{1}{4}$ HIGH FOR $\frac{1}{2}$ AND $\frac{5}{8}$ RIVETS			FLATTENED TO $\frac{3}{8}$ HIGH FOR $\frac{3}{4}$ TO 1" RIVETS			FULL HEADS	COUNTERSUNK AND CHIPPED		
BOTH SIDES	BOTH SIDES	NEAR SIDE	FAR SIDE	BOTH SIDES	NEAR SIDE	FAR SIDE	BOTH SIDES	NEAR SIDE	FAR SIDE	BOTH SIDES	NEAR SIDE	FAR SIDE	BOTH SIDES	BOTH SIDES	NEAR SIDE	FAR SIDE	BOTH SIDES

Fig. 25.12 Conventional symbols for rivets.

25.10

Detailing Information. The type of rivets and their treatment are indicated on structural drawings by the American National Standard conventional symbols shown in Fig. 25.12.

The holes for field rivets are indicated in solid black on a drawing, while shop rivets are shown by open circles having the same diameter as the rivet head (Fig. 25.13). Rivets should be drawn with either a drop pen or a bow pencil. In practice, the circles representing rivets are often drawn free-hand on pencil drawings.

25.11

Location of Dimension Lines. Since shopmen are never permitted to scale a drawing, all dimensions must be placed in such a manner that they will be easily understood. Principal dimensions are generally obtained from the design sheets; other dimensions necessary for detailing are found in tables or are determined by the detailer.

Dimensions for rivet spacing, minor location and size dimensions, and so on, are placed close to the view, while the longer dimensions, such as overall lengths, are placed farther away so that extension lines will not cross dimension lines (Figs. 25.14 and 25.15).

Dimension figures are generally placed above continuous (unbroken) dimension lines, which are made narrow and black. These lines usually should be placed off the view, but oftentimes added clarity may be obtained by putting a few dimensions in an open area on the view itself. Dimension lines ordinarily should not be placed less than $\frac{3}{8}$ in. apart or closer to the view than $\frac{1}{2}$ in. All of the above rules for the location of dimension lines may be modified to suit the available space (Fig. 25.16).

25.12

Dimensions and Notes in Structural Detailing.

1. Figures may be compressed without reducing their height, in order to place them in a limited space between arrowheads.

2. Figures can be placed to one side, with a leader to the dimension line, if the available space is very small.

3. Figures and notes must read from the bottom and the right side of the sheet because shopmen are accustomed to reading from these positions (Fig. 25.14).

4. For dimensions less than 1 ft, the inch marks (″) may be omitted (Fig. 25.15).

5. With the exception of widths of plates and depths of sections, all dimensions of 1 ft or more are expressed in feet and inches (Fig. 25.14).

BUTTON HEAD
(a)

COUNTERSUNK HEAD
(b)

Fig. 25.13 Dimensions of rivet heads.

Correct	Incorrect
$\frac{1}{4}$	$0\frac{1}{4}$
9	0′-9″
10	10″
1′-0	12″
2′-3$\frac{1}{4}$	2′-03$\frac{1}{4}$″
4′-0$\frac{1}{4}$	4$\frac{1}{4}$″

Fig. 25.14 Structural dimensioning.

Fig. 25.15 Structural dimensioning. (*Courtesy American Bridge Co.*)

Fig. 25.16 Column details. *(Courtesy American Bridge Co.)*

6. Usually, dimensions for rivet spacing are given in multiples of $\frac{1}{8}$ in. or, preferably, $\frac{1}{4}$ in. It is not desirable to use multiples of $\frac{1}{16}$ or $\frac{1}{32}$ in., except in rare cases.

7. Decimals found in tables should be converted to fractions to the nearest $\frac{1}{16}$ in. (except for machine-shop drawings for gears, shafts, and so forth).

8. To avoid complications that arise when corrections are made, dimensions shown on one view should not be repeated on another.

9. Rivets and holes are located by dimensions from center to center (Fig. 25.14).

10. Edge distances are frequently omitted, unless they are necessary to ensure clearances with connecting parts. (Shopmen understand that the distances on opposite edges are to be made equal.)

11. Dimensions *always* should be given to the center lines of I-beams and to the backs of angles and channels. (See Sec. 25.8, "gage" definition.)

12. When three or more rivet spaces for a line of rivets are equal, they should be dimensioned as a group (4 @ 3 = 1′-0). Staggered rivets are dimensioned as though they were on one gage line (Fig. 25.14).

13. Since a workman must use a rule or tape to lay off angles, a slope triangle should be shown to give the inclination of a working line (Fig. 25.15).

14. A person in the shop should never be compelled to add or subtract to obtain a necessary dimension.

15. A general note is usually placed on a detail drawing giving painting instructions, size of rivets, size of open holes, reaming instructions, and so on.

16. Members that are shipped separately for field erection are given a shipping mark of a letter and number that appears on the detail drawing and on the erection plan (Fig. 25.14).

17. The size of a member is indicated by a specification (in the form of a note) parallel to it (Fig. 25.14).

18. The width of plate is always given in inches (Fig. 25.14).

25.13

Machine-shop Work. The structural detailer will occasionally be called on to prepare detail drawings for castings for bridge shoes and roadway expansion joints and at times drawings for complicated gearing and shafting for movable bridges. Regular machine drawing practices apply, the principal difference being that bridge machinery in general is ponderous with single castings often weighing 10 or 15 tons.

25.14

Structural Welding. Since a large portion of structural work is either partially or completely welded, a working knowledge of the use of welding symbols (Chapter 23) is a requirement for a structural draftsman (Fig. 25.17).

Sizes and location of welds are usually given on the design drawing, but very often connections and minor details are left to the detailer.

An example of a welded truss of light construction is shown in Fig. 25.18.

25.15

Computer–plotter-prepared Structural Drawings. Programming systems have been prepared to produce complete shop (detail) drawings for the structural steel fabricating industry by automated drafting methods (see Chapter 17). At present, these systems will make shop drawings of beams, columns and bracing for most types of steel building frames. The Control Data Corporation (CDC) CONSTRUCTS System, which has been under constant development, now contains all of the specifications of the American Institute of Steel Construction and recognizes the standard practices for fitting and connecting structural steel shapes to carry the design loads indicated on the engineer's design drawings. The drawing shown in Fig. 25.19 was made using the Control Data Structural System.

In the application of programming systems that have been developed to prepare shop drawings to fabricate different types of structural members, a technician fills out an input form utilizing information furnished on the structural design drawing. This information, in turn, is fed into a computer for computation of the dimensions necessary for the fabrication of each structural member in a building framing plan.

The output from the initial run of the program is a complete set of tabulated computations and information. This print-out should be inspected for obvious errors. At this time, if there is a special

Fig. 25.17 Welded connections.

Fig. 25.18 Welded roof truss. (*Courtesy American Institute of Steel Construction, Inc.*)

reason, the person making the check may introduce nonstandard treatment of connections of members.

A second run through the computer produces a magnetic tape with the drafting instructions needed to direct the off-line plotter that will prepare the drawings.

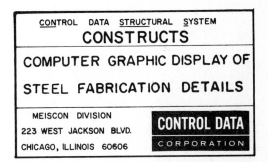

Fig. 25.19 Structural drawing prepared on a plotter. This drawing was made using the CONSTRUCTS System of automated drafting. (*Courtesy Meiscon Division, Control Data Corporation*)

Problems

The following problems are intended to furnish experience in the preparation of layouts and to emphasize the principles of structural drawing.

1. Make a pencil layout, plan, and elevation of joint A, Fig. 25.20. Follow carefully the steps outlined in Sec. 25.7 for making a layout of a gusset plate. The following requirements must be observed:

Use $\frac{3}{4}$-in. ϕ rivets.

Make the minimum allowable clearance between members $= \frac{1}{4}$ in.

Use minimum rivet spacing $= 3 \times$ diameter of rivets.

Use the preferred edge distance given in the table in Fig. 25.11.

Use standard gage distances, as given in the table in Fig. 25.10.

Make the width of the plate equal to the width of a standard plate. The variation of plate widths is by inches.

The bearing plate is a $9 \times \frac{3}{4} \times 1'-0$ steel plate. The rivet pitch in the plan view should be equal to the pitch of the rivets in the vertical leg,

and they must be so located that the distances c, f, and k are equal to or exceed the minimum values for these distances for a $\frac{3}{4}$-in. rivet as given in structural handbooks. The minimum value of f for a $\frac{3}{4}$-in. rivet is $1\frac{1}{4}$ in.; the minimum value of k is $1\frac{3}{4}$ in.; and the minimum value of c is $1\frac{1}{4}$ in. These three values represent minimum driving clearance for a $\frac{3}{4}$-in. rivet.

The holes for the anchor bolts are $\frac{13}{16}$ in. in diameter. Letter the correct specifications for the gusset plate.

2. Make a pencil drawing of the cross frame shown in Fig. 25.21. The following requirements must be observed:

Use $\frac{7}{8}$-in. rivets.

Use a minimum rivet pitch of $2\frac{1}{2}$ in.

The elevation of the cross frame is the only view that is required.

Use standard gage and preferred edge distances, as shown in Figs. 25.10 and 25.11.

Use Fig. 25.7 as a model for the placing of complete dimensions.

The open holes are spaced at $3\frac{1}{2}$-in. pitch.

Fig. 25.20 Roof truss layout.

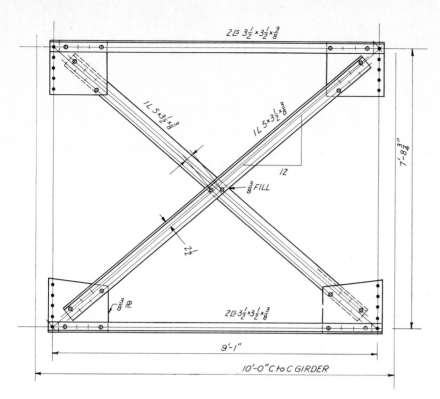

Fig. 25.21 Layout of interior cross frame for 100-ft railroad deck girder span.

Plan view for a portion of a highway project. Distances and size values are in meters and millimeters. (*Courtesy of Ohio Department of Transportation*)

topographic and engineering map drawings 26

26.1

Map Drawing.* A map is a drawing that represents a portion of the earth's surface area. Since it usually represents a relatively small part and the third dimension (the height) is not shown except in some cases by contour lines, a map may be thought of as a one-view orthographic projection (Fig. 26.1). Various forms of maps have been devised to satisfy different requirements. Land maps, plats, and

*See Figs. 17.28 and 17.29 for engineering maps prepared on a plotter.

so on, which fulfill their purpose by revealing only the natural and man-made features along with imaginary division lines and geometrical measurements, show only two dimensions. Others, such as topographical maps, show three dimensions, by representing height by means of contours

26.2

Engineering maps and plats—metric. Ideally, the scales used for preparing topographic maps, engineering maps, and plats should be from the 1, 2, and 5 series,

though the 2.5 scale is used to some extent. The U.S.G.S. recommends that the inch-foot scales be replaced by the 1, 2, 5 series as follows:

Inch-foot	SI
1″ = 80′	1:1000
1″ = 100′	1:1000
1″ = 200′	1:2000
1″ = 400′	1:5000
1″ = 500′	1:5000

U.S.G.S. maps are prepared to scales of 1:25,000, 1:50,000, and 1:100,000.

When a land area is small, say city lot size, the area should be given in square

Fig. 26.1 Section from an engineering map (metric).

meters. Large areas may be given in hectares or in cases of very large areas, such as state and national parks, in square kilometers.

For new residential subdivisions, where lengths, widths, and angles are unrestricted, roads should be made 10 m, 12 m, or 15 m wide. Arterial thoroughfares and streets should be laid out to a width of 20 m, 22 m, or 25 m as may be required for parking and expected traffic. Utility easements should be either 2 m, 3 m, or 5 m in width. Using full meters may be thought of as making a hard conversion to metric.

For metric maps of site plans, contour intervals can be either 0.1 m, 0.2 m, or 0.5 m. For topographic maps of rough terrain the U.S.G.S. commonly uses intervals of 1 m, 1.5 m, 2 m, 50 m, or 100 m.

26.3

Classification of Maps. Maps of interest to the engineer may be grouped for study in four general classes, in accordance with their purpose and method of preparation. The recognized classes are (1) topographical, (2) cadastral, (3) engineering, and (4) aerial photographic.

26.4

Topographic Maps. Topographic maps, although they are drawn to a relatively small scale, contain much detail. All natural features, such as lakes, streams, forests, fields, mines, and so on, and important permanent man-made creations, such as buildings, bridges, and houses,

may be represented if necessary to fulfill the purpose of a map. Topographic maps, prepared by the United States Geological Survey to a scale of 1:24000, naturally do not contain very much detail (Fig. 26.2). The form of the surface of the ground is represented by contour lines. Any one contour line passes through points of the same elevation and closes either on the map or beyond its limits. Closed contour lines represent either a hill or a depression. Figure 26.6 shows a topographic map.

26.5

Cadastral Maps. The major portion of this group consists of city plats, city development maps, town maps, county maps, and maps prepared to show ownership (particularly for the purposes of governmental control and taxation). These maps, although they show practically no detail, must be accurate and for this reason are drawn to a large scale. Property lines, political boundaries, and a few important features, such as streams, roads, and towns, may be given on township and county maps, to enable a reader to identify particular locations.

26.6

Engineering Maps (see Chapter opening facing page). Working maps prepared for engineering projects are known as engineering maps. They may be drawn for either reconnaissance or construction purposes. They usually are made to a large scale and accurately show the location of all property lines and important features. On maps of a topographic nature, practically all natural and man-made features along a right-of-way or on a site are shown, and the form of the surface of the ground is indicated by means of contours.

26.7

Plats of Land Surveys. A plat of a tract of land should contain a complete description of the land surveyed. It should show the lengths and bearings (or included angles) of the bounding sides and division lines, the included acreages, the locations of the monuments, and the names of the owners of the adjoining properties. Figure 26.3 shows a plat of a typical land survey. Note that a clear, concise title is lettered in the large open area. A certification of the survey is generally required by law. In most states, a plat must bear the seal of a licensed surveyor.

Fig. 26.2 Portion of a typical map prepared by the U.S. Geological Survey.

**26.8
Subdivision Plats.** A plat of a real-estate development should show the measurements and angles of the survey of the whole tract of land, the sizes of the included lots, the widths of the streets and drives, and the locations of all monuments. Plats of subdivisions must be complete and accurate, since they are filed as public records in the county recorder's office. Sufficient information must be given to enable a surveyor to locate the corners of any lot with precision when making a resurvey at a later date. Figure 26.4 is a plat of a city subdivision.

**26.9
Plats and Partial City Maps.** Plats made from subdivision plats or city maps are prepared by the engineering departments of cities and public utilities. The purpose of these partial maps is to record special

TRACT "A"

0.71±AC.

TRACT "B"

1.11±AC.

STATE ROAD 26

R/W LINE

SECTION LINE

L.E.&W.(NICKEL PLATE) R.R.

R/W LINE

PLAT OF PART OF THE WEST HALF (½) OF
SECTION 24, TOWNSHIP 23, RANGE 5 WEST.
I hereby certify that I am a registered engineer
of the State of Indiana, and that the above is a
true and correct plat.

John H. Matthews
Ind. Reg. No. 1579

HALF SECTION LINE

S.W. CORNER OF THE
N.W.¼ OF SEC. 24
T23N, R5W

Fig. 26.3 Plat of a land survey.

information concerning such things as proposed improvement projects, the location of lines of transportation, and the location of existing and proposed water mains, sewers, and so on. It is not necessary for such a map to contain all the information given on the subdivision plat from which it is made. The locations of monuments and angles generally are not shown. The widths of streets and the sizes of lots may or may not be shown, depending on the usefulness of such information. A few important buildings may be indicated for the sake of aiding the reader in orienting himself. Figure 26.5 shows the proposed locations of water mains and fire hydrants in a portion of a city.

26.10
Topographic Drawing. As previously stated, a topographic map is a reproduction, to scale, of a large area. On a complete topographic map, the natural and artificial features are represented by recognized conventional symbols and the form of the ground is shown by contours. Excessive detail should be avoided, and only necessary surface features should be shown. All names and required notes should be lettered in a position where they can be easily read; a complete title should be lettered, in the lower-right-hand corner. Ordinarily, single-stroke Reinhardt lettering is preferable on topographic maps prepared solely for construction projects,

Fig. 26.4 Plat of a city subdivision (also see Fig. 17.29).

while vertical modern Roman lettering is more desirable on finished maps where effect and pleasing appearance are important. Figure 26.6 shows a topographic map.

The scale to be used for a topographic map depends on the size of the area and the amount of the detail that must be shown. Scales range from 1 in. = 100 ft to 1 in. = 4 miles. Maps prepared by the United States Geological Survey (U.S.G.S.) have in the past usually been drawn to 1:62,500, which is almost 1 in. to a mile.

For drawings now being prepared in metric, the U.S.G.S. uses 1:1000 in place of 1″ = 100′. Other commonly used metric scales are: 1 = 25,000, 1:50,000, and 1:100,000.

26.11

Use of Standard Topographic Symbols (U.S.G.S.). Recognized signs and symbols are used to represent the natural and artificial features on a topographic map. Many of these symbols have been designed to bear some pictorial resemblance to the feature or object represented and, for convenience, may be grouped as follows: (1) physical features (buildings, highways, railroads, and so on); (2) ground formations (elevations and depressions); (3) water-surface features (rivers, lakes, and streams); and (4) vegetation growths (grass, trees, and cultivated crops).

Conventional symbols used by the United States Geological Survey for repre-

Fig. 26.5 Plat or partial city map showing the proposed location of water mains and fire hydrants.

senting the works and structures of man are shown in Fig. 26.7. Symbols for natural land formations, water features, and vegetation growths, both natural and cultivated, are shown in Figs. 26.8–26.10.

26.12

Topographic Symbols. Topographic symbols are drawn either freehand or mechanically, depending on the character of the features to be represented. For example, the symbols representing natural features are drawn freehand, while those representing artificial works are drawn mechanically. See Figs. 26.7–26.10. On topographic maps prepared for engineering projects, the symbols are drawn in India ink. When colors are used, as in finished maps, the artificial features (buildings, bridges, railroads, and so on) are drawn in black, the contours in brown, the water features in blue, and vegetation growths in green.

Although the size of symbols may vary somewhat with the size of the map, they are never to scale but are always exaggerations. The usual mistake of the beginner is to draw symbols too large or too close together. Either fault produces a disagreeable appearance and tends to attract the reader's attention away from more important features. The symbols representing prominent features are made to stand out from those of lesser importance by being drawn slightly larger and with heavier lines.

The beginner should study carefully the symbols as given in the various illustrations, so he will not miss some of the essential points in their construction. For instance, he should note that the symbol for a tree (Fig. 26.10) is composed of separate lines, irregularly located, and not of one closed line drawn without lifting the pen. The symbols for grass, corn, and other vegetation should be placed with the bases of the tufts and stocks parallel to the lower borderline.

Fig. 26.6 Topographic map. (A portion of a map taken from a manual prepared by the U.S. Department of the Interior.)

26.13

Drawing Water Lines. Water lining, used to indicate water surfaces, is done entirely freehand with an ordinary lettering pen. The starting line (shoreline) should be fairly heavy, and each successive line should decrease in width until the center of the body of water is reached. See Fig. 26.9. The line next to the shoreline should be drawn parallel to it throughout its entire length, and the space between should be equal to the width of the shoreline. The spacing between succeeding lines should increase gradually to the center, but the change should be so slight that no marked increase will be noticeable. Each added line should show fewer of the small irregularities of the shoreline, the last few following only the prominent ones.

If several bodies of water are to be indicated on the same map, a good way to obtain uniformity is to draw all the shorelines first, then all the lines next to the shorelines, and so on, working back and forth from one body of water to another until the representations are completed.

Excessive waviness gives these lines an unnatural appearance and should be avoided.

26.14

Contour Lines (Fig. 26.2). A contour line is a line through points of the same elevation on the surface of the ground. Theoretically, the contour lines on a map may be thought of as the lines of intersection of a series of horizontal planes and the ground surface. In practice, the imaginary planes are equally spaced vertically so that the contour intervals will be equal and the horizontal distances between contours on a map will indicate the steepness of the rise or descent of the surface. The closer together they are, the greater the slope and, conversely, the farther apart they are, the less the slope. An arrangement of contour lines that close indicates either a hill or a depression (Fig. 26.8). The case, whatever it is, usually can be determined by reading the values of the elevations of the contours. Usually each fifth contour is drawn heavier than the

Fig. 26.7 Standard symbols for works and structures. (*From Standard
Symbols, published by the U.S. Geological Survey*)

Fig. 26.8 Relief (natural land formations).

Fig. 26.9 Hydrographic symbols.

Fig. 26.10 Vegetation symbols.

Fig. 26.11 Contour map of a small area.

others and has a break in it where its elevation above a datum plane is recorded. If a U.S.G.S. bench mark is used, the datum plane will be at mean sea level.

The selection of the contour interval (vertical distance between contour planes) for a topographic survey is determined by the nature of the ground forms and the purpose for which the map will be prepared. For instance, if the area is relatively level, a 1- or 2-ft interval probably would be desirable, while, if the area is rugged, an interval of 50 or even 100 ft might be used. For a metric drawing, if the area is rather level, the contour interval may be 0.1 m, 0.2 m, or 0.5 m. For more rugged areas, the interval, could be 1 m, 2 m, 50 m, or in some cases 100 m in mountainous country.

Contour lines are plotted from survey notes made in the field. In the case of small areas, the usual method for locating contours is to divide the area into squares and take level readings at every intersection and at intermediate points where a pronounced change in slope takes place (Fig. 26.11). On the assumption that the slope of the ground is uniform between two points, the contours are sketched in by interpolating between the readings to establish the points at which the contours cross the survey lines. The interpolation may be done by eye or by calculation.

Frequently, contours are determined from level readings taken along known lines or from stadia survey notes. When extreme accuracy is necessary and the land is fairly level, contours are established by finding and locating points directly on each contour.

26.15
Profiles. The profile of a line is prepared to show the relative elevation of every point on the line. Theoretically, it may be thought of as a view showing the line of intersection of a vertical plane through the line and the ground surface. Figure 26.12 shows a plan view and a profile of a section of a proposed highway project. When the plan and profile are on the same sheet, as in the illustration, the horizontal scale for the profile should be the same as the scale for the plan view.

Profiles are drawn by plotting elevations computed from level readings taken at regular intervals on the ground and at points where the ground changes in slope. Since the slope between adjacent points is assumed to be uniform, a profile consists of a series of straight lines joining successive points.

Fig. 26.12   Plan and profile of a proposed highway project. (Distances and sizes are in meters and millimeters). This drawing was prepared manually. It is suggested that this drawing be compared with the one shown in Fig. 17.28. It can be observed that the drawing prepared by the plotter shows all needed information. (*Courtesy of Ohio Department of Transportation*)

tables

TRIGONOMETRIC **A**

APPENDIX

Table 1   Trigonometric functions

Angle	Sine	Cosine	Tan	Cotan	Angle
0°	0.0000	1.0000	0.0000	∞	90°
1°	0.0175	0.9998	0.0175	57.290	89°
2°	.0349	.9994	.0349	28.636	88°
3°	.0523	.9986	.0524	19.081	87°
4°	.0698	.9976	.0699	14.301	86°
5°	.0872	.9962	.0875	11.430	85°
6°	.1045	.9945	.1051	9.5144	84°
7°	.1219	.9925	.1228	8.1443	83°
8°	.1392	.9903	.1405	7.1154	82°
9°	.1564	.9877	.1584	6.3138	81°
10°	.1736	.9848	.1763	5.6713	80°
11°	.1908	.9816	.1944	5.1446	79°
12°	.2079	.9781	.2126	4.7046	78°
13°	.2250	.9744	.2309	4.3315	77°
14°	.2419	.9703	.2493	4.0108	76°
15°	.2588	.9659	.2679	3.7321	75°
16°	.2756	.9613	.2867	3.4874	74°
17°	.2924	.9563	.3057	3.2709	73°
18°	.3090	.9511	.3249	3.0777	72°
19°	.3256	.9455	.3443	2.9042	71°
20°	.3420	.9397	.3640	2.7475	70°
21°	.3584	.9336	.3839	2.6051	69°
22°	.3746	.9272	.4040	2.4751	68°
23°	.3907	.9205	.4245	2.3559	67°
24°	.4067	.9135	.4452	2.2460	66°
25°	.4226	.9063	.4663	2.1445	65°
26°	.4384	.8988	.4877	2.0503	64°
27°	.4540	.8910	.5095	1.9626	63°
28°	.4695	.8829	.5317	1.8807	62°
29°	.4848	.8746	.5543	1.8040	61°
30°	.5000	.8660	.5774	1.7321	60°
31°	.5150	.8572	.6009	1.6643	59°
32°	.5299	.8480	.6249	1.6003	58°
33°	.5446	.8387	.6494	1.5399	57°
34°	.5592	.8290	.6745	1.4826	56°
35°	.5736	.8192	.7002	1.4281	55°
36°	.5878	.8090	.7265	1.3764	54°
37°	.6018	.7986	.7536	1.3270	53°
38°	.6157	.7880	.7813	1.2799	52°
39°	.6293	.7771	.8098	1.2349	51°
40°	.6428	.7660	.8391	1.1918	50°
41°	.6561	.7547	.8693	1.1504	49°
42°	.6691	.7431	.9004	1.1106	48°
43°	.6820	.7314	.9325	1.0724	47°
44°	.6947	.7193	.9657	1.0355	46°
45°	.7071	.7071	1.0000	1.0000	45°
Angle	Cosine	Sine	Cotan	Tan	Angle

Table 2   Table of chords*

Degrees	0′	10′	20′	30′	40′	50′
0	.0000	.0029	.0058	.0087	.0116	.0145
1	.0174	.0204	.0233	.0202	.0291	.0320
2	.0349	.0378	.0407	.0436	.0465	.0494
3	.0523	.0553	.0582	.0611	.0640	.0669
4	.0698	.0727	.0756	.0785	.0814	.0843
5	.0872	.0901	.0930	.0959	.0988	.1017
6	.1047	.1076	.1105	.1134	.1163	.1192
7	.1221	.1250	.1279	.1308	.1337	.1366
8	.1395	.1424	.1453	.1482	.1511	.1540
9	.1569	.1598	.1627	.1656	.1685	.1714
10	.1743	.1772	.1801	.1830	.1859	.1888
11	.1917	.1946	.1975	.2004	.2033	.2062
12	.2090	.2119	.2148	.2177	.2206	.2235
13	.2264	.2293	.2322	.2351	.2380	.2409
14	.2437	.2466	.2495	.2524	.2553	.2582
15	.2610	.2639	.2668	.2697	.2726	.2755
16	.2783	.2812	.2841	.2870	.2899	.2927
17	.2956	.2985	.3014	.3042	.3071	.3100
18	.3129	.3157	.3186	.3215	.3243	.3272
19	.3301	.3330	.3358	.3387	.3416	.3444
20	.3473	.3502	.3530	.3559	.3587	.3616
21	.3645	.3673	.3702	.3730	.3759	.3788
22	.3816	.3845	.3873	.3902	.3930	.3959
23	.3987	.4016	.4044	.4073	.4101	.4130
24	.4158	.4187	.4215	.4243	.4272	.4300
25	.4329	.4357	.4385	.4414	.4442	.4471
26	.4499	.4527	.4556	.4584	.4612	.4641
27	.4669	.4697	.4725	.4754	.4782	.4810
28	.4838	.4867	.4895	.4923	.4951	.4979
29	.5008	.5036	.5064	.5092	.5120	.5148
30	.5176	.5204	.5232	.5261	.5289	.5317
31	.5345	.5373	.5401	.5429	.5457	.5485
32	.5513	.5541	.5569	.5596	.5624	.5652
33	.5680	.5708	.5736	.5764	.5792	.5820
34	.5847	.5875	.5903	.5931	.5959	.5986
35	.6014	.6042	.6069	.6097	.6125	.6153
36	.6180	.6208	.6236	.6263	.6291	.6318
37	.6346	.6374	.6401	.6429	.6456	.6484
38	.6511	.6539	.6566	.6594	.6621	.6649
39	.6676	.6703	.6731	.6758	.6786	.6813
40	.6840	.6868	.6895	.6922	.6950	.6977
41	.7004	.7031	.7059	.7086	.7113	.7140
42	.7167	.7194	.7222	.7249	.7276	.7303
43	.7330	.7357	.7384	.7411	.7438	.7465
44	.7492	.7519	.7546	.7573	.7600	.7627
45	.7654	.7680	.7707	.7734	.7761	.7788

*Tabulated values are chord lengths of arcs of unit radius subtending the specified angles. Use multiples for accuracy.

Chord AB is tabulated

tables

METRIC **B**
APPENDIX

Table 3 Inch–millimeter equivalents

Common fractions					Decimal			Millimeters
4ths	8ths	16ths	32nds	64ths	To 4 places	To 3 places	To 2 places	To 4 places
				1/64	.0156	.016	.02	0.3969
			1/32		.0312	.031	.03	0.7938
				3/64	.0469	.047	.05	1.1906
		1/16			.0625	.062	.06	1.5875
				5/64	.0781	.078	.08	1.9844
			3/32		.0938	.094	.09	2.3813
				7/64	.1094	.109	.11	2.7781
	1/8				.1250	.125	.12	3.1750
				9/64	.1406	.141	.14	3.5719
			5/32		.1562	.156	.16	3.9688
				11/64	.1719	.172	.17	4.3656
		3/16			.1875	.188	.19	4.7625
				13/64	.2031	.203	.20	5.1594
			7/32		.2188	.219	.22	5.5563
				15/64	.2344	.234	.23	5.9531
1/4					.2500	.250	.25	6.3500
				17/64	.2656	.266	.27	6.7469
			9/32		.2812	.281	.28	7.1438
				19/64	.2969	.297	.30	7.5406
		5/16			.3125	.312	.31	7.9375
				21/64	.3281	.328	.33	8.3344
			11/32		.3438	.344	.34	8.7313
				23/64	.3594	.359	.36	9.1281
	3/8				.3750	.375	.38	9.5250
				25/64	.3906	.391	.39	9.9219
			13/32		.4062	.406	.41	10.3188
				27/64	.4219	.422	.42	10.7156
		7/16			.4375	.438	.44	11.1125
				29/64	.4531	.453	.45	11.5094
			15/32		.4688	.469	.47	11.9063
				31/64	.4844	.484	.48	12.3031
					.5000	.500	.50	12.7000
				33/64	.5156	.516	.52	13.0969
			17/32		.5312	.531	.53	13.4938
				35/64	.5469	.547	.55	13.8906
		9/16			.5625	.562	.56	14.2875
				37/64	.5781	.578	.58	14.6844
			19/32		.5938	.594	.59	15.0813
				39/64	.6094	.609	.61	15.4781
	5/8				.6250	.625	.62	15.8750
				41/64	.6406	.641	.64	16.2719
			21/32		.6562	.656	.66	16.6688
				43/64	.6719	.672	.67	17.0656
		11/16			.6875	.688	.69	17.4625
				45/64	.7031	.703	.70	17.8594
			23/32		.7188	.719	.72	18.2563
				47/64	.7344	.734	.73	18.6531
3/4					.7500	.750	.75	19.0500
				49/64	.7656	.766	.77	19.4469
			25/32		.7812	.781	.78	19.8438
				51/64	.7969	.797	.80	20.2406
		13/16			.8125	.812	.81	20.6375
				53/64	.8281	.828	.83	21.0344
			27/32		.8438	.844	.84	21.4313
				55/64	.8594	.859	.86	21.8281
	7/8				.8750	.875	.88	22.2250
				57/64	.8906	.891	.89	22.6219
			29/32		.9062	.906	.91	23.0188
				59/64	.9219	.922	.92	23.4156
		15/16			.9375	.938	.94	23.8125
				61/64	.9531	.953	.95	24.2094
			31/32		.9688	.969	.97	24.6063
				63/64	.9844	.984	.98	25.0031
					1.0000	1.000	1.00	25.4000

Table 4 Inches to millimeters*

Inches	0	1	2	3	4	5	6	7	8	9
					Inches					
0–9	0	25.4	50.8	76.2	101.6	127.0	152.4	177.8	203.2	228.6
10–19	254.0	279.4	304.8	330.2	355.6	381.0	406.4	431.8	457.2	482.6
20–29	508.0	533.4	558.8	584.2	609.6	635.0	660.4	685.8	711.2	736.6
30–39	762.0	787.4	812.8	838.2	863.6	889.0	914.4	939.8	965.2	990.6
40–49	1016.0	1041.4	1066.8	1092.2	1117.6	1143.0	1168.4	1193.8	1219.2	1244.6
50–59	1270.0	1295.4	1320.8	1346.2	1371.6	1397.0	1422.4	1447.8	1473.2	1498.6
60–69	1524.0	1549.4	1574.8	1600.2	1625.6	1651.0	1676.4	1701.8	1727.2	1752.6
70–79	1778.0	1803.4	1828.8	1854.2	1879.6	1905.0	1930.4	1955.8	1981.2	2006.6
80–89	2032.0	2057.4	2082.8	2108.2	2133.6	2159.0	2184.4	2209.8	2235.2	2260.6
90–99	2286.0	2311.4	2336.8	2362.2	2387.6	2413.0	2438.4	2463.8	2489.2	2514.6

*Based on 1 in. = 25.4 mm.

Examples: To obtain the millimeter equivalent of 14 in., read the value below 4 in. in the horizontal line of values for 10–19 in. (355.6). The equivalent value of 52.4 in. is: 1320.8 *plus* 10.16 = 1330.96, the 10.16 value being obtained by moving the decimal point left in the equivalent value for 4 in.

Table 5 Metric equivalents—millimeters to decimal inches*

mm = in.	mm = in.	mm = in.	mm = in.	mm = in.
1 = 0.0394	21 = 0.8268	41 = 1.6142	61 = 2.4016	81 = 3.1890
2 = 0.0787	22 = 0.8661	42 = 1.6535	62 = 2.4409	82 = 3.2283
3 = 0.1181	23 = 0.9055	43 = 1.6929	63 = 2.4803	83 = 3.2677
4 = 0.1575	24 = 0.9449	44 = 1.7323	64 = 2.5197	84 = 3.3071
5 = 0.1969	25 = 0.9843	45 = 1.7717	65 = 2.5591	85 = 3.3465
6 = 0.2362	26 = 1.0236	46 = 1.8110	66 = 2.5984	86 = 3.3858
7 = 0.2756	27 = 1.0630	47 = 1.8504	67 = 2.6378	87 = 3.4252
8 = 0.3150	28 = 1.1024	48 = 1.8898	68 = 2.6772	88 = 3.4646
9 = 0.3543	29 = 1.1417	49 = 1.9291	69 = 2.7165	89 = 3.5039
10 = 0.3937	30 = 1.1811	50 = 1.9685	70 = 2.7559	90 = 3.5433
11 = 0.4331	31 = 1.2205	51 = 2.0079	71 = 2.7953	91 = 3.5827
12 = 0.4724	32 = 1.2598	52 = 2.0472	72 = 2.8346	92 = 3.6220
13 = 0.5118	33 = 1.2992	53 = 2.0866	73 = 2.8740	93 = 3.6614
14 = 0.5512	34 = 1.3386	54 = 2.1260	74 = 2.9134	94 = 3.7008
15 = 0.5906	35 = 1.3780	55 = 2.1654	75 = 2.9528	95 = 3.7402
16 = 0.6299	36 = 1.4173	56 = 2.2047	76 = 2.9921	96 = 3.7795
17 = 0.6693	37 = 1.4567	57 = 2.2441	77 = 3.0315	97 = 3.8189
18 = 0.7087	38 = 1.4961	58 = 2.2835	78 = 3.0709	98 = 3.8583
19 = 0.7480	39 = 1.5354	59 = 2.3228	79 = 3.1102	99 = 3.8976
20 = 0.7874	40 = 1.5748	60 = 2.3622	80 = 3.1496	100 = 3.9370

*To nearest fourth decimal place.

Table 6   Metric tolerance equivalents: conversion of tolerances—decimal-inch to millimeters*

Design practice		Calculated equivalents of decimal-inch tolerances†	Design practice		Calculated equivalents of decimal-inch tolerances°
Decimal, in.	Metric, mm		Decimal, in.	Metric, mm	
0.0001	0.003	0.00254	0.004	0.1	0.1016
0.0002	0.005	0.00508	0.005	0.13	0.1270
0.0003	0.008	0.00762	0.006	0.15	0.1524
0.0004	0.01	0.01016	0.007	0.18	0.1778
0.0005	0.013	0.01270	0.008	0.2	0.2032
0.0006	0.015	0.01524	0.009	0.23	0.2286
0.0007	0.018	0.01778	0.010	0.25	0.254
0.0008	0.02	0.02032	0.015	0.4	0.381
0.0009	0.023	0.02286	0.02	0.5	0.508
0.001	0.025	0.0254	0.03	0.8	0.762
0.0015	0.04	0.0381	0.04	1	1.016
0.002	0.05	0.0508	0.06	1.5	1.524
0.0025	0.06	0.0635	0.08	2	2.032
0.003	0.08	0.0762	0.10	2.5	2.540

* To be used for tolerance conversions only. Table gives commonly used inch tolerances.
† Calculated on the basis of 1 in. = 25.4 mm.
Additional information relating to the conversion of tolerances may be obtained from the SAE Standard-Dual Dimensioning-SAE J390.

Table 7   English–metric equivalents

Measures of length

1 Meter = 39.37 Inches = 3.281 Feet = 1.094 Yards
1 Centimeter = 0.3937 Inch
1 Millimeter = $\frac{1}{25}$ Inch (approximately)
1 Kilometer = 0.621 Mile
1 Inch = 2.540 Centimeters = 25.400 Millimeters
1 Foot = 0.305 Meter
1 Mile = 1.609 Kilometers

Measures of surface

1 Square Meter = 10.764 Square Feet = 1.196 Square Yards
1 Square Centimeter = 0.155 Square Inch
1 Square Millimeter = 0.00155 Square Inch
1 Square Yard = 0.836 Square Meter
1 Square Foot = 0.0929 Square Meter
1 Square Inch = 6.452 Square Centimeters = 645.2 Square Millimeters

Measures of volume

1 Cubic Meter = 35.314 Cubic Feet = 1.308 Cubic Yards
1 Cubic Decimeter = 61.023 Cubic Inches = 0.0353 Cubic Foot
1 Cubic Centimeter = 0.061 Cubic Inch
1 Liter = 61.223 Cubic Inches = 0.0353 Cubic Foot = 0.2642 (U.S.) Gallon
1 Cubic Foot = 28.317 Cubic Decimeters = 28.317 Liters
1 Cubic Inch = 16.387 Cubic Centimeters

Measures of weight

1 Kilogram = 2.2046 Pounds
1 Metric Ton = 2204.6 Pounds = .9842 Ton (2240 pounds)
1 Ounce (avoirdupois) = 28.35 Grams
1 Pound = 0.4536 Kilogram

Speed measurements

1 Kilometer per Hour = 0.621 Mile per Hour

Table 8 Metric units—symbols and formulas

567

Metric tables

Quantity	Unit	SI Symbol	Formula
General			
Length	millimeter	mm	—
	meter	m	—
	kilometer	km	—
Area	square meter	—	m^2
	hectare (10,000 square meters)	ha	—
Volume	cubic centimeter	—	cm^3
	cubic meter	—	m^3
	milliliter	ml	—
	liter	l	—
Mass	gram	g	—
	kilogram	kg	—
	ton (1000 kilograms)	t	—
Speed	meter per second	—	m/s
	kilometer per hour	—	km/h
Time	second	s	—
	minute, hour, day, month, etc.	—	—
Temperature	degree Celsius	°C	—
Thermodynamic temperature	kelvin	K	—
*Power	watt	W	J/s
	kilowatt	kW	—
*Energy, work	joule	J	N·m
	kilowatt hour	—	kW·h
Electric current	ampere	A	—
Electric potential difference	volt	V	W/A
Electric resistance	ohm	Ω	V/A
Space and Time			
Velocity	meter per second	—	m/s
Acceleration	meter per second squared	—	m/s^2
Angular velocity	radian per second	—	rad/s
Angular acceleration	radian per second squared	—	rad/s^2
*Frequency	hertz	Hz	cycle/s
Mechanics			
Density	kilogram per cubic meter	—	kg/m^3
Momentum	kilogram-meter per second	—	kg·m/s
Moment of inertia	kilogram-meter squared	—	$kg·m^2$
*Force	newton	N	$kg·m/s^2$
Torque or moment of force	newton meter	—	N·m
*Pressure and stress	pascal	Pa	N/m^2
Light			
*Luminous flux	lumen	lm	—
*Illuminance	lux	lx	lm/m^2

*Derived units with special names.

Table 9 English–metric conversion multipliers*

To convert from	To	Multiplier
Centigrade (°C)	Fahrenheit (°F)	$\frac{9}{5}$(°C) + 32
Centimeters (cm)	feet	0.0328
Centimeters	inches	0.3937
Centimeters	meters	0.01
Cubic centimeters (cm³)	cubic feet	0.00003531
Cubic centimeters	cubic inches	0.06102
Cubic centimeters	liters	0.0010
Cubic centimeters	cubic meter	0.0000010
Cubic inches (c in.³)	cubic centimeters	16.3872
Cubic inches	cubic meters	0.000016
Cubic feet	cubic centimeters	28,317.08
Cubic feet	cubic meters	0.0283
Cubic meters	cubic feet	35.3133
Cubic millimeters (mm³)	cubic centimeters	0.001
Cubic yards	cubic meters	0.7646
Degrees (arc)	radians	0.01745
Fahrenheit (°F)	centigrade (°C)	$\frac{5}{9}$(°F-32)
Feet (ft)	centimeters	30.4801
Feet	meters	0.3048
Foot-pound-force	joule (J)	1.3558
Foot²	meter²	0.0929
Foot³	meter³	0.0283
Foot/second	meter/second	0.3048
Gallons (U.S.)	cubic meters	0.0038
Gallons (U.S.)	liters	3.7878
Grams (g)	kilograms	0.0010
Grams	milligrams	1000
Grams/cubic centimeter	kilograms/cubic meter	1000
Hectometers (hm)	meters	100
Horsepower (hp)	kilogram-meters/second	76.042
Horsepower	metric horsepower	1.0139
Horsepower, metric	kilogram-meters/second	75.0
Inches (in.)	centimeters	2.54
Inches	meters	0.0254
Inches²	meters²	0.0006452
Inches³	meters³	0.00001639
Inches of mercury	grams/square centimeter	34.542
Kilogram-meters/second	horsepower	0.01305
Kilogram-meters/second	horsepower, metric	0.01333
Kilograms (kg)	grams	1000
Kilograms	tons	0.0011
Kilograms	tons, metric	0.001
Kilogram-force (kgf)	newton (N)	9.8066
Kiloliters (kl)	liters	1000
Kilometers (km)	meters	1000
Kilometers	statute miles	0.6214
Liters (l)	cubic centimeters	1000
Liters	cubic feet	0.035313
Liters	cubic inches	61.02398
Liters	U.S. gallons	0.2641
Megameters	meters	100,000
Meters (m)	U.S. miles	0.000622
Microns	meters	0.000001
Miles, statute (mi)	kilometers	1.6093
Miles, statute	meters	1,609.34
Miles/hour (mph)	kilometers/hour	1.6093
Miles/hour	meters/second	0.4470
Milligrams	grams	0.001
Millimeters (mm)	inches	0.03937
Millimeters	meters	0.001
Millimeters	microns	1000
Ounces, avoirdupois	grams	28.3495
Ounces, fluid	milliliters	29.57
Ounces, U.S. fluid	liters	0.0296

To convert from	To	Multiplier
Pints, liquid	liters	0.4732
Pounds, avoirdupois	grams	453.5924
Poundal	newton (N)	0.1383
Poundal/foot^2	pascal (Pa)	1.4882
Pound-force (avoirdupois)	newton (N)	4.4482
Pound-force-foot	newton-meter (Nm)	1.3558
Pound-force/foot^2	pascal (Pa)	47.88
Pound-force/inch^2 (psi)	pascal (Pa)	6894.8
Pound-mass (avoirdupois)	kilogram	0.4536
Pound-mass/foot^2	kilogram/meter^2	4.8824
Pound-mass/foot^3	kilogram/meter^3	16.0185
Pound-mass/inch^3	kilogram/meter^3	2767.99
Quarts, liquid	liters	0.9464
Quintals	grams	100,000
Radians	degrees, arc	57.2958
Square centimeters	square feet	0.001076
Square centimeters	square inches	0.1550
Square centimeters	square millimeters	100
Square decameters	square meters	100
Square decimeters	square meters	0.01
Square feet	square centimeters	929.0341
Square hectometers	square meters	10,000
Square inches	square centimeters	6.4516
Square inches	square millimeters	645.1625
Square kilometers	hectares	100
Square kilometers	square meters	1,000,000
Square kilometers	square miles	0.3861
Square meters	square feet	10.7639
Square meters	square yards	1.1960
Square miles	square kilometers	2.590
Square millimeters (mm^2)	square inches	0.00155
Square millimeters	square meters	0.000001
Square yards	square meters	0.8361
Tons, long	kilograms	1016.0470
Tons, metric	kilograms	1000
Tons, short	kilograms	907.18
Yards	meters	0.9144

*A complete listing of conversion factors prepared for use with the computer may be found in the ASTM–Metric Practice Guide (E380–72). Definitions of derived units of the International System (SI), such as the joule, the newton, and the pascal, will be found in this publication.

Energy—the joule is the meter-kilogram-second unit of work that is equal to the work done by a force of one newton when the point of application is displaced one meter in the direction of the force (Nm).

Force—the newton is the standard meter-kilogram-second unit of force. It is equal to the force that produces an acceleration of one meter per second per second when applied to a body having a mass of one kilogram (kg·m/s^2).

Stress or pressure—the pascal is the SI unit of pressure or stress. The pascal (Pa) is the pressure or stress of one newton per square meter (N/m^2).

Table 10 ISO Metric screw threads—fine and coarse series

Nominal size (mm)	Series with graded pitches*				Thread-diameter preference†		
	Coarse	Tap drill‡	Fine	Tap drill‡	1	2	3¶
1.6	**0.35**	1.25	—	—	1.6	—	—
1.8	**0.35**	1.45	—	—	—	1.8	—
2	**0.4**	1.60	—	—	2	—	—
2.2	**0.45**	1.75	—	—	—	2.2	—
2.5	**0.45**	2.05	—	—	2.5	—	—
3	**0.5**	2.50	—	—	3	—	—
3.5	**0.6**	2.90	—	—	—	3.5	—
4	**0.7**	3.30	—	—	4	—	—
4.5	**0.75**	3.75	—	—	—	4.5	—
5	**0.8**	4.20	—	—	5	—	—
5.5	—	—	—	—	—	—	5.5
6	**1**	5.00	—	—	6	—	—
7	**1**	6.00	—	—	—	—	7
8	**1.25**	6.75	1	7.00	8	—	—
9	**1.25**	7.75	—	—	—	—	9
10	**1.5**	8.50	**1.25**	8.75	10	—	—
11	**1.5**	9.50	—	—	—	—	11
12	**1.75**	10.00	**1.25**	10.50	12	—	—
14	**2**	12.00	**1.5**	12.50	14	—	—
15	—	—	—	—	—	—	15
16	**2**	14.00	**1.5**	14.50	16	—	—
17	—	—	—	—	—	—	17
18	**2.5**	15.50	**1.5**	16.50	—	18	—
20	**2.5**	17.50	**1.5**	18.50	20	—	—
22	**2.5**	19.50	**1.5**	20.50	—	22	—
24	**3**	21.00	2	22.00	24	—	—
25	—	—	—	—	—	—	25
26	—	—	—	—	—	—	26
27	**3**	24.00	2	25.00	—	27	—
28	—	—	—	—	—	—	28
30	**3.5**	26.50	2	28.00	30	—	—
32	—	—	2	—	—	—	32
33	**3.5**	29.50	2	31.00	—	33	—
35	—	—	—	—	—	—	35
36	**4**	32.00	3	33.00	36	—	—
38	—	—	3	—	—	—	38
39	**4**	35.00	3	36.00	—	39	—
40	—	—	—	—	—	—	40
42	**4.5**	37.50	3	39.00	42	—	—
45	**4.5**	40.50	3	42.00	—	45	—
48	**5**	43.00	3	45.00	48	—	—
50	—	—	—	—	—	—	50
52	**5**	47.00	3	49.00	—	52	—
55	—	—	—	—	—	—	55
56	**5.5**	50.50	4	52.00	56	—	—
58	—	—	—	—	—	—	58
60	**5.5**	54.50	4	56.00	—	60	—
62	—	—	—	—	—	—	62
64	**6**	58.00	4	60.00	64	—	—
65	—	—	—	—	—	—	65
68	**6**	62.00	4	64.00	—	68	—

*The pitches shown in bold type are those which have been established by ISO as a selected coarse- and fine-thread series for commercial threads and fasteners.

†Select thread diameter from columns 1, 2, or 3, with preference for selection being in that order.

‡For an approximate 75% thread, use the formula: nominal OD minus .97 × pitch.

¶For diameters listed in this column, see Table 11—Series With Constant Pitches.

Table 11 ISO Metric screw threads—series with constant pitches

Nominal size (mm)	Pitches (mm)											
	6	4	3	2	1.5	1.25	1	0.75	0.5	0.35	0.25	0.2
1.6	—	—	—	—	—	—	—	—	—	—	—	0.2
1.8	—	—	—	—	—	—	—	—	—	—	—	0.2
2	—	—	—	—	—	—	—	—	—	—	0.25	—
2.2	—	—	—	—	—	—	—	—	—	—	0.25	—
2.5	—	—	—	—	—	—	—	—	—	0.35	—	—
3	—	—	—	—	—	—	—	—	—	0.35	—	—
3.5	—	—	—	—	—	—	—	—	—	0.35	—	—
4	—	—	—	—	—	—	—	—	0.5	—	—	—
4.5	—	—	—	—	—	—	—	—	0.5	—	—	—
5	—	—	—	—	—	—	—	—	0.5	—	—	—
5.5	—	—	—	—	—	—	—	—	0.5	—	—	—
6	—	—	—	—	—	—	—	0.75	—	—	—	—
7	—	—	—	—	—	—	—	0.75	—	—	—	—
8	—	—	—	—	—	—	1	0.75	—	—	—	—
9	—	—	—	—	—	—	1	0.75	—	—	—	—
10	—	—	—	—	—	1.25	1	0.75	—	—	—	—
11	—	—	—	—	—	—	1	0.75	—	—	—	—
12	—	—	—	—	1.5	1.25	1	—	—	—	—	—
14	—	—	—	—	1.5	1.25	1	—	—	—	—	—
15	—	—	—	—	1.5	—	1	—	—	—	—	—
16	—	—	—	—	1.5	—	1	—	—	—	—	—
17	—	—	—	—	1.5	—	1	—	—	—	—	—
18	—	—	—	2	1.5	—	1	—	—	—	—	—
20	—	—	—	2	1.5	—	1	—	—	—	—	—
22	—	—	—	2	1.5	—	1	—	—	—	—	—
24	—	—	—	2	1.5	—	1	—	—	—	—	—
25	—	—	—	2	1.5	—	1	—	—	—	—	—
26	—	—	—	—	1.5	—	1	—	—	—	—	—
27	—	—	—	2	1.5	—	1	—	—	—	—	—
28	—	—	—	2	1.5	—	1	—	—	—	—	—
30	—	—	3	2	1.5	—	1	—	—	—	—	—
32	—	—	—	2	1.5	—	—	—	—	—	—	—
33	—	—	3	2	1.5	—	—	—	—	—	—	—
35	—	—	—	—	1.5	—	—	—	—	—	—	—
36	—	—	—	2	1.5	—	—	—	—	—	—	—
38	—	—	—	—	1.5	—	—	—	—	—	—	—
39	—	—	—	2	1.5	—	—	—	—	—	—	—
40	—	—	3	2	1.5	—	—	—	—	—	—	—
42	—	4	3	2	1.5	—	—	—	—	—	—	—
45	—	4	3	2	1.5	—	—	—	—	—	—	—
48	—	4	3	2	1.5	—	—	—	—	—	—	—
50	—	—	3	2	1.5	—	—	—	—	—	—	—
52	—	4	3	2	1.5	—	—	—	—	—	—	—
55	—	4	3	2	1.5	—	—	—	—	—	—	—
56	—	4	3	2	1.5	—	—	—	—	—	—	—
58	—	4	3	2	1.5	—	—	—	—	—	—	—
60	—	4	3	2	1.5	—	—	—	—	—	—	—
62	—	4	3	2	1.5	—	—	—	—	—	—	—
64	—	4	3	2	1.5	—	—	—	—	—	—	—
65	—	4	3	2	1.5	—	—	—	—	—	—	—
68	—	4	3	2	1.5	—	—	—	—	—	—	—
70	6	4	3	2	1.5	—	—	—	—	—	—	—
72	6	4	3	2	1.5	—	—	—	—	—	—	—
75	—	4	3	2	1.5	—	—	—	—	—	—	—
76	6	4	3	2	1.5	—	—	—	—	—	—	—

Table 12 Metric hex socket cap screws

| d | Head | | Socket | |
| Nominal screw diameter and thread pitch | D | H | W | S |
	Head diameter max	Head height max	Hex socket size nom.	Hex socket depth
M4 × 0.7	7.0	4.0	3.0	2.4
M5 × 0.8	8.5	5.0	4.0	3.1
M6 × 1	10.0	6.0	5.0	3.78
M8 × 1.25	13.0	8.0	6.0	4.78
M10 × 1.5	16.0	10.0	8.0	6.25
M12 × 1.75	18.0	12.0	10.0	7.5
M14 × 2	21.0	14.0	12.0	8.6
M16 × 2	24.0	16.0	14.0	9.7
M20 × 2.5	30.0	20.0	17.0	11.8
M24 × 3	36.0	24.0	19.0	14.0

Dimension values for drawing purposes only.

Table 13 Metric hex bolts and hex cap screws

Hex Bolt   Hex Cap Screw

| Hex bolts and hex cap screws | | | Bolts | Hex cap screws | | |
| | W | A | H | H | B | T |
Nominal size and thread pitch	Width across flats max	Width across corners max	Head height max	Head height max	Washer face diameter min	Washer face thickness max
M5 × 0.8	8.00	9.24	3.98	3.65	7.0	0.5
M6 × 1	10.00	11.55	4.38	4.15	8.9	0.5
M8 × 1.25	13.00	15.01	5.68	5.50	11.6	0.6
M10 × 1.5	16.00	18.49	6.85	6.63	14.6	0.6
M12 × 1.75	18.00	20.78	7.95	7.76	16.6	0.6
M14 × 2	21.00	24.25	9.25	9.09	19.6	0.6
M16 × 2	24.00	27.71	10.75	10.32	22.5	0.8
M20 × 2.5	30.00	34.64	13.40	12.88	27.7	0.8
M24 × 3	36.00	41.57	15.90	15.44	33.2	0.8
M30 × 3.5	46.00	53.12	19.75	19.48	42.7	0.8
M36 × 4	55.00	63.51	23.55	23.38	51.1	0.8
M42 × 4.5	65.00	75.06	27.05	26.97	59.8	1.0
M48 × 5	75.00	86.60	31.07	31.07	69.0	1.0
M56 × 5.5	85.00	98.15	36.20	36.20	78.1	1.0
M64 × 6	95.00	109.70	41.32	41.32	87.2	1.0
M72 × 6	105.00	121.24	46.45	46.45	96.3	1.2

All dimensions are in millimeters.
Dimension values are for drawing purposes only.

Table 14 Metric hex nuts style 1 and style 2

Style 1   Style 2

Nominal size and thread pitch	W Width across flats max	A Width across corners max	D Bearing face diameter min	H₁ Nut thickness Style 1 max	H₂ Nut thickness Style 2 max	Washer face thickness max
M1.6 × 0.35	3.20	3.70	2.3	1.3	—	0.4
M2 × 0.4	4.00	4.62	3.1	1.6	—	0.4
M2.5 × 0.45	5.00	5.77	4.1	2.0	—	0.4
M3 × 0.5	5.50	6.35	4.6	2.4	2.9	0.4
M3.5 × 0.6	6.00	6.93	5.1	2.8	3.3	0.4
M4 × 0.7	7.00	8.08	6.0	3.2	3.8	0.4
M5 × 0.8	8.00	9.24	7.0	4.7	5.1	0.5
M6 × 1	10.00	11.55	8.9	5.2	5.7	0.5
M8 × 1.25	13.00	15.01	11.6	6.8	7.5	0.6
M10 × 1.5	16.00	18.48	14.6	8.4	9.3	0.6
M12 × 1.75	18.00	20.78	16.6	10.8	12.0	0.6
M14 × 2	21.00	24.25	19.4	12.8	14.1	0.6
M16 × 2	24.00	27.71	22.4	14.8	16.4	0.8
M20 × 2.5	30.00	34.64	27.9	18.0	20.3	0.8
M24 × 3	36.00	41.57	32.5	21.5	23.9	0.8
M30 × 3.5	46.00	53.12	42.5	25.6	28.6	0.8
M36 × 4	55.00	63.51	50.8	31.0	34.7	0.8

Type 1: M1.6–M36
Type 2: M3–M36

All dimensions are in millimeters.
Values are for drawing purposes only.

Table 15 Metric hex slotted nuts

Style 1   Style 2

Nominal size and thread pitch	W Width across flats max	A Width across corners max	B Bearing face diameter min	H₁ Nut thickness Style 1 max	H₂ Nut thickness Style 2 max	T₁ Unslotted thickness Style 1 max	T₂ Unslotted thickness Style 2 max	S Width of slot max	For drawing only Depth of slot Style 1	For drawing only Depth of slot Style 2
M5 × 0.8	8.00	9.24	7.0	4.5	5.3	3.2	3.7	2.2	1.3	1.6
M6 × 1	10.00	11.55	8.9	5.6	6.5	3.9	4.5	2.8	1.7	2.0
M8 × 1.25	13.00	15.01	11.6	6.6	7.8	4.5	5.3	3.3	2.1	2.5
M10 × 1.5	15.00	17.32	13.6	9.0	10.7	6.0	7.1	3.6	3.0	3.6
M12 × 1.75	18.00	20.78	16.6	10.7	12.8	7.1	8.5	4.3	3.6	4.3
M14 × 2	21.00	24.25	19.4	12.5	14.9	8.2	9.8	4.3	4.3	5.1
M16 × 2	24.00	27.71	22.4	14.5	17.4	9.5	11.4	6.0	5.0	6.0
M20 × 2.5	30.00	34.64	27.6	18.4	21.2	12.1	13.9	6.0	6.3	7.3
M24 × 3	36.00	41.57	32.9	22.0	25.4	14.4	16.6	7.0	7.6	8.8
M30 × 3.5	46.00	53.12	42.5	26.7	31.0	17.3	20.1	9.3	9.4	10.9
M36 × 4	55.00	63.51	50.8	32.0	37.6	20.8	24.5	9.3	11.2	13.1

All dimensions are in millimeters.
Threads are general-purpose threads.

Table 16 Metric slotted, recessed, and hex head machine screws

Nominal size and thread pitch	Flat head slotted			Pan head slotted				Oval head slotted					Slot width	Hex head	
	A Actual	H Max	T Max	A Max	H Max	R Max	T Min	A Actual	H Max	H₁ Max	R Approx	T Max	S Max	W Max	H Max
M2 × 0.4	3.60	1.20	0.6	3.90	1.35	0.8	0.55	3.60	1.20	0.50	3.8	1.0	0.7	3.20	1.27
M2.5 × 0.45	4.60	1.50	0.7	4.90	1.65	1.0	0.73	4.60	1.50	0.60	5.0	1.2	0.8	4.00	1.40
M3 × 0.5	5.50	1.80	0.9	5.80	1.90	1.2	0.80	5.50	1.80	0.75	5.7	1.5	1.0	5.00	1.52
M3.5 × 0.6	6.44	2.10	1.0	6.80	2.25	1.4	0.95	6.44	2.10	0.90	6.5	1.7	1.2	5.50	2.36
M4 × 0.7	7.44	2.32	1.1	7.80	2.55	1.6	1.15	7.44	2.32	1.00	7.8	1.9	1.4	7.00	2.79
M5 × 0.8	9.44	2.85	1.4	9.80	3.10	2.0	1.35	9.44	2.85	1.25	9.9	2.3	1.5	8.00	3.05
M6 × 1	11.87	3.60	1.8	12.00	3.90	2.5	1.70	11.87	3.60	1.60	12.2	3.0	1.9	10.00	4.83
M8 × 1.25	15.17	4.40	2.1	15.60	5.00	3.2	2.20	15.17	4.40	2.00	15.8	3.7	2.3	13.00	5.84
M10 × 1.5	18.98	5.35	2.6	19.50	6.20	4.0	2.70	18.98	5.35	2.50	19.8	4.5	2.8	15.00	7.49
M12 × 1.75	22.88	6.35	3.1	23.40	7.50	4.8	3.20	22.88	6.35	3.00	23.8	5.3	2.8	18.00	9.50

Dimension values are for drawing purposes only.

Table 17 Metric round head square neck bolts

Nominal bolt size and thread pitch	A	H	R	W	P
	Head diameter max	Head height max	Head radius approx	Width of square max	Depth of square max
M5 × 0.8	11.3	3.0	7.2	5.48	3.0
M6 × 1	13.8	3.6	8.9	6.78	3.0
M8 × 1.25	17.0	4.5	11.2	8.58	3.0
M10 × 1.5	20.6	5.5	13.7	10.58	4.0
M12 × 1.75	24.7	6.5	16.5	12.70	4.0
M14 × 2	28.3	7.5	19.1	14.70	4.0
M16 × 2	33.1	8.8	22.6	20.84	5.0
M20 × 2.5	39.8	10.8	27.3	24.84	5.0

Dimension values are for drawing purposes only.

Table 18 Metric set screws

HALF DOG POINT   FULL DOG POINT   CONE POINT   FLAT POINT   CUP POINT

D	D_1	A	B	C	W	S
M3	2	2.4		2	1.5	0.4
M4	2.5	2.8		2.5	2	0.6
M5	3.5	2.8		3	2.5	0.8
M6	4.5	3.35		3.5	3	1
M8	6	4.7	Half-dog B = A/2	5	4	1.2
M10	7	4.9		6	5	1.6
M12	9	6.4		8	6	—
M16	12	8.15		—	8	—
M20	15	8.85		—	10	—
M24	18	9.65		—	12	—

Dimension values are for drawing purposes only.

Table 19 Metric plain and lock washers

Nominal size			M4	M5	M6	M8	M10	M12	M16	M20	M24
Diameters	*Plain*	D_1	4.3	5.3	6.4	8.4	10.5	13	17	21	25
		D_2	9	10	12.5	17	21	24	30	37	44
	Lock	D_3	4.1	5.1	6.1	8.1	10.2	12.2	16.2	20.2	24.5
		D_4	7.6	9.2	11.8	14.8	18.1	21.1	27.4	33.6	40
Thickness		T_1	0.8	1	1.6	1.6	2	2.5	3	3	4
		T_2	0.9	1.2	1.6	2	2.2	2.5	3.5	4	5

Dimension values are for drawing purposes only.

Table 20-A SI metric limits and fits-ISO
limits of tolerance for selected holes (upper and lower deviations)

Nominal size (mm)		H7		H8		H9		H11		G7		F8		E9		D10	
Over	To (incl.)	+		+		+		+		+	+	+	+	+	+	+	+
1	3	10	0	14	0	25	0	60	0	12	2	20	6	39	14	60	20
3	6	12	0	18	0	30	0	75	0	16	4	28	10	50	20	78	30
6	10	15	0	22	0	36	0	90	0	20	5	35	13	61	25	98	40
10	18	18	0	27	0	43	0	110	0	24	6	43	16	75	32	120	50
18	30	21	0	33	0	52	0	130	0	28	7	53	20	92	40	149	65
30	50	25	0	39	0	62	0	160	0	34	9	64	25	112	50	180	80
50	80	30	0	46	0	74	0	190	0	40	10	76	30	134	60	220	100
80	120	35	0	54	0	87	0	220	0	47	12	90	36	159	72	260	120
120	180	40	0	63	0	100	0	250	0	54	14	106	43	185	85	305	145
180	250	46	0	72	0	115	0	290	0	61	15	122	50	215	100	355	170

Values in micrometers.
Example: $\varnothing 50 + 0.025$ (H7).

Table 20-B   SI metric limits and fits-ISO
limits of tolerance for selected shafts (upper and lower deviations)

Nominal size (mm) Over	To (incl.)	c11 −	−	d9 −	−	e8 −	−	f7 −	−	g6 −	−	h6 0	−	h9 0	−	h11 0	−	k6 +	+	n6 +	+	p6 +	+	r6 +	+	s6 +	+
1	3	60	120	20	45	14	28	6	16	2	8	0	6	0	25	0	60	6	0	10	4	12	6	16	10	20	14
3	6	70	145	30	60	20	38	10	22	4	12	0	8	0	30	0	75	9	1	16	8	20	12	23	15	27	19
6	10	80	170	40	76	25	47	13	28	5	14	0	9	0	36	0	90	10	1	19	10	24	15	28	19	32	23
10	18	95	205	50	93	32	59	16	34	6	17	0	11	0	43	0	110	12	1	23	12	29	18	34	23	39	28
18	30	110	240	65	117	40	73	20	41	7	20	0	13	0	52	0	130	15	2	28	15	35	22	41	28	48	35
30	40	120	280	80	142	50	89	25	50	9	25	0	16	0	62	0	160	18	2	33	17	42	26	50	34	59	43
40	50	130	290																								
50	65	140	330	100	174	60	106	30	60	10	29	0	19	0	74	0	190	21	2	39	20	51	32	60	41	72	53
65	80	150	340																					62	43	78	59
80	100	170	390	120	207	72	126	36	71	12	34	0	22	0	87	0	220	25	3	45	23	59	37	73	51	93	71
100	120	180	400																					76	54	101	79
120	140	200	450	145	245	85	148	43	83	14	39	0	25	0	100	0	250	28	3	52	27	68	43	88	63	117	92
140	160	210	460																					90	65	125	100
160	180	230	480																					93	68	133	108
180	200	240	530	170	285	100	172	50	96	15	44	0	29	0	115	0	290	33	4	60	31	79	50	106	77	151	122
200	225	260	550																					109	80	159	130
225	250	280	570																					113	84	169	140

Values in micrometers.

Example: $\varnothing\,50\,^{-0.009}_{-0.025}$ or 49.991–49.975 (g6).

Table 21-A   Single-row radial bearings—type 3000*

See table below for explanation of letters

Bearing no.	Bore B		Diameter D		Width W		Balls		Radius r†
	mm	in.	mm	in.	mm	in.	Diam	No.	
3200	10	.3937	30	1.1811	9	.3543	$7/32$	7	.025
3300	10	.3937	35	1.3780	11	.4331	$1/4$	7	.025
3201	12	.4724	32	1.2598	10	.3937	.210	8	.025
3301	12	.4724	37	1.4567	12	.4724	$9/32$	7	.04
3202	15	.5906	35	1.3780	11	.4331	.210	9	.025
3302	15	.5906	42	1.6535	13	.5118	$5/16$	7	.04
3203	17	.6693	40	1.5748	12	.4724	$9/32$	8	.04
3303	17	.6693	47	1.8504	14	.5512	$11/32$	7	.04
3204	20	.7874	47	1.8504	14	.5512	$5/16$	8	.04
3304	20	.7874	52	2.0472	15	.5906	$13/32$	7	.04
3205	25	.9843	52	2.0472	15	.5906	$5/16$	9	.04
3305	25	.9843	62	2.4409	17	.6693	$13/32$	8	.04
3206	30	1.1811	62	2.4409	16	.6299	$11/32$	9	.04
3306	30	1.1811	72	2.8346	19	.7480	$15/32$	8	.04
3207	35	1.3780	72	2.8346	17	.6693	$7/16$	9	.04
3307	35	1.3780	80	3.1496	21	.8268	$17/32$	8	.06
3208	40	1.5748	80	3.1496	18	.7087	$15/32$	9	.04
3308	40	1.5748	90	3.5433	23	.9055	$19/32$	8	.06
3209	45	1.7717	85	3.3465	19	.7480	$15/32$	10	.04
3309	45	1.7717	100	3.9370	25	.9843	$21/32$	8	.06
3210	50	1.9685	90	3.5433	20	.7874	$15/32$	11	.04
3310	50	1.9685	110	4.3307	27	1.0630	$23/32$	8	.08
3211	55	2.1654	100	3.9370	21	.8268	$1/2$	11	.06
3311	55	2.1654	120	4.7244	29	1.1417	$25/32$	8	.08

* For radial or combined loads from either direction where thrust is to be resisted by a single bearing and is not great enough to require use of angular contact type.

† Radius *r* indicates maximum fillet radius in housing or on shaft, which bearing radius will clear.

Pages 140–143 from Vol. I, *New Departure Handbook.*

Table 21-B   Radax bearings—type 30,000*

See table below for explanation of letters

Bearing no.	Bore B		Diameter D		Width W		Balls		Radius r†
	mm	in.	mm	in.	mm	in.	Diam	No.	
30204			47	1.8504	14	.5512	$^{11}/_{32}$	10	
30304	20	.7874	52	2.0472	15	.5906	$^3/_8$	10	.04
30404			72	2.8346	19	.7480	$^9/_{16}$	8	
30205			52	2.0472	15	.5906	$^{11}/_{32}$	11	.04
30305	25	.9843	62	2.4409	17	.6693	$^7/_{16}$	10	.04
30405			80	3.1496	21	.8268	$^5/_8$	9	.06
30206			62	2.4409	16	.6299	$^3/_8$	12	.04
30306	30	1.1811	72	2.8346	19	.7480	$^1/_2$	10	.04
30406			90	3.5433	23	.9055	$^{11}/_{16}$	9	.06
30207			72	2.8346	17	.6693	$^7/_{16}$	12	.04
30307	35	1.3780	80	3.1496	21	.8268	$^9/_{16}$	11	.06
30407			100	3.9370	25	.9843	$^3/_4$	9	.06
30208			80	3.1496	18	.7087	$^1/_2$	12	.04
30308	40	1.5748	90	3.5433	23	.9055	$^5/_8$	11	.06
30408			110	4.3307	27	1.0630	$^{13}/_{16}$	10	.08
30209			85	3.3465	19	.7480	$^1/_2$	13	.04
30309	45	1.7717	100	3.9370	25	.9843	$^{11}/_{16}$	11	.06
30409			120	4.7244	29	1.1417	$^7/_8$	10	.08
30210			90	3.5433	20	.7874	$^1/_2$	14	.04
30310	50	1.9685	110	4.3307	27	1.0630	$^3/_4$	11	.08
30410			130	5.1181	31	1.2205	$^{15}/_{16}$	10	.08
30211			100	3.9370	21	.8268	$^9/_{16}$	14	.06
30311	55	2.1654	120	4.7244	29	1.1417	$^{13}/_{16}$	12	.08
30411			140	5.5118	33	1.2992	1	10	.08
30212			110	4.3307	22	.8661	$^5/_8$	14	.06
30312	60	2.3622	130	5.1181	31	1.2205	$^7/_8$	12	.08
30412			150	5.9055	35	1.3780	$1^1/_{16}$	10	.08

* Single-row angular contact; provide maximum capacity for one-direction thrust loads. Mounted two bearings opposed for combined loads or thrust from either direction.
† Radius *r* indicates maximum fillet radius in housing or on shaft, which bearing radius will clear.
Page 143 from Vol. I, *New Departure Handbook.*

Table 22-A Shaft mounting fits

Bearing and bore numbers	Bearing bore				Shaft revolving				Shaft stationary			
	Diameters inches°		Diameters millimeters		Diameters inches		Diameters millimeters		Diameters inches		Diameters millimeters	
	max	min	max	min	max	min	max	min	max	min	max	min
0	.3937	.3934	10.000	9.992	.3939	.3936	10.005	9.997	.3935	.3932	9.995	9.987
1	.4724	.4721	12.000	11.992	.4726	.4723	12.005	11.997	.4722	.4719	11.995	11.987
2	.5906	.5903	15.000	14.992	.5908	.5905	15.005	14.997	.5904	.5901	14.995	14.987
3	.6693	.6690	17.000	16.992	.6695	.6692	17.005	16.997	.6691	.6688	16.995	16.987
4	.7874	.7870	20.000	19.990	.7877	.7873	20.008	19.998	.7871	.7867	19.993	19.983
5	.9843	.9839	25.000	24.990	.9846	.9842	25.008	24.998	.9840	.9836	24.993	24.983
6	1.1811	1.1807	30.000	29.990	1.1814	1.1810	30.008	29.998	1.1808	1.1804	29.993	29.983
7	1.3780	1.3775	35.000	34.987	1.3784	1.3779	35.010	34.997	1.3776	1.3771	34.990	34.977
8	1.5748	1.5743	40.000	39.987	1.5752	1.5747	40.010	39.997	1.5744	1.5739	39.990	39.977
9	1.7717	1.7712	45.000	44.987	1.7721	1.7716	45.010	44.997	1.7713	1.7708	44.990	44.977
10	1.9685	1.9680	50.000	49.987	1.9689	1.9684	50.010	49.997	1.9681	1.9676	49.990	43.977
11	2.1654	2.1648	55.000	54.985	2.1659	2.1653	55.013	54.998	2.1649	2.1643	54.988	54.973
12	2.3622	2.3616	60.000	59.985	2.3627	2.3621	60.013	59.998	2.3617	2.3611	59.988	59.973
13	2.5591	2.5585	65.000	64.985	2.5596	2.5590	65.013	64.998	2.5586	2.5580	64.988	64.973
14	2.7559	2.7553	70.000	69.985	2.7564	2.7558	70.013	69.998	2.7554	2.7548	69.988	69.973

Table 22-B Housing mounting fits

lgt	med	hvy	Bearing outer diam.				Housing stationary				Housing revolving			
Series			Diameters inches°		Diameters millimeters		Diameters inches		Diameters millimeters		Diameters inches		Diameters millimeters	
lgt	med	hvy	max	min	max	min	max	min	max	min	max	min	max	min
0			1.1811	1.1807	30.000	29.990	1.1815	1.1810	30.010	29.997	1.1810	1.1805	29.997	29.984
1			1.2598	1.2593	32.000	31.987	1.2603	1.2597	32.013	31.998	1.2597	1.2591	31.997	31.982
2	0		1.3780	1.3775	35.000	34.987	1.3785	1.3779	35.013	34.998	1.3779	1.3773	34.997	34.982
	1		1.4567	1.4562	37.000	36.987	1.4572	1.4566	37.013	36.998	1.4566	1.4560	36.997	36.982
3			1.5748	1.5743	40.000	39.987	1.5753	1.5747	40.013	39.998	1.5747	1.5741	39.997	39.982
	2		1.6535	1.6530	42.000	41.987	1.6540	1.6534	42.013	41.998	1.6534	1.6528	41.997	41.982
4	3		1.8504	1.8499	47.000	46.987	1.8509	1.8503	47.013	46.998	1.8503	1.8497	46.997	46.982
5	4		2.0472	2.0466	52.000	51.985	2.0479	2.0471	52.018	51.998	2.0472	2.0464	52.000	51.980
6	5		2.4409	2.4403	62.000	61.985	2.4416	2.4408	62.018	61.998	2.4409	2.4401	62.000	61.980
7	6	4	2.8346	2.8340	72.000	71.985	2.8353	2.8345	72.018	71.998	2.8346	2.8338	72.000	71.980
8	7	5	3.1496	3.1490	80.000	79.985	3.1503	3.1495	80.018	79.998	3.1496	3.1488	80.000	79.980
9			3.3465	3.3457	85.000	84.980	3.3473	3.3463	85.020	84.995	3.3466	3.3456	85.003	84.978
10	8	6	3.5433	3.5425	90.000	89.980	3.5441	3.5431	90.020	89.995	3.5434	3.5424	90.003	89.978
11	9	7	3.9370	3.9362	100.000	99.980	3.9378	3.9368	100.020	99.995	3.9371	3.9361	100.003	99.978
12	10	8	4.3307	4.3299	110.000	109.980	4.3315	4.3305	110.020	109.995	4.3308	4.3298	110.003	109.978
13	11	9	4.7244	4.7236	120.000	119.980	4.7252	4.7242	120.020	119.995	4.7245	4.7235	120.003	119.978
14			4.9213	4.9203	125.000	124.975	4.9223	4.9211	125.025	124.995	4.9214	4.9202	125.003	124.973
15	12	10	5.1181	5.1171	130.000	129.975	5.1191	5.1179	130.025	129.995	5.1182	5.1170	130.002	129.972
16	13	11	5.5118	5.5108	140.000	139.975	5.5128	5.5116	140.025	139.995	5.5119	5.5107	140.002	139.972
17	14	12	5.9055	5.9045	150.000	149.975	5.9065	5.9053	150.025	149.995	5.9056	5.9044	150.002	149.972
18	15	13	6.2992	6.2982	160.000	159.975	6.3002	6.2990	160.025	159.995	6.2993	6.2981	160.002	159.972

* Inch values compiled from *New Departure Handbook.*

Table 23 Small drills—metric*

Metric drill diameter	Diameter (in.)	Metric drill diameter	Diameter (in.)	Metric drill diameter	Diameter (in.)	Metric drill diameter	Diameter (in.)
0.200	.0078	2.300	.0905	5.70	.2244	9.20	.3622
0.250	.0098	2.35	.0925	5.75	.2264	9.25	.3642
0.300	.0118	2.400	.0945	5.80	.2283	9.30	.3661
0.350	.0138	2.45	.0964	5.90	.2323	9.40	.3701
0.400	.0157	2.50	.0984	6.00	.2362	9.50	.3740
0.450	.0177	2.60	.1024	6.10	.2401	9.60	.3779
0.500	.0197	2.70	.1063	6.20	.2441	9.70	.3819
0.550	.0216	2.75	.1083	6.25	.2461	9.75	.3838
0.600	.0236	2.80	.1102	6.30	.2480	9.80	.3858
0.650	.0256	2.90	.1142	6.40	.2520	9.90	.3898
0.700	.0275	3.00	.1181	6.50	.2559	10.00	.3937
0.750	.0295	3.10	.1220	6.60	.2598	10.50	.4134
0.800	.0315	3.20	.1260	6.70	.2638	11.00	.4331
0.850	.0335	3.25	.1279	6.75	.2657	11.50	.4527
0.900	.0354	3.30	.1299	6.80	.2677	12.00	.4724
0.950	.0374	3.40	.1338	6.90	.2716	12.50	.4921
1.000	.0394	3.50	.1378	7.00	.2756	13.00	.5118
1.050	.0413	3.60	.1417	7.10	.2795	13.50	.5315
1.100	.0433	3.70	.1457	7.20	.2835	14.00	.5512
1.150	.0453	3.75	.1476	7.25	.2854	14.50	.5709
1.200	.0472	3.80	.1496	7.30	.2874	15.00	.5905
1.250	.0492	3.90	.1535	7.40	.2913	15.50	.6102
1.300	.0512	4.00	.1575	7.50	.2953	16.00	.6299
1.350	.0531	4.10	.1614	7.60	.2992	16.50	.6496
1.400	.0551	4.20	.1653	7.70	.3031	17.00	.6693
1.450	.0571	4.25	.1673	7.75	.3051	17.50	.6890
1.500	.0590	4.30	.1693	7.80	.3071	18.00	.7087
1.550	.0610	4.40	.1732	7.90	.3110	18.50	.7283
1.600	.0630	4.50	.1772	8.00	.3150	19.00	.7480
1.650	.0650	4.60	.1811	8.10	.3189	19.50	.7677
1.700	.0669	4.70	.1850	8.20	.3228	20.00	.7874
1.750	.0689	4.75	.1870	8.25	.3248	20.50	.8071
1.800	.0709	4.80	.1890	8.30	.3268	21.00	.8268
1.850	.0728	4.90	.1929	8.40	.3307	21.50	.8464
1.900	.0748	5.00	.1968	8.50	.3346	22.00	.8661
1.950	.0768	5.10	.2008	8.60	.3386	22.50	.8858
2.000	.0787	5.20	.2047	8.70	.3425	23.00	.9055
2.050	.0807	5.25	.2067	8.75	.3445	23.50	.9252
2.100	.0827	5.30	.2087	8.80	.3464	24.00	.9449
2.150	.0846	5.40	.2126	8.90	.3504	24.50	.9646
2.200	.0866	5.50	.2165	9.00	.3543	25.00	.9842
2.250	.0886	5.60	.2205	9.10	.3583	25.50	1.0039

* Drills beyond the range of this table increase in diameter by increments of 0.50 mm—26.00, 26.50, 27.00, etc.

**Table 24 Metric woodruff keys
selected sizes**

| Nominal size | | Shaft dia mm | Key data | | Depth of keyseat in shaft |
W	H mm		L mm	D Dia	
1	1.4	3 to 4	3.82	4	1
1.5	2.6	4 to 6	6.76	7	2
2	2.6	6 to 8	6.76	7	1.8
2	3.7		9.66	10	2.9
2.5	3.7	8 to 10	9.66	10	2.9
3	3.7		9.66	10	2.5
3	5		12.65	13	3.8
3	6.5		15.72	16	5.3
4	5	10 to 12	12.65	13	3.5
4	6.5		15.72	16	5
4	7.5		18.57	19	6
5	6.5	12 to 17	15.72	16	4.5
5	7.5		18.57	19	5.5
5	9		21.63	22	7
6	7.5	17 to 22	18.57	19	5.1
6	9		21.63	22	6.6
6	11		27.35	28	8.6
8	9	22 to 30	21.63	22	6.2
8	11		27.35	28	8.2
8	13		31.43	32	10.2
10	11	30 to 38	27.35	28	7.8
10	13		31.43	32	9.8
10	16		43.08	45	12.8

Not American Standard.

tables

INCH APPENDIX C

Table 25 Standard Unified thread series*

Present Unified thread Nominal size—diameter			Coarse (NC) (UNC)		Fine (NF) (UNF)		Extra-fine (NEF) (UNEF)	
Inch		Metric Equiv.‡	Threads per inch	Tap drill†	Threads per inch	Tap drill†	Threads per inch	Tap drill†
.060	0	1.52	—	—	80	$3/64$	—	—
.073	1	1.85	64	No. 53	72	No. 53	—	—
.086	2	2.18	56	No. 50	64	No. 50	—	—
.099	3	2.51	48	No. 47	56	No. 45	—	—
.112	4	2.84	**40**	No. 43	48	No. 42	—	—
.125	5	3.17	40	No. 38	44	No. 37	—	—
.138	6	3.50	**32**	No. 36	40	No. 33	—	—
.164	8	4.16	**32**	No. 29	36	No. 29	—	—
.190	10	4.83	**24**	No. 25	**32**	No. 21	—	—
.216	12	5.49	24	No. 16	28	No. 14	32	No. 13
.250	¼	6.35	**20**	No. 7	**28**	No. 3	32	No. 2
.3125	$5/16$	7.94	**18**	F	**24**	I	32	K
.375	$3/8$	9.52	**16**	$5/16$	**24**	Q	32	S
.4375	$7/16$	11.11	**14**	U	**20**	$25/64$	**28**	Y
.500	½	12.70	**13**	$27/64$	**20**	$29/64$	**28**	$15/32$
.5625	$9/16$	14.29	**12**	$31/64$	**18**	$33/64$	24	$17/32$
.625	$5/8$	15.87	**11**	$17/32$	**18**	$37/64$	24	$19/32$
.6875	$11/16$	17.46	—	—	—	—	24	$41/64$
.750	$3/4$	19.05	**10**	$21/32$	**16**	$11/16$	**20**	$45/64$
.8125	$13/16$	20.64	—	—	—	—	**20**	$49/64$
.875	$7/8$	22.22	**9**	$49/64$	**14**	$13/16$	**20**	$53/64$
.9375	$15/16$	23.81	—	—	—	—	**20**	$57/64$
1.000	1	25.40	**8**	$7/8$	**12**	$59/64$	**20**	$61/64$
1.0625	1 $1/16$	26.99	—	—	—	—	18	1
1.125	1 $1/8$	28.57	**7**	$63/64$	**12**	$1\ 3/64$	18	$1\ 5/64$
1.1875	1 $3/16$	30.16	—	—	—	—	18	$1\ 9/64$
1.250	1 $1/4$	31.75	**7**	$1\ 7/64$	**12**	$1\ 11/64$	18	$1\ 13/64$
1.3125	1 $5/16$	33.34	—	—	—	—	18	$1\ 17/64$
1.375	1 $3/8$	34.92	**6**	$1\ 13/64$	**12**	$1\ 19/64$	18	$1\ 5/16$
1.4375	1 $7/16$	36.51	—	—	—	—	18	$1\ 3/8$
1.500	1 $1/2$	38.10	**6**	$1\ 21/64$	**12**	$1\ 27/64$	18	$1\ 29/64$
1.5625	1 $9/16$	39.69	—	—	—	—	18	$1\ 1/2$
1.625	1 $5/8$	41.27	—	—	—	—	18	$1\ 9/16$
1.6875	1 $11/16$	42.86	—	—	—	—	18	$1\ 5/8$
1.750	1 $3/4$	44.45	**5**	$1\ 35/64$	—	—	**16**	$1\ 11/16$
2.000	2	50.80	**4½**	$1\ 25/32$	—	—	**16**	$1\ 15/16$
2.250	2 $1/4$	57.15	**4½**	$2\ 1/32$	—	—	—	—
2.500	2 $1/2$	63.50	**4**	$2\ 1/4$	—	—	—	—
2.750	2 $3/4$	69.85	**4**	$2\ 1/2$	—	—	—	—
3.000	3	76.20	**4**	$2\ 3/4$	—	—	—	—
3.250	3 $1/4$	82.55	**4**	3	—	—	—	—
3.500	3 $1/2$	88.90	**4**	$3\ 1/4$	—	—	—	—
3.750	3 $3/4$	95.25	**4**	$3\ 1/2$	—	—	—	—
4.000	4	101.60	**4**	$3\ 3/4$	—	—	—	—

* Adapted from ANSI B1.1–1960.
Bold type indicates Unified threads. To be designated UNC or UNF.
Unified Standard—Classes 1A, 2A, 3A, 1B, 2B, and 3B.
For recommended hole-size limits before threading, see Tables 38 and 39, ANSI B1.1–1960.
† Tap drill for a 75% thread (not Unified—American Standard).
Bold-type sizes smaller than ¼ in. are accepted for limited applications by the British, but the symbols NC or NF, as applicable, are retained.
‡ The values listed as metric equivalents of decimal inch values have been given to assist user in selecting the closest metric size to be found in Table 10. Adherence to diameter preference is recommended, if feasible. For a metric thread use the tap drill size recommended in Table 10.

Table 26 Standard Unified special threads*
(8-pitch, 12-pitch, an 16-pitch series)

Nominal size—diameter			Threads per inch			Nominal size—diameter			Threads per inch		
mm†	inch					mm	inch				
12.70	.500	1/2	—	**12**	—	44.45	1.750	1 3/4	**8**	**12**	**16**
14.29	.5625	9/16	—	**12**	—	46.04	1.8125	1 13/16	—	—	16
15.88	.625	5/8	—	**12**	—	47.62	1.875	1 7/8	**8**	**12**	**16**
17.46	.6875	11/16	—	**12**	—	49.21	1.9375	1 15/16	—	—	16
19.05	.750	3/4	—	**12**	16	50.80	2.000	2	**8**	**12**	**16**
20.64	.8125	13/16	—	**12**	16	52.39	2.0625	2 1/16	—	—	16
22.22	.875	7/8	—	**12**	16	53.98	2.125	2 1/8	**8**	**12**	**16**
23.81	.9375	15/16	—	**12**	16	55.56	2.1875	2 3/16	—	—	16
25.40	1.000	1	**8**	**12**	**16**	57.15	2.250	2 1/4	**8**	**12**	**16**
26.99	1.0625	1 1/16	—	**12**	16	58.74	2.3125	2 5/16	—	—	16
28.58	1.125	1 1/8	**8**	**12**	**16**	60.32	2.375	2 3/8	—	**12**	**16**
30.16	1.1875	1 3/16	—	**12**	**16**	61.91	2.4375	2 7/16	—	—	16
31.75	1.250	1 1/4	**8**	**12**	**16**	63.50	2.500	2 1/2	**8**	**12**	**16**
33.34	1.3125	1 5/16	—	**12**	**16**	66.68	2.625	2 5/8	—	**12**	**16**
34.92	1.375	1 3/8	**8**	**12**	**16**	69.85	2.750	2 3/4	**8**	**12**	**16**
36.51	1.4375	1 7/16	—	**12**	**16**	73.02	2.875	2 7/8	—	**12**	**16**
38.10	1.500	1 1/2	**8**	**12**	**16**	76.20	3.000	3	**8**	**12**	**16**
39.69	1.5625	1 9/16	—	—	16	79.38	3.125	3 1/8	—	**12**	**16**
41.28	1.625	1 5/8	**8**	**12**	**16**	82.55	3.250	3 1/4	**8**	**12**	**16**
42.86	1.6875	1 11/16	—	—	16	85.72	3.375	3 3/8	—	**12**	**16**

* Adapted from ANSI B1.1–1960. Bold type indicates Unified threads (UN).
† The values listed as metric equivalents of decimal inch values have been given to assist user in selecting the closest metric size to be found in Tables 8 and 9. Adherence to diameter preference is recommended, if feasible.

Table 27 Standard Acme and Stub Acme threads*

Nominal size—diameter			Threads per inch	Nominal size—diameter			Threads per inch
mm	inch			mm	inch		
6.35	.250	1/4	16	31.75	1.250	1 1/4	5
7.94	.3125	5/16	14	34.92	1.375	1 3/8	4
9.52	.375	3/8	12	38.10	1.500	1 1/2	4
11.11	.4375	7/16	12	44.45	1.750	1 3/4	4
12.70	.500	1/2	10	50.80	2.000	2	4
15.88	.625	5/8	8	57.15	2.250	2 1/4	3
19.05	.750	3/4	6	63.50	2.500	2 1/2	3
22.22	.875	7/8	6	69.85	2.750	2 3/4	3
25.40	1.000	1	5	76.20	3.000	3	2
28.58	1.125	1 1/8	5				

* In agreement with ANSI B1.5 and B1.8–1952.

Table 28 Buttress threads*
(suggested combinations of diameters and pitches)

Nominal size—diameter								Associated pitches (tpi)
mm	inch	mm	inch	mm	inch	mm	inch	
12.70	.500	14.29	.5625	15.88	.625	17.46	.6875	20 16 12
19.05	.750	22.22	.875	25.40	1.000	—	—	— 16 12 10
28.58	1.125	31.75	1.250	34.92	1.375	38.10	1.500	— 16 12 10 8 6
44.45	1.750	76.20	3.000	57.15	2.250	63.50	2.500	— 16 12 10 8 6 5 4
69.85	2.750	76.20	3.000	88.90	3.500	101.60	4.000	— 16 12 10 8 6 5 4

* In agreement with ANSI B1.9–1953.

Table 29 American National standard wrench-head bolts and nuts—regular series*

Bolt diameter, nominal size		Bolt heads				Nuts				
		Width across flats	Height of head			Width across flats	Thickness			
							Regular		Jam	
		Unfinished and semifinished square and hexagon†	Unfinished square	Unfinished hexagon	Semifinished hexagon	Unfinished hexagon, semifinished hexagon and hexagon jam	Unfinished square and hexagon	Semifinished hexagon and hexagon slotted	Unfinished hexagon	Semifinished hexagon
.250	1/4	3/8 sq. 7/16 hex.	11/64	11/64	5/32	7/16	7/32	7/32	5/32	5/32
.3125	5/16	1/2	13/64	7/32	13/64	1/2	17/64	17/64	3/16	3/16
.375	3/8	9/16	1/4	1/4	15/64	9/16	21/64	21/64	7/32	7/32
.4375	7/16	5/8	19/64	19/64	9/32	11/16	3/8	3/8	1/4	1/4
.500	1/2	3/4	21/64	11/32	5/16	3/4	7/16	7/16	5/16	5/16
.5625	9/16	13/16	—	—	23/64	7/8	31/64	31/64	5/16	5/16
.625	5/8	15/16	27/64	27/64	25/64	15/16	35/64	35/64	3/8	3/8
.750	3/4	1 1/8	1/2	1/2	15/32	1 1/8	21/32	41/64	7/16	27/64
.875	7/8	1 5/16	19/32	37/64	35/64	1 5/16	49/64	3/4	1/2	31/64
1.000	1	1 1/2	21/32	43/64	39/64	1 1/2	7/8	55/64	9/16	35/64
1.125	1 1/8	1 11/16	3/4	3/4	11/16	1 11/16	1	31/32	5/8	39/64
1.250	1 1/4	1 7/8	27/32	27/32	25/32	1 7/8	1 3/32	1 1/16	3/4	23/32
1.375	1 3/8	2 1/16	29/32	29/32	27/32	2 1/16	1 13/64	1 11/64	13/16	25/32
1.500	1 1/2	2 1/4	1	1	15/16	2 1/4	1 5/16	1 9/32	7/8	27/32
1.625	1 5/8	2 7/16	1 3/32	—	—	2 7/16	—	1 25/64	—	29/32
1.750	1 3/4	2 5/8	—	1 5/32	1 3/32	2 5/8	—	1 1/2	—	31/32
1.875	1 7/8	2 13/16	—	—	—	2 13/16	—	1 39/64	—	1 1/32
2.000	2	3	—	1 11/32	1 7/32	3	—	1 23/32	—	1 3/32
2.250	2 1/4	3 3/8	—	1 1/2	1 3/8	3 3/8	—	1 59/64	—	1 13/64
2.500	2 1/2	3 3/4	—	1 21/32	1 17/32	3 3/4	—	2 9/64	—	1 23/64
2.750	2 3/4	4 1/8	—	1 13/16	1 11/16	4 1/8	—	2 23/64	—	1 37/64
3.000	3	4 1/2	—	2	1 7/8	4 1/2	—	2 37/64	—	1 45/64

*ANSI B18.2.1–1965; ANSI B18.2.2–1965.
Thread-bolts: coarse-thread series, class 2A.
Thread-nuts: unfinished; coarse series, class 2B: semifinished; coarse, fine, or 8-pitch series.
†Square bolts in 1/4–1 5/8 in. sizes (nominal) only.

Table 30 American National standard finished hexagon castle nuts*

Nominal size— diameter D		Threads per inch		Thickness A	Width across flats, B	Slot		Diameter of cylindrical part min
		UNC	UNF			Depth, C	Width, E	
.250	1/4	20	28	9/32	7/16	.094	.078	.371
.3125	5/16	18	24	21/64	1/2	.094	.094	.425
.375	3/8	16	24	13/32	9/16	.125	.125	.478
.4375	7/16	14	20	29/64	11/16	.156	.125	.582
.500	1/2	13	20	9/16	3/4	.156	.156	.637
.5625	9/16	12	18	39/64	7/8	.188	.156	.744
.625	5/8	11	18	23/32	15/16	.219	.188	.797
.750	3/4	10	16	13/16	1 1/8	.250	.188	.941
.875	7/8	9	14	29/32	1 5/16	.250	.188	1.097
1.000	1	8	12	1	1 1/2	.281	.250	1.254
1.125	1 1/8	7	12	1 5/32	1 11/16	.344	.250	1.411

*ANSI B18.2.2–1965.
Thread may be coarse- or fine-thread series, class 2B tolerance; unless otherwise specified, fine-thread series will be furnished.

Table 31 American National standard machine screw and stove bolt nuts*

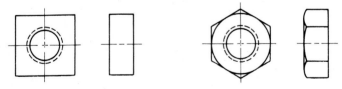

Diameter		Nominal size													
		0	1	2	3	4	5	6	8	10	12	1/4	5/16	3/8
Across flats (nominal)		5/32	5/32	3/16	3/16	1/4	5/16	5/16	11/32	3/8	7/16	7/16	9/16	5/8
Across corners (min)	Hexagonal	.171	.171	.205	.205	.275	.344	.344	.378	.413	.482	.482	.621	.692
	Square	.206	.206	.247	.247	.331	.415	.415	.456	.497	.581	.581	.748	.833
Thickness (nom)		3/64	3/64	1/16	1/16	3/32	7/64	7/64	1/8	1/8	5/32	3/16	7/32	1/4

*ANSI B18.6.3–1962.
Dimensions in inches.
Square nuts—threads are UNC, Class 2B.
Hexagonal nuts—threads are UNC or UNF, Class 2B.

Table 32 American National standard set screws*

FLAT POINT   CUP POINT   CONE POINT   OVAL POINT   DOG POINT   HALF DOG POINT

Nominal size—diameter, D		Slotted headless			Hexagonal socket (min)		Square-head F_{max}			Points								
										Cup and flat K_{max}			Cone	Oval	Full- and half-dog N_{max}			
		A	B	R_1	C	D†	F	H	R_2	K	W	X	Y	R_3	L	M	N	Z
.125	5	.023	.031	.125	1/16	.050	—	—	—	.067				.094	.060	.030	.083	
.138	6	.025	.035	.138	1/16	.050	—	—	—	.074				.109	.070	.035	.092	
.164	8	.029	.041	.164	5/64	.062	—	—	—	.087				.125	.080	.040	.109	
.190	10	.032	.048	.190	3/32	.075	.1875	9/64	15/32	.102				.141	.090	.045	.127	
.216	12	.036	.054	.216	3/32	.075	.216	5/32	35/64	.115				.156	.110	.055	.144	
.250	1/4	.045	.063	.250	1/8	.100	.250	3/16	5/8	.132				.188	.125	.063	.156	
.3125	5/16	.051	.078	.313	5/32	.125	.3125	15/64	25/32	.172				.234	.156	.078	.203	
.375	3/8	.064	.094	.375	3/16	.150	.375	9/32	15/16	.212				.281	.188	.094	.250	
.4375	7/16	.072	.109	.438	7/32	.175	.4375	21/64	1 3/32	.252				.328	.219	.109	.297	
.500	1/2	.081	.125	.500	1/4	.200	.500	3/8	1 1/4	.291				.375	.250	.125	.344	
.5625	9/16	.091	.141	.563	1/4	.200	.5625	27/64	1 13/32	.332				.422	.281	.140	.391	
.625	5/8	.102	.156	.625	5/16	.250	.625	15/32	1 9/16	.371				.469	.313	.156	.469	
.750	3/4	.129	.188	.750	3/8	.300	.750	9/16	1 7/8	.450				.563	.375	.188	.563	
.875	7/8	—	—	—	1/2	.400	.875	21/32	2 3/16	—				—	—	—	—	
1.000	1	—	—	—	9/16	.450	1.000	3/4	2 1/2	—				—	—	—	—	
1.125	1 1/8	—	—	—	9/16	.450	1.125	27/32	2 13/16	—				—	—	—	—	
1.250	1 1/4	—	—	—	5/8	.500	1.250	15/16	3 1/8	—				—	—	—	—	
1.375	1 3/8	—	—	—	5/8	.500	1.375	1 1/32	3 7/16	—				—	—	—	—	
1.500	1 1/2	—	—	—	3/4	.600	1.500	1 1/8	3 3/4	—				—	—	—	—	

W = 80°–90° (Draw as 90°.)
X = 118° ±5°
When L equals nominal diameter or less, Y = 118° ±2°. When L exceeds nominal diameter, Y = 90° ±2°.
Z = 100°–110°

* ANSI B18.6.2–1956; ANSI B18.3–1969.
† Dimensions apply to cup and flat-point screws 1 diam in length or longer. For screws shorter than 1 diam in length, and for other types of points, socket to be as deep as practicable.

Table 33 American National standard machine screws*

589

Inch tables

FLAT HEAD   ROUND HEAD   OVAL HEAD   FILLISTER HEAD

Size (number) and threads per inch			Standard dimensions (max)													
Nominal size Diameter, D	Thread		Head diameter			Height dimensions					Slot width	Slot depth			
	Coarse	Fine	A	B	C	E	F	G	H	K	J	L	M	N	O
.060	0	—	80	.119	.113	.096	.035	.053	.056	.045	.059	.023	.015	.039	.030	.025
.073	1	64	72	.146	.138	.118	.043	.061	.068	.053	.071	.026	.019	.044	.038	.031
.086	2	56	64	.172	.162	.140	.051	.069	.080	.062	.083	.031	.023	.048	.045	.037
.099	3	48	56	.199	.187	.161	.059	.078	.092	.070	.095	.035	.027	.053	.052	.043
.112	4	40	48	.225	.211	.183	.067	.086	.104	.079	.107	.039	.030	.058	.059	.048
.125	5	40	44	.252	.236	.205	.075	.095	.116	.088	.120	.043	.034	.063	.067	.054
.138	6	32	40	.279	.260	.226	.083	.103	.128	.096	.132	.048	.038	.068	.074	.060
.164	8	32	36	.332	.309	.270	.100	.120	.152	.113	.156	.054	.045	.077	.088	.071
.190	10	24	32	.385	.359	.313	.116	.137	.176	.130	.180	.060	.053	.087	.103	.083
.216	12	24	28	.438	.408	.357	.132	.153	.200	.148	.205	.067	.060	.096	.117	.094
.250	1/4	20	28	.507	.472	.414	.153	.175	.232	.170	.237	.075	.070	.109	.136	.109
.3125	5/16	18	24	.635	.590	.518	.191	.216	.290	.211	.295	.084	.088	.132	.171	.137
.375	3/8	16	24	.762	.708	.622	.230	.256	.347	.253	.355	.094	.106	.155	.206	.164
.4375	7/16	14	20	.812	.750	.625	.223	.328	.345	.265	.368	.094	.103	.196	.210	.170
.500	1/2	13	20	.875	.813	.750	.223	.355	.354	.297	.412	.106	.103	.211	.216	.190
.5625	9/16	12	18	1.000	.938	.812	.260	.410	.410	.336	.466	.118	.120	.242	.250	.214
.625	5/8	11	18	1.125	1.000	.875	.298	.438	.467	.375	.521	.133	.137	.258	.285	.240
.750	3/4	10	16	1.375	1.250	1.000	.372	.547	.578	.441	.612	.149	.171	.320	.353	.281

° ANSI B18.6.3–1962.
Thread length—screws 2 in. in length or less are threaded as close to the head as practicable. Screws longer than 2 in. should have a minimum thread length of 1¾ in.

Table 34 American National standard cap screws*

Nominal size—diameter		Head diameter					Height dimensions					Slot width	Slot depth			Socket dimensions	
		A	B	C	E	W	F_{av}	G	H_{nom}	K	M	J	O	P	Q	N_{min}	T_{min}
.250	1/4	.500	.437	.375	3/8	7/16	.140	.191	5/32	.172	.216	.075	.068	.117	.097	3/16	.120
.3125	5/16	.625	.562	.437	7/16	1/2	.177	.245	13/64	.203	.253	.084	.086	.151	.115	7/32	.151
.375	3/8	.750	.625	.562	9/16	9/16	.210	.273	15/64	.250	.314	.094	.103	.168	.142	5/16	.182
.4375	7/16	.8125	.750	.625	5/8	5/8	.210	.328	9/32	.297	.368	.094	.103	.202	.168	5/16	.213
.500	1/2	.875	.812	.750	3/4	3/4	.210	.354	5/16	.328	.413	.106	.103	.218	.193	3/8	.245
.5625	9/16	1.000	.937	.812	13/16	13/16	.244	.409	23/64	.375	.467	.118	.120	.252	.213	3/8	.276
.625	5/8	1.125	1.000	.875*	7/8	15/16	.281	.437	25/64	.422	.521	.133	.137	.270	.239	1/2	.307
.750	3/4	1.375	1.250	1.000	1	1 1/8	.352	.546	15/32	.500	.612	.149	.171	.338	.283	9/16	.370
.875	7/8	1.625	—	1.125	1 1/8	1 5/16	.423	—	35/64	.594	.720	.167	.206	—	.334	9/16	.432
1.000	1	1.875	—	1.312	1 5/16	1 1/2	.494	—	39/64	.656	.803	.188	.240	—	.371	5/8	.495
1.125	1 1/8	2.062	—	—	1 1/2	1 11/16	.529	—	11/16	—	—	.196	.257	—	—	3/4	.557
1.250	1 1/4	2.312	—	—	1 3/4	1 7/8	.600	—	25/32	—	—	.211	.291	—	—	3/4	.620
1.375	1 3/8	2.562	—	—	1 7/8	2 1/16	.665	—	27/32	—	—	.226	.326	—	—	3/4	.682
1.500	1 1/2	2.812	—	—	2	2 1/4	.742	—	15/16	—	—	.258	.360	—	—	1	.745

*ANSI B18.3–1969. ANSI B18.6.2–1956.
Basically, threads may be coarse, fine, or 8-thread series; class 2A for plain (unplated) cap screws.
Minimum thread length will be 2D + 1/4 in. for lengths up to and including 6 in.
Socket-head cap screws—thread coarse or fine, class 3A. Thread length: coarse, 2D + 1/2 in; fine, 1 1/2 D + 1/2 in.
Standard dimensions are maximum except as noted.

Table 35-A American National standard plain washers (type A)*

Washer size—nominal		Light-SAE (N = narrow)			Standard-plate (W = wide)		
		ID, A	OD, B	Thickness, H	ID, A	OD, B	Thickness, H
.250	1/4	.281	.625	.065	.312	.734	.065
.312	5/16	.344	.688	.065	.375	.875	.083
.375	3/8	.406	.812	.065	.438	1.000	.083
.438	7/16	.469	.922	.065	.500	1.250	.083
.500	1/2	.531	1.062	.095	.562	1.375	.109
.562	9/16	.594	1.156	.095	.625	1.469	.109
.625	5/8	.656	1.312	.095	.688	1.750	.134
.750	3/4	.812	1.469	.134	.812	2.000	.148
.875	7/8	.938	1.750	.134	.938	2.250	.165
1.000	1	1.062	2.000	.134	1.062	2.500	.165
1.125	1 1/8	1.250	2.250	.134	1.250	2.750	.165
1.250	1 1/4	1.375	2.500	.165	1.375	3.000	.165
1.375	1 3/8	1.500	2.750	.165	1.500	3.250	.180
1.500	1 1/2	1.625	3.000	.165	1.625	3.500	.180

*ANSI B27.2–1965.
Plain washers are specified as follows: ID × OD × thickness — .375 × .875 × .083 PLAIN WASHER.

SQUARE KEY

FLAT KEY

| Shaft diameter—nominal | | Square stock, W | Rectangular stock, W × H | Shaft diameter—nominal | | Square stock, W | Rectangular stock, W × H |
Over	To (inclusive)			Over	To (inclusive)		
$5/16$	$7/16$	$3/32$	—	$1\,3/4$	$2\,1/4$	$1/2$	$1/2 \times 3/8$
$7/16$	$9/16$	$1/8$	$1/8 \times 3/32$	$2\,1/4$	$2\,3/4$	$5/8$	$5/8 \times 7/16$
$9/16$	$7/8$	$3/16$	$3/16 \times 1/8$	$2\,3/4$	$3\,1/4$	$3/4$	$3/4 \times 1/2$
$7/8$	$1\,1/4$	$1/4$	$1/4 \times 3/16$	$3\,1/4$	$3\,3/4$	$7/8$	$7/8 \times 5/8$
$1\,1/4$	$1\,3/8$	$5/16$	$5/16 \times 1/4$	$3\,3/4$	$4\,1/2$	1	$1 \times 3/4$
$1\,3/8$	$1\,3/4$	$3/8$	$3/8 \times 1/4$	$4\,1/2$	$5\,1/2$	$1\,1/4$	$1\,1/4 \times 7/8$

*ANSI B17.1–1967.
All dimensions in inches.

Table 39 American National standard plain taper and gib-head keys†

W
Plain Taper

W
Alternate Plain Taper

Gib Head Taper

Plain and Gib Head Taper Keys Have a $\frac{1}{8}$" Taper in 12"

Plain taper and gib-head keys square and rectangular				Gib-head			
				Square		Rectangular	
Shaft diameter—nominal		Square type, W = H	Rectangular type, W × H	Head height, A	Length, B	Head height, A	Length, B
Over	To (inclusive)						
$5/16$	$7/16$	—	—	—	—	—	—
$7/16$	$9/16$	$1/8$	$1/8 \times 3/32$	$1/4$	$7/32$	$3/16$	$1/8$
$9/16$	$7/8$	$3/16$	$3/16 \times 1/8$	$5/16$	$9/32$	$1/4$	$3/16$
$7/8$	$1\,1/4$	$1/4$	$1/4 \times 3/16$	$7/16$	$11/32$	$5/16$	$1/4$
$1\,1/4$	$1\,3/8$	$5/16$	$5/16 \times 1/4$	$9/16$	$13/32$	$3/8$	$5/16$
$1\,3/8$	$1\,3/4$	$3/8$	$3/8 \times 1/4$	$11/16$	$15/32$	$7/16$	$3/8$
$1\,3/4$	$2\,1/4$	$1/2$	$1/2 \times 3/8$	$7/8$	$19/32$	$5/8$	$1/2$
$2\,1/4$	$2\,3/4$	$5/8$	$5/8 \times 7/16$	$1\,1/16$	$23/32$	$3/4$	$5/8$
$2\,3/4$	$3\,1/4$	$3/4$	$3/4 \times 1/2$	$1\,1/4$	$7/8$	$7/8$	$3/4$
$3\,1/4$	$3\,3/4$	$7/8$	$7/8 \times 5/8$	$1\,1/2$	1	$1\,1/16$	$7/8$

†ANSI B17.1–1967.
* For locating position of dimension H.
For longer sizes, see standard.
All dimensions in inches.

Table 40 American National standard woodruff keys*

Key no.	Nominal size, A × B	Height of key		Distance below center, E	Depth of keyseat in shaft +.005 −.000
		C_{max}	D_{max}		
202	$\frac{1}{16} \times \frac{1}{4}$	.109	.109	$\frac{1}{64}$	.0728
202.5	$\frac{1}{16} \times \frac{5}{16}$	.140	.140	$\frac{1}{64}$	.1038
203	$\frac{1}{16} \times \frac{3}{8}$	.172	.172	$\frac{1}{64}$	.1358
204	$\frac{1}{16} \times \frac{1}{2}$	.203	.194	$\frac{3}{64}$	.1668
302.5	$\frac{3}{32} \times \frac{5}{16}$	.140	.140	$\frac{1}{64}$	.0882
303	$\frac{3}{32} \times \frac{3}{8}$	.172	.172	$\frac{1}{64}$	.1202
304	$\frac{3}{32} \times \frac{1}{2}$	.203	.194	$\frac{3}{64}$	.1511
305	$\frac{3}{32} \times \frac{5}{8}$	.250	.240	$\frac{1}{16}$	.1981
403	$\frac{1}{8} \times \frac{3}{8}$	.172	.172	$\frac{1}{64}$	.1045
404	$\frac{1}{8} \times \frac{1}{2}$	.203	.194	$\frac{3}{64}$	.1355
405	$\frac{1}{8} \times \frac{5}{8}$	.250	.240	$\frac{1}{16}$	.1825
406	$\frac{1}{8} \times \frac{3}{4}$	.313	.303	$\frac{1}{16}$	.2455
505	$\frac{5}{32} \times \frac{5}{8}$	.250	.240	$\frac{1}{16}$	.1669
506	$\frac{5}{32} \times \frac{3}{4}$	.313	.303	$\frac{1}{16}$	.2299
507	$\frac{5}{32} \times \frac{7}{8}$	.375	.365	$\frac{1}{16}$	.2919
605	$\frac{3}{16} \times \frac{5}{8}$	.250	.240	$\frac{1}{16}$	.1513
606	$\frac{3}{16} \times \frac{3}{4}$	.313	.303	$\frac{1}{16}$	.2143
607	$\frac{3}{16} \times \frac{7}{8}$	.375	.365	$\frac{1}{16}$	.2763
608	$\frac{3}{16} \times 1$	.438	.428	$\frac{1}{16}$	.3393
609	$\frac{3}{16} \times 1\frac{1}{8}$	.484	.475	$\frac{5}{64}$	.3853
610	$\frac{3}{16} \times 1\frac{1}{4}$	.547	.537	$\frac{5}{64}$	.4483
707	$\frac{7}{32} \times \frac{7}{8}$	.375	.365	$\frac{1}{16}$	.2607
708	$\frac{7}{32} \times 1$	.438	.428	$\frac{1}{16}$	.3237
709	$\frac{7}{32} \times 1\frac{1}{8}$	.484	.475	$\frac{5}{64}$	.3697
710	$\frac{7}{32} \times 1\frac{1}{4}$	.547	.537	$\frac{5}{64}$	.4327
806	$\frac{1}{4} \times \frac{3}{4}$	.313	.303	$\frac{1}{16}$	.1830
807	$\frac{1}{4} \times \frac{7}{8}$	.375	.365	$\frac{1}{16}$	.2450
808	$\frac{1}{4} \times 1$	.438	.428	$\frac{1}{16}$	.3080
809	$\frac{1}{4} \times 1\frac{1}{8}$	.484	.475	$\frac{5}{64}$	.3540
810	$\frac{1}{4} \times 1\frac{1}{4}$	.547	.537	$\frac{5}{64}$	.4170
811	$\frac{1}{4} \times 1\frac{3}{8}$	.594	.584	$\frac{3}{32}$	.4640
812	$\frac{1}{4} \times 1\frac{1}{2}$	.641	.631	$\frac{7}{64}$	.5110
1008	$\frac{5}{16} \times 1$	.438	.428	$\frac{1}{16}$	.2768
1009	$\frac{5}{16} \times 1\frac{1}{8}$	.484	.475	$\frac{5}{64}$	.3228
1010	$\frac{5}{16} \times 1\frac{1}{4}$	.547	.537	$\frac{5}{64}$	.3858
1011	$\frac{5}{16} \times 1\frac{3}{8}$	.594	.584	$\frac{3}{32}$	.4328
1012	$\frac{5}{16} \times 1\frac{1}{2}$	.641	.631	$\frac{7}{64}$	.4798
1208	$\frac{3}{8} \times 1$	.438	.428	$\frac{1}{16}$	.2455
1210	$\frac{3}{8} \times 1\frac{1}{4}$	.547	.537	$\frac{5}{64}$	.3545
1211	$\frac{3}{8} \times 1\frac{3}{8}$	.594	.584	$\frac{3}{32}$	.4015
1212	$\frac{3}{8} \times 1\frac{1}{2}$	.641	.631	$\frac{7}{64}$	.4485

0.010 R.

* ANSI B17.2–1967.
All dimensions in inches. Key numbers indicate the nominal key dimensions. The last two digits give the nominal diameter in eighths of an inch and the digits preceding the last two give the nominal width in thirty-seconds of an inch.
Examples: No. 204 indicates a key $\frac{2}{32} \times \frac{4}{8}$ or $\frac{1}{16} \times \frac{1}{2}$.
No. 808 indicates a key $\frac{8}{32} \times \frac{8}{8}$ or $\frac{1}{4} \times 1$.

Table 41 Pratt & Whitney keys

SHAFT & KEY

Key no.	L*	W	H	D	Key no.	L*	W	H	D
1	1/2	1/16	3/32	1/16	22	1 3/8	1/4	3/8	1/4
2	1/2	3/32	9/64	3/32	23	1 3/8	5/16	15/32	5/16
3	1/2	1/8	3/16	1/8	F	1 3/8	3/8	9/16	3/8
4	5/8	3/32	9/64	3/32	24	1 1/2	1/4	3/8	1/4
5	5/8	1/8	3/16	1/8	25	1 1/2	5/16	15/32	5/16
6	5/8	5/32	15/64	5/32	G	1 1/2	3/8	9/16	3/8
7	3/4	1/8	3/16	1/8	51	1 3/4	1/4	3/8	1/4
8	3/4	5/32	15/64	5/32	52	1 3/4	5/16	15/32	5/16
9	3/4	3/16	9/32	3/16	53	1 3/4	3/8	9/16	3/8
10	7/8	5/32	15/64	5/32	26	2	3/16	9/32	3/16
11	7/8	3/16	9/32	3/16	27	2	1/4	3/8	1/4
12	7/8	7/32	21/64	7/32	28	2	5/16	15/32	5/16
A	7/8	1/4	3/8	1/4	29	2	3/8	9/16	3/8
13	1	3/16	9/32	3/16	54	2 1/4	1/4	3/8	1/4
14	1	7/32	21/64	7/32	55	2 1/4	5/16	15/32	5/16
15	1	1/4	3/8	1/4	56	2 1/4	3/8	9/16	3/8
B	1	5/16	15/32	5/16	57	2 1/4	7/16	21/32	7/16
16	1 1/8	3/16	9/32	3/16	58	2 1/2	5/16	15/32	5/16
17	1 1/8	7/32	21/64	7/32	59	2 1/2	3/8	9/16	3/8
18	1 1/8	1/4	3/8	1/4	60	2 1/2	7/16	21/32	7/16
C	1 1/8	5/16	15/32	5/16	61	2 1/2	1/2	3/4	1/2
19	1 1/4	3/16	9/32	3/16	30	3	3/8	9/16	3/8
20	1 1/4	7/32	21/64	7/32	31	3	7/16	21/32	7/16
21	1 1/4	1/4	3/8	1/4	32	3	1/2	3/4	1/2
D	1 1/4	5/16	15/32	5/16	33	3	9/16	27/32	9/16
E	1 1/4	3/8	9/16	3/8	34	3	5/8	15/16	5/8

*The length L may vary but should always be at least $2W$.

Table 42   American National standard taper pins*

Number of pin†	Diameter at large end		Maximum length, L	Approx. shaft diameter	Drill size‡
00000	.094	$3/32$	1	.250	No. 47 (.0785)
0000	.109	$7/64$	1	.312	No. 42 (.0935)
000	.125	$1/8$	1	.375	No. 37 (.1040)
00	.141	$9/64$	$1\frac{1}{4}$	.438	No. 31 (.1200)
0	.156	$5/32$	$1\frac{1}{2}$	.500	No. 28 (.1405)
1	.172	$11/64$	2	.562	No. 25 (.1495)
2	.193	—	$2\frac{1}{2}$	.625	No. 19 (.1660)
3	.219	$7/32$	3	.750	No. 12 (.1890)
4	.250	$1/4$	3	.812	No. 3 (.2130)
5	.289	—	3	.875	$1/4$ (.2500)
6	.341	—	4	1.000	$9/32$ (.2812)
7	.409	—	4	1.250	$11/32$ (.3438)
8	.492	—	4	1.500	$13/32$ (.4062)

* ANSI B5.20–1958.
† For 7/0, 6/0, 9, and 10 see the standard. Pins 11–14 are special sizes.
‡ Suggested sizes; not American National Standard.
Drill size is for reamer. The small diameter of a pin is equal to the large diameter minus .02083 × **L**, where **L** is the length.

Table 43   American National standard wrought-iron and steel pipe*

Nominal size	Outside diameter (all weights)	Threads per inch	Tap drill sizes†	Distance pipe enters fitting	Standard weight Nominal wall thickness Wrought iron	Standard weight Steel	Heavy Extra heavy Wrought iron	Heavy Extra heavy Steel	Heavy Double extra heavy Wrought iron	Heavy Double extra heavy Steel
$1/8$	.405	27	$11/32$	$5/16$	.069	.068	.099	.095	—	—
$1/4$	.540	18	$7/16$	$7/16$	.090	.088	.122	.119	—	—
$3/8$	.675	18	$19/32$	$7/16$	.093	.091	.129	.126	—	—
$1/2$	.840	14	$23/32$	$9/16$	.111	.109	.151	.147	.307	.294
$3/4$	1.050	14	$15/16$	$9/16$	.115	.113	.157	.154	.318	.308
1	1.315	$11\frac{1}{2}$	$1 5/32$	$11/16$	.136	.133	.183	.179	.369	.358
$1\frac{1}{4}$	1.660	$11\frac{1}{2}$	$1 1/2$	$11/16$	.143	.140	.195	.191	.393	.382
$1\frac{1}{2}$	1.900	$11\frac{1}{2}$	$1 23/32$	$11/16$	.148	.145	.204	.200	.411	.400
2	2.375	$11\frac{1}{2}$	$2 3/16$	$3/4$	.158	.154	.223	.218	.447	.436
$2\frac{1}{2}$	2.875	8	$2 5/8$	$1 1/16$	.208	.203	.282	.276	.567	.552
3	3.500	8	$3 1/4$	$1 1/8$	.221	.216	.306	.300	.615	.600
$3\frac{1}{2}$	4.000	8	$3 3/4$	$1 3/16$	.231	.226	.325	.318	—	—
4	4.500	8	$4 1/4$	$1 3/16$	.242	.237	.344	.337	.690	.674
5	5.563	8	$5 5/16$	$1 5/16$	.263	.258	.383	.375	.768	.750
6	6.625	8	$6 3/8$	$1 3/8$	.286	.280	.441	.432	.884	.864
8	8.625	8	—	—	.329	.322	.510	.500	.895	.875

* ANSI B36.10–1970, ANSI B2.1–1968.
All dimensions in inches.
† Not American Standard. See ANSI B36.10–1970 for sizes larger than 8 in.

Table 44 American National standard malleable-iron screwed fittings*

ELBOW · TEE · CROSS · 45° ELBOW · 45° Y-BRANCH · STREET ELBOW · 45° STREET ELBOW · STREET TEE · STRAIGHT REDUCING COUPLINGS

Nominal pipe size	A, Center to end, elbows, tees, and crosses	B_min, Length of thread	C, Center to end, 45° elbows	E_min, Width of band	F_min, Inside diameter of fitting	F_max	G, Metal thickness	H, Outside diameter of band	J, Center to male end, elbows and tees	K, Center to male end, 45° elbows	L_min, Length of external thread	M, Length of reducing couplings	N_max, Port diameter, male end	T, Center to end, inlet	U, Center to end, outlet	V, End to end	W, Length of straight couplings	Thickness of ribs of couplings
1/8	.69	.25	—	.200	.405	.435	.090	.693	1.00	—	.2638	1.00	.20	—	—	—	.96	.090
1/4	.81	.32	.73	.215	.540	.584	.095	.844	1.19	.94	.4018	1.13	.26	—	—	—	1.06	.095
3/8	.95	.36	.80	.230	.675	.719	.100	1.015	1.44	1.03	.4078	1.25	.37	.50	1.43	1.93	1.16	.100
1/2	1.12	.43	.88	.249	.840	.897	.105	1.197	1.63	1.15	.5337	1.44	.51	.61	1.71	2.32	1.34	.105
3/4	1.31	.50	.98	.273	1.050	1.107	.120	1.458	1.89	1.29	.5457	1.69	.69	.72	2.05	2.77	1.52	.120
1	1.50	.58	1.12	.302	1.315	1.385	.134	1.771	2.14	1.47	.6828	2.06	.91	.85	2.43	3.28	1.67	.134
1 1/4	1.75	.67	1.29	.341	1.660	1.730	.145	2.153	2.45	1.71	.7068	2.31	1.19	1.02	2.92	3.94	1.93	.145
1 1/2	1.94	.70	1.43	.368	1.900	1.970	.155	2.427	2.69	1.88	.7235	2.81	1.39	1.10	3.28	4.38	2.15	.155
2	2.25	.75	1.68	.422	2.375	2.445	.173	2.963	3.26	2.22	.7565	3.25	1.79	1.24	3.93	5.17	2.53	.173
2 1/2	2.70	.92	1.95	.478	2.875	2.975	.210	3.589	3.86	2.57	1.1375	3.69	2.20	1.52	4.73	6.25	2.88	.210
3	3.08	.98	2.17	.548	3.500	3.600	.231	4.285	4.51	3.00	1.2000	4.38	2.78	1.71	5.55	7.26	3.18	.231
3 1/2	3.42	1.03	2.39	.604	4.000	4.100	.248	4.843	—	—	—	—	—	—	—	—	—	—
4	3.79	1.08	2.61	.661	4.500	4.600	.265	5.401	5.69	3.70	1.3000	—	3.70	2.01	6.97	8.98	3.69	.265
5	4.50	1.18	3.05	.780	5.563	5.663	.300	6.583	6.86	—	1.4063	—	4.69	—	—	—	—	—
6	5.13	1.28	3.46	.900	6.625	6.725	.336	7.767	8.03	—	1.5125	—	5.67	—	—	—	—	—

* ANSI B16.3–1963.
For use under maximum working steam pressures of 150 lb per sq in.

Table 45   125-lb American National standard cast-iron screwed fittings*

Nominal pipe size	Center to end, elbows, tees, and crosses, A	Length of thread, B_{min}	Center to end, 45° elbows, C	Width of band, E_{min}	Inside diameter of fitting		Metal thickness, G_{min}	Outside diameter of band, H_{min}
					F_{min}	F_{max}		
¼	.81	.32	.73	.38	.540	.584	.110	.93
⅜	.95	.36	.80	.44	.675	.719	.120	1.12
½	1.12	.43	.88	.50	.840	.897	.130	1.34
¾	1.31	.50	.98	.56	1.050	1.107	.155	1.63
1	1.50	.58	1.12	.62	1.315	1.385	.170	1.95
1¼	1.75	.67	1.29	.69	1.660	1.730	.185	2.39
1½	1.94	.70	1.43	.75	1.900	1.970	.200	2.68
2	2.25	.75	1.68	.84	2.375	2.445	.220	3.28
2½	2.70	.92	1.95	.94	2.875	2.975	.240	3.86
3	3.08	.98	2.17	1.00	3.500	3.600	.260	4.62
3½	3.42	1.03	2.39	1.06	4.000	4.100	.280	5.20
4	3.79	1.08	2.61	1.12	4.500	4.600	.310	5.79
5	4.50	1.18	3.05	1.18	5.563	5.663	.380	7.05
6	5.13	1.28	3.46	1.28	6.625	6.725	.430	8.28
8	6.56	1.47	4.28	1.47	8.625	8.725	.550	10.63
10	8.08	1.68	5.16	1.68	10.750	10.850	.690	13.12
12	9.50	1.88	5.97	1.88	12.750	12.850	.800	15.47

*ANSI B16.4–1963.
All dimensions in inches.

Table 46 American National standard pipe plugs and caps*

PLUG   CAP

Nominal pipe size	Plug			Cap†						
	Length of thread	Height of square	Width across flats	Length of thread	Width of band	Inside diameter of fitting	Metal thickness	Outside diameter of band	Height	Thickness of ribs
	A_{min}	B_{min}	C_{nom}	D_{min}	E_{min}	F_{max}	G	H_{min}	P_{min}	
$\frac{1}{8}$	.37	.24	$\frac{9}{32}$	.25	.200	.435	.090	.693	.53	—
$\frac{1}{4}$	.44	.28	$\frac{3}{8}$	.32	.215	.584	.095	.844	.63	—
$\frac{3}{8}$	.48	.31	$\frac{7}{16}$	.36	.230	.719	.100	1.015	.74	—
$\frac{1}{2}$	.56	.38	$\frac{9}{16}$	.43	.249	.897	.105	1.197	.87	.105
$\frac{3}{4}$	.63	.44	$\frac{5}{8}$	.50	.273	1.107	.120	1.458	.97	.120
1	.75	.50	$\frac{13}{16}$	.58	.302	1.385	.134	1.771	1.16	.134
$1\frac{1}{4}$	.80	.56	$\frac{15}{16}$	.67	.341	1.730	.145	2.153	1.28	.145
$1\frac{1}{2}$	.83	.62	$1\frac{1}{8}$	.70	.368	1.970	.155	2.427	1.33	.155
2	.88	.68	$1\frac{5}{16}$	.75	.422	2.445	.173	2.963	1.45	.173
$2\frac{1}{2}$	1.07	.74	$1\frac{1}{2}$	.92	.478	2.975	.210	3.589	1.70	.210
3	1.13	.80	$1\frac{11}{16}$	.98	.548	3.600	.231	4.285	1.80	.231
$3\frac{1}{2}$	1.18	.86	$1\frac{7}{8}$	1.03	.604	4.100	.248	4.843	1.90	.248
4	—	—	—	1.08	.661	4.600	.265	5.401	2.08	.265
5	—	—	—	1.18	.780	5.663	.300	6.583	2.32	.300
6	—	—	—	1.28	.900	6.725	.336	7.767	2.55	.336

* ANSI B16.3–1963.
† The outside radius of top of cap is equal to 3 × F.

Table 47 American National standard cast-iron pipe flanges and flanged fittings*

90° ELBOW LONG RAD. ELBOW — 45° ELBOW — REDUCING ELBOW — SIDE OUTLET ELBOW — TEE — SIDE OUTLET TEE — CROSS — LATERAL — REDUCER — ECCENTRIC REDUCER — FLANGE — FLANGE

Nominal pipe size D	Center to face, elbow, tees, etc. A	Center to face, long radius elbow B	Center to face, 45° elbow C	Face to face, lateral D	Center to face, lateral E	Center to face, "Y" and lateral F	Face to face, reducer G	Diameter of flange H	Thickness of flange T	Diameter of hub X	Length of hub Y	Diameter of holes in flanges	Number of bolts for flanges	Diameter of bolts for flanges	Length of bolts for flanges	Diameter of bolt circle
1	3½	5	1¾	7½	5¾	1¾	—	4¼	7/16	1 15/16	11/16	5/8	4	1/2	1¾	3⅛
1¼	3¾	5½	2	8	6¼	1¾	—	4⅝	1/2	2 5/16	13/16	5/8	4	1/2	2	3½
1½	4	6	2¼	9	7	2	—	5	9/16	2 9/16	7/8	5/8	4	1/2	2	3⅞
2	4½	6½	2½	10½	8	2½	5	6	5/8	3 1/16	1	3/4	4	5/8	2¼	4¾
2½	5	7	3	12	9½	2½	5½	7	11/16	3 9/16	1⅛	3/4	4	5/8	2½	5½
3	5½	7¾	3	13	10	3	6	7½	3/4	4¼	1 3/16	3/4	4	5/8	2½	6
3½	6	8½	3½	14½	11½	3	6½	8½	13/16	4 13/16	1¼	3/4	8	5/8	2¾	7
4	6½	9	4	15	12	3	7	9	15/16	5 5/16	1 5/16	3/4	8	5/8	3	7½
5	7½	10¼	4½	17	13½	3½	8	10	15/16	6 7/16	1 7/16	7/8	8	3/4	3	8½
6	8	11½	5	18	14½	3½	9	11	1	7 9/16	1 9/16	7/8	8	3/4	3¼	9½
8	9	14	5½	22	17½	4½	11	13½	1⅛	9 11/16	1¾	7/8	8	3/4	3½	11¾
10	11	16½	6½	25½	20½	5	12	16	1 3/16	11 15/16	1 15/16	1	12	7/8	3¾	14¼
12	12	19	7½	30	24½	5½	14	19	1¼	14 1/16	2 3/16	1	12	7/8	3¾	17

* ANSI B16.1–1967.
For use under maximum working pressures of 125 lb per sq in.

600

Table 48A-1* American National standard running and sliding fits

Nominal size range (in.)† Over To	Class RC 1 Limits of clearance	Class RC 1 Standard limits Hole H5	Class RC 1 Standard limits Shaft g4	Class RC 2 Limits of clearance	Class RC 2 Standard limits Hole H6	Class RC 2 Standard limits Shaft g5	Class RC 3 Limits of clearance	Class RC 3 Standard limits Hole H7	Class RC 3 Standard limits Shaft f6	Class RC 4 Limits of clearance	Class RC 4 Standard limits Hole H8	Class RC 4 Standard limits Shaft f7
0–.12	.1 / .45	+.2 / 0	−.1 / .25	.1 / .55	+.25 / 0	−.1 / −.3	.3 / .95	+.4 / 0	−.3 / −.55	.3 / 1.3	+.6 / 0	−.3 / −.7
.12–.24	.15 / .5	+.2 / 0	−.15 / −.3	.15 / .65	+.3 / 0	−.15 / −.35	.4 / 1.12	+.5 / 0	−.4 / −.7	.4 / 1.6	+.7 / 0	−.4 / −.9
.24–.40	.2 / .6	.25 / 0	−.2 / −.35	.2 / .85	+.4 / 0	−.2 / −.45	.5 / 1.5	+.6 / 0	−.5 / −.9	.5 / 2.0	+.9 / 0	−.5 / −1.1
.40–.71	.25 / .75	+.3 / 0	−.25 / −.45	.25 / .95	+.4 / 0	−.25 / −.55	.6 / 1.7	+.7 / 0	−.6 / −1.0	.6 / 2.3	+1.0 / 0	−.6 / −1.3
.71–1.19	.3 / .95	+.4 / 0	−.3 / −.55	.3 / 1.2	+.5 / 0	−.3 / −.7	.8 / 2.1	+.8 / 0	−.8 / −1.3	.8 / 2.8	+1.2 / 0	−.8 / −1.6
1.19–1.97	.4 / 1.1	+.4 / 0	−.4 / −.7	.4 / 1.4	+.6 / 0	−.4 / −.8	1.0 / 2.6	+1.0 / 0	−1.0 / −1.6	1.0 / 3.6	+1.6 / 0	−1.0 / −2.0
1.97–3.15	.4 / 1.2	+.5 / 0	−.4 / −.7	.4 / 1.6	+.7 / 0	−.4 / −.9	1.2 / 3.1	+1.2 / 0	−1.2 / −1.9	1.2 / 4.2	+1.8 / 0	−1.2 / −2.4
3.15–4.73	.5 / 1.5	+.6 / 0	−.5 / −.9	.5 / 2.0	+.9 / 0	−.5 / −1.1	1.4 / 3.7	+1.4 / 0	−1.4 / −2.3	1.4 / 5.0	+2.2 / 0	−1.4 / −2.8
4.73–7.09	.6 / 1.8	+.7 / 0	−.6 / −1.1	.6 / 2.3	+1.0 / 0	−.6 / −1.3	1.6 / 4.2	+1.6 / 0	−1.6 / −2.6	1.6 / 5.7	+2.5 / 0	−1.6 / −3.2

Limits are in thousandths of an inch. Limits for hole and shaft are applied algebraically to the basic size to obtain the limits of size for the parts. Symbols H5, g5, etc., are hole and shaft designations used in ABC system.
*Tables 48A-1 through 48E are adapted from ANSI B4.1–1967.
† For diameters greater than those listed in Tables 48A-1 through 48E, see Standard.

Table 48A-2 American National standard running and sliding fits

Nominal size range (in.) Over To	Class RC 5 Limits of clearance	Class RC 5 Standard limits Hole H8	Class RC 5 Standard limits Shaft e7	Class RC 6 Limits of clearance	Class RC 6 Standard limits Hole H9	Class RC 6 Standard limits Shaft e8	Class RC 7 Limits of clearance	Class RC 7 Standard limits Hole H9	Class RC 7 Standard limits Shaft d8	Class RC 8 Limits of clearance	Class RC 8 Standard limits Hole H10	Class RC 8 Standard limits Shaft c9	Class RC 9 Limits of clearance	Class RC 9 Standard limits Hole H11	Class RC 9 Standard limits Shaft
0–.12	.6 / 1.6	+.6 / −0	−.6 / −1.0	.6 / 2.2	+1.0 / −0	−.6 / −1.2	1.0 / 2.6	+1.0 / 0	−1.0 / −1.6	2.5 / 5.1	+1.6 / 0	−2.5 / −3.5	4.0 / 8.1	+2.5 / 0	−4.0 / −5.6
.12–.24	.8 / 2.0	+.7 / −0	−.8 / −1.3	.8 / 2.7	+1.2 / −0	−.8 / −1.5	1.2 / 3.1	+1.2 / 0	−1.2 / −1.9	2.8 / 5.8	+1.8 / 0	−2.8 / −4.0	4.5 / 9.0	+3.0 / 0	−4.5 / −6.0
.24–.40	1.0 / 2.5	+.9 / −0	−1.0 / −1.6	1.0 / 3.3	+1.4 / −0	−1.0 / −1.9	1.6 / 3.9	+1.4 / 0	−1.6 / −2.5	3.0 / 6.6	+2.2 / 0	−3.0 / −4.4	5.0 / 10.7	+3.5 / 0	−5.0 / −7.2
.40–.71	1.2 / 2.9	+1.0 / −0	−1.2 / −1.9	1.2 / 3.8	+1.6 / −0	−1.2 / −2.2	2.0 / 4.6	+1.6 / 0	−2.0 / −3.0	3.5 / 7.9	+2.8 / 0	−3.5 / −5.1	6.0 / 12.8	+4.0 / 0	−6.0 / −8.8
.71–1.19	1.6 / 3.6	+1.2 / −0	−1.6 / −2.4	1.6 / 4.8	+2.0 / −0	−1.6 / −2.8	2.5 / 5.7	+2.0 / 0	−2.5 / −3.7	4.5 / 10.0	+3.5 / 0	−4.5 / −6.5	7.0 / 15.5	+5.0 / 0	−7.0 / −10.5
1.19–1.97	2.0 / 4.6	+1.6 / −0	−2.0 / −3.0	2.0 / 6.1	+2.5 / −0	−2.0 / −3.6	3.0 / 7.1	+2.5 / 0	−3.0 / −4.6	5.0 / 11.5	+4.0 / 0	−5.0 / −7.5	8.0 / 18.0	+6.0 / 0	−8.0 / −12.0
1.97–3.15	2.5 / 5.5	+1.8 / −0	−2.5 / −3.7	2.5 / 7.3	+3.0 / −0	−2.5 / −4.3	4.0 / 8.8	+3.0 / 0	−4.0 / −5.8	6.0 / 13.5	+4.5 / 0	−6.0 / −9.0	9.0 / 20.5	−7.0 / 0	−9.0 / −13.5
3.15–4.73	3.0 / 6.6	+2.2 / −0	−3.0 / −4.4	3.0 / 8.7	+3.5 / −0	−3.0 / −5.2	5.0 / 10.7	+3.5 / 0	−5.0 / −7.2	7.0 / 15.5	+5.0 / 0	−7.0 / −10.5	10.0 / 24.0	+9.0 / 0	−10.0 / −15.0
4.73–7.09	3.5 / 7.6	+2.5 / −0	−3.5 / −5.1	3.5 / 10.0	+4.0 / −0	−3.5 / −6.0	6.0 / 12.5	+4.0 / 0	−6.0 / −8.5	8.0 / 18.0	+6.0 / 0	−8.0 / −12.0	12.0 / 28.0	+10.0 / 0	−12.0 / −18.0

Limits are in thousandths of an inch. Limits for hole and shaft are applied algebraically to the basic size to obtain the limits of size for the parts. Symbols H8, e7, etc., are hole and shaft designations used in ABC system.

Table 48B-1 American National standard locational clearance fits

Nominal size range (in.) Over To	Class LC 1 Limits of clearance	Standard limits Hole H6	Standard limits Shaft h5	Class LC 2 Limits of clearance	Standard limits Hole H7	Standard limits Shaft h6	Class LC 3 Limits of clearance	Standard limits Hole H8	Standard limits Shaft h7	Class LC 4 Limits of clearance	Standard limits Hole H10	Standard limits Shaft h9	Class LC 5 Limits of clearance	Standard limits Hole H7	Standard limits Shaft g6
0–.12	0 / .45	+.25 / −0	+0 / −.2	0 / .65	+.4 / −0	+0 / −.25	0 / 1	+.6 / −0	+0 / −.4	0 / 2.6	+1.6 / −0	+0 / −1.0	.1 / .75	+.4 / −0	−.1 / −.35
.12–.24	0 / .5	+.3 / −0	+0 / −.2	0 / .8	+.5 / −0	+0 / −.3	0 / 1.2	+.7 / −0	+0 / −.5	0 / 3.0	+1.8 / −0	+0 / −1.2	.15 / .95	+.5 / −0	−.15 / −.45
.24–.40	0 / .65	+.4 / −0	+0 / −.25	0 / 1.0	+.6 / −0	+0 / −.4	0 / 1.5	+.9 / −0	+0 / −.6	0 / 3.6	+2.2 / −0	+0 / −1.4	.2 / 1.2	+.6 / −0	−.2 / −.6
.40–.71	0 / .7	+.4 / −0	+0 / −.3	0 / 1.1	+.7 / −0	+0 / −.4	0 / 1.7	+1.0 / −0	+0 / −.7	0 / 4.4	+2.8 / −0	+0 / −1.6	.25 / 1.35	+.7 / −0	−.25 / −.65
.71–1.19	0 / .9	+.5 / −0	+0 / −.4	0 / 1.3	+.8 / −0	+0 / −.5	0 / 2	+1.2 / −0	+0 / −.8	0 / 5.5	+3.5 / −0	+0 / −2.0	.3 / 1.6	+.8 / −0	−.3 / −.8
1.19–1.97	0 / 1.0	+.6 / −0	+0 / −.4	0 / 1.6	+1.0 / −0	+0 / −.6	0 / 2.6	+1.6 / −0	+0 / −1	0 / 6.5	+4.0 / −0	+0 / −2.5	.4 / 2.0	+1.0 / −0	−.4 / −1.0
1.97–3.15	0 / 1.2	+.7 / −0	+0 / −.5	0 / 1.9	+1.2 / −0	+0 / −.7	0 / 3	+1.8 / −0	+0 / −1.2	0 / 7.5	+4.5 / −0	+0 / −3	.4 / 2.3	+1.2 / −0	−.4 / −1.1
3.15–4.73	0 / 1.5	+.9 / −0	+0 / −.6	0 / 2.3	+1.4 / −0	+0 / −.9	0 / 3.6	+2.2 / −0	+0 / −1.4	0 / 8.5	+5.0 / −0	+0 / −3.5	.5 / 2.8	+1.4 / −0	−.5 / −1.4
4.73–7.09	0 / 1.7	+1.0 / −0	+0 / −.7	0 / 2.6	+1.6 / −0	+0 / −1.0	0 / 4.1	+2.5 / −0	+0 / −1.6	0 / 10	+6.0 / −0	+0 / −4	.6 / 3.2	+1.6 / −0	−.6 / −1.6

Limits are in thousandths of an inch. Limits for hole and shaft are applied algebraically to the basic size to obtain the limits of size for parts. Symbols H6, h5, etc., are hole and shaft designations used in ABC system.

Table 48B-2 American National standard locational clearance fits

Nominal size range (in.) Over To	Class LC 6 Limits of clearance	Standard limits Hole H9	Standard limits Shaft f8	Class LC 7 Limits of clearance	Standard limits Hole H10	Standard limits Shaft e9	Class LC 8 Limits of clearance	Standard limits Hole H10	Standard limits Shaft d9	Class LC 9 Limits of clearance	Standard limits Hole H11	Standard limits Shaft c10	Class LC 10 Limits of clearance	Standard limits Hole H12	Standard limits Shaft	Class LC 11 Limits of clearance	Standard limits Hole H13	Standard limits Shaft
0–.12	.3 / 1.9	+1.0 / 0	−.3 / −.9	.6 / 3.2	+1.6 / 0	−.6 / −1.6	1.0 / 3.6	+1.6 / −0	−1.0 / −2.0	2.5 / 6.6	+2.5 / −0	−2.5 / −4.1	4 / 12	+4 / −0	−4 / −8	5 / 17	+6 / −0	−5 / −11
.12–.24	.4 / 2.3	+1.2 / 0	−.4 / −1.1	.8 / 3.8	+1.8 / 0	−.8 / −2.0	1.2 / 4.2	+1.8 / −0	−1.2 / −2.4	2.8 / 7.6	+3.0 / −0	−2.8 / −4.6	4.5 / 14.5	+5 / −0	−4.5 / −9.5	6 / 20	+7 / −0	−6 / −13
.24–.40	.5 / 2.8	+1.4 / 0	−.5 / −1.4	1.0 / 4.6	+2.2 / 0	−1.0 / −2.4	1.6 / 5.2	+2.2 / −0	−1.6 / −3.0	3.0 / 8.7	+3.5 / −0	−3.0 / −5.2	5 / 17	+6 / −0	−5 / −11	7 / 25	+9 / −0	−7 / −16
.40–.71	.6 / 3.2	+1.6 / 0	−.6 / −1.6	1.2 / 5.6	+2.8 / 0	−1.2 / −2.8	2.0 / 6.4	+2.8 / −0	−2.0 / −3.6	3.5 / 10.3	+4.0 / −0	−3.5 / −6.3	6 / 20	+7 / −0	−6 / −13	8 / 28	+10 / −0	−8 / −18
.71–1.19	.8 / 4.0	+2.0 / 0	−.8 / −2.0	1.6 / 7.1	+3.5 / 0	−1.6 / −3.6	2.5 / 8.0	+3.5 / −0	−2.5 / −4.5	4.5 / 13.0	+5.0 / −0	−4.5 / −8.0	7 / 23	+8 / −0	−7 / −15	10 / 34	+12 / −0	−10 / −22
1.19–1.97	1.0 / 5.1	+2.5 / 0	−1.0 / −2.6	2.0 / 8.5	+4.0 / 0	−2.0 / −4.5	3.0 / 9.5	+4.0 / −0	−3.0 / −5.5	5 / 15	+6 / −0	−5 / −9	8 / 28	+10 / −0	−8 / −18	12 / 44	+16 / −0	−12 / −28
1.97–3.15	1.2 / 6.0	+3.0 / 0	−1.2 / −3.0	2.5 / 10.0	+4.5 / 0	−2.5 / −5.5	4.0 / 11.5	+4.5 / −0	−4.0 / −7.0	6 / 17.5	+7 / −0	−6 / −10.5	10 / 34	+12 / −0	−10 / −22	14 / 50	+18 / −0	−14 / −32
3.15–4.73	1.4 / 7.1	+3.5 / 0	−1.4 / −3.6	3.0 / 11.5	+5.0 / 0	−3.0 / −6.5	5.0 / 13.5	+5.0 / −0	−5.0 / −8.5	7 / 21	+9 / −0	−7 / −12	11 / 39	+14 / −0	−11 / −25	16 / 60	+22 / −0	−16 / −38
4.73–7.09	1.6 / 8.1	+4.0 / 0	−1.6 / −4.1	3.5 / 13.5	+6.0 / 0	−3.5 / −7.5	6 / 16	+6 / −0	−6 / −10	8 / 24	+10 / −0	−8 / −14	12 / 44	+16 / −0	−12 / −28	18 / 68	+25 / −0	−18 / −43

Limits are in thousandths of an inch. Limits for hole and shaft are applied algebraically to the basic size to obtain the limits of size for the parts. Symbols H9, f8, etc., are hole and shaft designations used in ABC system.

Table 48C American National standard locational transition fits

Nominal size range (in.) Over To	Class LT 1 Fit	LT1 Hole H7	LT1 Shaft js6	Class LT 2 Fit	LT2 Hole H8	LT2 Shaft js7	Class LT 3 Fit	LT3 Hole H7	LT3 Shaft k6	Class LT 4 Fit	LT4 Hole H8	LT4 Shaft k7	Class LT 5 Fit	LT5 Hole H7	LT5 Shaft n6	Class LT 6 Fit	LT6 Hole H7	LT6 Shaft n7
0–.12	−.10 / +.50	+.4 / −0	+.10 / −.10	−.2 / +.8	+.6 / −0	+.2 / −.2							−.5 / +.15	+.4 / −0	+.5 / +.25	−.65 / +.15	+.4 / −0	+.65 / +.25
.12–.24	−.15 / +.65	+.5 / −0	+.15 / −.15	−.25 / +.95	+.7 / −0	+.25 / −.25							−.6 / +.2	+.5 / −0	+.6 / +.3	−.8 / +.2	+.5 / −0	+.8 / +.3
.24–.40	−.2 / +.8	+.6 / −0	+.2 / −.2	−.3 / +1.2	+.9 / −0	+.3 / −.3	−.5 / +.5	+.6 / −0	+.5 / +.1	−.7 / +.8	+.9 / −0	+.7 / +.1	−.8 / +.2	+.6 / −0	+.8 / +.4	−1.0 / +.2	+.6 / −0	+1.0 / +.4
.40–.71	−.2 / +.9	+.7 / −0	+.2 / −.2	−.35 / +1.35	+1.0 / −0	+.35 / −.35	−.5 / +.6	+.7 / −0	+.5 / +.1	−.8 / +.9	+1.0 / −0	+.8 / +.1	−.9 / +.2	+.7 / −0	+.9 / +.5	−1.2 / +.2	+.7 / −0	+1.2 / +.5
.71–1.19	−.25 / +1.05	+.8 / −0	+.25 / −.25	−.4 / +1.6	+1.2 / −0	+.4 / −.4	−.6 / +.7	+.8 / −0	+.6 / +.1	−.9 / +1.1	+1.2 / −0	+.9 / +.1	−1.1 / +.2	+.8 / −0	+1.1 / +.6	−1.4 / +.2	+.8 / −0	+1.4 / +.6
1.19–1.97	−.3 / +1.3	+1.0 / −0	+.3 / −.3	−.5 / +2.1	+1.6 / −0	+.5 / −.5	−.7 / +.9	+1.0 / −0	+.7 / +.1	−1.1 / +1.5	+1.6 / −0	+1.1 / +.1	−1.3 / +.3	+1.0 / −0	+1.3 / +.7	−1.7 / +.3	+1.0 / −0	+1.7 / +.7
1.97–3.15	−.3 / +1.5	+1.2 / −0	+.3 / −.3	−.6 / +2.4	+1.8 / −0	+.6 / −.6	−.8 / +1.1	+1.2 / −0	+.8 / +.1	−1.3 / +1.7	+1.8 / −0	+1.3 / +.1	−1.5 / +.4	+1.2 / −0	+1.5 / +.8	−2.0 / +.4	+1.2 / −0	+2.0 / +.8
3.15–4.73	−.4 / +1.8	+1.4 / −0	+.4 / −.4	−.7 / +2.9	+2.2 / −0	+.7 / −.7	−1.0 / +1.3	+1.4 / −0	+1.0 / +.1	−1.5 / +2.1	+2.2 / −0	+1.5 / +.1	−1.9 / +.4	+1.4 / −0	+1.9 / +1.0	−2.4 / +.4	+1.4 / −0	+2.4 / +1.0
4.73–7.09	−.5 / +2.1	+1.6 / −0	+.5 / −.5	−.8 / +3.3	+2.5 / −0	+.8 / −.8	−1.1 / +1.5	+1.6 / −0	+1.1 / +.1	−1.7 / +2.4	+2.5 / −0	+1.7 / +.1	−2.2 / +.4	+1.6 / −0	+2.2 / +1.2	−2.8 / +.4	+1.6 / −0	+2.8 / +1.2

Limits are in thousandths of an inch. Limits for hole and shaft are applied algebraically to the basic size to obtain the limits of size for the mating parts. "Fit" represents the maximum interference (minus values) and the maximum clearance (plus values). Symbols H7, js6, etc., are hole and shaft designations used in ABC system.

Table 48D American National standard locational interference fits

Nominal size range (in.) Over To	Class LN 1 Limits of interference	LN1 Hole H6	LN1 Shaft n5	Class LN 2 Limits of interference	LN2 Hole H7	LN2 Shaft p6	Class LN 3 Limits of interference	LN3 Hole H7	LN3 Shaft r6
0–.12	0 / .45	+.25 / −0	+.45 / +.25	0 / .65	+.4 / −0	+.65 / +.4	.1 / .75	+.4 / −0	+.75 / +.5
.12–.24	0 / .5	+.3 / −0	+.5 / +.3	0 / .8	+.5 / −0	+.8 / +.5	.1 / .9	+.5 / −0	+.9 / +.6
.24–.40	0 / .65	+.4 / −0	+.65 / +.4	0 / 1.0	+.6 / −0	+1.0 / +.6	.2 / 1.2	+.6 / −0	+1.2 / +.8
.40–.71	0 / .8	+.4 / −0	+.8 / +.4	0 / 1.1	+.7 / −0	+1.1 / +.7	.3 / 1.4	+.7 / −0	+1.4 / +1.0
.71–1.19	0 / 1.0	+.5 / −0	+1.0 / +.5	0 / 1.3	+.8 / −0	+1.3 / +.8	.4 / 1.7	+.8 / −0	+1.7 / +1.2
1.19–1.97	0 / 1.1	+.6 / −0	+1.1 / +.6	0 / 1.6	+1.0 / −0	+1.6 / +1.0	.4 / 2.0	+1.0 / −0	+2.0 / +1.4
1.97–3.15	.1 / 1.3	+.7 / −0	+1.3 / +.7	.2 / 2.1	+1.2 / −0	+2.1 / +1.4	.4 / 2.3	+1.2 / −0	+2.3 / +1.6
3.15–4.73	.1 / 1.6	+.9 / −0	+1.6 / +1.0	.2 / 2.5	+1.4 / −0	+2.5 / +1.6	.6 / 2.9	+1.4 / −0	+2.9 / +2.0
4.73–7.09	.2 / 1.9	+1.0 / −0	+1.9 / +1.2	.2 / 2.8	+1.6 / −0	+2.8 / +1.8	.9 / 3.5	+1.6 / −0	+3.5 / +2.5

Limits are in thousandths of an inch. Limits for hole and shaft are applied algebraically to the basic size to obtain the limits of size for the parts. Symbols H7, p6, etc., are hole and shaft designations used in ABC system.

Table 48E American National standard force and shrink fits

Nominal size range (in.) Over To	Class FN 1 Limits of interference	Class FN 1 Standard limits Hole H6	Class FN 1 Standard limits Shaft	Class FN 2 Limits of interference	Class FN 2 Standard limits Hole H7	Class FN 2 Standard limits Shaft s6	Class FN 3 Limits of interference	Class FN 3 Standard limits Hole H7	Class FN 3 Standard limits Shaft t6	Class FN 4 Limits of interference	Class FN 4 Standard limits Hole H7	Class FN 4 Standard limits Shaft u6	Class FN 5 Limits of interference	Class FN 5 Standard limits Hole H8	Class FN 5 Standard limits Shaft x7
0–.12	.05 / .5	+.25 / −0	+.5 / +.3	.2 / .85	+.4 / −0	+.85 / +.6				.3 / .95	+.4 / −0	+.95 / +.7	.3 / 1.3	+.6 / −0	+1.3 / +.9
.12–.24	.1 / .6	+.3 / −0	+.6 / +.4	.2 / 1.0	+.5 / −0	+1.0 / +.7				.4 / 1.2	+.5 / −0	+1.2 / +.9	.5 / 1.7	+.7 / −0	+1.7 / +1.2
.24–.40	.1 / .75	+.4 / −0	+.75 / +.5	.4 / 1.4	+.6 / −0	+1.4 / +1.0				.6 / 1.6	+.6 / −0	+1.6 / +1.2	.5 / 2.0	+.9 / −0	+2.0 / +1.4
.40–.56	.1 / .8	+.4 / −0	+.8 / +.5	.5 / 1.6	+.7 / −0	+1.6 / +1.2				.7 / 1.8	+.7 / −0	+1.8 / +1.4	.6 / 2.3	+1.0 / −0	+2.3 / +1.6
.56–.71	.2 / .9	+.4 / −0	+.9 / +.6	.5 / 1.6	+.7 / −0	+1.6 / +1.2				.7 / 1.8	+.7 / −0	+1.8 / +1.4	.8 / 2.5	+1.0 / −0	+2.5 / +1.8
.71–.95	.2 / 1.1	+.5 / −0	+1.1 / +.7	.6 / 1.9	+.8 / −0	+1.9 / +1.4				.8 / 2.1	+.8 / −0	+2.1 / +1.6	1.0 / 3.0	+1.2 / −0	+3.0 / +2.2
.95–1.19	.3 / 1.2	+.5 / −0	+1.2 / +.8	.6 / 1.9	+.8 / −0	+1.9 / +1.4	.8 / 2.1	+.8 / −0	+2.1 / +1.6	1.0 / 2.3	+.8 / −0	+2.3 / +1.8	1.3 / 3.3	+1.2 / −0	+3.3 / +2.5
1.19–1.58	.3 / 1.3	+.6 / −0	+1.3 / +.9	.8 / 2.4	+1.0 / −0	+2.4 / +1.8	1.0 / 2.6	+1.0 / −0	+2.6 / +2.0	1.5 / 3.1	+1.0 / −0	+3.1 / +2.5	1.4 / 4.0	+1.6 / −0	+4.0 / +3.0
1.58–1.97	.4 / 1.4	+.6 / −0	+1.4 / +1.0	.8 / 2.4	+1.0 / −0	+2.4 / +1.8	1.2 / 2.8	+1.0 / −0	+2.8 / +2.2	1.8 / 3.4	+1.0 / −0	+3.4 / +2.8	2.4 / 5.0	+1.6 / −0	+5.0 / +4.0
1.97–2.56	.6 / 1.8	+.7 / −0	+1.8 / +1.3	.8 / 2.7	+1.2 / −0	+2.7 / +2.0	1.3 / 3.2	+1.2 / −0	+3.2 / +2.5	2.3 / 4.2	+1.2 / −0	+4.2 / +3.5	3.2 / 6.2	+1.8 / −0	+6.2 / +5.0
2.56–3.15	.7 / 1.9	+.7 / −0	+1.9 / +1.4	1.0 / 2.9	+1.2 / −0	+2.9 / +2.2	1.8 / 3.7	+1.2 / −0	+3.7 / +3.0	2.8 / 4.7	+1.2 / −0	+4.7 / +4.0	4.2 / 7.2	+1.8 / −0	+7.2 / +6.0
3.15–3.94	.9 / 2.4	+.9 / −0	+2.4 / +1.8	1.4 / 3.7	+1.4 / −0	+3.7 / +2.8	2.1 / 4.4	+1.4 / −0	+4.4 / +3.5	3.6 / 5.9	+1.4 / −0	+5.9 / +5.0	4.8 / 8.4	+2.2 / −0	+8.4 / +7.0
3.94–4.73	1.1 / 2.6	+.9 / −0	+2.6 / +2.0	1.6 / 3.9	+1.4 / −0	+3.9 / +3.0	2.6 / 4.9	+1.4 / −0	+4.9 / +4.0	4.6 / 6.9	+1.4 / −0	+6.9 / +6.0	5.8 / 9.4	+2.2 / −0	+9.4 / +8.0

Limits are in thousandths of an inch. Limits for hole and shaft are applied algebraically to the basic size to obtain the limits of size for the parts. Symbols H7, s6, etc., are hole and shaft designations used in ABC system.

Table 49 Twist drill sizes*

	Number sizes								Letter sizes			
No. size	Decimal equivalent	Metric equivalent	Closest metric drill (mm)	No. size	Decimal equivalent	Metric equivalent	Closest metric drill (mm)		Size letter	Decimal equivalent	Metric equivalent	Closest metric drill (mm)
1	.2280	5.791	5.80	41	.0960	2.438	2.45		A	.234	5.944	5.90
2	.2210	5.613	5.60	42	.0935	2.362	2.35		B	.238	6.045	6.00
3	.2130	5.410	5.40	43	.0890	2.261	2.25		C	.242	6.147	6.10
4	.2090	5.309	5.30	44	.0860	2.184	2.20		D	.246	6.248	6.25
5	.2055	5.220	5.20	45	.0820	2.083	2.10		E	.250	6.350	6.40
6	.2040	5.182	5.20	46	.0810	2.057	2.05		F	.257	6.528	6.50
7	.2010	5.105	5.10	47	.0785	1.994	2.00		G	.261	6.629	6.60
8	.1990	5.055	5.10	48	.0760	1.930	1.95		H	.266	6.756	6.75
9	.1960	4.978	5.00	49	.0730	1.854	1.85		I	.272	6.909	6.90
10	.1935	4.915	4.90	50	.0700	1.778	1.80		J	.277	7.036	7.00
11	.1910	4.851	4.90	51	.0670	1.702	1.70		K	.281	7.137	7.10
12	.1890	4.801	4.80	52	.0635	1.613	1.60		L	.290	7.366	7.40
13	.1850	4.699	4.70	53	.0595	1.511	1.50		M	.295	7.493	7.50
14	.1820	4.623	4.60	54	.0550	1.397	1.40		N	.302	7.671	7.70
15	.1800	4.572	4.60	55	.0520	1.321	1.30		O	.316	8.026	8.00
16	.1770	4.496	4.50	56	.0465	1.181	1.20		P	.323	8.204	8.20
17	.1730	4.394	4.40	57	.0430	1.092	1.10		Q	.332	8.433	8.40
18	.1695	4.305	4.30	58	.0420	1.067	1.05		R	.339	8.611	8.60
19	.1660	4.216	4.20	59	.0410	1.041	1.05		S	.348	8.839	8.80
19	.1610	4.089	4.10	60	.0400	1.016	1.00		T	.358	9.093	9.10
21	.1590	4.039	4.00	61	.0390	0.991	1.00		U	.368	9.347	9.30
22	.1570	3.988	4.00	62	.0380	0.965	0.95		V	.377	9.576	9.60
23	.1540	3.912	3.90	63	.0370	0.940	0.95		W	.386	9.804	9.80
24	.1520	3.861	3.90	64	.0360	0.914	0.90		X	.397	10.084	10.00
25	.1495	3.797	3.80	65	.0350	0.889	0.90		Y	.404	10.262	10.50
26	.1470	3.734	3.75	66	.0330	0.838	0.85		Z	.413	10.491	10.50
27	.1440	3.658	3.70	67	.0320	0.813	0.80					
28	.1405	3.569	3.60	68	.0310	0.787	0.80					
29	.1360	3.454	3.50	69	.0292	0.742	0.75					
30	.1285	3.264	3.25	70	.0280	0.711	0.70					
31	.1200	3.048	3.00	71	.0260	0.660	0.65					
32	.1160	2.946	2.90	72	.0250	0.635	0.65					
33	.1130	2.870	2.90	73	.0240	0.610	0.60					
34	.1110	2.819	2.80	74	.0225	0.572	0.55					
35	.1100	2.794	2.80	75	.0210	0.533	0.55					
36	.1065	2.705	2.70	76	.0200	0.508	0.50					
37	.1040	2.642	2.60	77	.0180	0.457	0.45					
38	.1015	2.578	2.60	78	.0160	0.406	0.40					
39	.0995	2.527	2.50	79	.0145	0.368	0.35					
40	.0980	2.489	2.50	80	.0135	0.343	0.35					

* Fraction-size drills range in size from one-sixteenth—4 in. and over in diameter—by sixty-fourths.

Table 50 Standard wire and sheet-metal gages*

Gage number	(A) Brown & Sharpe or American	(B) American Steel & Wire Co.	(C) Piano wire	(E) U.S. St'd.	Gage number
0000000	.6513	.4900	—	.5000	0000000
000000	.5800	.4615	.004	.4688	000000
00000	.5165	.4305	.005	.4375	00000
0000	.4600	.3938	.006	.4063	0000
000	.4096	.3625	.007	.3750	000
00	.3648	.3310	.008	.3438	00
0	.3249	.3065	.009	.3125	0
1	.2893	.2830	.010	.2813	1
2	.2576	.2625	.011	.2656	2
3	.2294	.2437	.012	.2500	3
4	.2043	.2253	.013	.2344	4
5	.1819	.2070	.014	.2188	5
6	.1620	.1920	.016	.2031	6
7	.1443	.1770	.018	.1875	7
8	.1285	.1620	.020	.1719	8
9	.1144	.1483	.022	.1563	9
10	.1019	.1350	.024	.1406	10
11	.0907	.1205	.026	.1250	11
12	.0808	.1055	.029	.1094	12
13	.0720	.0915	.031	.0938	13
14	.0641	.0800	.033	.0781	14
15	.0571	.0720	.035	.0703	15
16	.0508	.0625	.037	.0625	16
17	.0453	.0540	.039	.0563	17
18	.0403	.0475	.041	.0500	18
19	.0359	.0410	.043	.0438	19
20	.0320	.0348	.045	.0375	20
21	.0285	.0317	.047	.0344	21
22	.0253	.0286	.049	.0313	22
23	.0226	.0258	.051	.0281	23
24	.0201	.0230	.055	.0250	24
25	.0179	.0204	.059	.0219	25
26	.0159	.0181	.063	.0188	26
27	.0142	.0173	.067	.0172	27
28	.0126	.0162	.071	.0156	28
29	.0113	.0150	.075	.0141	29
30	.0100	.0140	.080	.0125	30
31	.0089	.0132	.085	.0109	31
32	.0080	.0128	.090	.0102	32
33	.0071	.0118	.095	.0094	33
34	.0063	.0104	.100	.0086	34
35	.0056	.0095	.106	.0078	35
36	.0050	.0090	.112	.0070	36
37	.0045	.0085	.118	.0066	37
38	.0040	.0080	.124	.0063	38
39	.0035	.0075	.130	—	39
40	.0031	.0070	.138	—	40

* Dimensions in decimal parts of an inch.

(A) Standard in United States for sheet metal and wire (except steel and iron).

(B) Standard for iron and steel wire (U.S. Steel Wire Gage).

(C) American Steel and Wire Company's music (or piano) wire gage sizes. Recognized by U.S. Bureau of Standards.

(E) U.S. Standard for iron and steel plate. However, plate is now generally specified by its thickness in decimals of an inch.

ANSI abbreviations and symbols for electrical diagrams

D

ANSI abbreviations
and symbols*

alternating current	AC
aluminum	AL
American Standard	AMER STD
approved	APPD
average	AVG
Babbitt	BAB
ball bearing	BB
brass	BRS
Brinell hardness number	BHN
bronze	BRZ
Brown & Sharpe	B & S
cast iron	CI
center line	CL or ₵
center to center	C to C
centimeter	CM
chemical	CHEM
circular	CIR
circular pitch	CP
cold-rolled steel	CRS
copper	COP
counterbore	CBORE
countersink	CSK
cubic	CU
cubic foot	CU FT
cubic inch	CU IN
cubic yard	CU YD
cylinder	CYL
degree	DEG or °
detail drawing	DET DWG
diagonal	DIAG
diameter	DIA
diametral pitch	DP
direct current	DC
drawing	DWG
drawn	DR
effective	EFF
electric	ELEC
engineer	ENGR
external	EXT
fabricate	FAB
fillister	FIL
finish	FIN.
foot	' or FT
gallon	GAL
galvanized iron	GI
grind	GRD
harden	HDN
hexagon	HEX
horsepower	HP
hour	HR
impregnate	IMPREG
inch	" or IN.
inside diameter	ID
internal	INT
lateral	LAT

left hand	LH
linear	LIN
long	LG
longitude	LONG.
machine	MACH
malleable iron	MI
material	MATL
maximum	MAX
meter	M
miles	MI
miles per hour	MPH
millimeter	MM
minimum	MIN
minute (angular measure)	' or MIN
minute (time)	MIN
outside diameter	OD
pattern	PATT
phosphor bronze	PH BRZ
piece	PC
pitch	P
pitch diameter	PD
plate	PL
pound	# or LB
pounds per square foot	PSF
pounds per square inch	PSI
Pratt & Whitney	P & W
quantity	QTY
radius	R or RAD
required	REQD
revolution per minute	RPM
right hand	RH
round	RD
round bar	ϕ
screw	SCR
second (angular measure)	"
second (time)	SEC
section	SECT
Society of Automotive Engineers	SAE
square	SQ
square foot	SQ FT
square inch	SQ IN
standard	STD
steel	STL
steel casting	STL CSTG
thousand	M
thread	THD
ton	TON
traced	TR
volt	V
watt	W
weight	WT
Woodruff	WDF
wrought iron	WI
yard	YD
year	YR

*Only those abbreviations that are commonly used on engineering drawings have been given in the list above. These have been selected from the long and comprehensive list given in ANSI Z32.13–1950. The professional draftsman should have this standard close at hand for use when needed.

AMPLIFIER General	**DEVICE, VISUAL SIGNALING** Annunciator	**PATH, TRANSMISSION** General
ANTENNA General	**ELEMENT, THERMAL** Thermal Cutout, Flasher	Crossing, Not Connected
	Thermal Relay	Junction
	or	Junction, Connected Paths
ARRESTER General Multigap		Pair
BATTERY Multicell		Assembled Conductors; Cable Coaxial
BREAKER, CIRCUIT General	**FUSE** General	2-Conductor Cable
CAPACITOR General	Fusible Element	Grouping Leads
COIL Blowout (Broken line not part of symbol) Operating	**GROUND**	or
	HANDSET General	**RECEIVER** General Headset
CONNECTOR Female Contact Male Contact	**INDUCTOR WINDING** General or	**RECTIFIER** General
CONTACT, ELECTRIC Fixed ○ or Locking Nonlocking Rotating Closed Open ☐ or	**LAMP** Ballast Lamp Incandescent	**REPEATER** 1-Way
		RESISTOR General
	MACHINE, ROTATING Basic Generator Motor Wound Rotor Armature	**SWITCH** General Single Throw Double Throw Knife
CORE Magnetic (General) Magnet or Relay	Winding Symbols 1-Phase 2-Phase 3-Phase Wye 3-Phase Delta	**THERMOCOUPLE** Temperature- Measuring
COUPLER, DIRECTIONAL General		**TRANSFORMER** Magnetic Core Shielded

*ANSI Y32.2.

american national and ISO standards*

E APPENDIX

*International Organization for Standardization

B16.4–1963	Cast-iron Screwed Fittings for Maximum WSP of 125 and 250 lb
B16.5–1968	Steel Pipe Flanges and Flanged Fittings
B16.9–1964	Steel Butt-welding Fittings
B17.1–1967	Shafting and Stock Keys
B17.2–1967	Woodruff Keys and Keyseats
B18.1–1965	Small Solid Rivets
B18.2.1–1965	Square and Hex Bolts and Screws
B18.2.2–1965	Square and Hex Nuts
B18.3–1969	Socket Set Screws and Socket-head Cap Screws
B18.4–1960	Large Rivets
B18.5–1959	Round-head Bolts
B18.6.2–1956	Hexagon-head Cap Screws, Slotted-head Cap Screws, Square-head Set Screws, and Slotted Headless Set Screws
B36.1–1956	Welded and Seamless Pipe (ASTM A53–44)
B36.2–1956	Welded Wrought-iron Pipe
B36.10–1970	Wrought-iron and Wrought-steel Pipe
B48.1–1933	Inch-Millimeter Conversion for Industrial Use
B94.6–1966	Knurling
Y14.1–1957	Size and Format
Y14.2–1973	Line Conventions, Sectioning, and Lettering
Y14.3–1957	Projections
Y14.4–1957	Pictorial Drawing
Y14.5–1973	Dimensioning and Notes
Y14.6–1957	Screw Threads
Y14.7–1958	Gears, Splines, and Serrations
Y14.7.1–1971	For Spur, Helical, Double Helical, and Rack
Y14.8– —	Casting (in preparation)
Y14.9–1958	Forgings
Y14.10–1959	Metal Stampings
Y14.11–1958	Plastics
Y14.12– —	Die Castings
Y14.13– —	Springs, Helical and Flat
Y14.14–1961	Mechanical Assemblies

(Y14.12 and Y14.13: in preparation)

Y14.15–1966	Electrical and Electronics Diagrams
Y14.15a–1971	Supplement— Interconnecting Diagrams
Y14.17–1966	Fluid Power Diagrams
Y14.18– —	Drawings for Optical Parts
Y14.22– —	Drawing Titles and Notes
Y14.24– —	Classifications of Drawings
Y14.25– —	Guide for Numbering Systems
Y14.26– —	Computer Graphics

(Y14.18 through Y14.26: in preparation)

Graphical and letter symbols*

Y1.1–1972	Abbreviations for Use on Drawings and in Text (Graphical)*
Y10.2–1958	Hydraulics (Letter)
Y10.3–1968	Mechanics for Solid Bodies (Letter)
Y10.4–1957	Heat and Thermodynamics (Letter)
Y10.7–1954	Aeronautical Sciences (Letter)
Y10.8–1962	Structural Analysis (Letter)*
Y10.9–1953	Radio (Letter)
Y10.11–1959	Acoustics (Letter)
Y10.12–1961	Chemical Engineering (Letter)
Y10.14–1959	Rocket Propulsion (Letter)
Y10.15–1958	Petroleum Reservoir Engineering and Electric Logging (Letter)
Y10.16–1964	Shell Theory (Letter)
Y10.17–1961	Guide for Selecting Greek Letters Used as Letter Symbols for Engineering Mathematics (Letter)
Y15.2–1960	Time-Series Charts, Manual of Design

*Y is the new letter assigned to standards for abbreviations, charts and graphs, drawings, graphical symbols, and letter symbols. The Z will be changed to Y as the standards are revised and reaffirmed.

Y32.2–1970	Electrical and Electronics Diagrams (Graphical)	
Y32.3–1969	Welding (Graphical)	
Y32.4–1955	Plumbing (Graphical)	
Y32.7–1972	Use on Railroad Maps and Profiles (Graphical)	
Y32.10–1967	Fluid Power Diagrams (Graphical)	
Y32.11–1961	Process Flow Diagrams in Petroleum and Chemical Industries (Graphical)	
Y32.12–1960	Metallizing Symbols (Graphical)	
Y32.17–1962	Nondestructive Testing Symbols (Graphical)	
Z10.1–1941	Abbreviations for Scientific and Engineering Terms (Letter)*	
Z10.6–1948	Physics (Letter)*	
Z15.3–1943	Engineering and Scientific Graphs for Publications	
Z32.2.3–1953	Pipe Fittings, Valves, and Piping (Graphical)*	
Z32.2.4–1953	Heating, Ventilating, and Air Conditioning (Graphical)*	
Z32.2.6–1956	Heat-Power Apparatus (Graphical)*	

*Y is the new letter assigned to standards for abbreviations, charts and graphs, drawings, graphical symbols, and letter symbols. The Z will be changed to Y as the standards are revised and reaffirmed.

International organization for standardization (ISO)

ISO/R 128	Engineering Drawing— Principles of Presentation
ISO/R 129	Engineering Drawing— Dimensioning
ISO/R 406	Inscription of Linear and Angular Tolerances
ISO/R 1101/I	Tolerances of Form and of Position—Part 1

ISO 1101/II	Tolerances of Form and of Position—Part 2
ISO 1302	Method of Indicating Surface Texture
ISO/R 1660	Tolerances of Form and of Position—Part 3 Dimensioning and Tolerancing of Profiles
ISO/R 1661	Tolerancing of Form and of Position—Part 4
ISO 2162	Representation of Springs
ISO 2203	Conventional Representation of Gears
ISO 3040	Dimensioning and Tolerancing Cones
ISO 3098	Lettering—Part 1
R225	Bolts, Screws, and Studs—Dimensioning
R278	Hexagon Bolts and Nuts
R288/1	Slotted and Castle Nuts with Metric Thread
R724	ISO General-Purpose Metric Screw Threads
R887	Washers for Hexagon Bolts and Nuts— Metric Series
R1207	Slotted Cheese (Fillister) Head Screws
R1234	Split Pins—Metric Series
R1478	Tapping Screw Thread
R1479	Hexagon Head Tapping Screws—Metric Series
R1481	Slotted Pan Head Tapping Screws
R1501	ISO Miniature Screw Threads
R1580	Slotted Pan Head Screws
ISO 2009	Slotted Countersunk (Flat) Head Screws— Metric Series
ISO 2010	Slotted Raised Countersunk (Oval) Head Screws
ISO 2306	Drills for Use Prior to Tapping Screw Threads
ISO 2339	Taper Pins, Unhardened—Metric Series
ISO 2340	Clevis Pins—Metric Series
ISO 2342	Slotted Headless Screws—Metric Series
ISO 2343	Hexagon Socket Set Screws

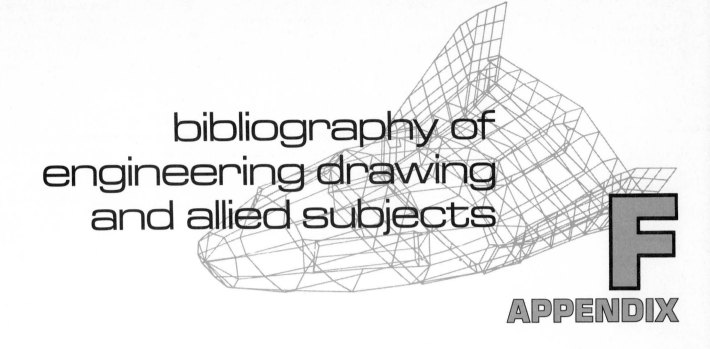

bibliography of engineering drawing and allied subjects

APPENDIX F

Aeronautical drafting and engineering

Anderson, N. H., *Aircraft Layout and Detail Design*, 2nd ed., New York: McGraw-Hill.

Katz, H. H., *Aircraft Drafting*, New York: Macmillan.

LeMaster, C. A., *Aircraft Sheet Metal Work*, Chicago: American Technical Society.

Blueprint reading

Bush, G. F., *Reading Engineering Drawings*, New York: Wiley.

DeVette, W. A., and D. E. Kellogg, *Blueprint Reading for the Metal Trades*, Milwaukee: Bruce.

Dick, A. A., *Blueprint Reading*, New York: Ronald.

Dwight, C., *Blueprint Reading in the Machine Industries*, New York: McGraw-Hill.

Heine, G. M., and C. H. Dunlap, *How to Read Electrical Blueprints*, Chicago: American Technical Society.

Hornung, W. J., *Blueprint Reading*, Englewood Cliffs, N.J.: Prentice-Hall.

Lincoln Electric Co., *Simple Blueprint Reading*, Cleveland.

Norcross, C., *Aircraft Blueprints and How to Read Them*, New York: McGraw-Hill.

Owens, A. A., and B. F. Slingluff, *How to Read Blueprints*, New York: Holt.

Spencer, H. C., and H. E. Grant, *Blueprint Language of the Machine Industries*, New York: Macmillan.

Thayer, H. R., *Blueprint Reading and Sketching*, New York: McGraw-Hill.

Weir, J. J., *Blueprint Reading*, New York: McGraw-Hill.

Castings

Campbell, H. I., *Metal Castings*, New York: Wiley.

Morris, J. L., *Metal Castings*, Englewood Cliffs, N.J.: Prentice-Hall.

Catalogues (instruments)

Alteneder, Theo., and Sons, Philadelphia.

B. K. Elliott Co., Pittsburgh.

Boston Gear Works, Inc., Chicago (gears, etc).

Crane Co., Chicago (pipe fittings).

Eugene Dietzgen Co., Chicago.

Frederic Post Co., Chicago.

Keuffel & Esser Co., Morristown, N.J.

New Departure, Bristol, Conn. (ball bearings).

Timken Roller Bearing Co., Canton, Ohio (roller bearings).

Computer graphics films

DAC-1 (*color*), General Motors Corporation, Public Relations Staff, Film Library, Detroit, Michigan 48202.

Graphic Data Processing, International Business Machines Corporation, available through local IBM representative.

SKETCHPAD, Lincoln Laboratories, Massachusetts Institute of Technology, Lexington, Massachusetts.

Computers—data processing systems

Bernstein, J., *The Analytical Engine*, New York: Random House.

Englebardt, S. L., *Computers*, New York: Pyramid.

Fetter, W. A., *Computer Graphics in Communication*, New York: McGraw-Hill.

Hagg, J. N., *Comprehensive Fortran Programming*, New York: Hayden.

International Business Machines Corporation, *1130 Presentation System*.

———, *Introduction to Data Processing Systems*.

Lohberg, R., and T. Lutz, *Electronic Brains*, New York: Sterling.

Thornhill, R. B., *Engineering Graphics and Numerical Control*, New York: McGraw-Hill.

Descriptive geometry

Johnson, L. O., and I. Wladaver, *Elements of Descriptive Geometry*, Englewood Cliffs, N.J.: Prentice-Hall.

Paré, E. G., R. O. Loving, and I. L. Hill, *Descriptive Geometry*, 2nd ed., New York: Macmillan.

Slaby, S. M., *Three-dimensional Descriptive Geometry*, New York: Harcourt.

Street, W. E., *Technical Descriptive Geometry*, New York: Van Nostrand.

Warner, F. M., and M. McNeary, *Applied Descriptive Geometry*, 5th ed., New York: McGraw-Hill.

Watts, E. F., and J. T. Rule, *Descriptive Geometry*, Englewood Cliffs, N.J.: Prentice-Hall.

Wellman, B. L., *Technical Descriptive Geometry*, 2nd ed., New York: McGraw-Hill.

Descriptive geometry problems

Paré, E. G., and others, *Descriptive Geometry Worksheets*, New York: Macmillan.

Wiedhaas, E. R., *Applied Descriptive Geometry Problems*, New York: McGraw-Hill.

Die casting

Chase, H., *Die Castings*, New York: Wiley.

Electrical drawing

Baer, C. J., *Electrical and Electronic Drawing*, New York: McGraw-Hill.

Bishop, C. C., C. T. Gilliam, and others, *Electrical Drafting and Design*, 3rd ed., New York: McGraw-Hill.

Carini, L. F. D., *Drafting for Electronics*, New York: McGraw-Hill.

Kuller, K. K., *Electronics Drafting*, New York: McGraw-Hill.

Mark, D., *How to Read Schematic Diagrams*, New York: Rider.

Raskhodoff, N. M., *Electronic Drafting Handbook*, New York: Macmillan.

Shiers, G., *Electronic Drafting*, Englewood Cliffs, N.J.: Prentice-Hall.

———, *Electronic Drafting Techniques and Exercises*, Englewood Cliffs, N.J.: Prentice-Hall.

Van Gieson, D. W., *Electrical Drafting*, New York: McGraw-Hill.

Engineering design

Alger, J. R., and C. V. Hayes, *Creative Synthesis in Design*, Englewood Cliffs, N.J.: Prentice-Hall.

Asimow, M., *Introduction to Design*, Englewood Cliffs, N.J.: Prentice-Hall.

Bartee, E. M., *Engineering Experimental Design Fundamentals*, Englewood Cliffs, N.J.: Prentice-Hall.

Crede, C. E., *Shock and Vibration Concepts in Engineering Design*, Englewood Cliffs, N.J.: Prentice-Hall.

Damon, A., H. W. Stoudt, and R. A. McFarland, *The Human Body in Equipment Design*, Cambridge, Mass.: Harvard University Press.

Dixon, J. R., *Design Engineering: Inventiveness, Analysis, and Decision-Making*, New York: McGraw-Hill.

Dreyfuss, H., *The Measure of Man*, New York: Whitney Library of Design.

Edel, D. H., *Introduction to Creative Design*, Englewood Cliffs, N.J.: Prentice-Hall.

Gibson, J. E., *Introduction to Engineering Design*, New York: Holt, Rinehart and Winston.

Graves, M. E., *The Art of Color and Design*, New York: McGraw-Hill.

Greenwood, D. C., *Mechanical Details for Product Design*, New York: McGraw-Hill.

Hill, P. H., *The Science of Engineering Design*, New York: Holt, Rinehart and Winston.

Klemm, F., *A History of Western Technology*, Cambridge, Mass.: M.I.T. Press.

Krick, E. V., *An Introduction to Engineering and Engineering Design*, New York: Wiley.

Luzadder, W. J., *Innovative Design*. Englewood Cliffs, N.J.: Prentice-Hall.

McCormick, E. J., *Human Factors Engineering*, New York: McGraw-Hill.

Middendorf, W. H., *Engineering Design*, Boston: Allyn and Bacon.

Mischke, C. R., *An Introduction to Computer-aided Design*, Englewood Cliffs, N.J.: Prentice-Hall.

Ray, W. S., *An Introduction to Experimental Design*, New York: Macmillan.

Ruskin, A. M., *Materials Considerations in Design*, Englewood Cliffs, N.J.: Prentice-Hall.

Spotts, M. F., *Design Engineering Projects*, Englewood Cliffs, N.J.: Prentice-Hall.

Starr, M. K., *Product Design and Decision Theory*, Englewood Cliffs, N.J.: Prentice-Hall.

Tuska, C. D., *An Introduction to Patents for Inventors and Engineers*, New York: Dover.

U.S. Patent Office, *Patent Laws*, Washington, D.C.: Superintendent of Documents.

————, *Patents and Inventions: An Information Aid for Inventors*, Washington, D.C.: Superintendent of Documents.

Von Fange, E., *Professional Creativity*, Englewood Cliffs, N.J.: Prentice-Hall.

White, W. J., and S. Schneyer, *Pocket Data for Human Factor Engineering*, Buffalo, N.Y.: Cornell University.

Engineering graphics

Black, E. D., *Graphical Communication*, New York: McGraw-Hill.

Earle, J. H., *Engineering Design Graphics*, Reading, Mass.: Addison-Wesley.

French, T. E., and C. J. Vierck, *Engineering Drawing*, 10th ed., New York: McGraw-Hill.

Giesecke, F. E., and others, *Technical Drawing*, 6th ed., New York: Macmillan.

Hammond, R. H., and others, *Engineering Graphics*, New York: Ronald.

Hoelscher, R. P., and C. H. Springer, *Engineering Drawing and Geometry*, 2nd ed., New York: Wiley.

Hornung, W. J., *Mechanical Drafting*, Englewood Cliffs, N.J.: Prentice-Hall.

Levens, A. S., *Graphics in Engineering and Science*, New York: Wiley.

Luzadder, W. J., *Basic Graphics for Design, Analysis, Communications, and the Computer*, 2nd ed., Englewood Cliffs, N.J.:: Prentice-Hall.

————, *Fundamentals of Engineering Drawing for Design, Communication, and Numerical Control*, 7th ed., Englewood Cliffs, N.J.: Prentice-Hall.

————, *Fundamentos de Dibujo Para Ingenieros*, 2nd ed., Mexico D.F.: Compañia Editorial Continental.

————, *Graphics for Engineers*, India: Prentice-Hall.

————, *Technical Drafting Essentials*, 2nd ed., Englewood Cliffs, N.J.: Prentice-Hall.

Rule, J. T., and S. A. Coons, *Graphics*, New York: McGraw-Hill.

Schneerer, W. F., *Programmed Graphics*, New York: McGraw-Hill.

Engineering graphics films
(see text films)

Engineering graphics problems

Earle, J. H., *Engineering Graphics and Design Problems* (Series 1, 2, and 3), Reading, Mass.: Addison-Wesley.

Giesecke, F. E., and others, *Technical Drawing Problems*, 3rd ed., New York: Macmillan.

Goss, L. D. and W. J. Luzadder, *Problems in Engineering Drawing*. Series 2, Englewood Cliffs, N.J: Prentice-Hall.

Johnson, L. O., and I. Wladaver, *Engineering Drawing Problems*, Englewood Cliffs, N.J.: Prentice-Hall.

Levens, A. S., and A. E. Edstrom, *Problems in Mechanical Drawing*, New York: McGraw-Hill.

Luzadder, W. J., and K. E. Botkin, *Problems in Engineering Drawing*, 7th ed., Englewood Cliffs, N.J.: Prentice-Hall.

————, K. E. Botkin, *Problems in Engineering Drawing*, 8th ed., Englewood Cliffs, N.J.: Prentice-Hall.

————, K. E. Botkin, and C. J. Rogers, *Engineering Graphics Problems for Design, Analysis and Communications*, Englewood Cliffs, N.J.: Prentice-Hall.

————, and R. E. Bolles, *Problems in Drafting Fundamentals*, Englewood Cliffs, N.J.: Prentice-Hall.

Spencer, H. C., and I. L. Hill, *Technical Drawing Problems*, New York: Macmillan.

Vierck, C. J., and R. I. Hang, *Engineering Drawing Problems*, New York: McGraw-Hill.

Graphical representation and computation

Davis, D. S., *Empirical Equations and Nomography*, New York: McGraw-Hill.

Douglass, R. D., and D. P. Adams, *Elements of Nomography*, New York: McGraw-Hill.

Levens, A. S., *Nomography*, 2nd ed., New York: Wiley.

Lipka, J., *Graphical and Mechanical Computation*, New York: Wiley.

Robinson, A. H., *Elements of Cartography*, New York: Wiley.

Handbooks

American Institute of Steel Construction, *Steel Construction*, New York.

Crocker, S., *Piping Handbook*, New York: McGraw-Hill.

Kent, W., *Mechanical Engineer's Handbook*, 12th ed., New York: Wiley.

Knowlton, A. E., *Standard Handbook for Electrical Engineers*, 9th ed., New York: McGraw-Hill.

Le Grand, R., ed., *New American Machinist's Handbook*, New York: McGraw-Hill.

McNeese, D. C., and A. L. Hoag, *Engineering and Technical Handbook*, Englewood Cliffs, N.J.: Prentice-Hall.

Marks, L. S., *Mechanical Engineers' Handbook*, 6th ed., New York: McGraw-Hill.

Oberg, E., and F. D. Jones, *Machinery's Handbook*, 17th ed., New York: Industrial Press.

O'Rourke, C. E., *General Engineering Handbook*, 2nd ed., New York: McGraw-Hill.

Perry, J. H., *Chemical Engineers' Handbook*, 4th ed., New York: McGraw-Hill.

Wilson, F. W., ed., *Tool Engineers' Handbook*, 2nd ed., New York: McGraw-Hill.

Jig and fixture design

Colvin, F. H., and L. L. Haas, *Jigs and Fixtures*, 5th ed., New York: McGraw-Hill.

Hinman, C. W., *Die Engineering Layouts and Formulas*, New York: McGraw-Hill.

Jones, F. D., *Jig and Fixture Design*, New York: Industrial Press.

Kinematics—machine design

Faires, V. M., *Design of Machine Elements*, 3rd ed., New York: Macmillan.

Hinkle, R. T., *Kinematics of Machines*, Englewood Cliffs, N.J.: Prentice-Hall.

Lent, D., *Analysis and Design of Mechanisms*, Englewood Cliffs, N.J.: Prentice-Hall.

Maleev, V. L., *Machine Design*, 3rd ed., Scranton, Pa.: International.

Norman, C. A., E. S. Ault, and I. F. Zarobsky, *Fundamentals of Machine Design*, New York: Macmillan.

Spotts, M. F., *Design of Machine Elements*, 5th ed., Englewood Cliffs, N.J.: Prentice-Hall.

Lettering

De Garmo, E. P., and F. Jonassen, *Technical Lettering*, New York: Macmillan.

Machine drawing

Svensen, C. L., *Machine Drawing*, 3rd ed., New York: Van Nostrand Reinhold.

Map and topographic drawing

Sloane, R. C., and I. M. Montz, *Elements of Topographic Drawing*, 2nd ed., New York: McGraw-Hill.

Metric System (Système International d'Unités—SI)

American Society for Testing Materials, *Metric Practice Guide*, New York.

Beloit Tool Co., *Discover Why Metrics*, Roscoe, Ill.: Swani Publishing Co.

————, *U.S.A. Goes Metric*, Roscoe, Ill.: Swani Publishing Co.

Foster, L. W., *Geo-metrics*, Reading, Mass.: Addison-Wesley.

Society of Automotive Engineers, *Dual-Dimensioning (SAE J390)*, New York.

Numerical control

International Business Machines Corp., *Automatic Programming of Machine Tools*, Poughkeepsie, N.Y.

Patent drawing (see engineering design)

Publications of U.S. Government Printing Office, Washington, D.C.:
Guide for Patent Draftsmen
Rules of Practice of the U.S. Patent Office

Pattern design

Hall, B. R., and H. E. Kiley, *Pattern Design*, Scranton, Pa.: International Textbook.

Perspective

Lawson, P. J., *Practical Perspective Drawing*, New York: McGraw-Hill.

Turner, W. W., *Simplified Perspective*, New York: Ronald.

Pipe

Crocker, S., *Piping Hand Book*, 4th ed., New York: McGraw-Hill.

Day, L. J., *Standard Plumbing Details*, New York: Wiley.

Thompson, C. H., *Fundamentals of Pipe Drafting*, New York: Wiley.

Shop practice

Begeman, M. L., *Manufacturing Processes*, 4th ed., New York: Wiley.

Boston, O. W., *Metal Processing*, 2nd ed., New York: Wiley.

Doyle, L. E., *Metal Machining*, Englewood Cliffs, N.J.: Prentice-Hall.

————, J. L. Leach, J. L. Morris, and G. F. Schrader, *Manufacturing Processes and Materials for Engineers*, Englewood Cliffs, N.J.: Prentice-Hall.

Lascoe, O. D., and others, Machine Shop, Chicago, American Technical Society.

Marek, C. T., *Fundamentals in the Production and Design of Castings*, New York: Wiley.

Stieri, E., *Fundamentals of Machine Shop Practice*, Englewood Cliffs, N.J.: Prentice-Hall.

Young, J. F., *Materials and Processes*, 2nd ed., New York: Wiley.

Sketching

Katz, H., *Technical Sketching and Visualization for Engineers*, New York: Macmillan.

Turner, W. W., *Freehand Sketching for Engineers*, New York: Ronald.

Zipprich, A. E., *Freehand Drafting for Technical Sketching*, 3rd ed., New York: Van Nostrand Reinhold.

Slide rule

Arnold, J. N., *The Slide Rule*, Englewood Cliffs, N.J.: Prentice-Hall.

Structural drafting

American Institute of Steel Construction, *Structural Steel Detailing*, New York.

Bishop, C. T., *Structural Drafting*, New York: Wiley.

Lothers, S. E., *Design in Structural Steel*, Englewood Cliffs, N.J.: Prentice-Hall.

Technical dictionary

Tweney, C. F., and L. E. C. Hughes, *Chamber's Technical Dictionary*, 3rd rev. ed., New York: Macmillan.

Technical illustration

Gibby, J. C., *Technical Illustration*, Chicago: American Technical Society.

Thomas, T. A., *Technical Illustration*, New York: McGraw-Hill.

Text films

Engineering Drawing, 7 films, 6 strips, New York: McGraw-Hill.

Tool design

Cole, C. B., *Tool Design*, Chicago: American Technical Society.

Donaldson, C., and G. H. LeCain, *Tool Design*, 2nd ed., New York: Harper & Row.

Doyle, L. E., *Tool Engineering, Analysis and Procedure*, Englewood Cliffs, N.J.: Prentice-Hall.

Jeffries, W. R., *Tool Design*, Englewood Cliffs, N.J.: Prentice-Hall.

Welding

Lincoln Electric Co., *Simple Blueprint Reading*, Cleveland.

————, *Procedure Handbook of Arc Welding Design and Practice*, Cleveland.

Morris, J. L., *Welding Processes and Procedures*, Englewood Cliffs, N.J.: Prentice-Hall.

————, *Welding Principles for Engineers*, Englewood Cliffs, N.J.: Prentice-Hall.

Index

A

Abscissa, 443
Acme thread, 314, 319, 322, 585 (table)
Acute angle, 62
ADAPT, 435
Addendum circle, 482, 483–84
Addition of vector forces, 171–72
Aero thread, 329, 330
Airbrush, 235
Alcoa Seaprobe, 128
Algebra, graphical, 467–69
ALGOL, 400
Aligned method of dimension placement, 283
Aligned views, 94
Alignment charts, 451–62
ALLADIN, 410
Allowance, 293
Alterations on working drawings, 357
Alternative positions, representation of, 96, 97
American Institute of Steel Construction, 539
American National Standards, 8, 610–12
  fasteners, 323–30, 586–94 (tables), 596 (table)
  fits, 292, 601–4 (tables)
  pipe, 596 (table)
  pipe fittings, 521, 597–600 (tables)
  pipe thread, 518–20
  screw threads, 315, 318–20
American National Standards Institute (ANSI), (see also American National Standards) 5, 21, 130, 134, 283
  electronic diagram abbreviations, 608
  electronic diagram symbols, 494, 609
  line symbols, 29
  material symbols, 138

American National Standards Institute (*Cont.*)
  resistance welding symbols, 514
  rivet symbols, 540
American Welding Society, 513
Ames lettering guide, 32, 33
Angle brackets, dimensioning, 284
Angles (*see also* Inclined lines), 62
  between a line and a given plane, 168–69
  between two intersecting oblique lines, 169
  between two nonintersecting (skew) lines, 169
  bisection of, 48
  construction of:
    chord method, 49
    equal to a given angle, 49
    tangent method, 49
  dihedral:
    auxiliary views of, 150
    between two planes, 170
  dimensioning, 285
  in isometric drawing, 215
  in multiview drawing, 83–84
  in oblique drawing, 223, 224
  trisection of, 48
Angular dimensions, 285
Angularity, tolerance specification for, 299, 301
Angular perspective, 228–29
  sketching, 119
Angular tolerances, 297
Aperture card, 370
Apollo Applications Program, 210
APT (Automatically Programmed Tools), 410, 430
Arcs, 62
  approximation of curve with, 55
  dimensioning, 285
  in isometric drawing, 216, 217
  laying off approximate length of, 55–56

619